Elwyn Simons: A Search for Origins

DEVELOPMENTS IN PRIMATOLOGY: PROGRESS AND PROSPECTS
Series Editor: Russell H. Tuttle, University of Chicago, Chicago, Illinois

---

This peer-reviewed book series melds the facts of organic diversity with the continuity of the evolutionary process. The volumes in this series exemplify the diversity of theoretical perspectives and methodological approaches currently employed by primatologists and physical anthropologists. Specific coverage includes: primate behavior in natural habitats and captive settings; primate ecology and conservation; functional morphology and developmental biology of primates; primate systematics; genetic and phenotypic differences among living primates; and paleoprimatology.

PRIMATE BIOGEOGRAPHY
Edited by Shawn M. Lehman and John Fleagle

REPRODUCTION AND FITNESS IN BABOONS: BEHAVIORAL, ECOLOGICAL, AND LIFE HISTORY PERSPECTIVES
Edited by Larissa Swedell and Steven R. Leigh

RINGTAILED LEMUR BIOLOGY: *LEMUR CATTA* IN MADAGASCAR
Edited by Alison Jolly, Robert W. Sussman, Naoki Koyama and Hantanirina Rasamimanana

PRIMATES OF WESTERN UGANDA
Edited by Nicholas E. Newton-Fisher, Hugh Notman, James D. Paterson and Vernon Reynolds

PRIMATE ORIGINS: ADAPTATIONS AND EVOLUTION
Edited by Matthew J. Ravosa and Marian Dagosto

LEMURS: ECOLOGY AND ADAPTATION
Edited by Lisa Gould and Michelle L. Sauther

PRIMATE ANTI-PREDATOR STRATEGIES
Edited by Sharon L. Gursky and K.A.I. Nekaris

CONSERVATION IN THE 21ST CENTURY: GORILLAS AS A CASE STUDY
Edited by T.S. Stoinski, H.D. Steklis and P.T. Mehlman

ELWYN SIMONS: A SEARCH FOR ORIGINS
Edited by John G. Fleagle and Christopher C. Gilbert

John G. Fleagle · Christopher C. Gilbert
Editors

Elwyn Simons: A Search for Origins

Springer

Editors
John G. Fleagle
Department of Anatomical Sciences
Health Sciences Center
Stony Brook University
Stony Brook, NY 11794-8081

Christopher C. Gilbert
Interdepartmental Doctoral Program in
Anthropological Sciences
Department of Anthropology
Stony Brook University
Stony Brook, NY 11794-4364

ISBN: 978-0-387-73895-6 e-ISBN: 978-0-387-73896-3

Library of Congress Control Number: 2007936873

Cover image: Photographs courtesy of Lewis Ladocsi and Christopher Gilbert.

Printed on acid-free paper.

9 8 7 6 5 4 3 2 1

springer.com

Preface

For over five decades, Elwyn Simons has dominated the field of primate evolution through his paleontological expeditions throughout the world, his exceptional record of publications and his role as mentor to hundreds of students in the classrooms at Yale and Duke as well as at the Duke University Primate Center or on his field expeditions. His boundless energy and larger than life personality have always left their mark on anyone who has worked with him.

As a researcher in the field of primate evolution, his record is unparalleled. He has worked on the entire span of the primate fossil record, from early prosimians of North America and Europe, to early anthropoids of Africa and possibly Asia, fossil apes and monkeys from Africa and Eurasia, and fossil hominids. It is no exaggeration to say that through his masterful review papers from the 1960's he invented the field of primate evolution as we know it today. Moreover his ongoing work in subsequent decades has provided much of the new material, the inspiration, and the manpower to make this one of the most exciting areas of research in the scientific world. His role in instigating, inspiring and facilitating research and conservation on the living and fossil primate fauna of Madagascar are no less remarkable.

To mark the occasion of Elwyn's 75th birthday, many of his friends, family, colleagues and students got together on September 16 and 17, 2005, in Durham, North Carolina for two days of scientific presentations, tributes, parties and conversations to celebrate and honor his life and work thus far. These festivities were organized by Freddie and John Oakley and Friderun Ankel-Simons. The events were made possible by donations from the Oakley family, The Anne and Gordon Getty Foundation, Herbert Simons, Sarah and Dan Hrdy, the Department of Biological Anthropology and Anatomy at Duke University and anonymous donors.

The papers in this volume are largely based on presentations delivered at the 2005 conference with a number of additional contributions by people who were unable to attend, but nonetheless wanted to participate in the tribute to Elwyn's life and work. Thus this volume offers not only a series of papers highlighting a small part of the breadth and personal influence of Elwyn's career on the field of paleontology and primatology, but also a record of the 2005 conference.

This volume reflects the efforts of many people. Each of the authors provided manuscripts more or less on schedule and many generously agreed to assist in reviewing other contributions. We also received reviews of the papers from numerous colleagues who did not contribute to the volume, among whom, Ann Yoder, James Rossie, Todd Rae, and Chris Heesy deserve special thanks. Marilyn Helms, Marty Meyer, Carl Vondra, and Jim Mead were especially helpful in our efforts to put together a photographic record of five decades of Fayum expeditions. At Springer, Andrea Macaluso, Lisa Tenaglia, and Cynthia Manzano have offered continue support along with the Series Editor Russ Tuttle and Sunayana Jain of Integra Software Services. Luci Betti-Nash designed the cover. Most of all, of course, this volume is the result of the tireless efforts of Elwyn Simons.

John Fleagle
Chris Gilbert

Contents

Contributors

Miloud M. Abugares
Division of Exploration
Petroleum Research Centre
Tripoli, Libya

Friderun Ankel-Simons
3603 Stoneybrook Drive
Durham, NC 27705

K. Christopher Beard
Section of Vertebrate
Paleontology
Carnegie Museum of Natural
History
Pittsburgh, PA 15213

Thomas M. Bown
Erathem-Vanir Geological, Inc.
10350 Dover Street, D-32
Westminster, CO 80021

Prithijit Chatrath
Division of Fossil Primates
Duke University Primate Center
1013 Broad Street
Durham, NC 27705

Russell L. Ciochon
Department of Anthropology
University of Iowa
114 Macbride Hall
Iowa City, IA 52242-1322

William C. Clyde
Department of Earth Sciences
University of New Hampshire
56 College Road
Durham, NH 03824

Herbert H. Covert
Department of Anthropology
University of Colorado at Boulder
Boulder, CO 80309-0233

Marian Dagosto
Department of Cell and
Molecular Biology
Feinberg School of Medicine
Northwestern University
Chicago, IL 60611

Brigitte Demes
Department of Anatomical Sciences
School of Medicine
Stony Brook University
Stony Brook, NY 11794-8081

John G. Fleagle
Department of Anatomical Sciences
Health Sciences Center
Stony Brook University
Stony Brook, NY 11794-8081

Daniel L. Gebo
Department of Anthropology
Northern Illinois University
DeKalb, IL 60115

Christopher C. Gilbert
Interdepartmental Doctoral Program in Anthropological Sciences
Department of Anthropology
Stony Brook University
Stony Brook, NY 11794-4364

Philip D. Gingerich
Museum of Paleontology and Department of Geological Sciences,
The University of Michigan
Ann Arbor, MI 48109-1079

Laurie R. Godfrey
Department of Anthropology
240 Hicks Way
University of Massachusetts
Amherst, MA 01003-9278

Michael D. Gottfried
Department of Geological Sciences
Michigan State University Museum
Michigan State University
East Lansing, MI 48824

Gregg F. Gunnell
Museum of Paleontology
University of Michigan
1109 Geddes Avenue
Ann Arbor, MI 48109-1079

Donna Hart
Department of Anthropology
University of Missouri - St. Louis
One University Blvd
St. Louis, MO 63121

Patricia A. Holroyd
Museum of Paleontology
University of California
Berkeley, CA 94720

Mitchell T. Irwin
Department of Biology
McGill University
1205 Docteur Penfield Avenue
Montreal, Quebec
Canada H3A 1B1

Jukka Jernvall
Developmental Biology Program
Institute of Biotechnology
University of Helsinki
PO Box 56, FIN-00014
Helsinki, Finland

William L. Jungers
Department of Anatomical Sciences
School of Medicine
Stony Brook University
Stony Brook, NY 11794-8081

Saidi Kapilima
Department of Geology
University of Dar es Salaam
P.O Box 35065
Dar Es Salaam, Tanzania

Richard F. Kay
Department of Biological Anthropology and Anatomy
Duke University
Durham, NC 27708

Xijun Ni
Institute of Vertebrate Paleontology and Paleoanthropology
Chinese Academy of Sciences
Post Office Box 643
Beijing, China 100044

Sifa Ngasala
Department of Geology
University of Dar es Salaam
P.O Box 35065
Dar Es Salaam, Tanzania

Fredericka B. Oakley
39598 Lupine Court
Davis, CA 95616

John B. Oakley
University of California at Davis
400 Mrak Hall Drive
Davis, CA 95616-5201

Patrick M. O'Connor
Department of Biomedical Sciences
Ohio University College of
Osteopathic Medicine
228 Irvine Hall
Ohio University
Athens, OH 45701

Rajeev Patnaik
Center of Advanced Study in Geology
Panjab University
Chandigarh-160014, India

David Pilbeam
Peabody Museum
11 Divinity Avenue
Cambridge, MA 02138-2019

Tao Qi
Institute of Vertebrate Paleontology
and Paleoanthropology
Chinese Academy of Sciences
Post Office Box 643
Beijing, China 100044

Le Khac Quyet
Fauna & Flora International –
Vietnam Programme
IPO Box 78
340 Nghi Tam
Hanoi, Vietnam

Eric M. Roberts
School of Geosciences
University of the Witwatersrand
2050 Wits, South Africa

D. Tab Rasmussen
Department of Anthropology
Washington University
St. Louis, MO 63130

Jennifer L. Ross
Department of Biological
Anthropology and Anatomy
Duke University
Durham, NC 27708

Gary T. Schwartz
Institute of Human Origins
School of Human Evolution
and Social Change
PO Box 872402
Arizona State University
Tempe, AZ 85287

Erik R. Seiffert
Department of Anatomical Sciences
Heleth Sciences Center
Stony Brook University
Stony Brook, NY 11794-8081

Joshua B. Smith
Department of Earth
and Planetary Sciences
Washington University
St. Louis, MO 63130 USA

Elwyn L. Simons
Division of Fossil Primates
Duke University Primate Center
Department of Biological
Anthropology and Anatomy
Duke University
1013 Broad Street
Durham NC 27705

Nancy J. Stevens
Department of Biomedical Sciences
Ohio University College of
Osteopathic Medicine
228 Irvine Hall
Ohio University
Athens, OH 45701

Suzanne G. Strait
Department of Biological Sciences
Marshall University
Huntington, WV 25755

R. W. Sussman
Department of Anthropology
Washington University
St. Louis, MO 63130

Ian Tattersall
Division of Anthropology
American Museum of Natural History
New York, NY 10024

Linda Taylor
Department of Anthropology
University of Miami
P.O. Box 248106
Coral Gables, FL 33124-2005

Chris Tilden
1121 Williamsburg Court
Lawrence, KS 66049

Sefau O. Tshakreen
Division of Exploration
Petroleum Research Centre
Tripoli, Libya

Bernard Wood
Center for the Advanced Study of Hominid Paleobiology
George Washington University
2110 G St NW
Washington, DC 20052

Barth W. Wright
Center for The Advanced Study of Hominid Paleobiology
Department of Anthropology
George Washington University
2110 G. St. NW
Washington, DC 20052

Patricia C. Wright
Department of Anthropology
Stony Brook University
Stony Brook, NY 11794-4364

Section 1
The Life of a Scientist

In the course of his first seventy-five years, Elwyn Simons has influenced the lives of many people in many ways. The papers in this section include a series of personal memories and tributes to Elwyn as a person as well as a short biography of one of his mentors. As such, they provide insight into some of the many facets of the man that are not evident to those who know Elwyn only through his many publications.

In the initial contribution, the "Introduction to the *Festschrift*", Fredericka ("Freddie") Oakley, the organizer of the 2005 Tribute, provides a brief outline of Elwyn's early years and his education at Rice as an undergraduate, Princeton as a graduate student, and a postdoctoral Marshall Fellowship at Oxford University under the renowned anatomist Sir Wilfrid Le Gros Clark. This is followed by a joyous essay, "Elwyn LaVerne Simons—a Very Personal View", by Friderun Ankel-Simons on life as Elwyn's wife. Dr. David Pilbeam, probably Elwyn's most distinguished student and long time colleague, recounts his days as a graduate and subsequently a faculty colleague in "A Personal Reminiscence of Elwyn". Pilbeam's contribution is followed by a short biography of one of Elwyn's most influential mentors, Le Gros Clark, by Bernard Wood. Wood's essay, "Sir Wilfrid Le Gros Clark: The Making of a Paleoanthropologist", focuses in particular on Le Gros's contributions to the study of human evolution, and especially the acceptance of *Australopithecus* as a human ancestor. The biography of Le Gros Clark is followed by an addendum, "Sir Wilfrid Le Gros Clark: Personal Recollections of Le Gros", in which Elwyn recounts some of his own memories of his studies at Oxford and interactions with Le Gros Clark both at Oxford and later at Yale University where Le Gros Clark was a visiting professor in 1962. The final contribution in this section is a transcript of the presentation made by Dr. John Oakley, Professor of Law and Philosophy at the University of California at Davis, spouse of the organizer Freddie Oakley, and a close friend of the Simons family for nearly forty years. In this personal appreciation of Elwyn, "Human Evolution and the Challenge of Creationism", Professor Oakley lays out his own views on the teaching of evolution in schools and the place of science and religion in human existence.

Introduction to the *Festschrift*

Fredericka B. Oakley

In the year of his 75th birthday, colleagues, friends, students and family of Elwyn Simons gathered near Duke University in Durham, North Carolina, for three days of conferences and gatherings celebrating his life and work. It was a wonderful time of scholarship and comradeship, and a great tribute to the man and scholar who has shaped many lives and much science in the more than a half century, to date, of his professional life.

Elwyn LaVerne Simons was born July 14, 1930, at Lawrence, Kansas, a son of Verne Franklin Simons and Verna Irene Cuddeback. A Kansas native and the descendent of pioneers, Verne Franklin Simons spent his career as a professor of accounting and financial advisor, first at the University of Kansas, and from 1929 onward, at Rice University in Houston, Texas, where Elwyn and his younger brother, Herbert were raised. Verna Cuddeback Simons, Elwyn's mother, was herself the descendant of a pioneering Kansas family, and was an art student when she met Elwyn's father. She devoted herself to family and beauty throughout her long life.

Elwyn's particular genius for original thinking about the natural world was evident very early in his life. His mother preserved many stories and artistic renderings centered on nature and animals that he began producing at around three years of age.

Reading these stories, one is struck by the precocious evidence for a natural-born natural scientist. Elwyn appreciated very early the importance of preserving the history of his family gleaned both from written records and from oral histories. These amazing and entertaining stories have been a great delight to hear told and retold, and the recordings Elwyn made as a youngster, of his grandparents singing folk songs on the porch of their farmhouse in their final years, are a true treasure of American history (see http://www.ils.unc.edu/dpr/archives/folksongs/).

Elwyn Simons attended public schools in the West University neighborhood of Houston. He received a B.S. degree from W. M. Rice University in 1953. He received a M.A. degree from Princeton University in 1955, a Ph.D.

Fredericka B. Oakley
39598 Lupine Court, Davis, California 95616

J. G. Fleagle, C. C. Gilbert (eds.), *Elwyn Simons: A Search for Origins.*

from Princeton University in 1956, a D.Phil. from University College, Oxford in 1959, an Honorary M.A. from Yale University in 1967 and a D.Sc. degree from University College, Oxford, in 1995. He has held too many academic appointments to detail, but his career has been spent chiefly at Princeton University (1953–1956), Oxford University (1956–1959), the University of Pennsylvania (1959–1961), and for many years at Yale University (1960–1977) and most recently at Duke University (1977-present).

Simons has, likewise, received nearly countless awards and fellowships, including a Fulbright Fellowship, the Alexander von Humboldt Senior Scientist's Award, membership in the U.S. National Academy of Sciences, the Charles R. Darwin award of the American Association of Physical Anthropology, the Founders Award of the American Society for the Prevention of Cruelty to Animals, and many, many more. Elwyn remembers with particular humor and affection his years as a Marshall Scholar at Oxford University.

Elwyn and Friderun Ankel-Simons have been married for more than 35 years. They have two children, Cornelia Simons Seiffert and Verne Simons. Elwyn is also the father of David Brenton Simons.

Elwyn Simons has conducted fieldwork in paleontology and primate conservation all over the globe, from Wyoming to Egypt to Madagascar, with many stops in between. The contributions in this volume speak to his broad range of interests and to the incredible appetite for science and the preservation of the natural world that he has communicated to his many distinguished

Fig. 1 The Oakley and Simons families in Wyoming 1985 in front of the Shasta. From left: John Oakley, Cornelia, Friderun, Verne, Elwyn, Freddie Oakley, Adélie and Antonia

students. As the sole author, and as a coauthor with many of his colleagues and students, Elwyn has written well over 300 scholarly books and articles.

It is a great honor to count myself among Elwyn's former students, and to know him and his wife, Friderun Ankel-Simons (an important scholar in her own right) as very dear friends of more than 35 years, and our families have been close for many years (Fig. 1). When I was inspired, in 2004, to organize this conference honoring the life and work of Elwyn LaVerne Simons in the year of his 75th birthday, I was confident that I would receive the cooperation of many busy people whose dedication to scholarship and conservation he inspired. I was not disappointed. In the year of planning and preparation for these events, I was helped and supported, in particular, by Friderun Ankel-Simons, by the faculty and staff of the Duke University Department of Anatomy and Biological Anthropology, and by the editors of this volume, John G. Fleagle and Christopher C. Gilbert.

Most particular thanks are due to the Ann and Gordon Getty Foundation, which provided extraordinary support for the conference, allowing many to attend who would otherwise not, and providing meals and meeting rooms for the entire conference. The gala dinner that concluded the conference was provided chiefly by the generosity of Herbert Simons and John Oakley.

It is a profound honor to introduce this *Festschrift* for Elwyn Simons. I conclude with this encouragement from Henry Wadsworth Longfellow, which Elwyn has often quoted, and which, I believe, says much about his character and his life.

> *"Let us then, be up and doing. With a heart for any fate;*
> *Still achieving, still pursuing, Learn to labor and to wait."*

Freddie Oakley
Davis, California
May 18, 2006

Elwyn LaVerne Simons

A Very Personal View

Friderun Ankel-Simons

This is the most unlikely story to ever have happened—a young German biologist went to Yale University in January of 1971 to work at the Division of Fossil Vertebrates of Yale University's Peabody Museum for one year—and one year only, not a day longer—and to learn as much as she could about fossil primates.

She learned much about vertebrate and primate fossils and in the bargain she unexpectedly found the man who would be the love of her life: Elwyn Simons, perhaps one of the last Renaissance men, who is the most complex, knowledgeable, caring, responsible and wonderful human being imaginable (Fig. 1). Some of his unusual qualities are: prudent consensus seeker, diplomat, visionary for the future, honorary Aye-aye—nobody else understands like Elwyn how to make a displaced lemur feel at home—*Exempla gratia*: By helping an Aye-aye, that just had arrived from its homeland Madagascar, to build its first nest in a strange new world at the Duke Primate Center, or grooming a lonely *Propithecus* named Nigel on a regular basis. Being married to Elwyn since December 2nd, 1972 I am likely to be biased. But this doesn't matter as I am trying to be realistic.

What does it mean to be Married to Elwyn?

It means continuously going to distant and daunting places where he is finding spectacular fossils. These may be expeditions to the stunning Badlands of Wyoming (Fig. 2), to the arid and overwhelmingly gorgeous and vast Fayum Desert in Egypt (Fig. 3), or to the magical Mecca of any naturalist, the island world of Madagascar. It means vehicles getting stuck in sand or mud, or running out of gas on the highway (Fig. 4); sleeping in tents (Fig. 5) in magical places and seeing more stars than one ever imagined existed; hearing

Friderun Ankel-Simons
3603 Stoneybrook Drive, Durham, NC 27705
friderun@aol.com

J. G. Fleagle, C. C. Gilbert (eds.), *Elwyn Simons: A Search for Origins.*

Fig. 1 Elwyn and Friderun Simons in Wyoming, 1972

Fig. 2 Simons Family in front of Goldtooth MacDonald's Cabin, Wyoming 1979. From left: David Brenton, Elwyn, Cornelia, Friderun holding Verne

Fig. 3 The Simons Family in Sakkara, Egypt 1985. From left: Egyptian Guide, Elwyn, Verne and Cornelia on the Camel, Friderun

Fig. 4 1995, the Simons Family on the Road to the Ankarana, northern Madagascar: fun and games!

the desert fox calling or the lemurs quarrel in the middle of the night. It means long evenings of lively discussions and conversations about the world, about fossils, about genealogy, the history of Christianity, the ear region of the

Fig. 5 1995, Ankarana Camp. Friderun emerging from her tent

primate genus *Tarsius*, about students and children, the Pharaohs of Egypt, about living prosimians, primates and wild life conservation, about bee keeping, the meaning of life, art, agonizing about politics gone wrong that are more often than not based in the uniquely human dilemma known as religion. It is an endless series of fond but hilarious anecdotes about Frank Goto, the paleontology preparator at Princeton University. It is mourning the loss of family members and friends. It is sharing the triumphs of receiving scientific awards, being knighted by the country of Madagascar, or being elected to scholarly societies. It is proudly seeing students succeed in their lives.

Being married to Elwyn means wonderful meals, gardening, restoring beautiful handmade quilts and paintings, trips to India or Europe, months living in Germany and Paris, France, vacations in the Caribbean, visits to Houston, Texas, pulling weeds, planting roses and palm trees, visits to Yellowstone Park, glaciers, mountains and geysers, Simons family reunion in the Ozarks, drives through New England or Kansas, all the way across the continent to the American West. It means Art and Natural History Museums, country music, being in Colorado, Montana and taking the road over the Beartooth Mountains, singing lullabies, camp songs and telling stories. Smelling flowers and watching humming birds, washing dishes, a kitchen sticky all over with delectable honey, watering plants, going for walks. Life with Elwyn is never, ever boring.

It also means limitless trust, love, warmth, responsibility and adding the next generation to our union, David Brenton, Cornelia and Verne. It means having wonderful, loving and trusting livelong friends and family members.

All of them true friends indeed. It means never ending support, generosity, understanding and allowing each other to grow. It means flower bouquets, orchids blooming, caterpillars and butterflies, pyramids and running creeks, picking up sticks, looking for fossils, sitting around camp fires and giving parties. It is writing papers and books, proofreading manuscripts and grant proposals, discussing ideas.

There have been complicated times, both familial and professional. There have been crosscurrents caused by envious, misleading or even devious people, uncomprehending administrators, senseless minions, misunderstandings, but Elwyn stands tall above any confusion and hurdle that life entails.

This is what it means to be married to Elwyn.

A Personal Reminiscence of Elwyn

David Pilbeam

I met Elwyn for the first time in the Natural History Museum in London sometime during the early spring of 1963. I had applied to Yale as a prospective graduate student in Geology and Geophysics and Elwyn interviewed me (as did, on another occasion that spring, John Buettner-Janusch who was then in the Yale Anthropology Department). I had applied to Yale to work with Elwyn because, one day almost a year previously, my teacher in physical anthropology at Cambridge, Jack Trevor, had said to me: "David, my boy, there's a brilliant young American paleontologist named Elwyn Simons who is interested in human origins; why don't you go and work with him." So I did.

I went (up) to Cambridge in 1959 to read (study) Natural Sciences as the first half of a medical degree (then, as now, the pre-clinical portion of medical training made up a significant part of a Cambridge undergraduate degree), but decided after two years that medicine and I were not for each other. I ended up taking the second part of my bachelors degree in physical anthropology. At Cambridge, physical anthropology was a quiet little backwater of a sub-department, with two instructors, one of them the aforementioned Jack Trevor. Jack had been born in Tanzania and had begun making a splash as a rising star before the Second World War. But by the time I met him, his star had faded and he had become something of a recluse, although still a brilliant one, and for me, an inspiring teacher. Jack did no lecturing but instead taught through weekly supervisions (tutorials) which involved (sometimes) critical reading of my essays, but as frequently meandered across a dozen different topics. This style of teaching gave me plenty of time to sit in libraries, and in particular to browse restlessly. It was during one session in the library of the Cambridge Philosophical Society, sometime in the 1961–1962 academic year, that I came across Elwyn's 1961 paper on *Ramapithecus* in what was to me an obscure journal, ***Postilla***, of the Yale Peabody Museum. I then tracked down his various papers published between 1959 and 1961 on the Fayum primates discovered earlier in the century by Richard Markgraf, and his report in 1962 of the first new primates from

David Pilbeam
Peabody Museum, 11 Divinity Avenue, Cambridge, MA 02138-2019
pilbeam@fas.harvard.edu

Egypt, fruits of the first full season in the Fayum. I became intrigued with hominoid evolution and hominid origins and with the Oligocene and (particularly) Miocene, to the point of obsession.

In the spring of 1962, Jack took me to a meeting at the London Zoo sponsored by the Zoological Society of London. Among the talks was one by a young primatologist who had just reported seeing chimps make tools—Jane Goodall. Few of us realized that we were present at the birth of studies of great ape behavioral ecology, which are now so important in making sense of the ape and human fossil record. I got to meet many of the then movers and shakers in the field—Neil Tappen, John Napier, Charles Oxnard, and George "Erik" Erikson among the morphologists and primatologists, along with several behaviorists and geneticists. Only much later, looking back at the publication that came from that meeting (Napier, 1963), did I realize that there were already strong hints from genetic comparisons that humans and apes were very similar.

I received a graduate fellowship in Geology and Geophysics that began in 1963. It paid $2000/year, which was enough to live on and even run a car (I bought Elwyn's fading VW Beatle). When I arrived at Yale, hominids were still thought to be very old by almost all morphologists and paleontologists: the great apes were believed monophyletic and the earliest human ancestors were traced back into the early Miocene or even earlier. The great phenotypic differences between humans and apes were thought to reflect great phylogenetic depth. More genetic comparisons among hominoids were just beginning to surface, but they were ignored by almost all morphologists unless they corresponded to the then conventional wisdom on branching times. The value of primate field studies was stated but not truly understood and they were certainly not valued in the ways they are today.

Once I had adjusted to Yale, the United States, and to Elwyn, whose teaching style, to be fair, was not that different from Jack's, and very compatible with my learning style, I fell into a regular rhythm. My office was in the newly built Kline Geology Laboratories, less than a minute's walk from Elwyn's (who was right next door to John Ostrom). Grant Meyer was the chief lab tech and preparator. I had a few courses to take in my first year there, but not enough to keep me from being a regular occupant of Elwyn's kingdom.

Rather quickly I was drawn into Elwyn's research projects, and in particular those involving Oligocene primates and the dryopithecines. After a brief visit to Egypt in 1960, Elwyn had been able to raise funds to begin in 1961 what has turned out to be a quite extraordinarily successful series of field seasons in the Oligocene and Eocene of the Fayum. I arrived just before Elwyn left for the third field season, which yielded important new primates including *Aegyptopithecus* mandibles. I was privileged to be able to participate as a very junior "learner" in the discussions and analyses of the new material. I was also becoming increasingly involved in another of Elwyn's major projects, taxonomic revision of the Neogene apes known collectively as the dryopithecines, after the type genus

Dryopithecus (described by Lartet in the mid-1800's and mentioned by Darwin in the *Descent of Man*).

At that time, the record of Miocene and Pliocene apes (the boundary between them still sat at what is now the Middle/Upper Miocene boundary) consisted almost entirely of scrappy jaws and isolated teeth (an exception being the *Proconsul* partial skeleton), to which were attached a multitude of names (over twenty generic names, as I recall). We attacked this morass with great zeal, which resulted in what can now (given a greatly expanded Miocene fossil ape record) be recognized as excessive nomenclatural parsimony. We ended up with just two Miocene genera: *Ramapithecus*, the hominid, and *Dryopithecus* for the rest, with barely more than a half dozen species. However, we did assign geographically-based subgeneric names, for the European, Asian, and African species groups. The study was completed by late 1964, and published in 1965 (see Simons and Pilbeam, 1965).

Elwyn's work patterns were simple: work constantly, and always have at least one and preferably two or three manuscripts in preparation at any one time. I did my best to keep up with the work rate, although I have never been able to match his level of productivity. All I can say is that this arrangement was enormous fun, exhilarating, and a terrific way to learn.

As a lowly graduate student I quickly became aware of the way in which Elwyn and my association with him opened doors for me. Through Elwyn I was able to meet many of the greats. A couple of months after my arrival at Yale, Elwyn took me to New York, to the American Museum of Natural History where I met Malcolm McKenna and Ned Colbert. I remember that day well because it was November 22, 1963, the day Kennedy was assassinated. I subsequently met George Gaylord Simpson. Louis Leakey came to Yale to deliver the Silliman Lectures in the spring of 1964. I believe he was invited mainly through contacts with John Buettner-Janusch, and Elwyn arranged for me to talk with the grand old man. I visited him in his lodgings (somewhere on the Yale Old Campus), as did Elwyn, and Louis was extremely generous in showing us casts of published as well as unpublished hominids. I remember seeing OH13, "Cinderella", and other *Homo habilis* material but have no recollection of the lectures themselves.

I went home to England for a few weeks in the summer of 1964 and while there took a side trip to Sabadell in Catalonia to see Miguel Crusafont Pairo's Miocene apes. I enjoyed his generosity in showing me the material as well as his great hospitality; with his wife, we enjoyed a subversive toast to "Free Catalonia" (Franco was still alive). After finishing the dryopithecine revision and sending the manuscript to Adolph Schultz in Zurich (editor of Folia Primatogica), I headed off to Cairo in November to join the fourth Fayum field season. I overlapped for several weeks with Elwyn, learning the ropes of desert collecting, and stayed in the field after he returned to the States. In January I moved on to Nairobi and was able to see the Miocene material there, thanks to the graciousness of Louis Leakey once again. Thence to South Africa where I was most kindly treated by Phillip Tobias and Raymond Dart, and from there on again to Zurich and a chance to meet the great Adolph Schultz. During my stay in Switzerland I had the pleasure of meeting Schultz's young research assistant, Friderun Ankel. Before returning to the States I had one last research stop in Paris, meeting one of the rising stars of French paleontology, Yves Coppens. All this was ultimately made possible by Elwyn, and through my association with him.

Elwyn's generosity made my PhD dissertation possible. Bill Bishop had recovered some interesting Early Miocene ape fossils in Uganda when he was with the Geological Survey there, and had asked Elwyn to describe them. With Bill's agreement, Elwyn arranged to turn the material over to me, and as I headed back to Cambridge I was able to take the specimens with me. During the summer of 1965, while preparing for what I had planned to be my third and final dissertation-writing year at Yale, I received an unexpected phone call from Jack Trevor in Cambridge offering me the junior physical anthropology position there. Describing those Ugandan fossil apes led me to meet Alan Walker and Mike Rose, and to join them in a couple of seasons of field work in Uganda. I also met Andrew Hill at around this time. Despite a very heavy teaching load at Cambridge I did manage to finish my dissertation. I also managed a semester's sabbatical at Yale in early 1968, and while there, was offered a job in the Anthropology Department. This coincided with my failure to be appointed to Jack Trevor's position at Cambridge (Jack had died a year earlier). I jumped at the chance, having found Cambridge after my return much less palatable than I had anticipated; to steal an old Mort Sahl joke, it was a place that didn't believe in doing anything for the first time—and I was spoiled by American attitudes and ways of getting things done.

Much was changing and had changed at Yale, with Kingman Brewster as President. Being able to work closely with Elwyn again was wonderful. The Egyptian field program had been stopped in 1967, and the famous red Dodge truck which had served so well in the Fayum was shipped for storage to Sabadell where Professor Crusafont Pairo generously gave it a home. Elwyn had also begun a field program in India, having decided to return to the Siwalik rocks from which *Ramapithecus* had been recovered many decades before. India became the venue because a future collaborator had responded to Elwyn's enquiry, while a similar request to the Geological Survey of Pakistan (GSP) had gone unanswered. Elwyn took me with him to India in early 1969 where we had a chance to visit the collections of the Geological Society of India in Calcutta. But the project was not going well and soon finished.

The vertebrate paleontology section of the Geology Department was extremely lively at that time. Phil Gingerich and Rich Kay were among Elwyn's graduate students, as was a little later John Barry, and John Fleagle was an undergraduate; Glenn Conroy was nominally my graduate student. I remember those hectic and stimulating days with enormous pleasure (Fig. 1). Soon after I had returned to Yale, I began fieldwork in Spain, and then in Kenya, greatly

Fig. 1 Elwyn Simons and David Pilbeam revising dryopithecine fossils circa 1971

helped in both places by Grant Meyer as well as by Prithijit Chatrath (who had worked on the Indian Siwalik project for Elwyn and who had then moved to Yale to join Elwyn's lab).

I planned a field season for the summer of 1973 in Kenya. But earlier that year Elwyn had received a letter from the Geological Survey of Pakistan inviting him to visit. Once again, Elwyn's generosity made it possible for me to respond to the invitation. And so it was that in the fall of 1973, Grant Meyer and Glenn Conroy, who had spent the summer in Kenya, flew to Spain, picked up the red Dodge, and drove it across Eurasia to Pakistan (not something one could do today). Phil Gingerich joined them, as did Mahmood Raza who was the young GSP officer assigned to the project (and who later completed a PhD in Geology and Geophysics at Yale). Thus began the Yale (later Harvard)-GSP project—and it still continues 32 years later. The success of the project owes a great debt to Elwyn and to those few Yale greats and future greats who got the project rolling.

There followed four more years of close contact with Elwyn—much discussion and argument, and much laughter. After Elwyn moved to Duke, our contacts inevitably became fewer and less intense (Fig. 2). But fourteen years of more-or-less continuous and intense intellectual contact have left their

Fig. 2 David Pilbeam and Elwyn Simons at a conference in 1982

mark in more than memory. I owe a very great deal to Elwyn. I value our forty plus years of friendship, and salute a very great paleontologist. We are all in his debt.

References

Simons, E. L., and Pilbeam, D. R. (1965). Preliminary revision of Dryopithecinae (Pongidae, Anthropoidea). *Folia Primatol.* 3:81–152.

Sir Wilfrid Le Gros Clark

The Making of a Paleoanthropologist

Bernard Wood

Introduction

One of Elwyn Simons' many talents is mimicry. Mimicry can be cruel and devastating or kindly and respectful; Elwyn's imitations of Sir Wilfrid Le Gros Clark are transparently of the second variety. For many these imitations are the closest they will ever come to "seeing" Le Gros Clark "in the flesh." For Sir Wilfrid Le Gros Clark (henceforth I will refer to him as "Le Gros" for that is how he was almost universally referred to when he was alive) died on the 28th June 1971 aged 76, having retired from the Dr. Lee's Professorship of Anatomy at Oxford in 1962.

Elwyn was a graduate student of Le Gros' from September 1956 to 1959. By that time Le Gros had been a Fellow of the Royal Society (i.e., an FRS) for just over twenty years and he enjoyed an international reputation as a neuroscientist, a primatologist and as a paleoanthropologist. He had been elected President of the International Anatomical Congress in 1950, he was elected President of the Anatomical Society of Great Britain and Ireland in 1951 and in 1955 the first edition of *The Fossil Evidence for Human Evolution: an Introduction to the Study of Palaeoanthropology* had appeared. He was also firmly embedded into the British scientific establishment, for a year before Elwyn's arrival in Oxford Le Gros had been knighted by Queen Elizabeth II. Anatomists are only rarely recommended for a knighthood, but Le Gros had evidently made sufficient impact as a scientist to be honored in this way. In other words Elwyn interacted with Le Gros at the height of the latter's powers and influence.

But despite Le Gros' eminence remarkably little has been written about him. He wrote in a guarded way about himself in the form of ten autobiographical essays collected together in a book with the whimsical title *Chant of Pleasant Exploration* (Le Gros Clark, 1968) (when I quote from this I will refer to it as "CPE"). This was published by the Scottish publisher Livingstone as part of a

Bernard Wood
Center for the Advanced Study of Hominid Paleobiology, George Washington University, 2110 G St NW, Washington, DC 20052
bernardawood@gmail.com

J. G. Fleagle, C. C. Gilbert (eds.), *Elwyn Simons: A Search for Origins.*

rather dry series of biographies of distinguished medical men with names like Sir George Buckston Browne and Sir Arbuthnot Lane. I would be surprised if the print run of CPE had exceeded a few hundred copies. In an introduction to the book Le Gros writes that three of the essays are based on lectures he had given and he aptly describes CPE as a "medley of autobiographical fragments" (CPE, p. vi).

Despite his urbanity, Le Gros was a shy man. For example, although he had been associated with Hertford College in Oxford for 35 years, initially in 1934 as a Professorial Fellow (for some reason lost in the mists of time, since the 17th century, Dr. Lee's Professors of Anatomy have always been made Professorial Fellows of Hertford College) and then as an Honorary Fellow from 1962 until his death, when I contacted the Fellows Librarian at Hertford College they could find no photograph, group or otherwise, that included Le Gros. There is information about his professional career in the official biography that is prepared for all recently deceased Fellows of the Royal Society, usually by another Fellow. Le Gros had the misfortune to have his obituary written by his nemesis, Solly Zuckerman. This was published two and half years after Le Gros' death and it is fair to say that Zuckerman made the very best of the opportunity to "damn with faint praise" and the bibliography included in Zuckerman's obituary is incomplete.

In this contribution I provide some biographical context for Le Gros' interest in, and contributions to, primatology and paleoanthropology. I address the following questions. When and how did Le Gros become interested in human evolution? How central was this interest professionally? What were Le Gros' views about human evolution and how did those views change over time?

Family Background

The family name "Le Gros Clark" is an amalgamation of the English family name "Clark" (it most likely originated because one, or more, of his ancestors had been "clerks") and the name of a family that probably originated in Guernsey, for there is evidence of several "Le Gros" lineages in the Guernsey records. Wilfrid (whose namesake was sanctified for helping in the 7th century to introduce the practices of the Roman Catholic Church into Northeastern England) was born on the 5th June 1895, the youngest of three sons of the Reverend Travers Le Gros Clark, a Church of England clergyman. The three brothers were evidently a close-knit group, sharing common interests in natural history and in exploring the outdoors. Le Gros claimed his middle brother Cyril was the "practical" one (CPE, p. 244) and credited the eldest brother Bill with "the most acute intellect of us all" (CPE, p. 243). Curiously, we are only told the nickname given to his eldest brother during the First World War. The name "Bill" came about because Le Gros' eldest brother sported a bushy moustache like the one worn by "Old Bill" a fictional character that featured in Bruce Bairnsfather's famous cartoons about life in the trenches in the First World War.

Le Gros' early childhood was spent in Hemel Hempstead, in Hertfordshire, where his father was the parish priest at St. Mary's Church in what was then still a small market town some 30 miles north of London. By the time he was seven years old the family had moved to a village called Newnham on the northern bank of the River Severn where for a short time has father was the vicar of the parish church. Newnham is in the county of Gloucestershire in England, but it is close to the Welsh border and is quite isolated. Two years later Le Gros' mother died and the family was uprooted once more when his father accepted the living as the rector of the church in the small village of Washfield, near Tiverton, in Devon. Washfield's proximity to Dartmoor (the wildest of three areas of moorland in the West of England; the others are Exmoor and Bodmin Moor) provided the Le Gros Clark boys with many opportunities for hikes and exploration. One of their favorite pastimes was to go off for several days at a time hiking along the banks of the rivers of South Devon tracing them from their source on Dartmoor to the sea.

We learn nothing about Le Gros' parents in CPE. In fact neither of his parents is mentioned by name in any of the ten autobiographical essays in the book.

Education

The education of the Le Gros boys followed the pattern of the sons of professional families in England in the early part of the 20th century. Le Gros went as a "boarder" to a "preparatory school" called Wells House (in his youth Sir Edward Elgar taught music there, but the school is now defunct) in Malvern Wells, a small town at the foot of the Malvern Hills. Later he followed his brothers to Blundell's School, which is a "public school" outside of Tiverton. Confusingly in England, "public schools" are actually fee-paying private schools. Blundell's is noted more for its sport, notably rugby, than for its academics. Le Gros excelled academically and apparently had neither interest in, nor aptitude for, sport.

Le Gros suggests his father had always planned for him to be medical doctor and in 1912, at the early age of 17, he had been admitted as a medical student to St. Thomas's Hospital Medical School in Lambeth, London. He had family connections there for his paternal grandfather, Frederick Le Gros Clark, who went on to become the President of the Royal College of Surgeons of England had been a distinguished consultant surgeon at St. Thomas's Hospital and his maternal grandfather, Edward Clapton, had been a consultant physician at St. Thomas's Hospital.

Like Blundell's School, St. Thomas's Hospital Medical School had a reputation for sport, especially rugby, but Le Gros seems to have been at the more studious end of the spectrum for he writes that while still a medical student he had read Charles Darwin's account of his travels in the *Voyage of the Beagle*, Thomas Henry Huxley's recollections of his voyage on H.M.S. Rattlesnake, Henry Walter Bates' account of his journeys in the Amazon between 1848 and 1862 and Alfred Russel Wallace's account of his even longer sojourn exploring the natural history of the Malaysian Archipelago. I somehow doubt whether this was the usual reading fare for a St. Thomas's Hospital Medical School student, even in the second decade of the 20th century. Le Gros wrote that his study of human anatomy led him to "further studies of the relationship with Man with lower animals, in particular his relationship to apes and monkeys that have their natural habitat in equatorial forests and jungles" (CPE, p. 8). His Anatomy Professor, F.G. Parsons, may also have played a role in shaping Le Gros' career for Le Gros wrote that Parson's "own interests were in the field of comparative anatomy and evolutionary morphology, subjects that particularly interested me as a student" (CPE, p. 38).

Physical anthropology and paleoanthropology had very different origins in the UK than in the USA. In the UK at this time there were no professional physical anthropologists other than Professors of Anatomy and if there was a center of human evolution research it was the Royal College of Surgeons of England. That had by far the largest collection of modern human and comparative skeletal material in the UK. When an archeologist uncovered the remains of, for example, Romano-British burials, these would be offered to either the Anatomy Department of the local Medical School or failing that to one of the Royal Colleges of Surgeons, and if the site was in England this would the Royal College of Surgeons of England in Lincoln's Inn Fields in London. Many Anatomy Departments, especially those like Edinburgh, which are close to major ports, had substantial collections of crania from the UK and from the British colonies. When I was last in the Edinburgh Department I went round the old "skull room". This was fitted out like a Victorian library with a gallery and ladders for accessing the shelves, but instead of books the shelves housed several hundred modern human skulls. Most of the substantial collections of modern human skeletons now held at the Natural History Museum in London are on "permanent loan" from Anatomy Departments or from the Royal College of Surgeons of England. A Professor of Anatomy in the first half of the 20th century was expected to be familiar with what at the time was still a very meager fossil record for human evolution.

The Great War: Service in the RAMC and its Sequelae

The First World War, also called by those who participated in it the Great War, had a major impact on Le Gros' generation and the Le Gros Clark boys were no exception. Unusually for a family with three sons they all survived the war, but all three were deeply affected by their experiences. His own wartime experiences, as well as those of his brothers in both the First and Second World

Wars, evidently greatly influenced Le Gros and I believe that these experiences are part of the reason for the eventual rift between Le Gros and Solly Zuckerman.

At the outbreak of the war in 1914 Le Gros was a second year medical student. Medical students were discouraged from enlisting in the armed forces - indeed they were "instructed" to continue their education so they could later be of use as doctors in one of the three services. Le Gros' two brothers, however, joined the army immediately, but under very different circumstances. His older brother, Bill, was "in the middle of his undergraduate career full of brilliant promise at Balliol College (in Oxford), and joined a Public Schools infantry battalion of the Middlesex Regiment as a private" (CPE, p. 42), whereas Le Gros' younger brother Cyril had joined the Officer's Training Corps at Blundell's School and so was commissioned as a junior officer in the Somerset Light Infantry. Somerset is the adjacent county to Devon, and the county town of Somerset and the base of the Somerset Light Infantry, Taunton, is relatively close to Tiverton and to Blundell's school.

Le Gros qualified as M.R.C.S. and L.R.C.P. (at this time most licenses to practice medicine in England were given by the Royal College of Surgeons of England and the Royal College of Physicians of London – a newly qualified doctor was a Member of the Royal College of Surgeons, but only a Licentiate of the Royal College of Physicians) in 1916 and immediately joined the Royal Army Medical Corps as a medical officer. After some basic training he was attached to a regiment that was sent to France early in 1918. However, he contracted and became seriously ill with diphtheria and had to be sent home to England to recover. But he was soon back in France posted to "No. 8 Stationary Hospital" at Wimereux in northern France, where he spent the rest of the war. Although the Great War was to end at the eleventh hour, of the eleventh day, of the eleventh month, of 1918, the last few months of the war saw no respite to the horrors of trench warfare on the "Western Front". Le Gros writes with typical reticence that "the work imposed a fairly heavy strain, particularly during the desperate fighting of the last German offensive of 1918 when convoys of wounded arrived in rapid succession day and night" (CPE, p. 44) and he admits with typical British understatement that my "memories of my work as a medical officer in France, where I had to deal with many terrible mutilations of the wounded during the last phase of fighting before the armistice, were not happy ones" (CPE, p. 8). He claims that "after the armistice there followed a short period of my life for which there is a curious blank in my memory" claiming that "I do not recall my return journey from Germany, nor have I any conscious memory of being demobilized in 1919" (CPE, p. 45).

But the armistice was a bitter relief for the Le Gros Clark brothers, for on the day the armistice was signed his much admired eldest brother was wounded and lost the sight in both eyes. Earlier in the war, in 1915, his middle brother Cyril had been badly affected by a syndrome called "shell-shock". After recovering in hospital Cyril was transferred to the Indian Army, but on the way out to India his troopship was torpedoed in the Mediterranean and he was lucky to escape with his life. Cyril later joined the Sarawak Civil Service (the Civil Service administered colonies like Sarawak) and eventually became the Chief Secretary, effectively the head of the Government and second only to the Rajah. But after the Japanese over-ran Sarawak Cyril along with other administrators was interned in prison camps. He was apparently responsible for saving the lives of several of his staff, but he was found guilty of a trumped-up charge and was transferred to a particularly harsh prison camp in North Borneo. The Australian Army invaded North Borneo just before the Japanese surrendered, but literally days before the Australians would have rescued them, all the prisoners, including Cyril, were killed. Thus, both of Le Gros' beloved brothers suffered terribly in the course of the two World Wars and as well shall see below Le Gros himself, while not physically wounded, was far from unscathed.

After his demobilization from the army in 1919, Le Gros spent the next year as a Demonstrator in the Department of Anatomy at St. Thomas's Hospital Medical School under his old Anatomy Professor, F.G. Parsons. In that year he took and passed the examinations to become qualified as surgeon (this transferred him from Membership to Fellowship of the Royal College of Surgeons of England and instead of being addressed as "Dr. Le Gros Clark" he would have been addressed as "Mr. Le Gros Clark"). He wrote that during that year he engaged himself "in the pursuit of anatomical and anthropological research" (CPE, p. 8). But he was far from content. He wrote enigmatically of "personal and intimate tragedies that affected me deeply" but he provides no clue as to what they were. He went that "it was in such a mood of moral perplexity that I was overcome with an intense longing to escape from the artificialities of civilization by losing myself somewhere in one of the remoter parts of the world" (CPE, p. 8). Within a few weeks of making this decision Le Gros, at the age of twenty-five, was appointed on a three years contract to the post of Principal Medical Officer to the Sarawak Government. The post involved substantial administrative responsibility "supervising the Government Medical Service of Sarawak" (CPE, p. 58). He left from Liverpool for Sarawak on the 10th October 1920 and he stayed at his post in Kuching until the end of his three-year contract in 1923.

An Anatomical Career

In the last year of his three-year contract in Sarawak Le Gros was encouraged by Professor Parsons to make what proved to be a successful application for the post of Professor of Anatomy at St. Bartholomew's Hospital Medical School ("Barts") in Charterhouse Square in the City of London. After six weeks leave which Le Gros used to court his future wife, Frida Giddey, he started at Barts in September 1923. Even by the standards of 1923 Le Gros' publication record was "thin" (a case report in the "Lancet" written when he was a student, a seven-page report about 18 Eskimo crania prepared when he was Demonstrator in Anatomy at St. Thomas's Hospital Medical School, and a

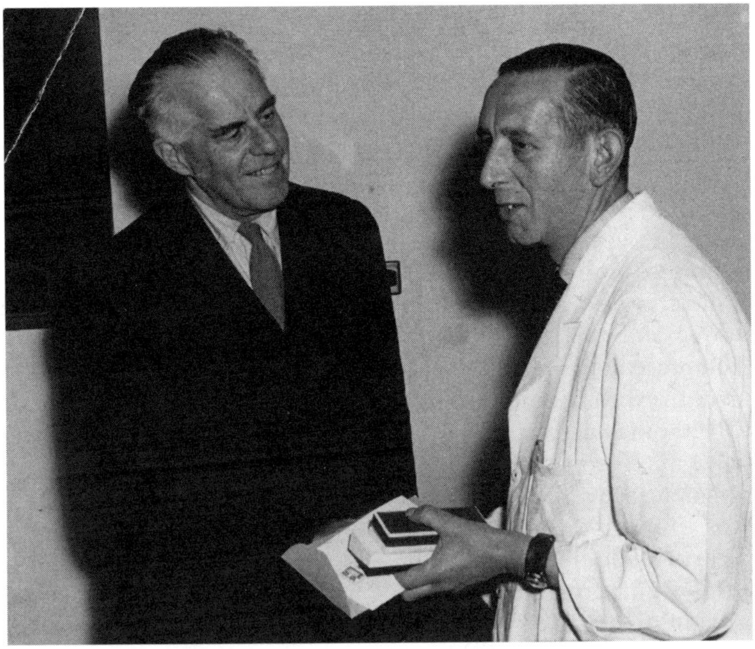

Fig. 1 Professor Le Gros Clark in 1934, the first year of his Oxford Professorship, making a retirement presentation to one of his staff. Thanks to John Morris and to the staff of the Oxford Anatomy Department

histological analysis of sensory skin receptors) is a modest publication record in anyone's book, but considering the latter two papers were written and published in the same year he managed to pass his Final F.R.C.S exam it is a considerable achievement.

The focus of the next eight years (1924–32) of his published research output was largely devoted to the comparative neuroanatomy of the primates (tree shrews and tarsiers) he had collected while in Sarawak. After six years at St. Bartholomew's Hospital Medical School in 1930 he was invited to succeed his mentor as Professor of Anatomy at St. Thomas's Hospital Medical School and then in 1934 he was invited to take over as the Dr. Lee's Professor of Anatomy (and effectively the Chair of the Anatomy Department) at Oxford University (Fig. 1).

Among the many advantages of Oxford University over a London Medical School was that Le Gros was surrounded by eminent scientists instead of eminent clinicians. This was a "heady" time for Oxford preclinical and medical science for his professorial colleagues included Charles Sherrington, Howard Florey, Robert Robinson and Cyril Hinshelwood; four of his professorial colleagues went on to win Nobel Prizes. The Professor of Geology was William Sollas and Le Gros friendship with Sollas may have helped regenerate Le Gros' interest in paleoanthropology. Despite all these academic advantages of

Oxford, and perhaps because of the brilliance of his fellow Professors, Le Gros and his family took some time to settle down. So much so that in 1939 he had been successfully courted by University College London (UCL) and just before the outbreak of the Second World War Le Gros had accepted the Chair of Anatomy at UCL. He and his wife were engaged in looking for a suitable house when the Second World War broke out, but by mutual consent Le Gros and the Provost of UCL agreed that the arrangements for him to move should be set aside in view of the exigencies of wartime.

By the time Le Gros moved to Oxford his studies of neuroanatomy had moved from the descriptive to the experimental and in a series of papers published between 1933 and 1935, including two in the Philosophical Transactions and one in the Proceedings of Royal Society, he had managed to generate a topological map of the ways the main sensory areas of the cerebral cortex map onto the thalamus, one of the deep nuclei of the brain. It was this work and his meticulous comparative microscopic morphology of the brain that earned him at the comparatively young age of 40 election to the Fellowship of the Royal Society in 1935.

Le Gros set about the task of modernizing the Oxford Anatomy Department and making new additions to the staff. One of the first of these was the appointment of Solly Zuckerman as a Junior Demonstrator. Le Gros acknowledged the important role played by Zuckerman in the process of modernization and nowhere in Le Gros' autobiographical essays is he other than generous about Solly Zuckerman. Indeed, it is worthwhile quoting one passage at length to make it clear that the antipathy shown by Zuckerman towards Le Gros Clark cannot the blamed on any lack of generosity on Le Gros' part. He wrote that Zuckerman "was of considerable help to me in the reorganizing and replanning of my Department" (CPE, p. 134) and later he states "Zuckerman remained a member of my staff for ten years: active in research himself, he had an unusual capacity for organizing teams of young research workers concentrating their attention on problems of common interest. In view of the success of his scientific career in later years, I feel happy that I was able to grant him the facilities he needed for his earlier endocrinological work in my Department, for it was this work that gave him a foothold in the long ladder of success that followed" (CPE, p. 134). This was written by the man whom Solly Zuckerman's biographer John Peyton described as being "gloomy and resentful by nature" (Peyton, 2001, p. 25). I suggest that what Zuckerman might have found objectionable about this quote was that he (Zuckerman) would not want to acknowledge that anyone other than himself was responsible for his success.

Interest in Human Evolution

Did Le Gros' undergraduate study of human anatomy kindle his interest in human evolution, or rekindle it? He entitled the eighth of his autobiographical essays "The Antiquity of Man" and he introduces it by recalling that "in my early youth my brothers and I visited Kent's Cavern at Torquay" (CPE, p. 185) and he writes of his fascination at the discovery of a hand axe embedded at the base of a six foot-high stalagmite, commenting that this must mean that the hand axe was very ancient. Kent's Cavern is now a tourist attraction and eminently accessible, but when Le Gros and his brothers explored the caves they had just been purchased by Francis Powe who used them as a carpentry shop and wood store. Even to be aware of the importance of the caves suggests that the Le Gros Clark boys were well informed about their local prehistory.

Le Gros' first published excursion into physical anthropology was the analysis of 16 adult crania taken from graves in North Greenland that had been donated to St. Thomas's Hospital Medical School (Le Gros Clark, 1920). It is well organized, clearly written and quantitative: in other words it had all the hallmarks of Le Gros' later more substantial publications. By 1928 he was evidently familiar enough with the literature to take Pycraft to task for assigning "Rhodesian Man" to a novel genus, *Cyphanthropus*. Le Gros showed that Pycraft's interpretation of the pelvis was erroneous, and went on the declare with respect to the cranium of "Rhodesian Man" that "I find it impossible to believe that a comparison between the Rhodesian skull and the skulls of Neanderthal man will justify the creation of a separate genus for the former" (Le Gros Clark, 1928, p. 207). This was the first sign of Le Gros' sensible sensitivity to the erection of unnecessary species and higher taxa.

In 1935 Le Gros presented a paper at The International Congress of Anthropological and Ethnological Sciences. This suggests that he had a parallel interest in human evolution in addition to the experimental neuroscience that in the same year had brought him an FRS. It is a *tour d'horizon* that extends from tree shrews to modern humans. He considers the possibility that "the resemblances between Man and the anthropoid apes are ... the result of parallel evolution" but suggests that "in the sum of his anatomical characters man is closely approached only by the anthropoid apes" and that a "searching analysis of human morphology leads to the inevitable conclusion that in the structure and form of the brain, the skull, the dentition and other systems, the human stem must have passed through a phase of evolution in which it so closely resembled the known anthropoid apes that it is necessary to postulate an anthropomorph ancestry for modern Man. This is not to say, of course, that in the line of human descent there was ever a form which showed the characteristic specializations of the modern anthropoid apes" (Le Gros Clark, 1935, p. 3). In this, and in other respects, Le Gros' analysis is remarkably "modern". There is little sign of "emphasis being placed on the now discredited idea of orthogenesis" which is one of the accusations leveled at Le Gros by Zuckerman in his Royal Society biography (Zuckerman, 1973, p. 222).

A year later we see further evidence that Le Gros is thinking deeply about the difficult task of interpreting the fossil record. In a paper entitled "Evolutionary parallelism and human phylogeny" he warns that because of parallelism "systematic position . . .can only be established by a complete anatomical survey, and the systematist is liable to fall into serious error if he confines his attention to one part of the body" (Le Gros Clark, 1936, p.4). In the same paper he nails his colors to the possibility of a more speciose interpretation the of the human fossil record because "there is no doubt that in the early stages of his evolution Man produced a number of different types, not all of which survived the struggle for existence" (*ibid*, p. 6). This clearly recognizes the possibility of cladogenesis. The first substantial review of the fossil evidence for human evolution came in a 1940 paper entitled "Palaeontological evidence bearing on human evolution." It is evident, even to the point of using the same illustrations, that this article was the starting-off point for his *Fossil Evidence for Human Evolution* which was published 15 years later. At this time the only South African fossil evidence with which he was familiar was the Taung child's skull. He refers to a 1936 paper in which he, Cooper and Zuckerman describe the difficulty of identifying the lunate sulcus of an endocranial cast of a chimpanzee (Le Gros Clark *et al*., 1936). He does not dismiss the possibility that creatures like Taung could be ancestral to Man, but suggests that if this is the case that "it must be accepted that the differentiation of man is to be referred back to a correspondingly early phase of the evolution of the higher primates" (Le Gros Clark, 1940a, p. 211). He also warned that the then published interpretations of the fossil evidence from South Africa may be based on "preliminary observations without an adequate comparative study" that takes "into account the variability shown in the teeth and skulls of modern anthropoid apes" (*ibid*, p. 210). Ten years later he repaired that deficiency by comparing the dentition of the australopiths from South Africa with a comparative sample of several hundred higher primate dentitions. In 1940 Le Gros also re-enters the debate about hominin taxonomy and unambiguously argues that *Sinanthropus* should be sunk into *Pithecanthropus* (Le Gros Clark, 1940b). In the same paper he warns that characters that were (and still are) being cited as characteristic of an early hominin taxon, such as the platymeric femora of *Pithecanthropus* that might have a nutritional rather than a genetic basis.

The last two of Le Gros' assessments of the significance of the South African fossil evidence before his conflict with Zuckerman were six years later, in the equivalent of a cross between a modern day "News and Views" article and an extended book review (Le Gros Clark, 1946) and three years after that, in 1949, in the first edition of the *History of the Primates* published by the British Museum (Natural History) (Le Gros Clark, 1949).

In his *Nature* review Le Gros puts Robert Broom's recently published monograph "The South African Fossil Ape-Men: The Australopithecinae" (Broom and Schepers, 1946) into context. Le Gros makes the point that the monograph provides the first opportunity for scholars to "assess independently the significance of these remarkable fossils" (Le Gros Clark, 1946, p. 863) for although many of the discoveries he refers to were made in the late 1930s the war effectively made them inaccessible to all but South African scientists. As a neuroscientist Le Gros is rightfully critical in his review of Scheper's claim that he (Schepers) could "delineate no less than twenty-six separate cyto-architectural areas and to compare them in their relative extent with those of modern apes and man (*ibid*, p. 864)." Le Gros goes on to suggest that "cortical physiologists may likewise feel inclined to demur at some of the inferences regarding function such as speech, abstract thought and motor skill which, it is suggested, can be drawn simply from the examination of a cast of the inside of the skull" (*ibid*, p. 864). But otherwise he unambiguously swings his considerable influence and reputation behind many, if not most, of Dart's earlier and Broom's recently published assessments. He concedes that perhaps if one only had access to the dental evidence then it might be possible to dismiss the South African fossil evidence as belonging to "a group of extinct apes, somewhat similar to the gorilla and chimpanzee, in which the (human) characters had developed (possibly independently) along lines almost identical to those of human evolution" (*ibid*, p. 863). However, he goes on to suggest "some most important fragments of limb bones .... allow, and even make probable, a much more startling interpretation of these fossil remains" (*ibid*, p. 863). Le Gros gently chides Broom that because "so much depends on this limb material for a proper assessment of the Australopithecinae" his description of the postcranial material of *Plesianthropus* "is tantalizingly brief (it is confined to 34 lines!)" (*ibid*, p. 864). Le Gros paid particular attention to the postcranial remains attributed to *Paranthropus* and made two main points, which to my shame, I was not previously aware of. First, he stressed the discrepancy in size between the modern human-liked sized upper limb remains and the diminutive talus which he described as "a remarkably small bone" that is "well short of the minimum recorded for modern ...mankind" (*ibid*, p. 863). Second, he stresses the medial extension of the articular surface of the head indicating "very considerable mobility of the sub-talar joint" (*ibid*, p. 863).

In *History of the Primates*, Le Gros summarizes his own assessment of the South African fossil evidence and this is worth quoting in full: -

"The question now arises, what is the proper place of the *Australopithecinae* in the classification of the higher primates? Clearly they belong to the Hominoidea, and clearly they show much closer resemblances to Man than does any of the living or fossil apes so far known; but should they be grouped with the *Hominidae* or the *Pongidae*? The answer to this question is still a matter for controversy, and no doubt depends ultimately on what criteria are employed for defining these two families. If the absolute size of the brain is regarded as the most important criterion, the *Australopithecinae* are to be regarded as apes of a very advanced type showing in the details of their anatomy a remarkable approximation to the *Hominidae*, and no close relationship to the modern anthropoid apes. On the other hand, if more emphasis is laid on the criteria of skull structure, dental anatomy and the details of the pelvis and limb bones, there can be no doubt that the *Australopithecinae* should be grouped with the *Hominidae* and regarded as exceedingly primitive types of mankind. Taxonomic difficulties of this sort, of course, and bound to arise as discoveries are made of fossils of a seemingly transitional type, and, with the increasing perfection of the fossil record, probably the differentiation of Man from ape will ultimately have to rest on a functional rather than an anatomical basis, the criterion of humanity being the ability to speak and make tools. So far as the *Australopithecinae* are concerned, no stone implements or other evidence of tool making have been found in association with their remains, and unfortunately, a study of the endocranial cast does not permit any legitimate conclusions regarding the power of speech. It may be noted that, at the original site at Taungs, where the first Australopithecine skull was discovered, a number of baboon skulls were found showing depressed fractures on the top, which suggests that they were killed by well-aimed blows with a weapon of some sort, and it has been surmised from this evidence that baboons were systematically hunted for food by the *Australopithecinae*. If this conclusion should be established as the result of further studies, it would indeed make it probable that the *Australopithecinae* were endowed with an intelligence superior to that of the anthropoid apes.

Finally the question arises whether the Australopithecine fossils so far discovered are likely to bear any direct or indirect ancestral relationship to *Homo sapiens*. A careful analysis of all the purely anatomical data brings to light no serious grounds for precluding such a possibility. Indeed it is possible to go further and to affirm that the anatomical characters of the *Australopithecinae* conform very closely to theoretical postulates for an intermediate stage of human evolution, which had been primarily based on the indirect evidence of comparative anatomy. But the place occupied by these fossils in the evolutionary history of Man will be precisely determined only when more adequate evidence is available for estimating their exact geological age. If they date from the latter Pliocene age, they may well represent the ancestral stock from which Man took his origin. On the other hand, if it should be established as the result of further systematic excavations that they are no older than the early Pleistocene, it would then appear probable that they represent late and but little modified survivors of such an ancestral stock." (Le Gros Clark, 1949, pp. 73–74).

This is a remarkably intelligent and insightful assessment of the possible significance of what was then known of the fossil hominins from southern Africa.

Conclusion

Sir Wilfrid Le Gros Clark was a complex man. In many ways he was the epitome of an English gentleman academic; intelligent, reserved, courteous and well liked by the vast majority of his colleagues and students (Fig. 2). I suspect that his youthful interest in things historic and prehistoric helped to whet his appetite for his later involvement in human evolution. From being a skeptic about whether the southern African hominins played a role in human evolution, he became a powerful advocate for paleoanthropological research in South Africa. His voice must have had an important influence on the decision to provide long term, if modest, research support for Robert Broom. Whatever the stimulus for it, Le Gros' interest in human evolution, his support of Dart

Fig. 2 Portrait of Sir Wilfrid Le Gros Clark taken in the mid 1950s around the time Elwyn Simons was his graduate student. Thanks to John Morris and to the staff of the Oxford Anatomy Department

and Broom and his subsequent support of Louis Leakey played an important role in the growth of paleoanthropology. His elegant and concise prose in the two editions of *The Fossil Evidence for Human Evolution: an Introduction to the Study of Palaeoanthropology* (Le Gros Clark, 1955, 1964) and in *Man-Apes or Ape-Men? The story of hominin discoveries in southern Africa* (Le Gros Clark, 1967) opened up the world of paleoanthropology to many students (including the writer) and they still make interesting reading today. Le Gros was also willing to "roll up his sleeves" and collect comparative data. To collect data from "the permanent dentition of 238 gorillas, 276 chimpanzees, and 39 orangutans, and the milk dentition of 89 gorillas, 105 chimpanzees and 29 orangutans" (*ibid*, p. 35) would be no mean achievement for a full-time graduate student, let alone the Head of the Oxford University Department of Anatomy and an FRS.

The long standing friendship between Le Gros and Elwyn, which began when Elwyn was Le Gros's graduate student, extended well into the latter's retirement. Elwyn would stop in England on his many journeys to and fro from the Fayum and would go and see Le Gros in the Department, and later on at his retirement home outside of Oxford. Elwyn was also host to Le Gros in Morse College when the latter taught a semester at Yale. On the face of it they were an "odd couple": one half the age of the other, one the epitome of urbanity, the other never wanting nor attempting to disguise his origins in the American West. But their common keen intelligence, humanity and shared interest in paleontology transcended these differences. Sir Wilfrid Le Gros Clark gained Elwyn Simon's respect and admiration, and he deserves ours.

Acknowledgment I am grateful to Elwyn Simons for arranging for me to attend and contribute to his Festschrift, and for sharing his memories of Le Gros Clark. I am only sorry that the serious illness of my father prevented me from attending and I am particularly grateful to Erik Seiffert for coping with this last minute change and to Alan Walker for standing in for me. I was not a student of Elwyn's nor did I have the good fortune to be his colleague (but I know many who have been), but I was pleased to be given this opportunity to thank Elwyn for his example as a scholar, for his fieldwork, for his book on "Primate Evolution", for demonstrating that enthusiasm is not the prerogative of youth, for his friendship and fellowship and for (despite his eminence) not being "all puffed up" as the Book of Common Prayer succinctly describes people who are all too conscious of their own importance. Thanks, too, to John Morris for helping me with photographs and other images of Le Gros Clark.

References

Broom, R. and Schepers, G. W. H. (1946). The South African fossil ape-men, the Australopithecinae. *Transvaal Museum Memoirs*: 1–283.

Le Gros Clark, W. E. (1920). On series of ancient Eskimo Skulls from Greenland. *J. R. Anth. Inst.* 50: 281–298.

Le Gros Clark, W. E. (1928). Rhodesian man. *Man* 28: 206–207.

Le Gros Clark, W. E. (1935). Man's place among the primates. *Man* 35: 1–6.

Le Gros Clark, W. E. (1936). Evolutionary parallelism and human phylogeny. *Man* 36: 4–8.

Le Gros Clark, W. E. (1940a). Palaeontological evidence bearing on human evolution. *Biol. Rev.* 15: 202–230le.

Le Gros Clark, W. E. (1940b). The relationship between pithecanthropus and sinanthropus. *Nature* 145: 70–71.

Le Gros Clark, W. E. (1946). Significance of the Australopithecinae. *Nature* 157: 863–865.

Le Gros Clark, W. E. (1949). *History of the primates*. British Museum (Natural History), London.

Le Gros Clark, W. E. (1955). *The fossil evidence for human evolution: An introduction to the study of paleoanthropology*. The Scientist's Library: Biology and Medicine. University of Chicago Press, Chicago.

Le Gros Clark, W. E. (1964). *The fossil evidence for human evolution: An introduction to the study of paleoanthropology*, 2nd ed. The Scientist's Library: Biology and Medicine. University of Chicago Press, Chicago.

Le Gros Clark, W. E. (1967). *Man-Apes or Ape-Men? The Story of Discoveries in Africa*. Holt, Rinehart and Winston, New York.
Le Gros Clark, W. E. (1968). *Chant of pleasant exploration*. E. and S. Livingstone, Edinburgh.
Le Gros Clark, W. E., Cooper, D. M. and Zuckerman, S. (1936). The endocranial cast of the chimpanzee. *J. Roy. Anthropol. Inst.* 66: 249–268.
Peyton, J. (2001). *Solly Zuckerman. A scientist out of the ordinary*. John Murray, London.
Zuckerman, S. (1973). Wilfrid Le Gros Clark. 1895–1971. *Biographical memoirs of fellows of the royal society* 19: 216–233.

Sir Wilfrid Le Gros Clark

Personal Recollections of Le Gros

Elwyn Simons

John Fleagle has kindly asked me to include in this volume some of my recollections of Sir Wilfrid Le Gros Clark, or Le Gros, as everyone called him. Much of the following is taken from a memoir of my life in science that I am preparing. Initially, I should say that when I was a graduate student at Oxford he was most affable and generous of his time with me although he was more than twice my age and we were from about as different backgrounds as imaginable. Nevertheless, among a lot of characteristics we had in common was our love of primates and of primate history. Some people thought that Le Gros was distant and stuffy because he was from the British establishment, but I had no such impression at all. Perhaps part of the problem was that he was slightly hard of hearing and once told me that he could not enjoy being at conferences because the "buzz" of many people talking at once made it hard for him to understand the particular person he was conversing with.

Years after Le Gros' death, Joe Weiner brought up with me the biographical note on Sir Wilfrid Le Gros Clark published in the Biographical Memoirs of Fellows of the Royal Society of London (1973, v. 19: 217–233). This had been written by Sir Solly Zuckerman who had been with Clark in the Oxford Department of Anatomy but later had gone on to a professorship at the University of Birmingham. This biographical note is unbelievably condescending and should actually be called "The Revenge of Lord Zuckerman." It is believed that Zuckerman wanted to have Le Gros help him to gain an Oxford professorship, an event that did not happen. Weiner hoped that I could write a biographical article about Sir Wilfrid that would set the record straight. Unfortunately, apart from the thoughts here, I've never been able to do this.

When I went to Oxford as a Marshall Scholar it was to work with Joseph Weiner. After I had made considerable progress on studies of early tarsioids and lemuroids that might be related to the origin of Anthropoidea, the topic chosen for my Oxford M. Sc. thesis, Dr. Weiner took my manuscript in to Sir Wilfrid's office to show him. Sir Wilfrid's immediate reaction was to say that

Elwyn Simons
Division of Fossil Primates, Duke University Primate Center, Department of Biological Anthropology, Duke University, 1013 Broad Street, Durham, NC 27705, USA

J. G. Fleagle, C. C. Gilbert (eds.), *Elwyn Simons: A Search for Origins.*

this project should be "put forward" for a D.Phil. This would become my second Ph.D. (D.Phil. + Ph.D.). In order to do the further studies necessary to complete a Ph.D. dissertation at Oxford, I submitted to the Marshall Scholarship authorities a request for a further (third) year of support that was granted. After that development Sir Wilfrid began to take a more active interest in my work as an advisor, as did Dr. John Napier (University of London), who was chosen as a third reader for my dissertation. From then on I began to meet regularly with Sir Wilfrid. As head of the department, he was very busy, but was always happy to make appointments to discuss aspects of primate evolution with me. Throughout he seemed to be completely available and at ease during these conversations, which were very wide ranging and cordial. In retrospect, it seems somewhat improbable that a Texan from a Kansas family and a distinguished Oxford professor would have very much in common, but our interests, mainly about primates, overcame our differences and I enjoyed very much the many conversations we had together. In these conversations he also told me much about himself and many of the points he made to me are also expanded on in his autobiography, *Chant of Pleasant Exploration* (E. and S. Livingstone Ltd., Edinburgh and London, 1968).

His father was the Reverend Travers Clark, who was a vicar in the Midlands (Hemel Hempstead). I asked Le Gros about how he became a scientist, and he said that he was already very interested in natural science as a boy, and that he didn't always find his father too helpful in understanding the world around us. Le Gros said his father was a "primitive." I asked him what he meant by that word. He said that on one occasion while young, he had said to his father, "What is the wind?" His father replied, "Only God knows." It seems that Le Gros was unsatisfied with this type of explanation. It was obvious that Le Gros and I were in total agreement that meteorological phenomena could and can be understood. One matter which Le Gros mentioned was that he, when young, had been a stutterer. He said that, on one occasion, he had taken the train down to London to see a very early psychiatrist who, Le Gros told me, had studied with Sigmund Freud in Vienna, in the hope that this doctor would have something to recommend about stuttering, but the London specialist was unable to help him. Returning by train, he said to himself, "Wilfrid, Wilfrid, you'll just have to do it yourself." Somehow, in fact, he did manage to suppress the stuttering, which by the time I knew him was completely eradicated. Zuckerman, in the aforementioned Royal Society biography, states that he did take psychiatric treatment for this stammer, which is evidently incorrect.

Sir Wilfrid Le Gros Clark, in 1920 when he was only 25, received an unusual appointment of Chief Medical Officer of British North Borneo. He told me that he had served in the First World War, in the RAMC in France from early in 1918, and had been demobilized in 1919. The violence of that war and loss of life had an extremely disillusioning effect on him, and he decided that he would like to "get right out of Europe" and the "artificialities of civilization." He noticed an advertisement for a competition to select a Chief Medical Officer for British North Borneo, and decided to apply. To his surprise, after interviews and

I believe written and/or oral examinations, Le Gros was chosen for this position. For one so young, the position of Chief Medical Officer in North Borneo was demanding, and led to extensive contact with all sorts of illnesses. Being in Sarawak, he also rekindled his childhood interests in natural history. I recall, for instance, on one occasion he said he had a tarsier living in his garden. During this period he also was able to study the full range of what were then considered primates, from the orangutan to the pen-tailed tree shrew.

He was particularly interested in the latter species, which was extremely rare (he afterwards wrote a monograph on the pen-tailed tree shrew). Through the medical services, which his department provided in Borneo, he became acquainted with many of the local tribesman, including members of the Sea Dyaks. He had developed a method of treatment for "yaws" (similar to syphilis in symptoms but not a venereal disease) by a solution of Neosalvarsan in distilled water, and the injections were effective in the short-term. This treatment led to Le Gros's acquiring a reputation as a "medicine man."

Le Gros also liked to go on long river explorations with a group of Dyak men that he had befriended. On one occasion he was traveling with them they paused at the riverside, which Le Gros referred to as a "green grassy bank," where he lay down for a few moments. He said to me that while relaxing on the grassy bank he had a very strong but mystifying feeling of fulfillment and achievement. Then he remembered that, as a child, he had been fascinated by books about exploration up the Amazon by 19th century naturalists, and he realized that the feeling of fulfillment was that he had now done something resembling his childhood dream of exploration. His admiration for the Dyak was strong, and at one point he joined with them in some ceremony which required a tattoo on the forearm. This he showed me one time at Oxford many years later by pulling up his sleeve.

Somewhat before the time of Le Gros's retirement, a group of admiring former students got up a collection and hired the noted sculptor, Sir Jacob Epstein (1880–1959), to prepare a commemorative bust of their distinguished teacher. The bust is a good likeness, but made larger than life with an expansive forehead and impressive pose and expression. This bust was, in due time, set up outside Sir Wilfrid's offices. One day when I was passing down the hall I noticed that Sir Wilfrid was showing the bust to a group of visiting scientists. As I passed by I overheard him say, "If it's not the way I am, it's the way 'GAWD' should have made me!"

Epstein made another bust at almost the same time of a prominent master at an Oxford College. Student wits told a story around Oxford that the two busts looked so much alike that they must have been made from the same original model in order to save time. It seems impossible that this could have been the case, but if so, Sir Wilfrid was the more fortunate of the two, because the bust does really look like him (see frontispiece in his autobiography *Chant of Pleasant Exploration,* Livingstone Ltd., 1968). I am reminded of a limerick that goes somewhat as follows:

A remarkable family named Stein,
There's "Gert", an there's "Ep" and there's "Ein"
"Gerts'" poetry's bunk,
"Eps'" statues are junk,
And nobody understands "Ein".

On another occasion I passed Sir Wilfrid in the hall and sensing that he had paused I turned around to look back at him. He had turned and was standing looking at me. Sir Wilfrid was imposing in his manner and spoke very clearly. While standing there he remarked, "I am going to 'lectua' on the middle ear." After this there was a long pause and then a second dictum: "I'm nothing. . .(long pause). . .if not versatile." Le Gros Clark was anything but arrogant, and in this case must have been thinking that the middle ear was a subject rather far afield from his more regular scholarly concerns.

Le Gros Clark's lecture style was remarkable. He spoke measuredly and very lucidly with a delivery timed to make him easy to understand. Le Gros was very emphatic that one should not search for and use complex scientific terms any more than one had to. My friend and classmate at University College, Bill Haddow, son of the noted British cancer researcher, Sir Alexander Haddow, told me on more than one occasion that the whole experience of Oxford would be worth it just to have heard one lecture of Sir Wilfrid's.

Sir Wilfrid had an interesting comment to make about the Piltdown forgery. Some years after his retirement, I visited him at his home in the Cotswold region of England, and during conversation he remarked that at least once he had woken up at night and sat up, wondering if the Piltdown evidence might not be true after all. He pointed out to me that, of course, it might have been regarded with suspicion much earlier because of the association of an ape-like jaw and teeth with fully modern cranial parts were it not for the fact that at a second site, Sheffield Park, at about two miles away from the Piltdown gravel pit (Piltdown I), similar thick skull fragments and teeth had been found-Piltdown II. Being logical, and of course not suspecting forgery, he stressed that the two odd associations tended to confirm each other. It must have been hard for him to thoroughly discard Piltdown man after its exposure as a forgery. He was operating on the basis of everyone's adhering to strict scientific honesty. Although it was conceivable that a human fossil with a thick cranial vault could be found at a site that also yielded an ape jaw, or ape teeth, this happening twice was far less likely to be a coincidence. Of course, at that time we knew that no late Tertiary ape had ever been found in the Plio-Pleistocene deposits of Europe. What we were not aware of then was that, as is commonly now agreed, the faker used other fragments of the thick skull and other teeth from the same orangutan jaw to make up the specimen from Sheffield Park.

Le Gros had an older brother Cyril whom he greatly admired. Cyril also went out to the Indonesian region where he eventually rose to be Chief Secretary of British North Borneo. However, during the Second World War he was captured by the Japanese and locked up with many other Europeans and Australians. In 1945 and in a completely barbaric manner, when a group of Australian forces were almost at the point of reaching the prison camp to free the captives, the Japanese officers of the camp ordered that all the prisoners be shot, including Le Gros' brother. Cyril, however, did find the chance, as Le Gros told me, to write a last message to his family, found scratched on a piece of bark and nailed to a tree.

Although Sir Wilfrid's papers in some bibliographies are listed under "Le Gros" he told me that it was not part of his surname but that for decades it was a tradition that males born in his family were given Le Gros as a middle name. Somewhere back in his ancestry a woman with the surname Le Gros had married into the Clark family. I believe that the family were members of the Salter's guild, and were from the island of Jersey or Gurnsey. Le Gros eventually became a Grand Master of the Salter's Company of London, a medieval guild, and he sat for a portrait in the robes of that office. He told me, I believe, that his connection with the Salter's was through the Le Gros family.

It seems that Le Gros found things American somewhat mystifying. One time he showed me a New Yorker Magazine and wanted to known why the various cartoons were funny. To the best of my abilities I told him but was not always sure myself. Another time he professed mystification about the frequency of music in America. He had traveled in the States and he said: "You have music playing in your restaurants, in your lifts, and even in your lavatories!!"

After I had been at Yale for some time John Buettner-Janusch of the anthropology department took a sabbatical leave and he suggested that Le Gros should be hired to teach in his place. To my considerable surprise Le Gros accepted. I don't recall exactly how it came about, but since at that time I was a bachelor resident fellow of Morse College it was decided that Sir Wilfrid would stay in my apartment there which had two bedrooms. By this circumstance we became room-mates! There were many long and detailed discussions about primate evolution. One incident I remember clearly involved a book edited by Sherry Washburn, *The Social Life of Early Man* that contained an article by the great French paleontologist Jean Piveteau. On page 13 of that volume Piveteau wrote: "It has been possible to reconstruct the general behavior of *Mesopithecus* (a Miocene fossil monkey from Greece). Their speed going forward must have been almost as high as their backward speed." Le Gros found this droll remark quite amusing. It must have been a mistranslation from the French. He was sitting in the living room of the Morse College apartment. With dry wit he remarked to me: "I find these sentences impossible to understand, but I see from the spine of the book that the editor is S.L.Washburn. I'm going to write Sherry to seek clarification." And he did. Days later he told me Washburn's reply stated only: "The error has been pointed out to me before."

While Le Gros was in the States we visited together the American Museum of Natural History in New York and went out to lunch with Malcolm McKenna, then curator of paleontology. Malcolm knew of the tensions that existed between Sir Wilfrid and Solly Zuckerman who by then had become an advisor to the British defense ministry. Kidding Le Gros, Malcolm remarked: "Sir Wilfrid do you believe that your country is well-defended? Le Gros replied: "I do not!"

While Sir Wilfrid was staying at Morse College with me I received a phone call in the middle of the night after Le Gros had gone to bed. The call was from Le Gros' family doctor saying that Lady Le Gros Clark had passed away from a sudden flu. Fortunately, I had never before had to deal with someone's death but I knew that Le Gros would have to return at once to England and was in a quandary as to whether to wake him up with the bad news. After phoning several people who all said I should wake him, I decided that doing so would be unkind and accomplish nothing and so I waited till morning to tell him. Several times afterwards he repeated his thanks to me that I had delayed giving him the sad news.

In later years, I stopped by to visit him more than once, going or coming from Egypt, and we continued many discussions. He once expressed to me that I, and in fact all younger colleagues, were fortunate because we would surely live to see many new discoveries of fossils concerning the course of human evolution. This has happened to an even greater extent than he could have imagined.

References

Le Gros Clark, W. E. (1968). Chant of Pleasant Exploration. E. and S. Livingstone, Edinburgh.

Zuckerman, S. (1973). Wilfrid Le Gros Clark. 1895–1971. Biographical memoirs of fellows of the royal society 19: 216–233.

Washburn, S. L. (1962). Social Life of Early Man. Aldine, Chicago.

Human Evolution and the Challenge of Creationism

A speech given at Duke University, September 16, 2005

John B. Oakley

Good morning. I'm delighted to be able to add to these proceedings greetings from the University of California, an institution that over the years has made more than a few contributions to the advancement of science, including the field of vertebrate paleontology and its sub-discipline, human evolution. Elwyn Laverne Simons is responsible for a great deal of what we know about mammalian evolution in the Eocene and Oligocene periods. On his shoulders stand many of my colleagues at the University of California who seek to advance our understanding of how *Homo sapiens* evolved from predecessor species. Elwyn has also been my friend for 36 years, and has allowed me to accompany him on many field expeditions to the Big Horn Basin in Wyoming, and twice to the Fayum in Egypt, where I have witnessed the art of collection and have even found a few fossils. Indeed, I see many of my favorite fossils among the audience today. By and large, however, what I found throughout the Fayum and the Big Horn Basin was an immense personal collection of gribblies and grodes. My disciplines are law and philosophy, not vertebrate paleontology, so I have a hard time telling a scute from a vertebrae, let alone a dusty rock. So why am I, a mere lawyer and philosopher, here today?

Why am I here today?

That, of course, is the question I seek to address. One can hardly hold a conference these days on primate evolution without acknowledging that many people, some in very high places, question the intellectual integrity of the whole theory of evolution. You're all familiar with the main strains of the counter-argument: whatever clever theories can be contrived to explain how organic molecules can evolve through photo-synthesis out of an inorganic molecular soup, the evolution of man requires a quantum leap in complexity that defies

John B. Oakley
University of California at Davis, 400 Mrak Hall Drive, Davis, CA 95616-5201, jboakley@ucdavis.edu

scientific explanation. If among the gribblies and grodes one finds evidence of biological complexity on the level of a watch, these anti-Darwinists say, that implies not a statistically implausible process of random mutation and adaptive selection but instead the purposeful work of a cosmic watchmaker. Should not we be teaching the future scientists of America about the truth claims of this counter-argument, this new species of creationism called "intelligent design?"

First of all, let me confess that I believe in intelligent design. In a small way. I'm not an evolutionary scientist. I'm not a scientist at all. But I'm speaking to you today. That's because I'm married to the organizer of this conference. QED. Humans are intelligent, and human systems are, or at least can be, intelligently designed. But of course that's not the claim that proponents of intelligent design make in seeking to introduce their concept of science into the curriculum of the public schools. They see intelligent design as a theory that implies the existence of divinity, by which I mean some intelligent agent prior to and independent of the material universe.

When we speak of divinity, we speak in ordinary language of the existence of God. Some of you don't believe in God, and some of you think that the existence of God is uncertain. Others, myself included, believe in God as a matter not of science, but of faith. There is thus a secular trinity of atheists, agnostics, and believers in some sort of God beyond and behind the known universe. Historically, this has been seen as a debate about metaphysics, about first causes, not as a debate about the fabric of the physical world that science seeks to explain. But the current debate about teaching intelligent design as a theory of science in the public schools has introduced a new element into contemporary science. That is a matter of philosophical as well as scientific import, and that is what I want to examine today. It is also a matter of fundamental importance to contemporary educational policy. Is there any reason why Elwyn Simons and other great scientists should concede that their work is inadequate to refute the skeptical claim that human evolution necessarily implies the existence of an intelligent designer? Should intelligent design be forced into the curriculum of public schools and universities?

I say no, emphatically and with profound conviction that the teaching of science needs to be defended against any such qualification or distortion. I shall seek to demonstrate that the theory of intelligent design is founded on a fundamental philosophical mistake quite independent of background beliefs about the existence or non-existence or fundamental unknowability of the existence of God. My argument turns on essential concepts that go to the root of all human knowledge: causation, consciousness, intelligence, and free will. What I seek to establish is that intelligent design falls on the far side of a proper divide between science, where empiricism is sovereign, and religion, where the Constitution demands accommodation of competing tenets of faith.

Let me start with the most important of the four concepts I invoked: causation. The philosopher John Mackie (1974) called causation "the cement of the universe," and used that phrase as the title of a famous book. The whole glorious history of scientific enlightenment, on which our material well being so conspicuously depends, can be summarized as an accelerating process of cultural acceptance of causal rather than supernatural explanations of natural phenomena. Proponents of intelligent design as a theory of human origins are enemies of scientific enlightenment, because they seek to insert into a causal account of biological evolution an indeterminate number of asterisks or ellipses, points at which uncaused events happened, and changed the course of prehistory.

When I speak of "uncaused events" that have ostensibly changed "the course of prehistory," I refer to events that break a scientifically verifiable sequence of material cause and effect unfolding from the origins of the material universe to what I might grandly call the beginning of history. History begins when one species—our species, homo sapiens—acquires the capacities of consciousness and intelligence and choice that afford it self-awareness and the ability through language to report experience across generations. I'll have more to say about causation and the human mind later. For now I want to limit my discussion to events in the prehistorical world, when questions of the causal nature of human action can be put to one side. To attribute prehistorical events to a supernatural cause, to a divine intelligence purposefully molding the universe to its ends, renders unintelligible what science seeks to do. What rational view of science can explain a prehistorical world in which miracles rather than mutations randomly occur—randomly in the sense of following a pattern or purpose beyond human comprehension? In my view, arguments for teaching students the viability of the theory of intelligent design threaten to dissolve the cement of the universe as we know it, and if widely accepted would return us to the world of superstition and chaos we dimly know as the dark ages.

The consequences of such a skeptical view of science extend far beyond theories of prehistory, of course. The modern world, the world of history as it has unfolded up to the present day, is founded on scientific knowledge of the material universe with which we humans interact. Imagine what our modern world would be like, what it would become, if, say, meteorology or medicine were qualified by general cultural acceptance of rain dancing and faith healing. Is this the course we want meteorology and medicine to take, because occasionally outcomes occur that defy statistical probabilities? What sort of lassitude would causal skepticism induce in forecasting and reacting to hurricanes and heart attacks? And of course these are only two among myriad examples that could be offered of the loss to humanity were causally-based science to be supplanted by a shared belief that the course of material events is haphazardly subject to causal interventions that we can perhaps observe but—short of knowing the mind of God—can never understand. A belief in material causation does not always allow us to control the forces of nature, but at least it allows us to make predictions that minimize harm and thus maximize human welfare. Today our nation mourns the tragedy of New Orleans, and in seeking to recover economically finds itself caught between Iraq and a wet place. Tomorrow threatens us with many more avoidable tragedies if we let down our causal guard, and become objects rather than subjects in molding our fates.

When I speak of proponents of intelligent design, I'm tempted to call them intelligent designers, but that seems far too kind. There is another term for this kind of causal skeptic that makes more clear their standing in the history of ideas. Let's call them shamans. I mean to be descriptive, not pejorative. Intelligent design invokes magic to make sense of the prehistorical universe. That's shamanism, or at least a passive cousin of it that does not make the conjurer's claim of occult control but still is willing to substitute magical processes for cause and effect in the shaping of prehistory and the natural world that history has inherited and inhabits.

Nothing I have said so far is likely to provoke much dissent from within the scientific community, or by and large from within the more general academic community. But some of my academic colleagues go farther, and equate religious faith in general with anti-intellectualism. I now want to talk briefly of consciousness, intelligence, and free will to urge caution in thinking that the rejection of creationism as a theory of prehistory, whether in the crude form debated in the Scopes trial or the sophisticated form of intelligent design, commits one to a rejection of religion in general. The study of causation in history rather than prehistory must take account of the capacity of humans to be, if we believe the evidence of our daily lives, uncaused causers.

At a very early age human beings, absent pathology, develop consciousness. This sense of self-awareness, whether or not we can eventually map its potentiality to a particular sequence of the human genome, is and is likely long to remain a philosophical puzzle of the first order. Add in intelligence—the ability not only to perceive ourselves in the context of the material universe but also to develop explanatory concepts such as causation that allow us to understand the natural world sufficiently to manipulate its processes—and causal theories of human action become more matters of faith than science. There are determinists, to be sure, who proclaim us to be self-conscious billiard balls careening through lives of cause and effect internally as well as externally. But they do so only as a matter of professional theory. Such folks, like the rest of us, live lives of constant eternal debate about what to have for breakfast and for whom to vote. Regardless of what we profess to think, we all act as if we have free will, and I think we do. Our consciousness and intelligence give us the capacity to make choices, and so to shape history in ways that no theory of cause and effect can yet quite explain. We are uncaused causes that make history fundamentally different from prehistory. That seems to me to make us all, in a very small but vastly significant way, divine.

Something, and I can say ecumenically "who knows what?", had to cause the causal universe. An uncaused cause. And if human beings have the free will that we all assume we have in our personal if not our academic lives, the historical world cannot be explained solely in causal terms. Here, in our self-aware perception that we have a capacity for choice that is vaguely divine, and in our intelligent efforts to understand how we can be both of and apart from the causality of the material world, there seems to me ample room for at least the rational contemplation and perhaps conviction of religious belief. But that room exists at the back end of human evolution, not among its precursors. The challenge of creationism to human evolution is ultimately a challenge to our scientific understanding of the world in which we live, not to our humanistic understanding of ourselves. It is we who are challenged to be intelligent designers of a world we can reshape to suit our own purposes. God help us if we reject science in the exercise of this divinely human prerogative.

So thank you, Elwyn, for living a life dedicated to science, and for doing so much to help history understand and learn from the causal processes that created prehistory and are the hinge upon which human welfare so vitally depends every day. The world would be a much better place if everyone consciously, intelligently, and freely chose to live lives such as yours.

References

Mackie, J.L. (1974). *The Cement of the Universe: A Study of Causation.* Oxford University Press.

Section 2
The Fayum and Other Fossil Adventures

Although he is probably best known for his ongoing expeditions over five decades in Egypt, Elwyn Simons' paleontological work extends to many other places and many other continents. From his days as a graduate student at Princeton University he has conducted fieldwork in Paleocene and Eocene of Wyoming and built up major collections of early Eocene mammals at Yale. It was his field expeditions to Wyoming in the early 1970's that led to the ongoing work by Ken Rose and Tom Bown in the same area. During his tenure at Yale, he organized two seasons of fieldwork (1967–1969) in the Indian Siwaliks that led to the discovery of *Gigantopithecus bilaspurensis* as well as a brief expedition to Iran and an aborted effort in Libya. In subsequent years, as noted by David Pilbeam in the preceding section, he was instrumental in setting up expeditions in Kenya and Spain as well. The papers in this section reflect much of this diversity in Elwyn's paleontological history and interests.

Simons' five decades of paleontological expeditions in the Fayum region of Egypt are probably unmatched in the history of vertebrate paleontology in both the duration of work and the discovery of new primate taxa. It is thus fitting that this section begins with a photographic record of the many field crews that have been involved in Fayum fieldwork from 1959 through 2006. In the second contribution of this section, "Geology, Paleoenvironment, and Age of Birket Qarun Locality (BQ-2), Fayum Depression, Egypt", Erik Seiffert and colleagues describe the latest fieldwork in Egypt that extends the record of fossil primates and other mammals back in time to the early late Eocene. This is followed by a review by Elwyn Simons of the mammalian fossil record from the Fayum, "Eocene and Oligocene Mammals of the Fayum, Egypt", much of which is the result of his own expeditions. In the third contribution, "Early Evolution of Whales: A Century of Research in Egypt", Phil Gingerich provides a fascinating and thorough historical review of the extraordinary record of whale evolution that has come from work in Egypt over the past 125 years and continues today. Some of the most spectacular primate fossils recovered in Egypt over the past have century have been skulls. Rich Kay and colleagues describe aspects of the cranial anatomy of *Aegyptopithecus* and what this tells us about cranial evolution in early anthropoids of Africa and South America in their contribution, "The Basicranial Anatomy of African Eocene/Oligocene Anthropoids. Are There Any Clues for Platyrrhine Origins?".

Two contributions are devoted to paleontological work in other parts of Africa. Paying homage to Simons' extraordinary paleontological discoveries in Egypt, in "Paleontological Exploration in Africa: A View from the Rukwa Rift Basin of Tanzania" Nancy Stevens and colleagues describe the results of their recent work in Tanzania. In the Rukwa Basin they have found both Cretaceous deposits yielding dinosaurs and Paleogene deposits yielding rodents, primates and a variety of other vertebrates. Similarly, in "Return to Dor Al-Talha: Paleontological Reconaissance of the Early Tertiary of Libya" Tab Rasmussen and colleagues note Simons' initial attempts at fieldwork in Libya that was thwarted by the revolution of 1969 and then describe the results of their own recent surveys in 2005.

Throughout his career, Elwyn has been involved with paleontological research in Asia in one way or another. In "Revisiting Haritalyangar, the Late Miocene Ape Locality of India", Rajeev Patnaik discusses the history of paleontological research at this site, including Yale –Chandigarh Expeditions of 1967–1969 that yielded the jaw of *Gigantopithecus bilaspurensis* and other fossil primates. He also summarizes recent dating results that bring the dates of this site more in line with results from similar deapoits in Pakistan. For many years Elwyn has been was involved in the debates over the status of the fossil primates *Amphipithecus* and *Pondaungia* from Myanmar (formerly Burma). This controversy continues, and in their paper entitled "Revisiting Primate Postcrania from the Pondaung Formation of Myanmar: the Purported Anthropoid Astragalus" Gunnell and Ciochon reevaluate a fossil talus recently described from Myanmar and question its anthropoid status. In another paper devoted to fossil primate foot bones, "A Haplorhine First Metatarsal from the Middle Eocene of China", Dan Gebo and colleagues describe a fossil first metatarsal from the middle Eocene of China. Although not clearly associated with any particular taxon, it offers clues to the evolution of grasping abilities in Eocene haplorhines from Asia.

As noted above, Elwyn has maintained a tradition of fieldwork in the Paleocene and Ecocene of Wyoming ever since his days as a graduate student. He has described numerous fossil mammals from the Western early Tertiary and written extensively on the affinities of various Eocene primate groups. Following in this tradition, Pat Holroyd and Suzanne Strait describe some newly recovered fossils of the omomyid primate *Loveina* and discuss phylogenetic relationship among the poorly understood washakiine omomyoids in "New Data on *Loveina* (Primates: Omomyidae) from the Early Eocene Wasatch Formation and Implications for Washakiin Relationships".

Although best known today for the work on early anthropoids from the Fayum, in the 1960's and early 1970's Elwyn was at the center of research on early hominid evolution. This aspect of Elwyn's career is well-documented in the contribution by Robert Sussman and Donna Hart, entitled "The Behavioral

Ecology of Our Earliest Hominid Ancestors". In this paper derived from their recent book, *Man the Hunted*, the authors review efforts at reconstructing the behavioral ecology of early hominids, calling into question many attempts to find vicious killer apes in our evolutionary lineage.

Five Decades in the Fayum

Elwyn Simons, Prithijit Chatrath, Christopher C. Gilbert and John G. Fleagle

Introduction

For nearly fifty years now, Elwyn Simons has been leading expeditions to the fossiliferous deposits north of Birket Qarun in the large area known as the Fayum Depression. As documented in hundred of publications (see Appendix), including many in this volume, these expeditions have led to remarkable and unparalleled collections of fossil mammals from the Eocene and Oligocene of Africa and a spectacular and unimagined diversity of fossil primates and many other groups of mammals.

In addition, these expeditions have been a training ground for several generations of students from all over the world. Many have gone on to successful careers in paleontology, anthropology, zoology, or geology. Others have become teachers, doctors, lawyers, executives or other professionals. All look back on their weeks in the Fayum as a special time in their life when they learned the techniques of fossil collecting, learned the bureaucratic and logistic difficulties of running a field camp in a remote site in the African desert, learned a bit about geology, learned a lot about the history of the field as well as the Simons and Cuddeback families, ate a lot of eggplant, and made many new friends from many cultures. They learned about the ability of the desert wind to collapse a tent of any size, and the treacherous nature of soft sand with the ability to stop any vehicle.

On the following pages, we have put together a gallery of field crews from the initial exploratory trip in 1959 through the 2006 field season. Those from 1977 onward are "real" photos. For some of the earlier expeditions in the 1960's there were few formal group photos, so we have used digital technology and Elwyn's well-known artistic skills, to patch together images of as many members of the crews as possible.

This is DLC publication #1023.

Elwyn Simons
Division of Fossil Primates, Duke University Primate Center, Department of Biological Anthropology, Duke University, 1013 Broad Street, Durham, NC 27705, USA

J. G. Fleagle, C. C. Gilbert (eds.), *Elwyn Simons: A Search for Origins.*

Fig. 1 1961/62: From left to right, Frans Sami Faheem, Abu Hassan Ali Issa, Donald E. Russell, Ibrahim, unidentified, Mahmood Bishir Mahmood, Donald E. Savage, Yussef Shawki Moustafa, Paul Lemke. Inset: Elwyn Simons

Fig. 2 1962/63: From left to right, Mahmoud Abdel Rahman, James G. Mead, Elwyn Simons, Arnold D. Lewis, Abdul Mohammed Ahmed, Sayed Moustafa

Fig. 3 1963/64: From left to right, Charles Seymour III, Reis Unis, Grant E. Meyer, Elwyn Simons. Inset: Grant and Marty Meyer

Fig. 4 1964/65: From left to right, **Bottom Frame:** Baher El Khashab, Reis Unis Abdulla, Elwyn Simons, Ibrahim Kalifa Fadel, David Pilbeam, Mohammed Suliman. **Upper left inset:** Grant Meyer and Thomas Walsh at Wadi Hitan (Zeuglodon Valley). **Upper right inset:** Jeffrey A. Smith, Darwish el-Farr (partly obscured), Grant E. Meyer

Fig. 5 1965/66: From left to right, **Back Row:** Reis Unis Abdulla, Grant Meyer, Lloyd Tanner, Baher el Khashab, unidentified; **Front Row:** Sayed Moustafa Osman, Bruce Bowen

Fig. 6 1966/67: From left to right, **Back Row:** E. L. Simons, Baher el Khashab, Carl Vondra, Grant Meyer, John Boyer, Lloyd Tanner, Dennis Powers; **Front Row:** Reis Unis Abdulla, unidentified, unidentified

Fig. 7 1967/68: From left to right, Mohammed (assistant to Baher el Khashab), Wayne Meyer, Grant Meyer, Ken Rose, Tony Gaston, John Fleagle

Fig. 8 1977: From left to right, Baher el Khashab, Lloyd Tanner, Scott Wing, E. L. Simons, Debbie Meinke, Steve Knisley, John Fleagle, Mohammed Abdulla

Fig. 9 1978: From left to right, **Back Row:** Baher el Khashab, Karen Messenger, Lloyd Tanner, Charles Messinger, Elwyn Simons, Achmed el-Awadi Qandil, Wahid Ibrahim; **Front Row:** Mohammed Abdulla, unidentified

Fig. 10 1979: From left to right, **Back Row:** John Fleagle, John Kappelman, Mohammed Abdulla, Mary L. Tanner, Baher el Khashab, Lloyd G. Tanner, Susan N. Tanner, Abdel Ghani Ibrahim, Fathi Abed Rabouh, Kenneth D. Rose; **Front Row:** unidentified, unidentified, unidentified, unidentified

Fig. 11 1980: From left to right, **Back Row:** Bruce McKenna (on top of mattresses), Rich Kay (on top of truck); **Middle Row:** Wahid Ibrahim, Mary Kraus, Elwyn Simons, Halsey Frank, Mohammed Abdulla, Lloyd Tanner, Abdel Ghani Ibrahim, Herbert Covert; **Front Row:** Radwan, Achmed el Awadi Kandil, Abdel Azim, Tom Bown (seated), unidentified

Fig. 12 1981: From left to right, **Back Row:** Scott Wing (seated on truck); **Middle Row:** Michael Stuart, Lloyd Tanner, Elwyn Simons, Herbert Covert, Tom Bown, Prithijit Chatrath.; **Front Row:** John Fleagle

Fig. 13 1982: From left to right, **Back Row:** Said Moustafa Osman, Elwyn Simons, Achmed el Awadi Kandil, Rich Kay (standing); **Middle Row:** unidentified, unidentified, Tab Rasmussen, Mohammed Achmed Abdulla, Daniel Gebo, Abdel Ghani Ibrahim Ibrahim, unidentified, Tom Churchill; **Front Row:** Rick Madden

Fig. 14 1983: From left to right, **Back Row**: Elwyn Simons, Rick Madden, Casey McKinney, Dan Gebo, Gregg Gunnell, Prithijit Chatrath; **Middle Row:** Maroof Goma Abdulla, Refaay Ibrahim, Said Moustafa Osman, Marc Godinot; **Front Row:** John Fleagle, Bedawi Abdelbaki Ali, Bishir Hassan Ali, Mohammed el Soghair, Gaber Achmed Awaga, Tom Bown, Ali

Fig. 15 1984: From left to right, **Back Row:** Prithijit Chatrath, Bedawi Abdel Baki Ali, Ali Faisel, Marc Wolf, Said Mohammed, Said Faisel, Tab Rasmussen, Baher el Khashab, Mohammed Hassan Taha; Front **Row:** Maroof Goma Abdulla, Bruce McKenna, Abdul Ghany Ibrahim, Casey McKinney, Mohammed Hassan, Elwyn Simons

Fig. 16 1985: From left to right, **Back Row:** Said Mohammed, Bedawi Abdel Baki Ali, J. P. Watters, Elwyn Simons, Friederun Ankel-Simons, Tom Bown, Yousry Attia, Yasser El Safori, Prithijit Chatrath; **Middle Row:** Cornelia Simons, Verne Simons; **Front Row:** Ali Faisal, Bert Covert, Mohammed Hassan Taha, Achmed Mohammed, Said Faisal, John Fleagle

Fig. 17 1986: From left to right, **Back Row:** Casey McKinney, Ali Faisel, Said Faisel, Saber Yahaya, John Fleagle; **Middle Row:** Jeff Brown, Elwyn Simons, Prithijit Chatrath, Alex Van Nievelt, Mohammed Hassan Taha, Tab Rasmussen; **Front Row:** Yousry Attia, Magdy Zakaria, Mohammed Hassanter, Goda Fasal Satah

Fig. 18 1987: From left to right, **Back Row:** Said Fasal, Mario Gagnon, Chris Tilden, Sylvia Cornero, Tom Bown, Mohammed Hassanter, Elwyn Simons, Yousry Attia, Mark Brown, Prithijit Chatrath; **Front Row:** Ali Fasal, Verne Simons, Goda Farag, Mohammed Maud, Lewis Ladocsi, Mohammed Hassan Taha

Fig. 19 1988: From left to right, **Back Row:** Abdullah Hassan Taha, Ali Fasal, Said Fasal; **Middle Row:** Mario Gagnon, Prithijit Chatrath, Mohammed Hassan Taha, Thomas Naiman, unidentified, Yousry Attia, Elwyn Simons, Pat Holroyd, Said Ramadan, Nick Court; **Front Row:** Callum Ross, John Fleagle, Tom Bown, Sylvia Cornero

Fig. 20 1989: From left to right, **Back Row:** Mario Gagnon, Pat Holroyd, Blythe Williams, Adelie Oakley, Elwyn Simons, Verne Simons; **Middle Row:** Prithijit Chatrath, Said Faisal, Abdullah Hassan Taha, Ali Fasal, Yousry Attia, Bill Sanders, Hassan Ali, Ragab Mohammed Achmed, Holly Smith, Phillip Gingerich, Ali Barakat; **Front Row:** Ramadan Achmed Mohammed, Mohammed Hassan Taha, Gamal Goma Abdel Baky

Fig. 21 1990: From left to right, **Back Row:** Baday Mohamed, Ali Faisel, Abdullah Hassan, Said Faisal, Ramadan Ahmed Mohammed, Callum Ross; **Front Row:** Myron Shekelle, Todd Rae, Hassan Ali, Yousry Attia, Issa Achmed Mohammed, Elwyn Simons, Mohammed Hassan Taha, Medhat Said Abdel Ghani, Prithijit Chatrath, Ted Roese, Mark Hamrick. Photo by Todd Rae

Fig. 22 1991: From left to right, **Back Row:** Ali Faisel, Abdullah Hassan Taha, Gumah, Said Faisal, Ramadan Achmed Mohammed, Mohammed Hassan Taha, Chia Tan, Karen Messenger; **Middle Row**: Fred Hays, Bill Sanders, Philip Gingerich, Elwyn Simons, Prithijit Chatrath, Holly Smith, Charles Messenger, Yousry Attia; **Front Row**: Issa Achmed Mohammed, Hassan, Said Hassan Sheraf, Casey McKinney, Sam Senturia, Robert Whiting, John Fleagle

Fig. 23 1992: From left to right, **Back Row:** Hassan Ali, Ragab Shaban, Ragab Ali, Achmed Mahmood, Kwame Said, Hassan Ali, "Sammy" Issa Zakri, Ellen Miller; **Middle Row:** Mohammed Daood, Said Hashem, Ali Barakat, Abdel Latif, David Froehlich, Jorge Genise, Tom Bown, Mohammed Hassan Taha, Elwyn Simons, Yousry Attia, Fritz Hertel, Issa Achmed Mohammed, Laura MacLatchy, Prithijit Chatrath, Paige Vinson, David Hobbs; **Front Row:** Casey McKinney

Fig. 24 1993: From left to right, **Back Row:** Elwyn Simons, Mohammed Attia, Maureen O'Leary, Don DeBlieux, Will Clyde; **Middle Row:** Abdel Galeel, Mohammed Hassan Taha, Casey McKinney, "Sammy" Issa Zakri, Ali Bakarat, Hassan Ali, Issa Achmed Mohammed, Mark Hamrick, Prithijit Chatrath, Abdel Latif, Jonathan Bloch; **Front Row:** Ragab Shaban, Ragab Ali, Ellen Miller

Fig. 25 1994: From left to right, **Back Row:** Tony Friscia; **Middle Row:** Ragab Shaban, Hassan Ali, "Sammy" Issa Zakri, Elwyn Simons, Abdul Es Salaam, Khalid Hachmood, Yousry Attia, Devon McLennan, John Fleagle, Abdel Latif, Issa Achmed Mohammed, Abdel Galeel, Ellen Miller, Prithijit Chatrath, Don DeBlieux; **Front Row:** Bert Covert, Casey McKinney, Mohammed Hassan Taha

Fig. 26 1995: From left to right, **Back Row:** Sylvia Cornero, Ellen Miller; **Middle Row:** Hassan Ali, Marcelo Sanchez-Villagra, Abdel Galeel, Aieed Ali Ibrahim, Tom Bown, Mohammed Hassan Taha, Ragab Ali Abbas, Elwyn Simons, "Sammy" Issa Zakri, Abdel Latif, Yousry Attia, John Fleagle, Prithijit Chatrath; **Front Row:** David Hobbs, Don DeBlieux

Fig. 27 1996: From left to right, **Back Row:** Ellen Miller, David Hobbs; **Middle Row:** Ayeed el Ali, John Fleagle, Tom Bown, Friederun Ankel-Simons, Mohammed Hassan Taha, Elwyn Simons, "Sammy" Issa Zakri, Hassal Ali, Goma Kemel, Abdel Latif, Mahmood Abdulla, Prithijit Chatrath; **Front Row:** Don DeBlieux, Abdel Galeel, Issa Achmed Mohammed, Beth Townsend

Fig. 28 1997: From left to right, **Back Row:** Don DeBlieux, Abdel Galeel, Kari Wright, Natalia Rybczynski, Achmed Moustafa, Hassan Ali, Tab Rasmussen; **Front Row**: Glenn Conroy, Magdy Zakaria, Abdel Latif, Issa Achmed Mohammed, unidentified, Ellen Miller, Elwyn Simons, Goma Kemel, Hassan Rabyia, Tom Bown, Mohammed Hassan Taha, Yousry Attia, Prithijit Chatrath

Fig. 29 1998: From left to right, **Back Row:** Hassan Ali, Susan Williams, Said Mohammed Said, Lea Ann Jolley, Nancy Stevens, Michael Wilson, Elwyn Simons; **Middle Row:** Prithijit Chatrath, Issa Achmed Mohammed, Yousry Attia, Medhat Said Abdel Ghani, Jacob Hogarth, Goma Kemel, "Sammy" Issa Zakri, Abdel Latif, Verne Simons, Mohammed Hassan Taha, Magdy Zakaria, Mahmood Abdulla; **Front Row:** Don DeBlieux, Ali Barakat, Ayeed el Ali

Fig. 30 1999: From left to right, **Back Row:** Hussein Mohammed Ramadan, Nancy Stevens, Amy Judd, Said Mohammed Said, Elwyn Simons; **Middle Row:** Soher Mohammed, Ashraf Achmed, Marc Godinot, Abdel Bayuomi, Yousry Attia, Tom Bown, Medhat Said Abdel Ghani, Patrick Lewis, Jacob Hogarth, Mohammed Hassan Taha, Jay Norejko, Abdel Latif, Prithijit Chatrath; **Front Row:** Don DeBlieux, Ali Hamida Gebril, Goma Kemel, Ayeed el Ali, Moustafa Said

Fig. 31 2000: From left to right, **Back Row:** Yousry Attia, Erik Seiffert, Elwyn Simons; **Middle Row:** Patrick Lewis, James Rossie, Said Mohammed Said, Hussein Mohammed Ramadan, Mike Pasin, unidentified, Jacob Hogarth, Medhat Said Abdel Ghani, Ashraf Achmed, Hassan Ali, Prithijit Chatrath, Abdul Zedan; **Front Row:** John Fleagle, Mahmood Abdulla, Goma Kemmel, Don DeBlieux, Mohammed Hassan Taha

Fig. 32 2001: From left to right, **Back Row:** unidentified, Lea Ann Jolley Burger, Ahmed Abdel Ghafar, Mark Mathison, Said Mohammed Said, Erik Seiffert; **Middle Row:** Patrick Lewis, Joseph English, Prithijit Chatrath, Gomma Mohammed, Mahmood Abdulla, Yousry Attia, Magdy Zakaria, Tom Bown, Elwyn Simons, Chris Heesy, Goma Kemel, John Fleagle, Mohammed Hassan Taha; **Front Row:** Ashraf Achmed, Rajeev Patnaik

Fig. 33 2002: From left to right, **Back Row:** Elwyn Simons, Ashraf Achmed, Khamis Fikry, Erik Seiffert, Abdul Zedan, Achmed Mohammed Hassan**; Middle Row:** Yousry Attia, Tom Bown, Kimberly Nichols Bown, Mohammed Hassan Taha, Mahmood Abdulla, Ayeed el Ali, Goma Kemel, Yasser Abdel Razik, Prithijit Chatrath, Adel Bayoumi, Medhat Said Abdel Ghani, Lea Ann Jolley Burger, Benjamin Burger; **Front Row:** Verne Simons, Mark Mathison

Fig. 34 2003: From left to right, **Back Row:** Kristian Carlson, Rabia Mohammed, Alison Murray, Elwyn Simons; **Middle Row:** Prithijit Chatrath, Yasser Abdel Razik, Aid Ali Ibrahim, Ali Hassan, Hani Abdel Ghafar, Mark Mathison, Goma Kemel, Abu Zedan, Yousry Attia, Mohammed Hassan Taha, "Rada" Mohammed Hassan Ibrahim; **Front Row:** John Fleagle, Rajeev Patnaik, Ashraf Achmed, Biren Patel

Fig. 35 2004: From left to right, **Back row:** Achmed Mohammed Hassan, El Said Abdel Raba, Sobhi Aziz Abu Seif, Hussien Mohammed Ramadan, Elwyn Simons, Hani Abdel Ghafar, Erik Seiffert; **Middle Row:** Mohammed Hassan Taha, Yousry Attia, Prithijit Chatrath, Abu Zedan, Jacob Hogarth, Jonathan Perry, Ashraf Achmed, John Bennett, Gebyle Abu el Khair, Medhat Said Abdel Ghani, "Rada" Mohammed Hassan Ibrahim, Patrick Lewis; **Front row:** Samuel Turvey, Chris Gilbert, Yasser Abdel Razik, Mark Mathison

Fig. 36 2005: From left to right, **Back row:** Mohammed Abdel Ghany Osman, Suzanne Cote, Patrick Lewis, Said M. Said, Erik Seiffert, Elwyn Simons, Ahmed Mohammed Hassan, Michael Steiper; **Middle row:** Rabia Eid Abdel Twab, Hamid Abdel Waheed, Ashraf M. el Said, "Rada" Mohammed Hassan Ibrahim, Magdi Zakaria, Osama Ahmad Abdullah, Mohammed Hassan Taha, Prithijit Chatrath, Joan Jach, Rachel Dunn; **Front row:** Mark Mathison

Fig. 37 2006: From left to right, **Back row:** Brenda Frazier, Mercedes Gutierrez, Jennifer Ross, Chloe Brindley, Erik Seiffert, Abu Zedan, Alexander Liu, Biren Patel, Mark Coleman; **Middle row:** Yassir Abdelrazik, Mohammed Hassan Taha, Tom Bown, Ashraf Rifaey, Elwyn Simons, Goma Kamel, Hesham Sallam, Yousry Attia, Unidentified, Prithijit Chatrath; **Front row**: Mark Mathison, Badawi Ali Badawi, Ahmed Hassan Taha, Hussain, unidentified

Geology, Paleoenvironment, and Age of Birket Qarun Locality 2 (BQ-2), Fayum Depression, Egypt

Erik R. Seiffert, Thomas M. Bown, William C. Clyde and Elwyn Simons

Introduction

Vertebrate paleontological research in the Fayum Depression began in 1879, with Georg Schweinfurth's recovery of whale and fish fossils on the island Geziret el-Qarn in Birket Qarun (Dames, 1883; Schweinfurth, 1886). In later years Schweinfurth worked north of the lake, in part within the uppermost levels of the Birket Qarun Formation that are exposed to the south of (and stratigraphically below) the site of the Qasr el-Sagha Temple, but he is not known to have collected any vertebrate fossils from those beds. Subsequent exploration by Hugh Beadnell, and later Richard Markgraf, led to the discovery of fragmentary remains of the cetacean *Basilosaurus isis* from near this stratigraphic horizon on the northwest side of Birket Qarun (Andrews, 1904; Stromer, 1908; see also Gingerich, 2008), but no vertebrate fossils were reported from sediments exposed on the northeast side of the lake until late in the 20th century, when a single premolar of the proboscidean *Moeritherium* was described (Holroyd et al., 1996).

In the year 2000, paleontological reconnaissance in the sediments exposed along the "plain of Dimê" led to the identification of a number of new vertebrate fossil localities (Fig. 1), most of which preserve fragmentary remains of proboscideans (*Barytherium* and *Moeritherium*), sirenians, and whales. One locality situated on the northeast side of Birket Qarun, now called Birket Qarun Locality-2 or BQ-2, initially produced surface finds of creodont postcrania, a basicranium and partial mandible of *Barytherium*, a partial mandible and postcrania of *Moeritherium*, and a small placental petrosal. Subsequent quarrying and dry screening at BQ-2 from 2001 to 2005 has revealed craniodental and postcranial remains of numerous small mammals, including primates, hyracoids, herodotiines, ptolemaiids, anomaluroid and hystricognathous rodents,

Erik R. Seiffert
Department of Anatomical Sciences, Stony Brook University, Stony Brook, New York, 11794
erik.seiffert@stonybrook.edu

J. G. Fleagle, C. C. Gilbert (eds.), *Elwyn Simons: A Search for Origins.*

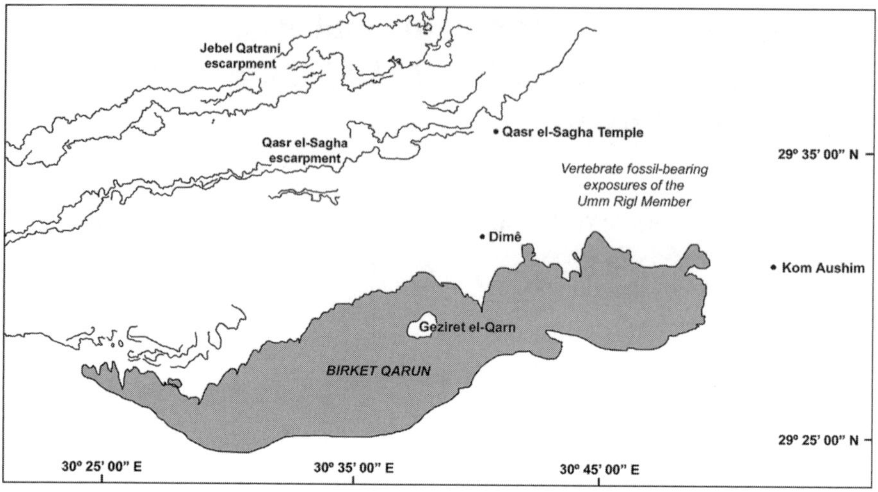

Fig. 1 Map of the Fayum area, showing major landmarks mentioned in the text

creodonts, chiropterans, and insectivoran-grade placentals. The mammalian fauna from BQ-2 is currently the most diverse from the entire Afro-Arabian Paleogene. Primates from BQ-2 include a diverse array of adapiform and djebelemurine stem strepsirrhines, the lorisiform crown strepsirrhines *Karanisia* and *Saharagalago* (Seiffert et al., 2003), and two new species of the primitive parapithecoid anthropoid *Biretia* (Seiffert et al., 2005). The purpose of this contribution is to provide a more detailed review of the geology, taphonomy, paleoenvironment, and age of Locality BQ-2.

Geology of Vertebrate Fossil Locality BQ-2 and Adjacent Rocks

Locality BQ-2 occurs near the escarpment exposed north of the far eastern end of Birket Qarun. This horizon was originally placed in the "Birket el Qarûn Series" (now the Birket Qarun Formation) by Beadnell (1905), who stated that the former designation is "…convenient and applicable to these beds, which form the escarpment immediately overlooking the lake on the north side throughout its length" (p. 41). Beadnell's placement of these beds in his "Birket el Qarûn Series" is confirmed both by his maps and his statement that "between Tamia and Dimê, near the eastern end of the Birket el Qurûn, the lowest ground, consisting of poor sandy land with tamarisk scrub, bordering the lake and cultivation, is bounded by a low escarpment of the Birket el Qurûn series." (p. 72). Beadnell's basis for placing these beds into the "Birket el Qurûn series" is not entirely clear, but it might have been influenced by an assumption that the immense, oblate spheroidal sandstone nodules exposed in the beds

Fig. 2 On left, $CaCO_3$-cemented oblate spheroidal sandstone nodules (or "biscuits") exposed north of the eastern end of Birket Qarun in the Umm Rigl Member of the Birket Qarun Formation. These structures are likely to be bioclastic infillings of large-scale interference ripple marks formed under very shallow marine conditions. On right, nodules from Unit 5, showing vertical borings (the ichnofossil *Psilolithes*) left by Eocene bivalves. Picture on right courtesy of M. Mathison

between Kom Aushim and Dimê (Fig. 2) are lithologically equivalent to the more globular structures of the Birket Qarun Formation exposed farther to the west. We agree with Gingerich (1992) that the broadly oblate, biscuit-shaped nodules exposed on the northeast side of the lake, and the globular concretions exposed to the west, are of different origin – the former likely being bioclastic infillings of large-scale interference ripple marks formed under very shallow marine conditions [but not "bioherms" as suggested by Gingerich (1992)].

In part on the basis of the lithological distinctiveness of the "biscuit"-like concretions, Gingerich (1992) removed the eastern beds from the Birket Qarun Formation and placed them in a new Umm Rigl Member of the Qasr el-Sagha Formation. We see no need for such a transfer, and here retain the Umm Rigl Member in the Birket Qarun Formation. The lithology of the Umm Rigl beds—which by any definition, are lithosomal (essentially discontinuous) in character—correspond neither to the lithology of the underlying beds of the Birket Qarun Formation nor to that of the overlying Qasr el-Sagha Formation, and no units, newly-named or not, lie between rocks included by Beadnell in his Birket Qarun Formation and those included by Gingerich in his Umm Rigl Member of the Qasr el-Sagha Formation. Macrofossils also appear to be of little use in placing the Umm Rigl Member in either the Birket Qarun or Qasr el-Sagha Formation; as noted by Beadnell (1905, p. 48), "the exact junction between the Birket el Qurûn series and the overlying Qasr el Sagha beds is naturally perfectly arbitrary, many of the fossils being common to both groups. *Carolia placunoides*, which is perhaps the most distinctive fossil in the Qasr el Sagha series, is sometimes very common in the upper beds of the underlying group, and, as shown before, is common enough in the still lower *Nummulites gizehensis* beds of Wadi Rayan."

Retaining the Umm Rigl beds in the Birket Qarun Formation is consistent with previous work, as well as the suggestion that that formation provides evidence for the first Egyptian coastal deposits since the Cretaceous (Bown and Kraus, 1988; also advocated by Gingerich, 1992). Gingerich's (1992) new, overlying Harab Member, being a succession of embayment mudstones with very little surface exposure, belongs to the base of the Qasr el-Sagha Formation, because it has virtually no lithologic characteristics which separate it from the base of the Temple Member of the Qasr el-Sagha except, perhaps, the absence of "hardgrounds."

Thus in what follows we employ the following nomenclature (from top to bottom):

QASR EL-SAGHA FORMATION (Beadnell, 1905)
  Dir Abu Lifa Member (Bown and Kraus, 1988)
  Temple Member (Bown and Kraus, 1988; including Harab Member of Gingerich, 1992, at its base)
BIRKET QARUN FORMATION (Beadnell, 1905)
  Umm Rigl Member (Gingerich, 1992, as he applied it to the Qasr el-Sagha Formation)
  Unnamed lower member

The remaining rocks of the Birket Qarun Formation are those that Gingerich (1992) retained in that formation, and so are "restricted to the cliff-forming yellow sandstones that are so conspicuous above the northwest shore of Birket Qarun, at Garet Gehannam, in Wadi Hitan, and at Minqar el-Hut" (Gingerich, 1992, p. 47). These sandstones reach a maximum thickness of about 70 m at Garet Gehannam and Minqar el-Hut, but are only about 10 m thick in Beadnell's (1905) type section, 7 m thick on Geziret el-Qarn, and even thinner on the lake shore cliffs north of Geziret el-Qarn (Gingerich, 1992).

The oblate spheroidal biscuit-like sandstone nodules exposed over a large area north of the east side of Birket Qarun, and extending as far east as the Roman town of Karanis (east of the modern village of Kom Aushim), are the most distinctive geological feature of the area, and are worthy of more detailed description. There are as many as five levels of these nodules. They are commonly bioclastic and contain either a transported hash of fragmentary mollusc remains, or in places, as at Kom Aushim, also whole valves of the oyster *Carolia*. As noted previously, we consider it likely that these structures represent the largely detrital bioclastic infillings of large-scale interference ripple marks formed under very shallow marine conditions. In many places, the nodules contain no fossil material whatsoever, and consist only of $CaCO_3$-indurated medium-to-coarse sand. In other locales, the nodules are seen to subtly conjoin into a more-or-less continuous sand sheet, with or without associated invertebrate remains. Moreover, unfossiliferous nodules commonly contain cross-stratification coincident with that of the less indurated encasing rocks. These data, combined with the gently downward conical curvature of the bases of the nodules, strongly suggest that the nodules represent the calcite cemented clastic infillings of depressions developed upon an undulating seafloor bottom, probably (by super- and substratigraphic evidence) in exceedingly shallow water (on the order of a few meters at most). This inference is also supported by the presence of numerous vertical burrows in the nodules, representing several generations of boring bivalve activity resulting in the trace fossil *Psilolithes* (Häntzschel, 1975). These burrows could not have resulted from activity during a Pliocene marine transgression (Beadnell, 1905; Gingerich, 1992) because they are present in hundreds of buried *in situ* nodules uncovered by recent nearby mining activity.

A geological section through sediments of the Umm Rigl Member exposed at and near fossil locality BQ-2 was measured in 2001, and lithological details are provided in Fig. 3. The section was started about 1,500 m to the south of the locality (bearing of section = 351 degrees; dip 4 degrees). The combined thickness of that part of the Umm Rigl Member of the Birket Qarun Formation lying

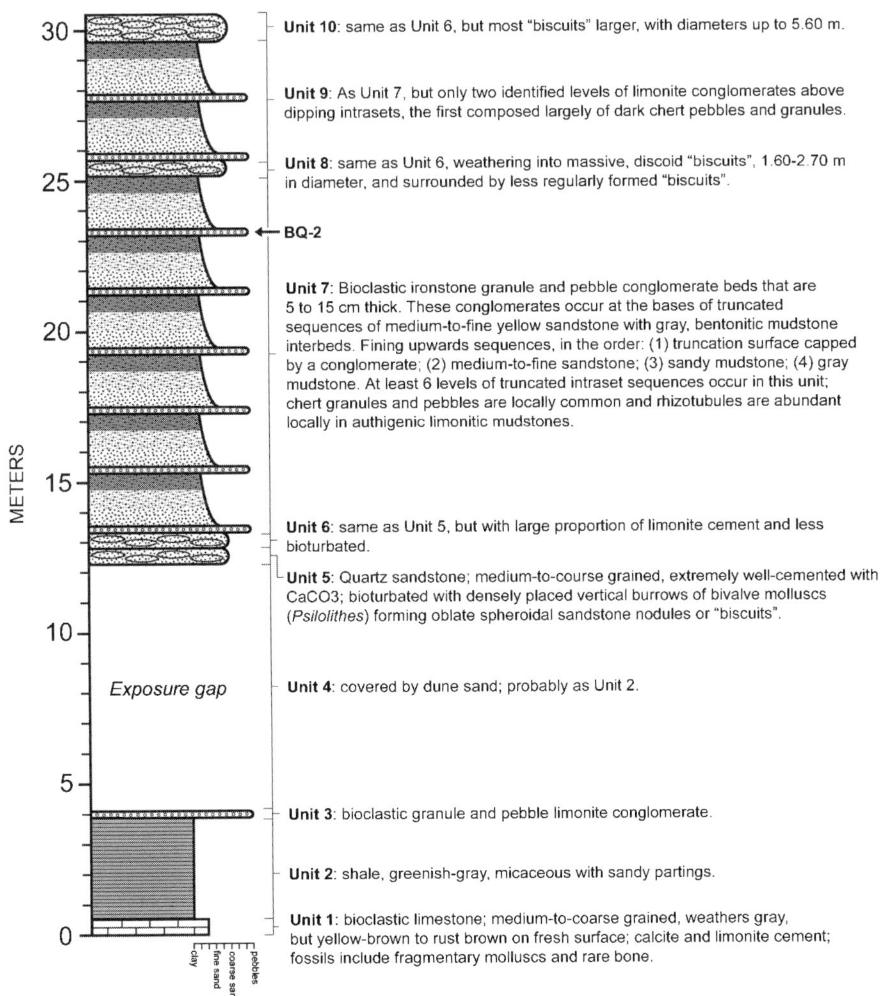

Fig. 3 Lithological column for stratigraphic section measured near Locality BQ-2

above locality BQ-2, plus that of the overlying Temple Member of the Qasr el-Sagha Formation is 94.4 m. The thickness of the Dir Abu Lifa Member of the Qasr el-Sagha Formation is 77.0 m (Bown and Kraus, 1988), and the next-oldest primate-bearing fossil vertebrate locality in Egypt, Quarry L-41, lies 46.0 m above the base of the Jebel Qatrani Formation. Thus localities BQ-2 and L-41 are separated by 217.4 m of sediment. Seiffert et al. (2003) reported that BQ-2 lies 183 m below the contact of the Jebel Qatrani and Qasr el-Sagha Formations, but it has since been recognized that a small part of that section was repeated due to a fault.

Paleoenvironment of Locality BQ-2

A massive, 10 to12 m thick sandbody of loosely indurated, pure, white, fine-to-coarse quartz sandstone has been exposed by industrial quarrying in a large pit about one kilometer west of locality BQ-2 (Fig. 4). With the exception of one small exposure north of the ancient town of Dimê, this unit is not known to occur elsewhere in the Umm Rigl Member, and was not encountered in any of the measured sections. In fact, this unit does not occur naturally in any of the

Fig. 4 Massive sandbody of fine-to-coarse grained quartz sandstone, interpreted here as being of fluvial origin, exposed by industrial quarrying about one kilometer west of locality BQ-2. Humans standing near the escarpment (E.L. Simons, on right, and J.G. Fleagle, on left) provide scale. Cross-stratification in this sandbody consists of medium scale planar-tabular sets and medium-large-scale trough cross-stratification; set boundaries often contain authigenic limonite

south-facing exposures of the Birket Qarun Formation that we have examined in the Dimê-Kom Aushim region. The top of this sandbody occupies a stratigraphic position about two meters above locality BQ-2 (but does not occur at that locality), and the axis of the sandbody appears to trend to the NW and WNW, the same directions as suggested for stream flow by paleocurrent azimuths. Cross-stratification consists of medium scale planar-tabular sets and medium-large-scale trough cross-stratification, with set boundaries commonly containing accumulations of authigenic limonite, a material that also occurs in diffuse "seams" throughout the unit. These data strongly indicate a fluvial origin for the sandbody.

The presence of such features in sandstones associated with the fossil-bearing beds at BQ-2, and in the massive sandbody exposed to the west of the quarry, provide compelling evidence for a fluvial origin of the deposits. This sedimentologic evidence for a fluvial origin is supplemented by the abundance of rhizoliths in both sandstones and ironstones directly associated with the fossil-bearing beds at BQ-2 (Fig. 5). Other evidence for a fluvial origin is provided by the presence of limonitic ironstone—both as authigenic and transported granule-pebble-sized rip-up clasts from other authigenic deposits—which indicates terrestrial subaqueous deposition in bogs subject to periodic flushing during floods. During these floods, in which water depth probably never exceeded one meter, in-channel cross-bedded sand and mud units were truncated. The newly formed surface of truncation was then buried (at high "flood") with coarser material (chert, ironstone granules and pebbles, and vertebrate fossils), which then formed the base of a successive, overlying fining-upwards sequence (pebbles, granules, and fossils at the base, followed above by coarse sand, finer sand, and mud), that was, in turn, truncated by a new flood event, and so on. There are at least six such cycles at BQ-2. This evidence is supplemented by abundant carbonaceous debris in the muds associated with BQ-2, any occurrence of which is very rare in purely marine environments. Finally, the

Fig. 5 On left, fragments of authigenic iron-oxyhydrate-cemented quartz sandstone exhibiting numerous rhizoliths (fossil root traces), from near locality BQ-2. On right, rhizotubules (fossil root traces), also recovered near locality BQ-2. Pictures courtesy of M. Mathison

presence of abundant, unabraded remains of terrestrial mammals indicates little transportation of the fossils (i.e., they are from a nearby source). Terrestrial vertebrate remains overwhelmingly dominate the BQ-2 fauna, and the few shark's teeth, skate and ray mouthparts, and teeth of the few marine fish in the fauna are probably those of outfaunal elements that are able to survive in brackish water, and to invade continental streams to the point where the fresh-water levels become too extreme (see discussion of same in Bown and Kraus, 1988).

In sum, the mammalian fossils at BQ-2 probably accumulated in near-coastal shallow streams or wide rivulets of larger perennial rivers—perhaps represented by the sandbody exposed to the west—that were ponded and stagnant for most of their existence but which were flooded occasionally by shallow yet rapidly flowing water, perhaps during seasonal effluences. Absolutely no evidence was found that would suggest a nearshore marine origin for the BQ-2 deposits. The BQ-2 sediments are terrestrial, they are alluvial, and they constitute the earliest known wholly terrestrial sediments in Egypt since the Cretaceous. As the fossil bones and teeth show little evidence of transport, the presence of abundant arboreal primates at BQ-2 also indicates the presence of forested conditions at, or very close to, the site of the fossil deposit.

Age of Locality BQ-2

Three primary lines of evidence have been brought to bear on the age of the Umm Rigl sediments – sequence stratigraphic information (Gingerich, 1992), biostratigraphic information, and paleomagnetic reversal stratigraphy (Seiffert et al., 2005). The mammals from the Qasr el-Sagha and Jebel Qatrani Formations are of limited biostratigraphic use due to the largely endemic nature of Paleogene Afro-Arabian faunas and the poorly constrained ages of other mammal-bearing localities in Afro-Arabia, but there is now general agreement that the Jebel Qatrani Formation contains both late Eocene and early Oligocene beds (Kappelman et al., 1992; Rasmussen et al., 1992; Simons and Rasmussen, 1994; Holroyd et al., 1996; Seiffert, 2006), while the Qasr el-Sagha and Birket Qarun Formations are later Eocene in age (Bown and Kraus, 1988; Van Couvering and Harris, 1991; Gingerich, 1992, 1993; Seiffert, 2006). Planktic foraminifera from the underlying Gehannam Formation indicate that those beds are of late middle Eocene age (Bartonian, Zone P14, approximately 38 to 39.5 Ma) (Strougo and Haggag, 1984; Haggag, 1985). Nummulitids identified in the Birket Qarun Formation include *Nummulites gizehensis* (Beadnell, 1905), *Nummulites striatus* and *Nummulites* aff. *pulchellus* (Strougo, 1992), and, according to Abdel-Kireem et al. (1985) and Beadnell (1905), *Nummulites beaumonti*. It is unclear whether these fossils were recovered from below the Umm Rigl Member; Strougo (1992) placed *N. striatus* and *N.* aff. *pulchellus* approximately 87 m above the base of a 92.5 m-thick section of "Birket Qarun Formation" sediment (measured at Guta, at the western end of Birket Qarun) that is composed almost entirely of clays at the base and top (60 m of the section are not figured). As noted earlier, Gingerich argued that the Birket Qarun Formation is composed almost entirely of cliff-forming sandstones, reaches a maximum thickness of 70 m, and occurs much further to the west. As such it is possible that the top of Strougo's section contains Umm Rigl Member sediments. *N. striatus* ranges from the late Bartonian to the early Priabonian, *N. pulchellus* is entirely Priabonian; the latest occurrence of *N. beaumonti* is early Bartonian and that of *N. gizehensis* is Lutetian (Serra-Kiel et al., 1998).

Kappelman et al. (1992) produced a paleomagnetic reversal stratigraphy of the Jebel Qatrani Formation and the upper 60 m of the Dir Abu Lifa Member of the Qasr el-Sagha Formation, and concluded that the lower sequence of the Jebel Qatrani Formation was of late Eocene age. In 2002, paleomagnetic sampling by W.C.C. was extended down-section through the rest of the Dir Abu Lifa Member, the Temple Member of the Qasr el-Sagha Formation, and the Umm Rigl Member of the Birket Qarun Formation (Seiffert et al., 2005). This work revealed that the newly sampled section is entirely of normal polarity. Within the context of Kappelman et al.'s (1992) preferred magnetostratigraphic correlation for the Jebel Qatrani Formation, this long normal would likely correlate either with C16n.2n alone, or both C16n.2n and C17n.1n. Correlating this long normal with C16n.2n alone would require that BQ-2 was, at most, only about 700,000 years older than Quarry L-41, and would imply exceedingly high sedimentation rates through the Qasr el-Sagha Formation. This correlation would also necessitate that the Umm Rigl Member was deposited during a time of relatively high sea level, which would make little sense of BQ-2's fluvial origin and abundant terrestrial mammals situated above and below near-shore marine sediments. A more likely scenario is that at least part of the long normal zone correlates both with Chron C16n.2n and Chron C17n.1n, and that the short reversal separating these chrons (C16r.2r) was lost in the erosional unconformity demarcating the contact of the Temple Member and the Dir Abu Lifa Member (Seiffert et al., 2005). Under this correlation, BQ-2 would likely fall close to the base of Chron C17n.1n, which is currently considered to last from 36.51 to 37.24 Ma (Ogg and Smith, 2004), and thus the locality would probably be about 37 Ma or earliest Priabonian in age. As the Bartonian-Priabonian boundary is marked by a global sea-level lowstand, this correlation also makes sense of BQ-2's fluvial sediments and terrestrial mammals intercalated within a succession of otherwise near-shore marine sediments.

Seiffert (2006) has recently argued for a revised interpretation of the Fayum magnetostratigraphy that places the Eocene-Oligocene boundary (upper part of Chron C13r) 48 meters above the contact between the Qasr el-Sagha Formation and Jebel Qatrani Formation (i.e., just above Quarry L-41). Seiffert's argument for this new correlation was based largely on mammalian biostratigraphy, however we show here that this hypothesis is further supported by graphic correlation (Shaw, 1964), which is a method that compares the pattern and spacing of stratigraphic events in a local record (e.g., local polarity reversals) to that known from a standard record (e.g., Geomagnetic Polarity Timescale [GPTS]; Ogg and Smith, 2004). Without *a priori* knowledge of how sediment accumulation rates vary through a local stratigraphic section, it is reasonable to assume relatively uniform accumulation rates. Comparisons of different correlations between the local record and the standard record can therefore be accomplished by comparing the significance of their respective coefficients of determination (r^2; Davis, 1986, p.66–67). Evaluating competing magnetostratigraphic correlations of the Fayum section in this way indicates that the new correlation proposed by Seiffert (2006) exhibits the best fit to the GPTS (Fig. 6). The statistical fit of the Seiffert correlation ($p = 4.9*10^{-7}$) is two orders of magnitude better than the correlations of Kappelman et al. (1992; $p = 1.6*10^{-5}$) and Gingerich (1993; $p = 1.7*10^{-5}$). This is particularly noteworthy since the new correlation also explains a greater number of the reversals in the Fayum section (11) than do the other correlations (7 For Kappelman et al., 1992 and 9 for Gingerich, 1993). This means that the new correlation requires fewer extra, unexplained reversals in the local magnetostratigraphic record compared to the previous correlations. Although there are many reasons why a local magnetostratigraphic section might not sample a given polarity reversal from the GPTS (e.g., unconformity, low sampling resolution, etc.), extra reversals in a local section are very difficult to justify and should be minimized when choosing among alternative correlations.

Although this new correlation does not affect the age assignment of BQ-2, which would still correlate to the lower part of Chron C17n.1n, it does have significant implications for the ages of important localities further up-section.

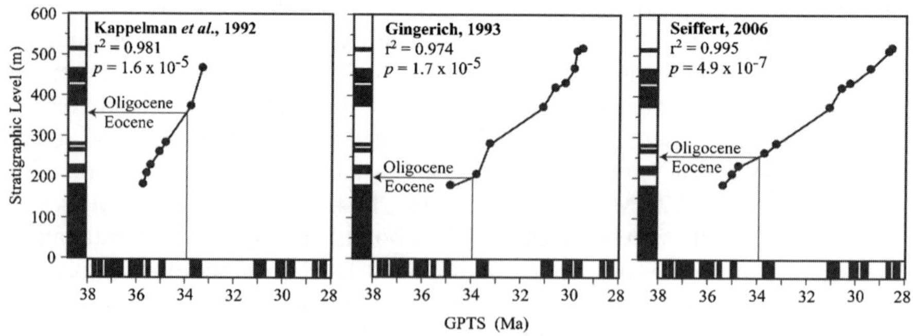

Fig. 6 Comparison of three hypothesized correlations of the local Fayum magnetostratigraphic record to the Geomagnetic Polarity Time Scale (GPTS) of Ogg and Smith (2004) using graphic correlation (Shaw, 1964). In the absence of *a priori* knowledge about stratigraphic changes in sediment accumulation rate, the fit of the Fayum polarity record to the GPTS can be measured by the significance (p) of the coefficient of determination (r^2). The correlation proposed by Seiffert (2006; right) exhibits a fit that is two orders of magnitude more significant than the correlations proposed by Kappelman et al. (1992; left) and Gingerich (1993; middle)

For instance the long interval of normal polarity at the base of the Fayum magnetostratigraphy would correlate with Chrons C16n.1n, C16n.2n, and C17n.1n, making BQ-2 approximately 3 million years older than Quarry L-41 and as much as 7 million years older than quarries I and M. No other succession in Africa documents such a long record of Paleogene mammalian evolution.

Correlation with Other Mammal-Bearing Localities in Afro-Arabia

The mammalian faunas from Gour Lazib and Glib Zegdou in Algeria (Sudre, 1979), and Chambi in Tunisia (Hartenberger et al., 1985; Hartenberger et al., 1997) are not well-dated, but appear to document a much older phase in Afro-Arabian mammalian evolution than that which is recorded at BQ-2 or younger Fayum localities (Hartenberger et al., 1997; Gheerbrant et al., 1998). The recently described mammalian fauna from Aznag in Morocco (Tabuce et al., 2005) is estimated to be between 45.8 and 43.6 Ma in age based on associated foraminifera. These Maghrebi sites do not share any mammalian species with sites in the Fayum succession, and lack anthracotheriid artiodactyls and hystricognathous and anomaluroid rodents; the first appearance datum for anthracotheriids is either at Bir el Ater or in the Dir Abu Lifa Member of the Qasr el Sagha Formation (Holroyd et al., 1996), while that for the rodent clades is either at Bir el Ater in Algeria (Jaeger et al., 1985) or at BQ-2 (Fig. 7). Despite extensive collecting, anthracotheriids have not been recovered from the Umm Rigl Member. BQ-2, Bir el Ater, and younger sites also lack the peculiar zegdoumyid rodents that are present at Glib Zegdou and Chambi (Vianey-Liaud et al., 1994), and the azibiid primates that occur at Glib Zegdou and Gour Lazib (Tabuce et al., 2004). The hyracoid *Microhyrax* from Gour Lazib is much more primitive than any hyracoid known from the Jebel Qatrani Formation or from BQ-2 (Tabuce et al., 2001; Seiffert, 2003), while *Seggeurius*—remains of which have recently been recovered from Glib Zegdou (Adaci et al., 2006)—is arguably even more primitive than *Microhyrax* and is otherwise only known from the early Eocene localities of El Kohol and Ouled Abdoun (Court and Mahboubi, 1993; Gheerbrant et al., 2003). Finally, charophytes support an early or early middle Eocene age for the Hammada du Dra localities (Mebrouk et al., 1997).

The mammalian faunas from BQ-2 and Bir el Ater are more similar to each other than they are to any other Afro-Arabian Paleogene fauna. These sites share common occurrences of the anthropoid *Biretia*, primitive anomaluroid rodents, hystricognathous rodents of *Protophiomys* grade, and a primitive species of *Moeritherium*. All of these taxa are absent from the Jebel Qatrani Formation. The only hyracoid species known from BQ-2 is morphologically similar to poorly known "*Bunohyrax*" *matsumotoi* from Bir el Ater (Tabuce et al., 2000), but the much more complete sample of

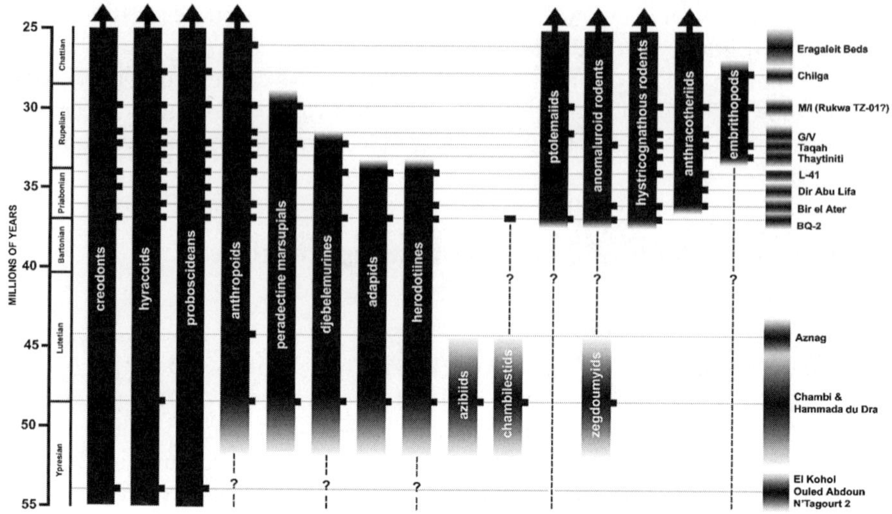

Fig. 7 Patterns in the evolution and extinction of major terrestrial Afro-Arabian mammal lineages during the Eocene and Oligocene, and suggested ages for major Afro-Arabian fossil mammal localities. Ticks on the right-hand side of each column indicate localities from which the taxa have been recovered. Dashed line with "?" at the base of the "anthropoids" column takes into account uncertainty surrounding the possible anthropoid status of late Paleocene *Altiatlasius* from Morocco (Hooker et al., 1999; Seiffert et al., 2005); djebelemurines are likely a close sister group of crown strepsirrhine primates (Seiffert et al., 2005), and the dashed line with "?" at the base of the "djebelemurines" column takes into account the fact that molecular estimates for the origin of crown Strepsirrhini often extend back into the Paleocene or Cretaceous (Yoder and Yang, 2004). Herodotiines and embrithopods are probably members of the endemic Afro-Arabian clade Afrotheria, which has likely been evolving on that landmass for at least 80 Ma (Springer et al., 2003), but their earliest records are from Chambi (Hartenberger, 1986) and just above L-41 in the Jebel Qatrani Formation, respectively. Despite only a few undoubted occurrences, it has been argued that Ptolemaiida likely had a long evolutionary history in Afro-Arabia prior to their first appearance in the fossil record (Simons and Bown, 1995); early Miocene *Kelba* appears to be a late-surviving ptolemaiid (Cote et al., 2007). An insectivoran-grade placental from BQ-2 may be a late-surviving chambilestid. Dashed line with "?" connecting the "zegdoumyids" column with the "anomaluroid rodents" column reflects the possibility that zegdoumyids were ancestral to anomaluroids (Vianey-Liaud and Jaeger, 1996). In the far right hand column, "Eragaleit Beds" refers to late Oligocene sediments exposed in the Lothidok Range west of Lake Turkana in Kenya (Leakey et al., 1995); "Rukwa TZ-01" refers to a recently discovered Oligocene mammal site in Tanzania that preserves remains of the rodent *Metaphiomys* that are very similar to those from the Fayum quarries I and M (Stevens et al., 2006); "Dhofar" refers to the Omani localities Taqah and Thaytiniti (Thomas et al., 1999); "Hammada du Dra" refers to the Algerian localities Gour Lazib and Glib Zegdou (Sudre, 1979)

craniodental and postcranial remains documenting the BQ-2 species indicate that it is intermediate in morphology between *Seggeurius* and *Microhyrax* and the more specialized geniohyid, saghatheriid, and titanohyracid species that are so well documented the Jebel Qatrani Formation (Seiffert, 2003), and is not a congener of *Bunohyrax*. The presence of an anthracotheriid at

Bir el Ater might support an age slightly younger than BQ-2 (Fig. 7), but together the mammalian data from BQ-2 lend strong support to Coiffait et al. (1984) hypothesis that Bir el Ater is late Eocene in age. Sites such as M'Bodione Dadere in Senegal (Gorodiski and Lavocat, 1953) and Tamaguilelt and In Tafidet in Mali (Lavocat, 1953) are argued to be Lutetian in age, but the only identifiable mammalian specimens from these localities are of the presumably semi-aquatic *Moeritherium*. Thus there may still be a gap of 8-to-15 million years in the middle Eocene record of terrestrial mammalian evolution in Africa.

Acknowledgment Thanks to Freddie Oakley for inviting E.R.S. and T.M.B. to participate in the 2005 conference honoring E.L.S.' 75th birthday. P. Chatrath manages our fieldwork in the Fayum region. Research in Egypt has been facilitated by the staff of the Egyptian Mineral Resources Authority and the Egyptian Geological Museum, and we are grateful for the support provided by Y. Attia, K. Soliman, and the field team provided by the Egyptian Geological Museum – particularly M. Abdel Ghany, Y. Abdel Razik, O. Ahmad, A. Hassen, and M. Zakaria. Thanks also to the other participants in recent Fayum field seasons who have assisted with excavation at BQ-2, particularly R. Patnaik and M. Mathison. P. Holroyd and D.T. Rasmussen provided valuable comments on the manuscript. Our research in Egypt has been supported by the U.S. National Science Foundation, The Leakey Foundation, and by Gordon Getty. This is DLC publication #1024.

References

Abdel-Kireem, M. R., Blondeau, A., and Shamah, K. M. (1985). Les nummulites de la localité-type de la Formation de Gehannam (Le Fayoum, Egypte). *Bull. Soc. Hist. Nat. Toulouse* 121:65–71.

Adaci, M., Bensalah, M., Mahboubi, M., Mebrouk, F., Tabuce, R., Jaeger, J.-J., Lazzari, V., and Otero, O. (2006). Les mammifères des formations Paleogenes du sud-ouest Algerien (region de Gour Lazib et de Glib Zegdou). Apports phylogenetique et paleobiogeographique. *8th International Conference on Geology of the Arab World*:203.

Andrews, C. W. (1904). Further notes on the mammals of the Eocene of Egypt, III. *Geol. Mag.* 5:211–215.

Beadnell, H. J. L. (1905). The topography and geology of the Fayum Province of Egypt. Survey Department of Egypt, Cairo.

Bown, T. M., and Kraus, M. J. (1988). Geology and paleoenvironment of the Oligocene Jebel Qatrani Formation and adjacent rocks, Fayum Depression, Egypt. *U.S. Geol. Surv. Prof. Paper* 1452:1–64.

Coiffait, P.-É., Coiffait, B., Jaeger, J.-J., and Mahboubi, M. (1984). Un nouveau gisement à Mammifères fossiles d'âge Éocène supérieur sur le versant sud des Nementcha (Algérie orientale): Découverte des plus anciens Rongeurs d'Afrique. *C. R. Acad. Sci. II* 299:893–898.

Cote, S., Werdelin, L., Seiffert, E. R., and Barry, J. C. (2007). Additional material of the early Miocene mammal Kelba and its relationship to the order Ptolemaiida. *P. Natl. Acad. Sci. U. S. A.* 104: 5510–5515.

Court, N., and Mahboubi, M. (1993). Reassessment of Lower Eocene *Seggeurius amourensis*: Aspects of primitive dental morphology in the mammalian order Hyracoidea. *J. Paleontol.* 67:889–893.

Dames, W. B. (1883). Über eine tertiäre Wirbelthierfauna von der westlichen Insel des Birket-el-Qurun im Fajum (Aegypten). *Sitzungsberichte der Königlich Preussischen Akademie der Wissenschaften zu Berlin* 1883:129–153.

Davis, J. C. (1986). Statistics and Data Analysis in Geology (2nd Edition). John Wiley and Sons, New York.

Gheerbrant, E., Sudre, J., Sen, S., Abrial, C., Marandat, B., Sigé, B., and Vianey-Liaud, M. (1998). Nouvelles données sur les mammifères du Thanétien et de l'Yprésian du Bassin d'Ouarzazate (Maroc) et leur contexte stratigraphique. *Palaeovertebrata* 27:155–202.

Gheerbrant, E., Sudre, J., Capetta, H., Mourer-Chauviré, C., Bourdon, E., Iarochene, M., Amaghzaz, M., and Bouya, B. (2003). Les localités à mammifères des carrières de Grand Daoui, bassin des Ouled Abdoun, Maroc, Yprésien: premier état des lieux. *B. Soc. Géol. Fr.* 174:279–293.

Gingerich, P. D. (1992). Marine mammals (Cetacea and Sirenia) from the Eocene of Gebel Mokattam and Fayum, Egypt: Stratigraphy, age, and paleoenvironments. *University of Michigan Papers on Paleontology* 30:1–84.

Gingerich, P. D. (1993). Oligocene age of the Gebel Qatrani Formation, Fayum, Egypt. *J. Hum. Evol.* 24:207–218.

Gingerich, P. D. (2008). Early evolution of whales: A century of research in Egypt. In: Fleagle, J. G. and Gilbert, C. C. (eds.), *Elwyn Simons: A Search for Origins*. Springer, New York, pp. 107–124.

Gorodiski, A., and Lavocat, R. (1953). Première découverte de mammifères dans le Tertiaire (Lutétien) du Sénégal. *C. R. Somm. Soc. Geol. France* 15:314–316.

Haggag, M. A. Y. (1985). Middle Eocene planktonic foraminifera from Fayoum area, Egypt. *Rev. Esp. Micropaleont.* 17:27–40.

Häntzschel, W. (1975). Trace fossils and problematica. In: Moore, R. C., (ed.) *Treatise on Invertebrate Paleontology, Supplement 1*. University of Kansas Press, Lawrence, Kansas, pp. W177–W245.

Hartenberger, J.-L. (1986). Hypothèse paléontologique sur l'origine des Macroscelidea (Mammalia). *C. R. Acad. Sci. II* 302:247–249.

Hartenberger, J.-L., Martinez, C., and Ben Saïd, A. (1985). Découverte de mammifères d'âge Éocène inférieur en Tunisie centrale. *C. R. Acad. Sci. II* 301:649–652.

Hartenberger, J.-L., Crochet, J.-Y., Martinez, C., Feist, M., Godinot, M., Mannai Tayech, B., Marandat, B., and Sigé, B. (1997). Le gisement de mammifères de Chambi (Éocène, Tunisie centrale) dans son contexte géologique. Apport à la connaissance de l'évolution des mammifères en Afrique. *Mém. Trav. E.P.H.E., Inst. Montpellier* 21:263–274.

Holroyd, P. A., Simons, E. L., Bown, T. M., Polly, P. D., and Kraus, M. J. (1996). New records of terrestrial mammals from the upper Eocene Qasr el Sagha Formation, Fayum Depression, Egypt. *Palaeovertebrata* 25:175–192.

Hooker, J. J., Russell, D. E., and Phélizon, A. (1999). A new family of Plesiadapiformes (Mammalia) from the Old World lower Paleogene. *Palaeontology* 42:377–407.

Jaeger, J.-J., Denys, C., and Coiffait, B. (1985). New Phiomorpha and Anomaluridae from the late Eocene of north-west Africa: Phylogenetic implications. In: Luckett, W. P., and Hartenberger, J.-L., (eds.), *Evolutionary Relationships among Rodents – A Multidisciplinary Analysis* (pp. 567–588). Plenum Press, New York.

Kappelman, J., Simons, E. L., and Swisher, C. C., III (1992). New age determinations for the Eocene-Oligocene boundary sediments in the Fayum Depression, Northern Egypt. *J. Geol.* 100:647–668.

Lavocat, R. (1953). Sur la présence de quelques restes de mammifères dans le bone-bed éocène de Tamaguilelt (Soudan français). *C. R. Somm. Soc. Geol. France* 7:109–110.

Leakey, M. G., Ungar, P. S., and Walker, A. (1995). A new genus of large primate from the Late Oligocene of Lothidok, Turkana District, Kenya. *J. Hum. Evol.* 28:519–531.

Mebrouk, F., Mahboubi, M., Bessedik, M., and Feist, M. (1997). L'apport des charophytes à la stratigraphie des formations continentales paléogènes de l'Algérie. *Geobios* 30:171–177.
Ogg, J. G., and Smith, A. G. (2004). The geomagnetic polarity time scale. In: Gradstein, F. M., Ogg, J. G., and Smith, A. G., (eds.), *A Geological Time Scale 2004* (pp. 63–86). Cambridge University Press, Cambridge.
Rasmussen, D. T., Bown, T. M., and Simons, E. L. (1992). The Eocene-Oligocene transition in continental Africa. In: Prothero, D. R., and Berggren, W. A., (eds.), *Eocene-Oligocene Climatic and Biotic Evolution* (pp. 548–566). Princeton University Press, Princeton.
Schweinfurth, G. A. (1886). Reise in das Depressionsgebiet im Umkreise des Fajum im Januar 1886. *Z. Gesellschafte Erdkurde Berlin* 21:96–149.
Seiffert, E. R. (2003). A Phylogenetic analysis of living and extinct afrotherian placentals. Ph.D., Duke University, Durham, North Carolina.
Seiffert, E. R. (2006). Revised age estimates for the later Paleogene mammal faunas of Egypt and Oman. *P. Natl. Acad. Sci. U.S.A.* 103:5000–5005.
Seiffert, E. R., Simons, E. L., and Attia, Y. (2003). Fossil evidence for an ancient divergence of lorises and galagos. *Nature* 422:421–424.
Seiffert, E. R., Simons, E. L., Clyde, W. C., Rossie, J. B., Attia, Y., Bown, T. M., Chatrath, P., and Mathison, M. (2005). Basal anthropoids from Egypt and the antiquity of Africa's higher primate radiation. *Science* 310:300–304.
Serra-Kiel, J., Hottinger, L., Caus, E., Drobne, K., Ferràndez, C., Jauhri, A. K., Less, G., Pavlovec, R., Pignatti, J., Samsó, J. M., Schaub, H., Sirel, E., Strougo, A., Tambareau, Y., Tosquella, J., and Zakrevskaya, E. (1998). Larger foraminiferal biostratigraphy of the Tethyan Paleocene and Eocene. *B. Soc. Géol. Fr.* 169:281–299.
Shaw, A. B. (1986). Time is Stratigraphy. McGraw Hill, New York.
Simons, E. L., and Rasmussen, T. (1994). A whole new world of ancestors: Eocene anthropoideans from Africa. *Evol. Anthropol.* 3:128–139.
Simons, E. L., and Bown, T. M. (1995). Ptolemaiida, a new order of Mammalia – with description of the cranium of *Ptolemaia grangeri*. *P. Natl. Acad. Sci. U.S.A.* 92:3269–3273.
Springer, M. S., Murphy, W. J., Eizirik, E., and O'Brien, S. J. (2003). Placental mammal diversification and the Cretaceous-Tertiary boundary. *P. Natl. Acad. Sci. U.S.A.* 100:1056–1061.
Stevens, N. J., O'Connor, P. M., Gottfried, M. D., Roberts, E. M., Ngasala, S., and Dawson, M. R. (2006). *Metaphiomys* (Rodentia: Phiomyidae) from the Paleogene of southwestern Tanzania. *J. Paleontol.* 80:407–410.
Stromer, E. (1908). Die Archaeoceti des ägyptischen Eozäns. *Beiträge zur Paläontologie und Geologie Österreich-Ungarns und des Orients, Vienna* 21:106–178.
Strougo, A. (1992). The Middle Eocene/Upper Eocene transition in Egypt reconsidered. *N. Jb. Geol. Paläont. Abh.* 186:71–89.
Strougo, A., and Haggag, M. A. Y. (1984). Contribution to the age determination of the Gehannam Formation in the Fayum Province, Egypt. *N. Jb. Geol. Paläon. Monat.* 1:46–52.
Sudre, J. (1979). Nouveaux mammifères éocènes du Sahara occidental. *Palaeovertebrata* 9:83–115.
Tabuce, R., Mahboubi, M., and Sudre, J. (2001). Reassessment of the Algerian Eocene hyracoid *Microhyrax*. Consequences on the early diversity and basal phylogeny of the Order Hyracoidea (Mammalia). *Eclogae Geologicae Helvetiae* 94:537–545.
Tabuce, R., Mahboubi, M., Tafforeau, P., and Sudre, J. (2004). Discovery of a highly-specialized plesiadapiform primate in the early-middle Eocene of northwestern Africa. *J. Hum. Evol.* 47:305–321.

Tabuce, R., Coiffait, B., Coiffait, P.-E., Mahboubi, M., and Jaeger, J.-J. (2000). A new species of *Bunohyrax* (Hyracoidea, Mammalia) from the Eocene of Bir el Ater (Algeria). *C. R. Acad. Sci. II A* 331:61–66.

Tabuce, R., Adnet, S., Cappetta, H., Noubhani, A., and Quillevere, F. (2005). Aznag (bassin d'Ouarzazate, Maroc), nouvelle localité à sélaciens et mammifères de l'Eocène moyen (Lutétien) d'Afrique. *B. Soc. Géol. Fr.* 176:381–400.

Thomas, H., Roger, J., Sen, S., Pickford, M., Gheerbrant, E., Al-Sulaimani, Z., and Al-Busaidi, S. (1999). Oligocene and Miocene terrestrial vertebrates in the southern Arabian peninsula (Sultanate of Oman) and their geodynamic and palaeogeographic settings. In: Whybrow, P. J., and Hill, A., (eds.), *Fossil Vertebrates of Arabia* (pp. 430–442). Yale University Press, New Haven.

Van Couvering, J. A., and Harris, J. A. (1991). Late Eocene age of Fayum mammal faunas. *J. Hum. Evol.* 21:241–260.

Vianey-Liaud, M., and Jaeger, J.-J. (1996). A new hypothesis for the origin of African Anomaluridae and Graphiuridae (Rodentia). *Palaeovertebrata* 25:349–358.

Vianey-Liaud, M., Jaeger, J.-J., Hartenberger, J.-L., and Mahboubi, M. (1994). Les rongeurs de l'Eocène d'Afrique nord-occidentale [Glib Zegdou (Algérie) et Chambi (Tunisie)] et l'origine des Anomaluridae. *Palaeovertebrata* 23:93–118.

Yoder, A. D., and Yang, Z. (2004). Divergence dates for Malagasy lemurs estimated from multiple gene loci: Geological and evolutionary context. *Mol. Ecol.*13:757–773.

Eocene and Oligocene Mammals of the Fayum, Egypt

Elwyn Simons

Introduction

Fossil mammals from the Eocene/Oligocene badlands north of Birket Qarun, Fayum Province, Egypt have long been known. In the 19th century Georg Schweinfurth, a German geologist collected whale fossils from Gizeret el-Qorn, an island in Lake Qarun and later also collected in the Qasr el Sagha Formation north of the Lake. The whale fossils he collected were called *Zeuglodon* (now *Basilosaurus*) *osiris* by Dames (1894).The detailed history of these early expeditions is discussed in greater detail by Gingerich (2008). At the turn of the last century, first Dr. Hugh Beadnell of the Egyptian Geological Survey beginning in 1898 and second, Dr. Charles Andrews of the British Museum of Natural History, who was surveying the distribution of living Fayum mammals, became involved in collecting the first large mammals from these sediments and these were described in detail by Andrews (1906) and the geology by Beadnell (1905). Andrews' monograph constituted the first analysis of the Fayum's Eocene/Oligocene mammalian fauna. Later the German geologists E. Fraas and Stromer worked in Egypt and the latter trained an amateur collector Richard Markgraf who eventually took up residence in Sinnuris in the northeastern Fayum and who, operating from there, collected many fossils for European and American museums until his death in 1916. Unlike other early collectors in the Fayum, Markgraf took great pains to locate small mammal fossils and actually collected all the original type specimens of primates that were to focus the attention of paleoanthropologists, and especially my interest, on the Fayum. He provided some of these materials, including the type of *Apidium*, to the American Museum group in 1907 and later sent the first discovered Fayum anthropoid frontal, which turned out to be of *Apidium* to that museum. Many of the larger vertebrate fossils Markgraf found were sent to

Elwyn Simons
Division of Fossil Primates, Duke University Primate Center, Department of Biological Anthropology, Duke University, 1013 Broad Street, Durham, NC 27705, USA

J. G. Fleagle, C. C. Gilbert (eds.), *Elwyn Simons: A Search for Origins.*

European museums and much of this material, which also included the types of *Parapithecus fraasi* and *Propliopithecus haeckeli*, came to the attention of Max Schlosser of Munich who published a second monograph on the Fayum land mammals (1911). Earlier, in 1907, Markgraf had joined the celebrated expedition to the Fayum from the American Museum of Natural History led by Walter Granger (see Fig. 1). Markgraf, somewhat a man of mystery, told Granger that he had a daughter living in Italy. He had gone to Egypt for health and according to Ralph von Koenigswald, Markgraf earned money by playing the piano in Cairo before he opened a fossil sales business in the Fayum village of Sennouris. While shopping there in 1961, Yousef Shawki Moustafa and I were shown "the house of the foreigner" and it was Shawki's opinion that this was Markgraf's home. Today, the center of Sennouris is where this house was entirely rebuilt. The collections Markgraf made at the beginning of the last century led to a number of early publications from the US and Europe concerning Fayum mammals, and his pioneering work in the Fayum was of the greatest importance to science.

Fig. 1 Richard Markgraf in the Fayum 1907. Courtesy of the American Museum of Natural History. Markgraf discovered how to collect mammalian microfauna in the Fayum

Because European and American academicians normally are committed to teaching in fall and spring, and because the Fayum badlands are too hot to collect in during summer, there was little attention to fossil collecting there during the two world wars and the period between them. Also it may have seemed that the monographs of Andrews and Schlosser had been definitive of the land mammal faunas of the region, but both of their collections were deficient in the recovery or description of small mammalian fossils. Together, these monographs referred to less than a dozen specimens of extinct mammals the size of a cat or smaller, and most of these few fossils were primates.

In 1947, Wendell Phillips (from the University of California at Berkeley), organizer of the Pan-African Expedition of that year, led a varied group of scientists to collect in the Fayum and some of the material found reached the collections of the Berkeley museum. Because this group worked most of the time they were in Egypt out of the former Governor General's headquarters located near Kom Aushim, they had to drive each day about 38 km both ways in and out of the region with continental exposures in order to collect fossils. These collections, consequently, were limited and were mainly from the marine sediments (see Gingerich, 2008).

In 1961 I held a position as Head Curator of Vertebrate Paleontology at the Yale University Peabody Museum and had negotiated an arrangement without teaching duties in the fall that I have continued throughout my career. In consequence, research groups I have directed on expeditions to the region have carried out our field programs during the fall months when the temperatures in the Fayum badlands are equitable. The many expeditions I have led, first in cooperation with the Geological Survey of Egypt and now with the Egyptian Mineral Resources Authority were originally sponsored by the Yale University Peabody Museum and afterwards launched from the Duke University Primate Center. The first field work from Yale was between 1961 and 1968 (7 years) and the second series of field projects from Duke University has continued under my direction from 1977 until this fall of 2006 (29 years).

My primary professional objective in the Fayum has been to clarify the nature and extent of early primate history in Africa, a continent which has certainly been the site of all the major events in higher primate evolution. Nevertheless, the remainder of the vertebrate fauna as well as the plants recovered are intrinsically important to science as well as they enable reconstruction of paleoenvironments and their biological communities. In what follows I will try to briefly outline what kinds of mammals—principally other than primates—inhabited Egypt during the Eocene and Oligocene epochs. The most complete fossils documenting African Paleogene continental land faunas come from Egypt. There are also early Tertiary sites in Morocco and Algeria that yield fossils of importance but, in general, Paleogene fossils from those regions are more incomplete or less well preserved than are those from Egypt. Hence, Egyptian studies are of primary importance to understanding of the development of mammals in Africa.

Eocene and Oligocene Mammals

It is believed that Africa was isolated as an island during the Paleogene, separated from Laurasia by the Tethys Sea, and that therefore there was little faunal interchange between the region and the continents of Europe and Asia. Because of this, many orders of Paleogene mammals that appeared early or arose in the northern continents did not reach Africa during the Paleogene. Those groups that did prosper at that time in this southerly continent belong to three divisions or sorts of animals:

I. **Those that could either fly or swim into the region:** Orders Chiroptera, Cetacea, and perhaps the anthracotheres of Order Artiodactyla.

II. **Those orders belonging to a superordinal cohort:** Afrotheria, or the afrotheres, are believed to have had an early unitary origin on the African continent and consequently are all more or less closely related to each other. These groups, as far as the Fayum is concerned, include Proboscidea (the elephants and allies such as *Moeritherium* and *Barytherium*), Sirenia (sea cows, *"arusha el Nar"*), Hyracoidea (the hyracoids or hyraxes, "*arnab sacri*"), Embrithopoda (the extinct arsinoitheres or "beasts of Queen Arsinoe"), Ptolomaiida (the extinct ptolomaiids or "beasts of King Ptolomy"), and Macroscelidea or elephant-shrews. All these six orders together with three other orders those of the Afrosoricida, Chrysochloridae (golden moles) and Tenricidae (the tenrecs) as well as Tubulidentata (the aardvarks), although not occurring in the Fayum, also belong in Afrotheria.

III. **The remaining Fayum mammals belong in five other Orders:** Presumably, these mammals reached the continent later than the basal stock of afrotheres and each at different times. There is good evidence in the Fayum and elsewhere in Africa that the first of these Orders to diversify there was the Primates. Four others, Rodentia (the rodents), Creodonta (the creodonts or false carnivores), Artiodactyla (represented by anthracotheres) and Marsupalia (represented by *Peritherium*-like marsupials) are less diversified and presumably got into the African continent either by rafting, swimming or crossing intermittently existing land bridges.

In addition to the Primates, Fayum mammals belong to at least eleven other orders. These will be described sequentially below.

The Chiroptera (Bats)

Bat jaws and/or teeth have been found at three localities of quite different ages in the Fayum, low in the marine section in continental riverine sediments at Birket Qarun locality 2 (BQ-2, age about 37 myr), at L-41 (age about 35 myr) near the base of the Jebel Qatrani Formation and at Quarry I (age about

32/33 myr) high in the Jebel Qatrani, for the latest information on dates see Seiffert et al. (2005). Sigé (1985) assigned one of the bats found at Quarry I to the microbat genus *Philisis*, a genus that appears to belong to the vespertillionid family of bats. This is a widespread family of bats ranging throughout the world from Eocene times. The various subgroups of these bats have many names. The bats from BQ-2, currently under study by Gunnell, appear to also include both nycterid and rhinolophoid bats and Gunnell has kindly supplied the following comment on Fayum bats:

> "The earliest records of bats in the Fayum sequence come from BQ-2 which is approximately latest middle Eocene (Bartonian-Priabonian boundary) in age. The endemic and extinct bat family Philisidae is represented at BQ-2 by *Philisis sphingis* along with a much larger philisid with cheek teeth in the size range of the large extant microchiropteran, *Macroderma gigas* (Australian Ghost Bat). In addition to the two philisids, a single upper molar from BQ-2 represents a new nycterid rhinolophoid. This is the earliest known record of Nycteridae, a family now widely distributed throughout Africa and southeastern Asia (Particular modern species are called Slit-faced or Hollow-faced bats). Somewhat higher in the Fayum sequence, three new genera of bats are represented from Quarry L-41 in the lower part of the Jebel Qatrani Formation (late Priabonian). One genus represents a philisid similar to *P. sphingis* but it is smaller and retains P/3 which is apparently lost in *P. sphingis*. A second taxon represents a moderate sized megadermatid rhinolophoid [living rhinolophoid bats are called Horseshoe-bats]. A third taxon is represented by a single, tiny dentary with cheek teeth rivaling those of the smallest known extant bat, *Craseonycteris*, in size. This small L-41 species is of unknown affinities. In the upper Jebel Qatrani Formation (early Oligocene, Rupelian) two microchiropterans are known, *P. sphingis* described by Sigé (1985) and the enigmatic *Provampyrus* described by Schlosser (1911). *P. sphingis* is represented by good upper and lower dentitions and a partial humerus from Fayum Quarry I while *Provampyrus* is known only from a single humerus from an uncertain locality in the Lower Fossil Wood Zone."

Since there is little else from the Fayum to compare the latter, right humerus with, it has remained little studied and is of uncertain affinities, but may also be a vespertillionid. There is nothing to prevent volant Chiroptera from reaching Africa from the Eurasian land mass and consequently members of this order provide no significant zoogeographic information about early mammalian migrations. Holroyd (1994) discussed the possible interrelationships of Fayum mammals with groups outside the continent and possible faunal exchange and Gagnon (1993) divided the Fayum succession into sequential faunal zones.

The Cetacea (Whales)

This order of mammals is abundant in the Fayum region, particularly throughout the Qasr el Sagha Formation and at Wadi Hitan, about 60 km west of the classic Fayum fossil mammal sites north of Birket Qarun. Although this group of mammals did not originate in Africa, whale fossils are an important element of the Fayum mammalian fauna. Gingerich et al. (1990) first documented the

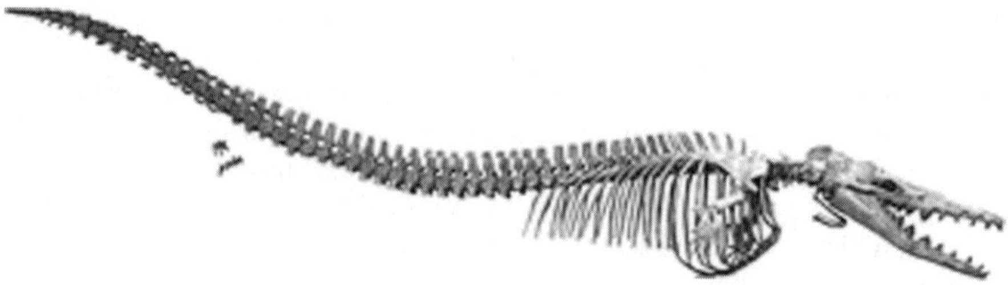

Fig. 2 Reconstructed skeleton of the Fayum whale *Dorudon* from Wadi Hitan as mounted in the University of Michigan Museum of Paleontology. Photo: P. D. Gingerich

nature and existence of hind limbs in whales on the basis of fossils found at Wadi Hitan. In this region, located to the west of Garet Gahennom, and formerly known as "*Zeuglodon* valley" skeletons of around 500 whales have been located. The region has recently been made a protected World Heritage site, mainly because of this extraordinary occurrence. These fossil whales are not the result of mass strandings but are scattered through different levels of a considerable sequence of beds. The evidence for whale origins has been discussed by Gingerich et al. (1983) and by Gingerich and Uhen (1998). In 1992 Gingerich proposed the name *Saghacetus osiris* for a small whale often found in the Qasr el Sagha Formation. The whales of Wadi Hitan belong mainly to the genera *Dorudon* and *Basilosaurus* (see Fig. 2), the latter genus having been named in the mistaken assumption that it was a reptile, Harlan (1834). *Dorudon* is a medium sized archaeocete while *Basilosaurus* is extremely long with elongated vertebrae. For a detailed account of Fayum whales, see the chapter by Gingerich in this volume.

The Artiodactyla (The Cloven-hoofed Ungulates)

Recent evidence is that whales are derived from the mammalian order Artiodactyla (Gingerich and Uhen, 1998; Gingerich et al., 2001). The first record for the appearance of this order of presumed holarctic origin in Africa is in Algeria (Bir el Ater). In the Fayum, earliest evidence is in the form of limb elements from the Dir abu Lifa member of the Qasr el Sagha Formation (Holroyd et al. 1996). The earliest well preserved craniodental and skeletal remains belong to a species of anthracothere found at Quarry L-41 near the base of the Jebel Qatrani section (see Fig. 3). The first of the anthracotheres (literally "coal beasts") was named "*Anthracotherium*" because the specimens on which this generic name was based were originally found in Eocene coal beds in France. It has long been thought that anthracotheres were semi-aquatic probable ancestors of the hippopotamus (in Arabic, *Sayed Eshta* or *Faras el Nil*) (but see also Boisserie et al., 2005). This aquatic habitat would help explain why the

Fig.3 Reconstruction of anthracotheres, similar to those that appeared in the Fayum, Egypt about 35 million years ago. These animals are related to both the whales and the hippopotamus. Illustration by R. Bruce Horsfall from W. B. Scott's *A History of Land Mammals in the Western Hemisphere*, figure 196

anthracotheres are the earliest members of Order Artiodactyla to reach Africa. Fayum anthracotheres were studied by Schmidt (1913) and analyzed again by Black (1978) but there is still much to be done to clarify their taxonomy. Recent studies, summarized in Boisserie et al. (2005) reaffirm that the hippopotamus is derived from the anthracotheres, a family which may have branched from the stock which produced whales at the end of the Paleocene or early Eocene Epoch. Analyses of blood proteins and DNA have supported a relationship between the hippopotamus and whales. When the anthracotheres first arrived in Egypt and Africa, the hyraxes or hyracoids (a group of afrotheres) were the dominant land living plant eaters and had already radiated into many genera and species. As one samples mammalian faunas from successive levels in the Fayum sediments, documented by fossils from the members of the lower and then upper sequence of the Jebel Qatrani Formation, anthracotheres become more common and the abundance of hyrax genera and species decreases.

The Ptolemaiida

The most recently described order of Fayum mammals Ptolemaiida was named by Simons and Bown (1995). The several species belonging within this order that have been described to date are based on specimens from three Fayum localities: Quarries V, A and M which are all contained in a circle with a radius of about one kilometer. In recent seasons isolated teeth of undescribed ptolemaiids have been located at the BQ-2 locality situated in the Fayum about 30 km southeast of these sites. Nevertheless this group of mammals must have one of the most limited geographic distributions of any mammalian Order, even Embrithopoda (the arsinoitheres) are much more widely distributed in the Fayum and have also been found in Ethiopia and Turkey. The name *Ptolemaia* was originally proposed by Osborn (1908) who was unable to place it in any then existing order but he did suggest that recovery of additional material might necessitate the definition of a new mammalian order. The very limited sample of species so far described for this group, mainly from Quarry V, belong to three genera: *Ptolemaia*, *Quarunavus* and *Cleopatrodon*. The latter genus, *Cleopatrodon,* was proposed by Bown and Simons (1987). It may be that ptolemaiids are derived from Eocene and Oligocene mammals of the Fayum such as the pantolestid insectivores, but that is far from certain. Placement as insectivores, *sensu lato*, might well ally these mammals with the African insectivorans. Schlosser (1923) referred *Ptolemaia*, together with the material named later by Simons and Gingerich (1974) as *Quarunavus* to the Pantolestidae. Another suggestion by Simons and Gingerich (1974) is that ptolemaiids may be related to the aardvarks: Order Tubulidentata. If this is so, parsimony analyses recently prepared by Seiffert (2003) firmly place the latter group into the Afrotheria as a sister group of the elephant shrews, golden moles and tenrecs. The molars of ptolemaiids appear to wear off flat as if the preferred foodstuffs were abrasive which could be a resemblance to aardvark teeth and their insectivorus diet. The only site where members of this order are relatively common is at Quarry V. Bown (*pers.com.*) believes that this quarry may represent a riverine quicksand accumulation and the abundance of ptolemaiids there is perhaps due to their having been semi-aquatic. Nevertheless, no limb bones for any member of this order have been positively identified and consequently, speculation as to their lifestyle is limited. In the year 2000, Seiffert and Simons described a tiny placental insectivore that they named *Widanelfarasia* from Quarry L-41 of late Eocene age in the Fayum. The species concerned is of uncertain affinities but a strong possibility is that it could be allied with the tenrecs and golden moles.

The Sirenia (Sea cows)

The possibly earliest occurring fossils belonging to the group Afrotheria in Egypt are the sirenians or sea cows of Order Sirenia. Having developed an aquatic lifestyle these animals were the first afrotheres to leave the African continent and the oldest still quadrupedal form is a species of the genus *Pezosiren* from the early Eocene of Jamaica (Domning, 2001). The first sea cows in Egypt come from the Lower Building stone member of the Mokattam

Limestone, Cairo thought to be of early middle Eocene age. This species was named *Protosiren fraasi* by Abel (1907) on the basis of material first described by Andrews (1906). In 1994 Domning et al. described additional material of this species. Also occurring in the Mokattam deposits is the genus *Eotheroides* and these early species appear to range from Lutetian to Priabonian in age. Another species of *Protosiren* from the Fayum, *P. smithae* was named by Domning and Gingerich (1994) and it is larger and more advanced than the species from the Mokattam limestone. This second species occurs only in the Gehannam and Birket Qarun formations of Wadi Hitan, western Fayum Depression. Its presumed age is latest middle to early late Eocene. Although *Protosiren* was fully acquatic it was not only marine but also entered fresh water rivers in late Eocene/Oligocene times in the Fayum and hence occurs in both the Qasr el Sagha and Jebel Qatrani formations. From the latter formation a well preserved skull and ribs of a new species of genus *Eosiren* has been described by Domning et al. (1994a,b).

The Embrithopoda (The Arsinoitheres)

Order Embrithopoda was first described in detail by Andrews (1906) who, however, placed them in a group he called Barypoda. This group, which also belongs in the Afrotheria, has been found mainly in the Fayum. They are the most characteristic animals of the fauna, being giant rhinoceros-like mammals with a four horned head (see Figs. 4–5). The youngest known specimens of this group come from the Chilga beds of Ethiopia. Arsinoitheres appear to be another group which found its way out of Africa since Sen and Heinz (1979) have described an Eocene arsinoithere from Romania.

The Proboscidea (Elephants)

Perhaps the most striking afrotherian group to characterize the Fayum is the elephants or Proboscidea. In fact, the discovery of fossil proboscideans of the genus later described as *Paleomastodon* in the Fayum in 1901 first drew attention to the continental vertebrate faunas of the region (Beadnell, 1905; Andrews, 1906; see Fig. 6).

Early discovered material of two primitive elephantids *Paleomastodon* and *Phiomia* are extensively discussed in Andrews (1906) and earlier references there cited. Andrews' monograph also discusses in detail material of two other genera of early elephant relatives: *Moeritherium* and *Barytherium*. An axial skeleton of the former was also discovered in the Qasr el Sagha formation during the Yale expedition of 1962 and specimens of the reconstructed skeleton of *Moeritherium* modeled by Grant E. Meyer and Arnold Lewis are on exhibit at the Geological Museum Cairo and at Yale (see Fig. 7). To the

Fig. 4 Lateral view of the cranium of *Arsinoitherium zitteli* from Andrews 1906. Note the enormous horn cores over the nostrils. Actual horns would have been much longer

axial skeleton have been added limb bones from other sites and specimens, but better carpals, tarsals and digital elements still need to be discovered. *Moeritherium*, in its degree of limb reduction, is reminiscent of the earliest sirenian *Pezosiren*.

Both of these ancient afrotheres have strikingly reduced, but apparently functional limbs with external independent digits on hands and feet, not having reached the condition of flippers seen in the more extreme aquatic adaptations of modern sea cows, seals, dolphins and whales. This limb reduction gives them both long otter-like bodies with an expanded number of lumbar vertebrae and it is thought that they could barely stand when unsupported by water. *Barytherium* is a very large land mammal that is extremely rare in the Fayum, all known material coming only from the Birket Qarun and Qasr el Sagha formations. It does not appear that this large proboscidean survived into Oligocene times in the Fayum but *Moeritherium*, like *Protosiren* entered the fresh water streams of the Fayum and is sometimes found in the Jebel Qatrani continental sediments. The Fayum sediments are being weathered only very slowly by wind and

Fig. 5 Reconstruction of arsinoitheres attacked by creodonts in the Fayum. Painting by Charles R. Knight. Courtesy of the American Museum of Natural History

water, and Bown and I have estimated that the surface of the *sourir* and unconsolidated rock there is only being removed by erosion at about a centimeter a century. In consequence, the largest Fayum mammals, including many of these proboscideans, which were found in such abundance by the early workers at the turn of the last century, are no longer frequently discovered weathering out. Those that were exposed have been found and fortunately collected. For continued research, it is the microfossils that now dominate our attention.

The Hyracoidea (Hyraxes)

An early sister-group that branched early from the Proboscidea, as did Order Sirenia, is the order Hyracoidea, many genera and species of which were described by Andrews (1906), Andrews and Beadnell (1902), and by Schlosser (1911). Apart from these contributions, many later papers present ideas about

Fig. 6 Cranium of *Paleomastodon beadnelli* as exhibited in the Egyptian Geological Museum. Teeth are upward as the skull is lying on its top. Photograph by Patrick Lewis

Fig. 7 Mounted skeleton of *Moeritherium lyonsi*, a primitive relative of the elephants. Note the closeness of the nasal opening and the orbital opening, characteristic of an aquatic animal. Photograph by E. L. Simons

the admittedly confusing classification of Fayum hyraxes or describe new genera and species. These are *Hyracoidea* by Meyer (1978) and references there cited as well as several more recent studies by either Rasmussen (1989, 2000), Rasmussen and Simons (1988, 1991), Rasmussen *et al.* (1990) and Schwartz *et al.* (1995). Finally, a phylogenetic analysis of this group is included in Seiffert (2003). Although (because of the rhinoceros-like tooth structure) hyraxes have been suggested as being related to even-toed ungulates (Perissodactyla), it is now clear that they belong with the Afrotheria. Andrews and Beadnell (1902) named a relatively small form that is not a great deal larger than the present day hyraxes of the Eastern Desert of Egypt, *Saghatherium*; a year later Andrews (1903) described a species of a giant hyrax, *Megalohyrax*, an animal presumably about the size of a small donkey. Later Matsumoto (1922, 1926) reviewed earlier discovered Fayum material and described a new genus perhaps even larger than the latter, *Titanohyrax*. This animal may have been about as large as a small rhinoceros. This remarkable range of absolute size variation among hyraxes has been reviewed by Schwartz *et al.* (1995), see Fig. 8.

Recently discovered skulls of these hyraxes at Quarry L-41 are as large and long as is an entire skeleton of a living hyrax. Meyer (1973) proposed a new genus, *Thyrohyrax*, that is not very different is size from *Saghatherium* but has a remarkable specialization in that there is an internal mandibular fenestra or chamber that is greatly inflated so that the interior of each horizontal ramus of

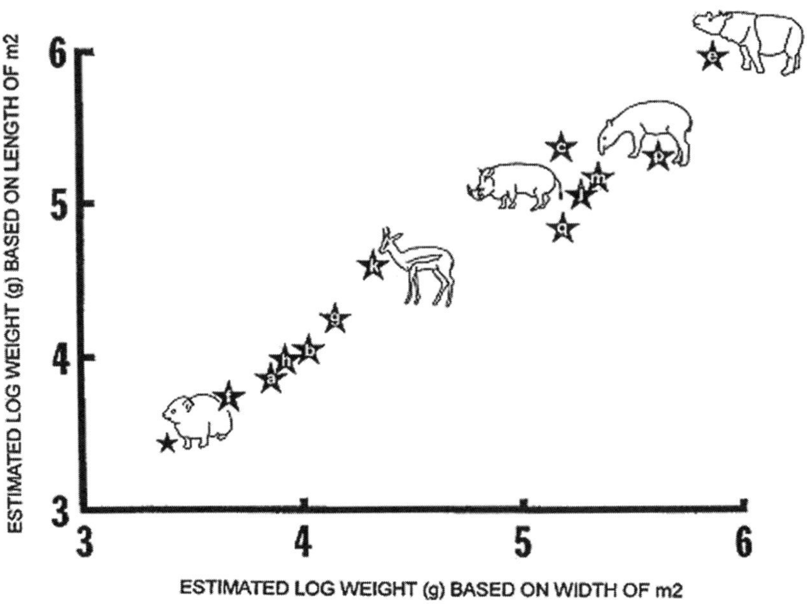

Fig. 8 Comparison of relative size in Fayum Hyracoidea, from Schwartz *et al.* (1995), modified from table 1 there (p.1089) and figure 1 (p.1092)

the jaw is hollowed out like an egg and, even more strangely, this condition is seen in only one sex of the species. It is restricted to approximately one half of specimens that cannot otherwise be distinguished in dental characteristics save that the lateral incisors of the chambered individuals are larger as in living and later male hyraxes.

Working with earlier found and far fewer samples, Meyer thought these chambered mandibles were of females, but recent research by De Blieux *et al.* (2006) presents confirmatory evidence that these individuals are males. The function of this chamber is unknown and is found in no other group of fossil mammals and in no living mammals. A possible speculation is that an air chamber filled these cavities and that it was used to amplify mating calls (see Fig. 9 for illustration of the internal mandibular fenestra). Rasmussen and Simons (2000) previously named a relatively long legged cursorial form *Antelohyrax pectidens*. This species has a most remarkable dental specialization in the central pair of lower incisors which are structured like tiny combs with each tooth variably having eight or nine tines. Presumably this is a feeding adaptation but its exact use remains enigmatic. Late in the Eocene at Quarry L-41 there are perhaps seven or eight genera of hyraxes but by the late Oligocene and in Miocene times the larger-sized hyraxes become more and more rare and today only three small species survive. The many hyraxes of the Fayum are divided into about four major kinds (see Schwartz *et al.*, 1995), and their size ranges are indicated in Fig. 8. Group 1 includes pig-like forms with bunodont, rounded tooth cusps and these are species of the genera *Geniohyus* and *Pachyhyrax*. Group 2 includes smaller species with more rhino-like teeth, similar to present-day hyraxes, genera *Saghatherium, Selenohyrax* and *Thyrohyrax*.

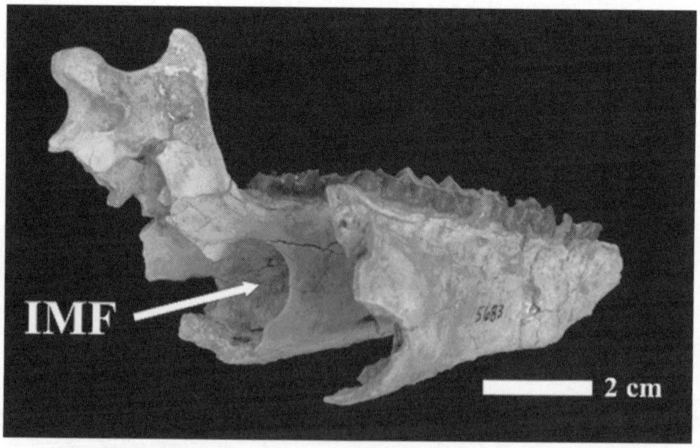

Fig. 9 Lower jaws of *Thyrohyrax domorictus*, a 33 million year old hyrax from the Fayum, Quarry I. Note the large internal mandibular fenestra (IMF) at arrow. The enlarged mandibular chamber extends forward under the teeth and also toward the back under the damaged ascending branch of the mandible. This condition, with an inflated chamber in the mandibular ramus is not known in any other mammal. Figure from DeBleux *et al.* (2006)

Group 3 contains the tapir-like hyraxes: *Megalohyrax*. Group 4 includes hyraxes with crescentic tooth cusps: *Titanohyrax*, perhaps as large as a Sumatran rhinoceros, and *Antelohyrax*, as stated above a running form with pectinate teeth. The latest African fossil hyraxes are from the Chilga beds of Ethiopia with an age of ~ 27 Ma. During the Miocene hyraxes left Africa and are found from Greece to the Bugti Beds of Pakistan.

The Rodentia (Rodents)

Several remaining orders did have or may have had their origins outside Africa. The first of these is the Order Rodentia. Very few specimens of rodents were recovered by early collectors in the Fayum. This order was reviewed by Wood (1968) who discussed and described several new species of the uniquely Fayum family Phiomyidae. Holroyd (1994) analyzed the possible origins outside Africa for this group and suggested that a probable stock for them were the Laurasian chappatemyids. Recently we have discovered an additional family of rodents, consisting of scattered teeth of anomalurid rodents, at the BQ 2 locality and these are now under study. It seems that the phiomyid rodents gave rise to two surviving groups: the rock rats of South Africa and the West African rodents known as "cutting grass" or cane rats.

The creodonts are a characteristic Fayum order of mammals. Formerly they were thought to belong in the Order Carnivora but are now considered a separate order, sometimes called "false carnivores". They were meat eating, small brained and had clawed feet. The commonest Fayum genera are *Apterodon* and *Pterodon* both genera that were originally described from specimens found in French Eocene deposits. Consequently, this is another group that has members both inside and outside Africa. Some scientists think, however, that the creodonts may have originated in Africa. Other genera reported from the Fayum are *Hyaenodon* and *Metasinopa*. Simons and Gingerich (1974) described a very small Fayum hyaenodont *Masrasector* (see Fig. 10). Creodonts are also

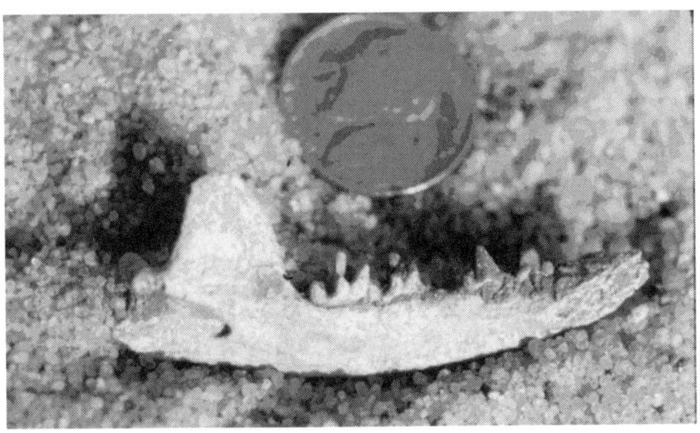

Fig. 10 Example of the lower jaw or mandible of a small creodont from the Fayum, genus *Masrasector*, soon after discovery in the Jebel Qatrani formation. Photograph by E. L. Simons

discussed in Holroyd (1995, 1999) and in Holroyd et al. (1996). Holroyd (1999) transferred Fayum *Hyaenodon* to *Metapterodon* and described a new Fayum creodont genus *Akenatenavus*.

The Primates

Probably the most ancient order of Mammalia known to be in Africa, other than the afrothere group, is the Primates. New discoveries made by a cooperative project between the Egyptian Mineral Resources Authority and Duke University clarify that the place of origin of the common ancestor of all living anthropoids (monkeys, apes and humans) was presumably in Africa. Seiffert *et al.* (2005) and references there cited, is most important because, together with earlier work by my group, this recent paper demonstrates a family tree linking *Algeripithecus* from the early Eocene or middle Eocene site of Glib Zegdou in Algeria as members of the Parapithecoidea, a stem anthropoid superfamily also represented at Fayum Quarry BQ-2 by much more complete dental material. BQ-2 finds include Egypt's oldest anthropoids. This new and expanded material included a species *Biretia megalopsis* that provides structural craniofacial evidence in the orbit of its being nocturnal. This is the world's earliest evidence of a nocturnal anthropoid. Even so, later and better known parapithecids are diurnal, as presumably was the last common anthropoid ancestor. In turn, the parapithecids are extensively exemplified by materials of yet more advanced members of the group from the Fayum Oligocene: *Apidium* and *Parapithecus*. Both of the latter genera are known from skulls and skeletal material which shows that these primates were active, arboreal leaping and springing animals (e.g., see Fleagle and Simons, 1995). Seiffert et al. (2005) also provide paleomagnetic evidence that dates BQ-2, located 229 m below Quarry L-41, at least as far back as the earliest late Eocene ~ 37 Myr in age (or Priabonian) and it is believed that the early or early middle Eocene *Algeripithecus* probably dates Eocene and Oligocene Mammals of the Fayum, Egypt back to greater than 45 Myr. Other anthropoids, yet to be described, occur at BQ-2 and, consequently, there is an emerging picture of diversity of early anthropoids in Africa. A consensus seems to be growing that *Altiatlasius* from the Paleocene of Morocco (~ 57 Myr), although poorly known, may well be a basal anthropoid and, if so, the oldest known stem anthropoid. Its presence in north Africa at so early a date could represent a separate immigration into Afro-Arabia or may relate to the origin of all later African anthropoids, including crown Anthropoidea. In any event the clade that later gave rise to parapithecoids, proteopithecids, platyrrhines, and catarrhines seems to have had an important early diversification in the African continent and it is here that the derivation of all extant groups of higher primates is documented. Ultimately crown Primates and the most primitive stem anthropoids may have had a Laurasian origin but, this is still a matter of debate and evidence to date is shaky or not compelling. One inference is that since tree shrews, flying lemurs and tarsiers are distributed in Asia from Eocene to Recent times and since these are sister groups to either Anthropoidea or to Primates, that the latter Order may have arisen in Asia (e.g., Beard, 2006). In addition, plesiadapiformes, thought by many to be a sister-group of Primates, occur solely on northern continents, supporting a Laurasian and not African origin of Primates.

Acknowledgment This chapter was originally presented at the First International Conference on the Geology of the Tethys, 2005, at Cairo University. I thank Prof. El Sayed Abd El Aziz Aly Youssef for allowing a somewhat expanded version of that contribution to be published here. Nevertheless, the contribution is only introductory since complete coverage of all the literature about Fayum mammals cannot be included in a single chapter such as this. Credits for the illustrations are indicated in each figure caption. I thank Gregg Gunnell for information concerning Fayum Chiroptera, Friderun Ankel-Simons for reviewing the manuscript and Erik Seiffert for discussions concerning afrotheres. I also wish to thank Drs. Sharif Sousa and Hassan Hamouda, both Chairmen of the Egyptian Mineral Resources Authority (EMRA) for support and assistance to our cooperative research project in Egypt. Special thanks also go to Yousry Attia, General Director of the Egyptian Geological Museum and his staff. Present research in Egypt is funded by NSF 0416164. This is DLC publication #1025.

References

Abel, O. (1907). Die Morphologie der Hüftbeinrudimente der Cetaceen. *Denkschriften der Mathematisch-Naturwissenschaftlichen Klasse der Kaiserlichen Akademie der Wissenschaften*, Wien, 81: 139–195.

Andrews, C. W. (1903). Notes on an Expedition to the Fayûm, Egypt, with descriptions of some new Mammals. *Geol. Mag.* 4(10): 337–343.

Andrews, C. W. (1906). A descriptive catalogue of the Tertiary Vertebrata of the Fayum, Egypt. 324 pp., plates 1–26, *Brit. Mus. (Nat. Hist) monograph*, Cromwell Rd., London.

Andrews, C. W. and Beadnell, H. J. L. (1902). A preliminary note on some new Mammals from the Upper Eocene of Egypt. *Surv. Dept. Cairo., Public Works Min.*, Cairo, pp. 1–9.

Beadnell, H. J. L. (1905). The topology and geology of the Fayum Province of Egypt. *Nat. Printing Off.*, Cairo. pp. 1–101.

Beard, K. C. (2006). Mammalian biogeography and anthropoid origins. In: Lehman, S. M. and Fleagle, J. G. (eds.), *Primate Biogeography*. Springer, New York, pp. 439–467.

Black, C. C. (1978). Anthracotheriidae. In: Maglio, V. J. and Cooke, H. B. S. (eds.), *Evolution of African Mammals*, Chapter 21. Harvard University Press, Cambridge, pp. 423–434.

Boisserie, J. R., Lihoreau, F. and Brunet, M. (2005). The position of Hippopotamidae within Cetartiodactyla. *P. Natl. Acad. Sci. USA* 102(5): 1537–1541.

Bown, T. M. and Simons, E. L. (1987). New Oligocene Ptolemaiidae (Mammalia:? Pantolesta) from the Jebel Qatrani Formation, Fayum Depression, Egypt. *J. Vertebr. Paleontol.* 7(3): 311–324.

Dames, W. (1894). Ueber Zeuglodonten aus Aegypten und die Beziehungen der Archaeoceten zu den übrigen Cetaceen. *Palaeont. Abh. NS* 1: 189.

De Blieux, D. E., Baumrind, M. R., Simons, E. L., Chatrath, P. S., Meyer, G. E. and Attia, Y. S. (2006). Sexual dimorphism of the internal mandibular chamber in Fayum Pliohyracidae (Mammalia). *J. Vertebr. Paleontol.* 26(1): 160–169.

Domning, D. P. (2001). The earliest known fully quadrupedal sirenian. *Nature* 413: 625–627.

Domning, D. P. and Gingerich, P. D. (1994). Protosiren smithae, new species (Mammalia, Sirenia), from the late middle Eocene of Wadi Hitan, Egypt, *Contrib. Mus. Paleonto, Univ. Michigan* 29: 69–87.

Domning, D. P., Gingerich, P. D., Simons, E. L. and Ankel-Simons, F. A. (1994a). A new early Oligocene dugongid (Mammalia, Sirenia) from Fayum Province Egypt. *Contrib. Mus. Paleont., Univ. Michigan* 29(4): 89–108.

Domning, D. P., Blane, C. E. and Uhen, M. D. (1994b). Cranial morphology of *Protosiren fraasi* (Mammalia, Sirenia) from the middle Eocene of Egypt. *Contrib. Mus. Paleont., Univ. Michigan* 29: 41–67.

Gagnon, M. (1993). Mammalian paleoecology of the Fayum, Egypt. Doctoral dissertation, Duke University, Dept. of Biological Anthropology, p. 393.

Gingerich, P. D., Haq, M., Zalmout, I. S., Khan, I. H., and Malkani, M. S. (2001). Origin of whales from early artiodactyls: hands and feet of Eocene Protocertidae from Pakistan. *Science* 292: 2239–2242.

Gingerich, P. D., Smith, B. H. and Simons, E. L. (1990). Hind limbs of Eocene *Basilosaurus*: Evidence of feet in whales. *Science* 249: 154–157.

Gingerich, P.D. (1992). Marine mammals (Cetacea and Sirenia) from the Eocene of Gebel Mokattam and Fayum, Egypt: Stratigraphy, age and paleoenvironments. *Contrib. Mus. Paleont., Univ. Michigan* 30: 1–84.

Gingerich, P.D. (2008). In: Fleagle, J. G. and Gilbert, C. C. (eds.), Elwyn Simons: *A Search for Origins*. Springer, New York, pp. 107–124.

Gingerich, P. D. and Uhen, M. D. (1998). Likelihood estimation of the time of origin of Cetacea and the time of divergence of Cetacea and Artiodactyla. *Palaeontologica Electronica* 1(2): http://wwwodp.tamu.edu/paleo/1998 2/ging uhen/issue2.htm.

Gingerich, P. D., Wells, N. A., Russell, D. E. and Shah, S. M. I. (1983). Origin of whales in epicontinental remnant seas: New evidence from the early Eocene of Pakistan. *Science* 220: 403–406.

Fleagle, J. G. and Simons, E. L. (1995). Limb skeleton and locomotor adaptations of *Apidium phiomense*, an Oligocene anthropoid from Egypt. *Am. J. Phys. Anthropol.* 97: 235–289.

Harlan, R. (1834). Notice of fossil bones found in the Tertiary formation of the state of Louisiana. *T. Am. Philos. Soc., Philadelphia* 4: 397–403.

Holroyd, P. A. (1994). An examination of dispersal origins for Fayum Mammalia. Doctoral dissertation, Duke University, Dept. of Biological Anthropology, p. 328.

Holroyd, P.A. (1999). New Pterodontinae (Creodonta: Hyaenodontidae) from the late Eocene early Oligocene Jebel Qatrani Formation, Fayum province, Egypt. *PaleoBios* 19(2):1–18.

Holroyd, P. A., Simons, E. L., Bown, T. M., Polly, P. D. and. Kraus, M. J. (1996). New records of terrestrial mammals from the upper Eocene Quasr el Sagha Formation, Fayum Depression, Egypt. *Paleovertebrata* 23: 175–192.

Matsumoto, H. (1922). *Megalohyrax* Andrews and *Titanohyrax* gen. nov. A revision of the genera of hyracoids from the Fayum, Egypt. *P. Zool. Soc. Lond.* 1921: 839–850.

Matsumoto, H. (1926). Contribution to the knowledge of the fossil Hyracoidea of the Fayum, Egypt, with description of several new species. *B. Am. Mus. Nat. Hist.* 56: 253–350.

Meyer, G. E. (1973). A new Oligocene hyrax from the Jebel el Qatrani Formation, Egypt. *Postilla* 163:1–11.

Meyer, G. E. (1978). Hyracoidea. In: Maglio, V. J. and Cooke, H. B. S. (eds.), *Evolution of African Mammals*, Chapter 14. Harvard University Press, Cambridge, pp. 284–314.

Osborn, H. F. (1908). New fossil mammals from the Fayum Oligocene, Egypt. *B. Am. Mus. Nat. Hist.* 24: 265–272.

Rasmussen, D. T. (1989). The evolution of the Hyracoidea: A review of the fossil evidence. In: Prothero, D. R. and Schoch, R. M. (eds.), *The Evolution of Perissodactyls*. Oxford University Press, New York, pp. 57–78.

Rasmussen, D. T. (2000). Hyraxes. In: Singer, R. (ed.), *Encyclopedia of Paleontology*. Fitzroy Dearborn Publishers, Chicago.

Rasmussen, D. T., Gagnon, M. and Simons, E. L. (1990). Taxeopody in the carpus and tarsus of Oligocene Pliohyracidae (Mammalia, Hyracoidea) and the phyletic position of hyraxes. *P. Nat. Acad. Sci. USA* 87: 4788–4691.

Rasmussen, D. T. and Simons, E. L. (1988). New Oligocene hyracoids from Egypt. *J. Vertebr. Paleontol.* 8: 67–83.

Rasmussen, D. T. and Simons, E. L. (1991). The oldest Egyptian hyracoids (Mammalia: Pliohyracidae): New species of *Saghatherium* and *Thyrohyrax* from the Fayum. *Neues Jahrbuch für Geol. und Paläontol., Abhandlungen* (Stuttgart) 182: 187–209.

Rasmussen, D. T. and Simons, E. L. (2000). Ecomorphological diversity among Paleogene hyracoids (Mammalia): A new cursorial browser from the Fayum, Egypt. *J. Vertebr. Paleontol.* 20(1): 167–176.

Schlosser, M. (1910). Uber einige fossil Säugetiere aus dem Oligocan vom Äegypten. *Zool. Anzeiger* 35:500–508.

Schlosser, M. (1911). Beiträge zur Kenntnis der Oligozänen Landsäugetiere aus dem Fayum: Ägypten. *Beitr. Paläont. Geol.* Osterreich-Ungarns 24: 167 pp., plates 1–16 Wien und Leipzig.

Schlosser, M. (1923). Mammalia. In: von Zittel, K. A. (ed.), *Grundzuege der Palaeontologie II Abteilung: Vertebrata,* 4th ed. Munich, Oldenburg press, pp. 402–689.

Schmidt, M. (1913). Ueber Paarhufer der fluviomarinen Schichted des Fajum. *Geol. Paläont. Abh.* 15: 153–264.

Schwartz G. T., Rasmussen, D. T. and Smith, R. J. (1995). Body size diversity and community structure of fossil hyracoids. *J. Mammal.* 76: 1088–1099.

Seiffert, E. R. (2003). A phylogenetic analysis of living and extinct afrotherian placentals. Doctoral dissertation, Duke University, Dept. of Biological Anthropology, 239 pp.

Seiffert, E. R., Simons, E. L., Clyde, W. C., Rossie, J. B., Attia, Y., Bown, T. M., Chatrath, P. and Mathison, M. E. (2005). Basal Anthropoids from Egypt and the Antiquity of Africa's Higher Primate Radiation. *Science* 310: 300–304.

Sen, S. and Heintz., E. (1979). *Palaeoamasia kansui* Ozansoy 1966, Embrithopode (Mammalia) de l'Éocène d'Anatolie: Annales de Paléontologie (Vértébres). 65:73–91.

Sigé, B. (1985). Les chriopères oligoctènes du Fayum, Egypte. *Geologica et Palaeontologica* 19: 161-189.

Simons, E. L. (1968). African Oligocene Mammals: Introduction, History of Study, and Faunal Succession. Part I in Early Cenozoic Mammalian Faunas, Fayum Province, Egypt. *Bull. Peabody Mus. Nat. Hist. Yale University* 28: 1–21.

Simons, E. L. and Bown, T. M. (1995). Ptolemaiida, a New Order of Mammalia—with description of the cranium of *Ptolemaia grangeri. P. Natl. Acad. Sci. USA* 92: 3269–3273.

Simons, E. L. and Gingerich, P. D. (1974). New carnivorous mammals from the Oligocene of Egypt. *Ann. Geol. Surv. Egypt.* 4: 157–166.

Sudre, J. (1979). Nouveaux mammifères éocènes du Sahara occidental. *Paleovertebrata* 9: 83–115.

Wood, A. E. (1968). The African Oligocene Rodentia. Part II in Early Cenozoic Mammalian Faunas, Fayum Province, Egypt, *Bull. Peabody Mus. Nat. Hist. Yale University* 28: 23–105.

Early Evolution of Whales

A Century of Research in Egypt

Philip D. Gingerich

Introduction

Living whales are fully aquatic and belong to two suborders of Cetacea: Mysticeti (baleen whales) and Odontoceti (toothed whales). Both of these modern suborders appeared when Earth changed from a 'greenhouse' earth to an 'icehouse' earth at in about the beginning of the Oligocene epoch (Zachos et al., 2001). Early whales, from the 'greenhouse' Eocene, all belong to a distinct paraphyletic suborder Archaeoceti. Archaeocetes differ from later modern whales in retaining many characteristics of land mammals, including complexly occluding cheek teeth, ear bones well integrated with the rest of the cranium, longer necks, less specialized forelimb flippers, and hind limbs with feet and toes. Archaeocetes are, in essence, the transitional forms documenting the origin of whales from an earlier land-mammal ancestry (Gingerich, 2005).

The first archaeocete fossil to be studied scientifically was a very large vertebral centrum collected in 1832 near the Ouachita River in Caldwell Parish, Louisiana. This measured some 35 cm in length and was but one of a series of 28 vertebrae found together there. The animal represented was named *Basilosaurus* or 'king lizard' because of its size and presumed reptilian heritage (Harlan, 1834). At the time the British anatomist Richard Owen was busy studying the large reptiles he eventually called dinosaurs. To solve the mystery of *Basilosaurus*, Owen secured additional remains and showed that it was a mammal because its cheek teeth are double-rooted. Owen (1839) deemed the name *Basilosaurus* to be inappropriate and proposed *Zeuglodon* or 'yoked teeth' as a replacement name. Relationship to whales was codified when Owen (1841) added the specific epithet *Zeuglodon cetoides*.

Additional archaeocete fossils were found in North America later in the nineteenth century, including the type specimen of *Dorudon serratus* Gibbes

Philip D. Gingerich
Museum of Paleontology and Department of Geological Sciences, The University of Michigan, Ann Arbor, MI 48109-1079
gingeric@umich.edu

J. G. Fleagle, C. C. Gilbert (eds.), *Elwyn Simons: A Search for Origins.*

(1845) and the type of *Zygorhiza kochii* (Reichenbach, 1847). These species were first classified in Basilosauridae by Cope (1868). Later all specimens known at the time were reviewed in a great monograph on Archaeoceti (Kellogg, 1936). This new monograph was important for the anatomical information it summarized, but also, on close reading, for the substantial gaps it revealed in our knowledge of archaeocetes: (1) no archaeocete was known from a complete axial skeleton, and it was impossible to say how many vertebrae any archaeocete had; (2) no archaeocete had the carpus and manus represented, and hence, conformation of the forelimb flipper was unknown (Gidley, 1913, and Kellogg, 1936, fabricated these using hands of sea lions as models); and (3) no archaeocete was known to have retained hind limbs with feet, and the innominates of *Basilosaurus* were mounted as if they still articulated with the vertebral column. Kellogg (1936) correctly inferred that advanced archaeocetes like *Dorudon* and *Zygorhiza* swam using upward and downward strokes of a powerfully-muscled and fluked tail (Uhen, 1996, 2004; Gingerich, 2003).

In recent years many of the gaps in our knowledge of archaeocetes have been filled by recovery of new and better archaeocete specimens from Egypt (Fig. 1)

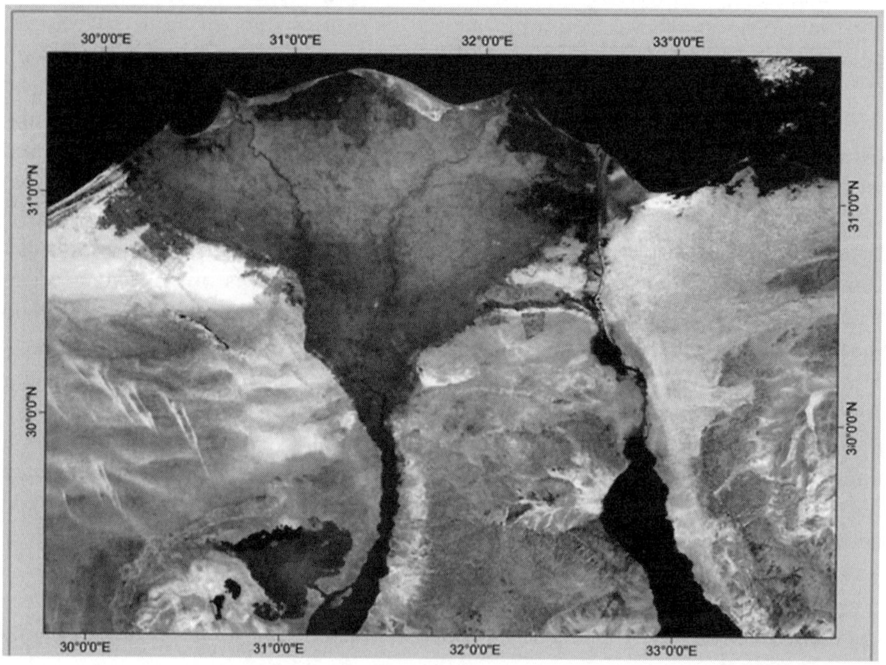

Fig. 1 Landsat image of present-day Egypt showing the location of Eocene archaeocete whale sites at Gebel Mokattam (site 1), Geziret el-Qarn (site 2), Qasr el-Sagha escarpment (site 3), Birket Qarun escarpment (site 4), Wadi Hitan (site 5), and Khashm el-Raqaba in Wadi Tarfa (site 6)

and from Pakistan. These include specimens with complete skulls, axial skeletons, forelimbs, hind limbs, and tails. The Egyptian archaeocetes include specimens classified as Basilosauridae (*Basilosaurus*, *Dorudon*, *Saghacetus*) and specimens classified as Protocetidae (*Protocetus*, *Eocetus*). The taxonomic history and stratigraphic distribution for each Egyptian archaeocete species is summarized in Table 1 and Table 2. Protocetidae and Basilosauridae represent different grades of adaptation to life in water, with early protocetids being semiaquatic foot-powered swimmers, while later basilosaurids were fully aquatic tail-powered swimmers (Gingerich, 2003). Of relevance here, the Egyptian and Pakistan field work was encouraged and facilitated, directly or indirectly, by Professor Elwyn Simons, and he was an active participant in recovery of the first whale hind limbs when these were found on *Basilosaurus*.

Table 1 Summary of the history of names for the genera and species of archaeocete whales found in Egypt. Stratigraphic distribution of each species is shown in Table 2

Egyptian form	History of names	Author	Geographical distribution
	Basilosaurus (no species)	Harlan (1834)	North America
	Zeuglodon (no species)	Owen (1839)	North America
	Zeuglodon cetoides	Owen (1841)	North America
	Dorudon serratus	Gibbes (1845)	North America
Saghacetus osiris	*Zeuglodon osiris*	Dames (1894)	Egypt
	Zeuglodon osiris	Slijper (1936)	Egypt
	Dorudon osiris	Kellogg (1936)	Egypt
	Saghacetus osiris	Gingerich (1992)	Egypt
Protocetus atavus	*Protocetus atavus*	Fraas (1904a)	Egypt
Eocetus schweinfurthi	*Mesocetus schweinfurthi*	Fraas (1904a)	Egypt
	Eocetus schweinfurthi	Fraas (1904b)	Egypt
Basilosaurus isis	*Zeuglodon isis*	Beadnell in Andrews (1904)	Egypt
	Zeuglodon isis	Slijper (1936)	Egypt
	Prozeuglodon isis (in part)	Kellogg (1936)	Egypt
	Basilosaurus isis	Gingerich et al. (1990)	Egypt
Dorudon atrox	*Prozeuglodon atrox*	Andrews (1906)	Egypt
	Prozeuglodon isis (in part)	Kellogg (1936)	Egypt
	Prozeuglodon isis	Moustafa (1954)	Egypt
	Dorudon atrox	Gingerich and Uhen (1996)	Egypt
Dorudon stromeri	*Prozeuglodon stromeri*	Kellogg (1928)	Egypt
	Dorudon stromeri	Kellogg (1936)	Egypt
Ancalecetus simonsi	*Ancalecetus simonsi*	Gingerich and Uhen (1996)	Egypt

Table 2 Stratigraphic distribution of middle and late Eocene archaeocete whales from Egypt. Taxonomic history for each genus and species is summarized in Table 1

Formation	Taxon	Locality (Figure 1)
Priabonian (late Eocene)		
Qasr el-Sagha Formation	*Saghacetus osiris* (Dames, 1894)	3
	Dorudon stromeri (Kellogg, 1928)	3
Birket Qarun Formation	*Basilosaurus isis* (Beadnell in Andrews, 1904)	2, 4, 5
	Dorudon atrox (Andrews, 1906)	2, 4, 5
Bartonian (middle Eocene)		
Giushi Formation	*Eocetus schweinfurthi* (Fraas, 1904a)	1
Gebel Hof Formation	Undescribed skeletons	6
Lutetian (middle Eocene)		
Mokattam Formation	*Protocetus atavus* Fraas, 1904a	1

Chronology of Egyptian Research

Egyptian archaeocetes were not the first archaeocetes to be found, but they include a number of early species (Table 1). Further, new specimens from Egypt are proving to be so complete and important for our understanding of early whale evolution that the history of their discovery deserves review.

Schweinfurth and Dames

Georg August Schweinfurth [1836–1925] was a German botanist who spent three years in the Eastern Desert of Egypt and Sudan from 1863 to 1866 while compiling his botanical dissertation *Beitrag zur flora Äthiopiens* (1867). In 1868 he was commissioned by the Alexander von Humboldt Foundation of Berlin to explore more of Central Africa, which he did from 1869 to 1871. Publication of *Im Herzen von Afrika* in 1874 secured Schweinfurth's reputation as an African explorer.

Schweinfurth lived in Cairo for 14 years from 1875 through 1889, and it was during this time that the first Eocene mammal fossils were described from Gebel Mokattam (site 1 in Fig. 1) near Cairo (the first was the sirenian *Eotherium aegyptiacum*, now *Eotheroides aegyptiacum*; Owen, 1875). While living in Cairo, Schweinfurth continued to explore locally, and in 1879 discovered the first vertebrate fossils in Fayum and the first archaeocete whales in Egypt. These were found on the island of Geziret el-Qarn (site 2 in Fig. 1) in the middle of lake Birket Qarun. In 1884 and again in 1886, Schweinfurth visited Qasr el-Sagha and the Qasr el-Sagha escarpment (site 3 in Fig. 1) on the north side of Birket Qarun. This is the time, too, when Schweinfurth published two studies

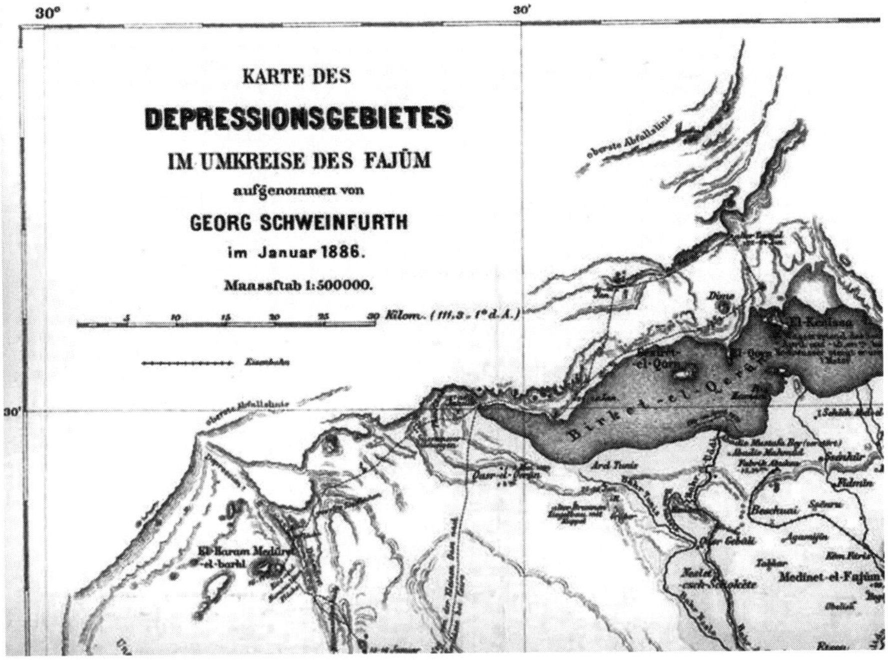

Fig. 2 Georg Schweinfurth's map of Fayum showing the path of his 1886 expedition through Wadi Rayan and north of Birket Qarun. Schweinfurth found the type specimen of *Zeuglodon osiris* (now *Saghacetus osiris*) at the site near his January 21–22 camp (Garet el Esh). He camped at Garet Gehannam on January 16–17, but never found the archaeocete skeletons there nor those in Wadi Hitan several kilometers to the west

important for Egyptian stratigraphy and archaeocete paleontology: 'On the geological stratification of Mokattam near Cairo' (Schweinfurth, 1883), and 'Travel in the depression circumscribing Fayum in January, 1886' (Schweinfurth, 1886). On the expedition described in the latter (Fig. 2), Schweinfurth came within a few kilometers of discovering Zeuglodon Valley, or what is now called Wadi Hitan, but turned back because of difficulties with his camels and staff.

Schweinfurth's fossils from both expeditions, the 1879 expedition to Geziret el Qarn and the 1886 expedition to Wadi Rayan and the north side of Birket Qarun, are conserved in the Humboldt University *Museum für Naturkunde* in Berlin, where they were studied by Wilhelm Barnim Dames [1843–1898]. Fossils from Geziret el Qarn include a diversity of isolated vertebrae from what are now called uppermost Gehannam and lower Birket Qarun formations (see also Seiffert, this volume). Preliminary notices of these were published by Dames (1883a, b). Fossils from north of Birket Qarun include a well preserved dentary from the Qasr el-Sagha excarpment that became the type of the first archaeocete named from Egypt: *Zeuglodon osiris* (or now *Saghacetus osiris*; Fig. 3).

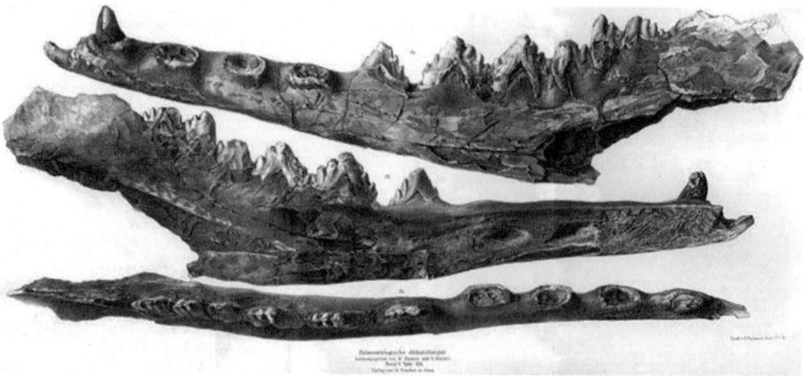

Fig. 3 Type specimen of *Zeuglodon osiris* Dames, 1894 [now *Saghacetus osiris* (Dames)] found by Georg Schweinfurth at a site near Garet el Esh (see Fig. 2), north of lake Birket Qarun in the Fayum Depression

Schweinfurth's specimens, including the type dentary of *Saghacetus osiris*, were reviewed and illustrated by Dames (1894).

Blanckenhorn, Fraas, Markgraf, and Stromer

The next group to investigate the marine Eocene north of Birket Qarun was also German. Max Blanckenhorn [1861–1947] started work for the Geological Survey of Egypt in 1897–1898, shortly after its founding, and he published several important reports on the stratigraphy of whale-bearing formations (Blanckenhorn, 1900, 1903). Eberhard Fraas [1862–1915] visited Egypt too, starting in 1897, where he engaged Richard Markgraf [1856–1916] as a private collector working first in the stone quarries of Gebel Mokattam, and later in Fayum. Markgraf was a Bohemian living in Egypt because of poor health.

During the winter of 1901–1902 Blanckenhorn was joined by Ernst Stromer von Reichenbach [1871–1952] of the *Königlich Bayerischen Akademie der Wissenschaften* in Munich. In January, 1902, they made an 11-day traverse starting at the temple ruin at Qasr Qarun near the west end of Birket Qarun. They went around the west end of Birket Qarun and then, following Schweinfurth's path, traveled eastward to Dimeh and Qasr el-Sagha on the north side of Birket Qarun. Stromer von Reichenbach (1903a) was generally disappointed by the results of this expedition, but he was able to describe a new skull and lower jaw of *Saghacetus osiris* (Dames) from what is now upper Qasr el-Sagha Formation. Later Stromer von Reichenbach (1903b) described a new species *Zeuglodon zitteli* from the type locality of *Saghacetus osiris*, but this has proven indistinguishable from Dames' species.

Markgraf made an important discovery in 1903 when he found the cranium and associated postcranial remains of a small archaeocete from the Mokattam Limestone of Lutetian, early middle Eocene, age, near the base of the Gebel Mokattam section (site 1 in Fig. 1). This was described by Fraas (1904a) as a new genus and species *Protocetus atavus*. The skull is primitive, with upper molars retaining protocones. Associated vertebrae include a partial sacrum with auricular processes indicating retention of articulation with a pelvis (the sacrum was originally interpreted as having consisted of a single centrum, but later comparisons indicate that the inferred reduction is an artifact of breakage).

A second skull of a different whale from Gebel Mokattam, recovered by Markgraf in 1904, was added to Fraas' 1904a publication, possibly at the proof stage. Fraas' name for this, *Mesocetus*, proved to be preoccupied, and the name *Eocetus* had to be substituted (Fraas, 1904b). The type of *Eocetus* came from what is now called Giushi Formation and is Bartonian, or late middle Eocene in age, from a level higher in the Gebel Mokattam section than that yielding *Protocetus* (site 1 in Fig. 1). The older and more primitive form, *Protocetus*, subsequently became the type of a new family of archaeocetes called Protocetidae (Stromer von Reichenbach, 1908). Protocetidae originally included just *Protocetus* and *Eocetus*, but the family now includes additional genera named in later years.

Stromer made a second trip to Egypt to collect Eocene whales and other vertebrates, starting in November 1903. This was a three-month expedition employing Markgraf and ranging widely. Two large archaeocete vertebrae were collected at Gebel Mokattam, and an archaeocete skull with jaws and vertebrae was collected north of Birket Qarun. Markgraf continued to collect fossil whales in Fayum, and later in 1904 he sent another specimen from the Qasr el-Sagha escarpment to Munich that Stromer von Reichenbach (1908) first identified as *Zeuglodon osiris*. This eventually became the type of *Dorudon stromeri* (Kellogg, 1928).

Eberhard Fraas made a final trip to Egypt to work with Markgraf in 1906. The trip started on March 11, and they reached Qasr el-Sagha on March 13. Camels were sent for water and fodder, and within days they moved to the west end of Birket Qarun (site 4 in Fig. 1). Here they excavated much of the skeleton of a large '*Zeuglodon*' (*Basilosaurus isis*) with a 1.3-meter skull and a 10-meter section of the following skeleton comprising vertebrae and ribs (Fraas, 1906). This specimen, described by Slijper (1936), is in the *Staatliches Museum für Naturkunde* in Stuttgart.

Beadnell and Andrews

Hugh J. L. Beadnell [1874–1944] was British, and was employed by the Egyptian Geological Survey from the time it was founded in 1896 until 1906. Beadnell worked on various projects, and then started work in the Fayum in October, 1898. Beadnell started in eastern Fayum, and then extended his mapping first to the escarpments north of Birket Qarun and then westward to Garet Gehannam during the spring of 1899. Charles W. Andrews [1866–1925] of the British Museum (Natural History) published the first note on Beadnell's paleontological discoveries (Andrews, 1899), and Beadnell himself (1901) provided a summary of Fayum stratigraphy that has guided most subsequent work. In April of 1901, Andrews joined Beadnell in the field for the first time to investigate bone beds discovered in 1898, and many new specimens including archaeocetes were found (Andrews, 1901). Collecting then continued during the winters of 1901–1902, 1902–1903, and 1903–1904. Sometime in this interval, Beadnell collected a very large dentary for the Cairo Geological Museum specimen that became the type of *Zeuglodon isis* Beadnell in Andrews, 1904 (now *Basilosaurus isis*). According to Beadnell (1905), this came from the Birket Qarun escarpment near the west end of the lake (site 4 in Fig. 1).

A second phase of mapping was carried out in the winter of 1902–1903, when Beadnell made a traverse from Garet Gehannam west and southwest 12 kilometers to a valley where large skulls and other remains of fossil whales were abundant (site 5 in Fig. 1). Very little was collected, but one skull recovered for the Cairo Geological Museum was made the type of *Prozeuglodon atrox* by Andrews (1906). This is now properly called *Dorudon atrox*. Andrews diagnosed *Prozeuglodon* as "intermediate between *Protocetus* and *Zeuglodon* proper," but he seemingly did not recognize that the type is a juvenile with deciduous premolars. This led to confusion when it was recognized later, and for some time *Prozeuglodon atrox* was thought to be the juvenile form of *Zeuglodon isis*. Kellogg (1936), for example, combined these, and referred to both as *Prozeuglodon isis* (later recovery of skeletons of mature *Dorudon atrox*, see below, showed that they are clearly different from contemporary *Basilosaurus isis*). Surprisingly, following Beadnell's initial work in the valley 12 kilometers WSW of Garet Gehannam in 1902–1903, no serious collection and study of the whales there was carried out for eighty years.

Osborn and Granger

Henry Fairfield Osborn [1857–1935] of the American Museum of Natural History in New York organized a 1907 collecting expedition to Fayum to follow in the footsteps of Blanckenhorn, Stromer, Beadnell, and Andrews. Osborn interpreted publication of Andrews' 1906 *Descriptive Catalog of the Tertiary Vertebrata of the Fayum, Egypt* to indicate that the area was now open for study by others, and Andrews himself (*in litt.*) encouraged Osborn to carry out further studies. Osborn's first published report on the expedition was submitted from Cairo and dated February 25. In this, Osborn coined the term 'Zeuglodon Valley' for Beadnell's valley 12 kilometers WSW of Garet Gehannam (Osborn, 1907a; site 5 in Fig. 1). He later called this "the most famous fossil locality in the Fayum" (Osborn, 1907b). Zeuglodon Valley has since been renamed Wadi

Hitan or 'Valley of Whales' to acknowledge synonymy of *Zeuglodon* with *Basilosaurus* and to provide an alternative in Arabic (Gingerich, 1992).

Osborn did some prospecting for fossils during the time he was in Fayum, but he was accompanied by his wife and children and did no real collecting. The Egyptian Geological Survey assigned Hartley T. Ferrar [1879–1932], recently returned from Robert F. Scott's British National Antarctic Expedition of 1901–1903, to accompany and assist Osborn. The area that the American Museum party worked included Beadnell and Andrews' principal localities, in continental beds above the Qasr el-Sagha escarpment (north of site 3 in Fig. 1). Here American Museum quarries A, B, and later C were developed. From here, Osborn and Ferrar made a three-day camel march west to Garet Gehannam and Wadi Hitan on February 14–16 (site 5 in Fig. 1). As Osborn (1907b) wrote:

> We found [Wadi Hitan] strewn with the remains of monster zeuglodonts, including heads, ribs and long series of vertebrae, most tempting to the fossil hunter, yet too large and difficult of removal from this very remote and arid point.

Osborn left the Fayum on February 18 to return to Cairo and New York, leaving Walter Granger [1872–1941] and Granger's assistant George Olsen [d. 1939] of the American Museum as the paleontologists responsible for all of the fossil collections to be made in Fayum. Granger and Olsen started work in Fayum on February 5 with a large team of Egyptian workers. They continued work at and near quarries A, B, and C through April 21, when they moved camp for three days to Qasr el-Sagha.

February 16, 1907, Granger had a chance meeting with Richard Markgraf, who was collecting fossils in the same area. Osborn met Markgraf on February 17 after his return from Wadi Hitan. At this time negotiations started for Markgraf to work the remainder of the season for the American Museum team. Two archaeocete specimens in the American Museum collection, a braincase and frontal of *Basilosaurus isis* and a fine skull of *Saghacetus osiris*, were collected by Markgraf. The latter is illustrated in Kellogg (1936). These Markgraf specimens came from the Birket Qarun escarpment (site 4 in Fig. 1) and the Qasr el-Sagha escarpment (site 3 in Fig. 1), respectively.

The only archaeocete that the American Museum party found themselves was a partial skull of *Saghacetus osiris* collected on April 23 from 1–2 kilometers west of Qasr el-Sagha temple. Granger left Fayum for Cairo on April 25, and then returned to the field in Fayum again from May 2 through May 23.

Phillips and Denison

Wendell Phillips [1925–1975], namesake and descendent of the nineteenth century abolitionist orator, was a young American, an enthusiastic Princeton undergraduate and University of California, Berkeley, graduate student in paleontology who became an anthropologist, archaeologist, Arabian and African explorer, and later founder of Wendell Phillips Oil Company based in London. He participated in the first Pan-African Congress on Prehistory in 1947 and then organized the University of California African Expedition from Cairo to Cape Town, which was active from 1947–1950 (Phillips, 1948). Phillips also founded the American Foundation for the Study of Man, in Washington, D.C., in 1949.

Robert H. Denison [1911–1985], then of Dartmouth College, was the principal paleontologist during the Egyptian phase of the University of California African Expedition. He was assisted by Paul Deraniyagala of Ceylon, V. L. VanderHoff of Stanford University, and H. B. S. Cooke of the University of Witwatersrand. This group worked in Fayum from September 24 through December 14, 1947. Most of their time was spent prospecting and excavating in the vicinity of the earlier Beadnell-Andrews and American Museum excavations and prospecting on the Qasr el-Sagha excarpment (site 3 in Fig. 1).

On November 14, Denison and Deraniyagala drove westward with two U.S. Marine soldiers and reached Garet Gehannam and Wadi Hitan on November 15 (site 5 in Fig. 1). The group camped for two nights, and collected the well preserved left half of a *Basilosaurus isis* skull for the University of California at Berkeley. Deraniyagala (1948) published an illustrated account of the field work, including the trip to Wadi Hitan. Deraniyagala reported the whale skeletons in Wadi Hitan as Oligocene in age, which they are not, and recorded that some twenty skeletons lay with a radius of 1–2 kilometers of their camp, which is likely. However, his interpretation that such a density of skeletons "shows that even 35 million years ago schools of whales committed race suicide by stranding themselves" would require more detailed documentation than he provided.

Moustafa

Y. Shawki Moustafa [n.d.] of Cairo University collected the cranium of a subadult archaeocete in 1950 from the lower part of the Birket Qarun Formation west of lake Birket Qarun (Moustafa, 1954, 1974; the specimen was found somewhere in or between sites 4 and 5 in Fig. 1). Following Kellogg (1936), Moustafa identified this as *Prozeuglodon isis*. However, with the experience of many seasons collecting archaeocetes in the Birket Qarun Formation (see below), the only archaeocete species found west of Birket Qarun for which numerous subadult individuals are known is *Dorudon atrox*, and the specimen described by Moustafa has the size and morphology of this species.

Simons and Meyer

Elwyn Simons [b. 1930] was an American professor at Yale University when he started field work in Egypt. Simons carried out his first season of field work in

Fayum during the winter of 1961–1962, working north of the Qasr el-Sagha escarpment (Simons, 1968; north of site 3 in Fig. 1). Simons is now at Duke University, and the project continues to the present day with emphasis on recovery of late Eocene and Oligocene fossil primates. The manager of Simons' field work during the 1960s was Grant E. Meyer [d. 2004], who organized two trips to Wadi Hitan in 1965–65 and 1966–1967. During the first, Meyer was assisted by Jeff Smith and Tom Walsh, and during the second he was assisted by John Boyer and Lloyd Tanner. Collections from these expeditions are in the Cairo Geological Museum and the Yale University Peabody Museum of Natural History.

Simons and Gingerich

Another American, Philip D. Gingerich [b. 1946], started working with Simons in 1983 to collect representative archaeocetes from the Qasr el-Sagha escarpment (site 3 in Fig. 1) and from Wadi Hitan (site 5 in Fig. 1). Initially the purpose was simply to acquire comparative specimens to facilitate research on archaeocetes being found in Pakistan, but both sites in Egypt proved to be so productive of exceptionally complete specimens that field research was focused here (Fig. 4). Field work was carried forward during six field seasons, in odd-numbered years from 1983 through 1993. Assistance was provided by Ali Barakat, Tom Bown, Will Clyde, Gregg Gunnell, Abd el-Latif, Alex van Nievelt, Bill Sanders, Holly Smith, and others in various years. Collections from these expeditions are in the Cairo Geological Museum and the University of Michigan Museum of Paleontology, and have been studied by Gingerich et al. (1990), Gingerich and Smith (1990), Gingerich (1992), Gingerich and Uhen (1996), and Uhen (1998, 2004).

The principal results of this work can be enumerated here:

(1) The only archaeocete species found commonly in the late Priabonian Qasr el-Sagha Formation of the Qasr el-Sagha escarpment (site 3 in Fig. 1) is the small ca. 3 meter-long *Saghacetus osiris*, which is known from skulls and several fairly complete axial skeletons.

(2) There are two archaeocete genera and species found commonly in the early Priabonian Birket Qarun Formation of Wadi Hitan (site 5 in Fig. 1), which are the large ca. 18 meter-long *Basilosaurus isis* and the medium-sized ca. 5 meter-long *Dorudon atrox*. Both are known from virtually complete skeletons (e.g., Fig. 5).

(3) Approximately 500 skeletons or partial skeletons of archaeocetes have been mapped in Wadi Hitan. Most specimens are *Basilosaurus isis* or *Dorudon atrox*. *B. isis* is a little more common than *D. atrox*, but skeletons of *B. isis* are also larger and therefore easier to map. All known *B. isis* specimens are mature adults, while *D. atrox* is approximately equally represented by

Fig. 4 Elwyn Simons helping to excavate the hind limb of a specimen of an 18-meter-long individual of *Basilosaurus isis* in Wadi Hitan in 1989. The vertebral column is articulated, and each individual centrum shown here measures about 30–33 cm in length. In contrast, a complete tibia, shown in situ here, is less than 15 cm long. Hind limbs and feet of Wadi Hitan *Basilosaurus isis*, the first for a cetacean, were described by Gingerich et al., 1990)

mature adult and immature specimens. Some immature *D. atrox* skulls exhibit bite marks made by a large predator, possibly *B. isis*.

(4) A third archaeocete species, *Ancalecetus simonsi*, is the size of *Dorudon atrox* and represented by a single partial skeleton (Gingerich and Uhen, 1996). There are seemingly two or three additional archaeocete genera and species present in Wadi Hitan.

(5) *Basilosaurus isis* and *Dorudon atrox* skeletons both retain flexible elbows but have wrists with blocky, tightly-packed carpals. The carpus consists of a separate scaphoid, lunar, and cuneiform in the proximal row; and a reduced trapezium, fused trapezoid-magnum, and large unciform in an alternating distal row. A small superficial centrale articulated with the scaphoid. The pollex is greatly reduced but not lost, and the pisiform is large, flat, and projecting (Gingerich and Smith, 1990; Uhen, 2004). This pattern differs somewhat from that of *Zygorhiza kochii* described by Kellogg (1936), which has relatively smaller carpals.

(6) *Basilosaurus isis* has 7 cervical vertebrae, 18 thoracics, and about 42 lumbocaudals (Gingerich et al., 1990), while the vertebral formula for *Dorudon atrox* was 7.17.20.21 (Uhen, 2004; lumbars and caudals are distinguished

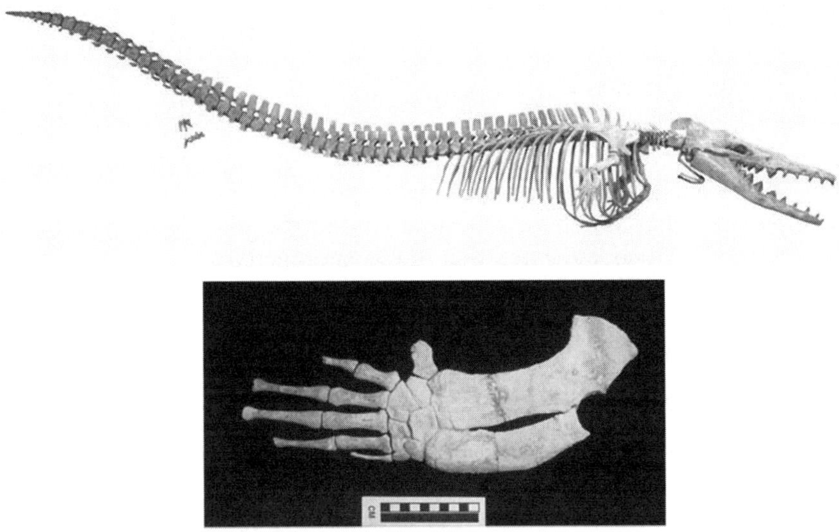

Fig. 5 Virtually complete skeleton of *Dorudon atrox* collected in 1991from Wadi Hitan. Note the forelimbs modified as flippers, expanded and robust lumbocaudal vertebral series, and reduced *Basilosaurus*-like hind limbs separated from the axial skeleton and embedded in the ventral body wall. All are characteristics of modern whales with a tail fluke and associated with aquatic propulsion by caudal undulation and oscillation (Uhen, 1996, 2004; Gingerich, 2003)

by the presence or absence of chevron facets). Skeletons of both genera and species lack any clear evidence of an intervening sacrum in the lumbocaudal vertebral series; both evidently retained pelves well separated from their vertebral column, and both were fully aquatic tail-powered swimmers. Both retained well-formed hind limbs and feet that were much reduced in size compared to the rest of the skeleton.

Gingerich, Fouda, and Attia

A new cooperative project was initiated in 2004 involving the Egyptian Environmental Affairs Agency (represented by Moustafa M. A. Fouda [b. 1950]), the Egyptian Mineral Resources Authority (including the Egyptian Geological Survey and Geological Museum; represented by Yousry Attia [n.d.]), and the University of Michigan (represented by Philip D. Gingerich). Field work was restarted to substantiate the importance of Wadi Hitan for understanding early whale evolution, which contributed to the site being designated a UNESCO World Heritage Site in 2005. New fossils have been collected, but these have not as yet been prepared or studied, and there are as yet no new scientific results to report.

Bianucci and Others

Giovanni Bianucci [n.d.] and others in Italy illustrated and exhibited the skeleton of a protocetid archaeocete recovered by stonecutters in Italy while cutting decorative limestone imported from Egypt (Bianucci et al., 2003). There are additional unpublished reports of archaeocetes and primitive sirenians being found in limestones imported to Europe from Egypt. Investigation in 2006 indicates that these specimens are coming from the Gebel Hof Formation or equivalent, of Bartonian late middle Eocene age, from a site at 28E 27' N latitude and 31E 50' E longitude, north of Khashm el-Raqaba in Wadi Tarfa, Eastern Desert of Egypt (Fig. 6). More work will be required to devise a way to find, recover, and prepare such specimens for scientific study. They are potentially very important for filling a gap between reasonably well known Lutetian early middle Eocene protocetids and classic Priabonian late Eocene basilosaurids. This gap is the time of transition from foot-powered to tail-powered swimming in archaecetes.

Fig. 6 Truck hauling middle Eocene limestone from a major quarry complex at 28E 27' N latitude and 31E 50' E longitude, north of Khashm el-Raqaba in Wadi Tarfa, Eastern Desert of Egypt. This limestone is exported to Europe and elsewhere as decorative building stone. Some is archaeocete-bearing, as indicated by the specimen illustrated by Bianucci et al. (2003)

Significance

Egyptian fossil localities and specimens are fundamental for understanding the early evolution of whales. Egyptian archaeocetes are found in four stratigraphic intervals spanning much of the middle and late Eocene, and all four intervals have been known for fully 100 years. Likewise, five of the six major sites in Fig. 1 have been known for fully 100 years. Wadi Hitan has been particularly important during the past twenty years because it is here that we have been able to answer many of the questions left unanswered by Kellogg's (1936) classic monograph on Archaeoceti. We now have several archaeocetes with virtually complete axial skeletons, and we know their vertebral formulae. We know the size, proportions, and articulation of the forelimb, wrist, and hand for *Basilosaurus* and *Dorudon*, and we know much about the size, proportions, and articulation of the hind limb, tarsus, and foot for both genera.

Further, the success of our research on skeletons of late Eocene archaeocetes in Wadi Hitan inspired a return to investigation of protocetid archaeocetes in Pakistan, where protocetids are now represented by equally complete skeletons with skulls, axial skeletons complete to the end of the tail, forelimbs with wrists, hands, fingers, and hooves, and hind limbs with ankles, feet, and toes. Recovery of ankle bones associated with protocetid skeletons demonstrated the artiodactyl ancestry of whales, solving a major mystery and controversy in mammalian phylogeny (Gingerich et al., 2001).

Finally, to close, this is an appropriate place to acknowledge and thank Elwyn Simons for encouraging and facilitating research on early whale evolution in Egypt. Simons said several times during his career that what surprised him about fieldwork in Egypt and elsewhere was that new discoveries rarely conform to expectation. This has been my experience studying early whale evolution too. I never expected that archaeocetes living 15–20 million years after the origin of whales would still retain well formed hind limbs and feet, long after they ceased to walk; and I never expected that whales would prove to be the direct descendants of artiodactyls. Field work in Egypt and Pakistan has shown both to be true. This means two things: first, we are still at a stage where emphasis on exploration and discovery now will greatly enhance reliable interpretation in the future; and second, it is probably futile to try to infer the paths of morphological transformation, adaptation, and phylogenetic relationships of whales (or primates) in much detail beyond what we can trace in the fossil record.

Inference from the present works in some instances, as in prediction that whales and artiodactyls should be related (starting with Boyden and Gemeroy, 1950), but animals living today are a small subset of those that have lived in the past, limiting what we can learn of the past from the present. Van Valen (1966) explained Boyden and Gemeroy's inference of a whale-artiodactyl relationship in terms consistent with what we knew of the fossil record at the time, but here, to my surprise, new fossils showed he was wrong. If inferential methods had much power, we would more often find what we expect in the fossil record.

Acknowledgment I thank Elwyn Simons for encouraging and facilitating research on early whale evolution in Egypt, and for the same, less directly, in Pakistan. John G. Fleagle provided important encouragement too when we started research in 'Zeuglodon Valley.' The editor and two anonymous reviewers provided comments improving the manuscript. Field research in Egypt was made possible by a succession of grants from the Committee for Research and Exploration of the National Geographic Society, and more recently by the U.S.-Egypt Joint Science and Technology Program and the U.S. National Science Foundation (OISE-0513544).

References

Andrews, C. W. (1899). Fossil Mammalia from Egypt, Part I. *Geol. Mag.* 6: 481–484.

Andrews, C. W. (1904). Further notes on the mammals of the Eocene of Egypt, Part III. *Geol. Mag.* 1: 211–215.

Andrews, C. W. (1906). A descriptive catalogue of the Tertiary Vertebrata of the Fayum, Egypt. British Museum (Natural History), London. p. 324.

Beadnell, H. J. L. (1901). The Fayum depression: A preliminary notice of the geology of a district in Egypt containing a new Palaeogene vertebrate fauna. *Geol. Mag., London* 8: 540–546.

Beadnell, H. J. L. (1905). The topography and geology of the Fayum Province of Egypt. Survey Department of Egypt, Cairo. p. 101.

Bianucci, G., Nocchi, C., Sorbini, C., and Landini, W. (2003). L'Archeoceto nella Roccia alle origini dei Cetacei. Museo di Storia Naturale e del Territorio, Università degli Studi di Pisa, Pisa. pp. 1–17

Blanckenhorn, M. (1900). Neues zur Geologie und Paläontologie Aegyptens. II. Das Palaeogen. *Z. Deut. Morganland. G.* 52: 403–479.

Blanckenhorn, M. (1903). Neue geologisch-stratigraphische Beobachtungen in Aegypten. *Sitzungsberichte der Mathematisch-physikalischen Classe der Königlichen bayerischen Akademie der Wissenschaften, München* 32: 353–433.

Boyden, A. A., and Gemeroy, D. G. (1950). The relative position of the Cetacea among the orders of Mammalia as indicated by precipitin tests. *Zoologica- New York* 35: 145–151.

Cope, E. D. (1868). An addition to the vertebrate fauna of the Miocene period, with a synopsis of the extinct Cetacea of the United States. *P. Acad. Nat. Sci. Phila.* 19: 138–156.

Dames, W. B. (1883a). Ein Epistropheus von *Zeuglodon* sp. *Sitzungsberchte der Gesellschaft naturforschender Freunde, Berlin* 1883: 3.

Dames, W. B. (1883b). Ueber eine tertiär Wirbelthierfauna von der westlichen Insel der Birket-el-Qurun. *Sitzungsberichte der Königlich-preussischen Akademie der Wissenschaften, Berlin* 1883: 129–153.

Dames, W. B. (1894). Über Zeuglodonten aus Aegypten und die Beziehungen der Archaeoceten zu den übrigen Cetaceen. *Geologische und Paläontologische Abhandlungen, Jena* 5: 189–222.

Deraniyagala, P. E. P. (1948). Some scientific results of two visits to Africa. *Spolia Zeylanica* 25: 1–42.

Fraas, E. (1904a). Neue Zeuglodonten aus dem unteren Mitteleocän vom Mokattam bei Cairo. *Geologische und Paläontologische Abhandlungen, Jena* 6: 197–220.

Fraas, E. (1904b). Neue Zeuglodonten aus dem unteren Mitteleocän vom Mokattam bei Cairo. *Geologisches Zentralblatt, Leipzig* 5: 374.

Fraas, E. (1906). Wüstenreise eines Geologen in Ägypten. *Kosmos, Handweiser für Naturfreunde, Stuttgart* 3: 263–269.

Gibbes, R. W. (1845). Description of the teeth of a new fossil animal found in the green-sand of South Carolina. *P. Acad. Nat. Sci. Phila.* 2: 254–256.

Gidley, J. W. (1913). A recently mounted *Zeuglodon* skeleton in the United States National Museum. *Proceedings of the U.S. National Museum* 44: 649–654.

Gingerich, P. D. (1992). Marine mammals (Cetacea and Sirenia) from the Eocene of Gebel Mokattam and Fayum, Egypt: stratigraphy, age, and paleoenvironments. *University of Michigan Papers on Paleontology* 30: 1–84.

Gingerich, P. D. (2003). Land-to-sea transition of early whales: Evolution of Eocene Archaeoceti (Cetacea) in relation to skeletal proportions and locomotion of living semiaquatic mammals. *Paleobiology* 29: 429–454.

Gingerich, P. D. (2005). Cetacea. In: Rose, K. D. and Archibald, J. D. (eds.), *Placental mammals: origin, timing, and relationships of the major extant clades*. Johns Hopkins University Press, Baltimore, pp. 234–252.

Gingerich, P. D., Haq, M., Zalmout, I. S., Khan, I. H., and Malkani, M. S. (2001). Origin of whales from early artiodactyls: hands and feet of Eocene Protocetidae from Pakistan. *Science* 293: 2239–2242.

Gingerich, P. D., and Smith, B. H. (1990). Forelimb and hand of *Basilosaurus isis* (Mammalia, Cetacea) from the middle Eocene of Egypt. *J. Vertebr. Paleontol.* 10A: 24.

Gingerich, P. D., Smith, B. H., and Simons, E. L. (1990). Hind limbs of Eocene *Basilosaurus isis*: Evidence of feet in whales. *Science* 249: 154–157.

Gingerich, P. D., and Uhen, M. D. (1996). *Ancalecetus simonsi*, a new dorudontine archaeocete (Mammalia, Cetacea) from the early late Eocene of Wadi Hitan, Egypt. *Contrib. Mus. Paleontol., Univ. Mich.* 29: 359–401.

Harlan, R. (1834). Notice of fossil bones found in the Tertiary formation of the State of Louisiana. T. *Am. Phil. Soc.* 4: 397–403.

Kellogg, R. (1928). The history of whales—their adaptation to life in the water. *Q. Rev. Biol.* 3: 29–76, 174–208.

Kellogg, R. (1936). A review of the Archaeoceti. *Carnegie Institution of Washington Publications* 482: 1–366.

Moustafa, Y. S. (1954). Additional information on the skull of Prozeuglodon isis and the morphological history of the Archaeoceti. *Proceedings of the Egyptian Academy of Science* 9: 80–88.

Moustafa, Y. S. (1974). Critical observations on the occurrence of Fayum fossil vertebrates. *Annals of the Geological Survey of Egypt* 4: 41–78.

Osborn, H. F. (1907a). The Fayum expedition of the American Museum. *Science* 25: 513–516.

Osborn, H. F. (1907b). Hunting the ancestral elephant in the Fayum desert. *Century Magazine* 74: 815–835.

Owen, R. (1839). Observations on the teeth of the *Zeuglodon*, *Basilosaurus* of Dr. Harlan. Proceedings *of the Geological Society of London* 1839: 24–28.

Owen, R. (1841). Observations on the *Basilosaurus* of Dr. Harlan (*Zeuglodon cetoides* Owen). *Transactions of the Geological Society of London* 2: 69–79.

Owen, R. (1875). On fossil evidences of a sirenian mammal (*Eotherium aegyptiacum*) from the nummulitic Eocene of the Mokattam cliffs, near Cairo. *Quarterly Journal of the Geological Society of London* 31: 100–105.

Phillips, W. (1948). Recent discoveries in the Egyptian Fayum and Sinai. *Science* 107: 666–670.

Reichenbach, H. G. L. (1847). Systematisches. In: Carus, C. G. (ed.), *Resultate geologischer, anatomischer und zoologischer Untersuchungen über das unter dem Namen* Hydrarchos *von Dr. A. C. Koch zuerst nach Europa gebrachte und in Dresden angestellte grosse fossile Skelett*. Arnoldische Buchlandlung, Dresden and Leipzig, pp. 13–15.

Schweinfurth, G. A. (1883). Ueber die geologische Schichtengliederung des Mokattam bei Cairo. *Zeitschrift der Deutschen Geologischen Gesellshaft, Stuttgart* 35: 709–737.

Schweinfurth, G. A. (1886). Reise in das Depressionsgebiet im Umkreise des Fajum im Januar 1886. *Zeitschrift der Gesellschaft für Erdkunde zu Berlin* 21: 96–149.

Seiffert, E. R. (this volume). Geology, paleoenvironment, and age of Birket Qarun Locality 2 (BQ-2), Fayum Depression, Egypt. In: Fleagle, J. G. (ed.), *Elwyn L. Simons: Mapping the Progress of Primate Evolution*. Plenum Press, New York.

Simons, E. L. (1968). Early Cenozoic mammalian faunas, Fayum Province, Egypt. Part 1. African Oligocene mammals: introduction, history of study, and faunal succession. *Bulletin of the Peabody Museum of Natural History, Yale University* 28: 1–21.

Slijper, E. J. (1936). Die Cetaceen, Vergleichend-Anatomisch und Systematisch. *Capita Zoologica* 6–7: 1–590.

Stromer von Reichenbach, E. (1903a). Bericht über eine von den Privatdozenten Dr. Max Blanckenhorn und Dr. Ernst Stromer von Reichenbach ausgeführte Reise nach Aegypten. Einleitung. Ein Schädel und Unterkiefer von Zeuglodon osiris Dames. *Sitzungsberichte der Mathematisch-physikalischen Classe der Königlichen Bayerischen Akademie der Wissenschaften, München* 32: 341–352.

Stromer von Reichenbach, E. (1903b). *Zeuglodon*-Reste aus dem oberen Mitteleocän des Fajum. *Beiträge zur Paläontologie und Geologie Österreich-Ungarns und des Orients, Vienna* 15: 65–100.

Stromer von Reichenbach, E. (1908). Die Archaeoceti des ägyptischen Eozäns. *Beiträge zur Paläontologie und Geologie Österreich-Ungarns und des Orients, Vienna* 21: 106–178.

Uhen, M. D. (1996). *Dorudon atrox* (Mammalia, Cetacea): Form, function, and phylogenetic relationships of an archaeocete from the late middle Eocene of Egypt. Ph.D. dissertation, University of Michigan, Ann Arbor, p. 608.

Uhen, M. D. (1998). Middle to late Eocene basilosaurines and dorudontines. In: Thewissen, J. G. M. (ed.), *The Emergence of Whales: Evolutionary Patterns in the Origin of Cetacea*. Plenum Press, New York, pp. 29–61.

Uhen, M. D. (2004). Form, function, and anatomy of *Dorudon atrox* (Mammalia, Cetacea): An archaeocete from the middle to late Eocene of Egypt. *University of Michigan Papers on Paleontology* 34: 1–222.

Van Valen, L. M. (1966). Deltatheridia, a new order of mammals. *B. Am. Mus. Nat. Hist*. 132: 1–126.

Zachos, J. C., Pagani, M., Sloan, L. C., Thomas, E. and Billups, K. (2001). Trends, rhythms, and aberrations in global climate 65 Ma to present. *Science* 292: 686–693.

The Basicranial Anatomy of African Eocene/Oligocene Anthropoids. Are There Any Clues for Platyrrhine Origins?

Richard F. Kay, Elwyn Simons and Jennifer L. Ross

Introduction

One of the more contentious issues in anthropoid evolution is clarifying the phylogenetic position of late Eocene and early Oligocene anthropoids from Egypt relative to Oligocene-to-Recent 'crown' Anthropoidea. There is general agreement that *Aegyptopithecus* and other Propliopithecidae are members of a stem catarrhine clade, but do any of the other African Eocene-Oligocene anthropoids represent stem platyrrhines? Related to this, do any of the late Eocene taxa, such as the Oligopithecidae (*Catopithecus* and *Oligopithecus*), also represent stem catarrhines, or are they stem anthropoids with a few characters convergent on the catarrhine condition?

Up until the mid-1990s, these issues were debated mainly on the basis of dental anatomy. Early Oligocene propliopithecids were regarded as catarrhine owing to a suite of dental similarities with Miocene-Recent catarrhines the most salient of which is the loss of a premolar, reducing the dental formula to two premolars above and below (Simons, 1965; Kay et al., 1981; Simons, 1989; Kay and Williams, 1994). Analyses of more anatomical systems also generally support this view (Fleagle and Kay, 1987; Kay et al., 2004; Seiffert et al., 2004). Because they also have lost a premolar, late Eocene Oligopithecidae are thought to represent an otherwise more primitive version of a catarrhine (Simons, 1962; Kay, 1978; for contrasting views, see Szalay and Delson, 1979; Rasmussen and Simons, 1992; Kay and Delson, 2000).

Having already acknowledged the existence of a catarrhine lineage presupposes that platyrrhines must also have existed contemporaneously. Leading candidates for platyrrhine status have been the late Eocene- early Oligocene parapithecoids with special attention given to *Apidium* (Hoffstetter, 1977) and, more recently to *Proteopithecus* (Takai et al., 2000). All sides have acknowledged that the case for a link between platyrrhines and any African taxon is

Richard F. Kay
Box 3170, Duke University Medical Center, Durham, NC, 27710. Ph: (919) 684-2143, FAX: (919) 684-8034
rich_kay@baa.mc.duke.edu

J. G. Fleagle, C. C. Gilbert (eds.), *Elwyn Simons: A Search for Origins.*

weak. The argument runs that a platyrrhine ancestor *must* have existed at this time (because catarrhines already were present) and one or another African taxon is sufficiently primitive anatomically to have given rise to platyrrhines. At the same time, no convincing synapomorphies have been proposed to link platyrrhines with *Apidium*, *Proteopithecus*, or some other African taxon.

In the present contribution, we examine these issues in light of the anatomy of the bony ear. Specifically the temporal (petrosal, mastoid, and squamous temporalcomponents) and tympanic (equivalent to the ectotympanic of comparative anatomy). We build on the descriptive work on the ear region of Omomyidae, *Tarsius*, and the Anthropoidea (catarrhines, platyrrhines and the Fayum taxa *Apidium* and *Aegyptopithecus* (Cartmill and Kay, 1978; Cartmill et al., 1981; MacPhee and Cartmill, 1986; Beard and MacPhee, 1994; Ross, 1994; Horovitz, 1997, 1999; Ross and Covert, 2000)). We begin with a general description of the anatomy of the ear region in haplorhines, including digressions concerning the development of intracranial venous sinuses, the auditory tube, and tympanic cavity with its associated paratympanic spaces.

Following on this general review, we present the distribution of traits of the ear regions of living and fossil anthropoids and pose the following questions:

1. What are the characters of the ear region that distinguish crown platyrrhines from crown catarrhines? Amongst these features, which are primitive retentions and which represent synapomorphies of one or the other group?
2. Are there features of the ear regions of any late Eocene and early Oligocene anthropoid that depart from the crown clade pattern and, if so, do they help resolve the question of whether any of these taxa are catarrhine, platyrrhine, or stem anthropoids?

New data evaluated in this contribution are CT scans of the temporal regions of a comparative sample of extant haplorhines as well as the Egyptian late Eocene *Catopithecus* and *Proteopithecus* and early Oligocene taxa *Parapithecus grangeri* (often called *Simonsius*), *Apidium*, and *Aegyptopithecus*. Also, comments are added concerning the ear regions of several early Miocene platyrrhines, based on new CT scans. This method of visualizing anatomy has the potential to offer new insights into anatomy that was hitherto concealed.

Taxonomic Considerations

For terminological convenience, we refer to larger collections of taxa, most of which are monophyletic but a few of which may prove to be paraphyletic. All 'crown groups' are considered monophyletic, as established by the preponderance of molecular genetic data. Of course, stem groups, made up exclusively of extinct taxa, and known often only by fragmentary remains, are less likely to prove monophyletic. For example, the Omomyidae may be a monophyletic sister taxon to crown Haplorhini, or one or several of its included genera may prove to be more closely related to the ancestry of *Tarsius* (Beard and MacPhee, 1994), or Anthropoidea (Kay, 1980), or of crown Haplorhini as a whole (Cartmill and Kay, 1976; Cartmill and Kay, 1978). Likewise, in addition to the Propliopithecidae, one or several of the late Eocene-to-early Oligocene African anthropoids may prove to be more closely related to crown catarrhines or crown platyrrhines.

Omomyidae, includes the following taxa for which data have been published about the ear region and surrounding structures: *Shoshonius*, *Omomys*, *Necrolemur*, and *Tetonius* (Szalay, 1976; Ross, 1994; for a summary, see Ross and Covert, 2000). At least two of these taxa *Necrolemur* (Simons, 1961; Rosenberger and Szalay, 1980; Rosenberger, 1985) and *Shoshonius* (Beard et al., 1991; Beard and MacPhee, 1994) have been advocated as being sister to *Tarsius* alone.

Crown haplorhines include all genera, living and extinct, that descend from the most recent common ancestor of *Tarsius*, New World monkeys (Platyrrhini), and Old World monkeys and apes (Catarrhini). The following taxa from the late Eocene and early Oligocene preserve significant parts of the ear region and are definitely crown haplorhines: *Proteopithecus sylviae*, *Apidium phiomense*, *Parapithecus grangeri*, *Catopithecus browni*, and *Aegyptopithecus zeuxis*. An isolated petrosal from the middle Eocene of China has been allocated to the eosimiid anthropoids (MacPhee et al., 1995) but structurally it is virtually indistinguishable from omomyids (Kay et al., 1997; Ross et al., 1998; Ross and Covert, 2000; Kay et al., 2004)

Amongst the late Eocene and early Oligocene genera mentioned above, *Proteopithecus sylviae*, *Apidium phiomense*, and *Parapithecus grangeri* are considered stem anthropoids, although it is distinctly possible that one of these is related to the ancestry of platyrrhines. *Apidium* and *Proteopithecus* have been variously proposed as such (Hoffstetter, 1977; Takai et al., 2000). Likewise *Catopithecus browni* and its close relative *Oligopithecus*, may be a stem anthropoid (Kay, 1977; Szalay and Delson, 1979; Kay and Williams, 1994; Kay and Delson, 2000) or a stem catarrhine (see below).

Living catarrhine genera *Miopithecus*, *Presbytis*, and *Hylobates* are analyzed here as representative crown catarrhines. Among the late Eocene and early Oligocene genera mentioned above, *Aegyptopithecus* is the only certain catarrhine (e.g., Simons, 1965; Simons and Rasmussen, 1989; Simons, 1992; Simons, 1995), and it is distinctly possible that *Catopithecus* is as well (Rasmussen and Simons, 1992; Simons, 1992; Simons, 1995; Gunnell and Miller, 2001; Miller et al., 2005).

Ear regions of 16 commonly recognized genera of extant crown platyrrhines were examined. These are divided into three families: Atelidae (*Alouatta*, *Brachyteles*, *Lagothrix*, and *Ateles*), Pitheciidae (*Callicebus*, *Pithecia*, *Cacajao* and *Chiropotes*), and Cebidae (*Aotus*, *Saimiri*, *Cebus*, *Callithrix*, *Callimico*, *Cebuella*, *Saguinus*, and *Leontopithecus*). Additionally, mention is made of several early Miocene platyrrhines that are stem platyrrhines (Hershkovitz, 1974; Hershkovitz, 1982; Fleagle and Bown, 1983; Fleagle and Kay, 1989; Kay et al., 2005; Kay et al., in press): *Homunculus*, *Tremacebus*, and *Dolichocebus*.

The Ear Region of Haplorhini

What follows is not intended as a complete description of the extant anthropoid and haplorhine ear region. Our intent is to introduce some of the more important features to which we will be referring later when we discuss the phylogenetic placement of late Eocene and early Oligocene African anthropoids. A more complete description is found in MacPhee and Cartmill (1986)

As among other Primates, and mammals generally, the haplorhine auditory apparatus is composed of an inner, middle, and outer ear. The inner ear houses the interconnected sensory apparati that include 1) the Organ of Corti within the spiral cochlea that detects the intensity and frequency of sound waves; 2) the semicircular canals and their ampullae that detect of rotational movement of the head; and 3) the maculae within the utricle and saccule that contain tonic gravitational receptors. In development, these interconnected sensory systems first appear as invaginations of the sides of the head which separate to become the two auditory vesicles (Patten, 1968). Later, they become enclosed within cartilaginous otic capsules that ossify into the petrosal bones (petrous temporal of human anatomy).

Early in development, an invagination of ectoderm on the side of the head and an evagination of endoderm from the pharynx (the first pharyngeal pouch between the first and second branchial arches) grow together to form a thin membrane on the side of the head, the ear drum (tympanic membrane). The invagination from the side of the head forms external auditory meatus, which, with the pinna, composes the external ear.

Expansion of the distal portion of the pharyngeal pouch becomes the tympanic cavity, an air-filled tympanic cavity in the adult connected by a narrow auditory tube to the pharynx. The tympanic cavity is situated lateral and ventral to the otic capsule (Figs. 1–3). The cochlea of the inner ear bulges into the tympanic cavity to form the promontorium. Depending on the orientation of the underlying cochlear coils and the size of the tympanic cavity, either one or two coils evaginate the promontorium. In the first case, a single bulge appears whereas in the second case there are two bulges separated by a shallow sulcus (Horovitz and MacPhee, 1999) (Fig. 4). The ventral aspect of the tympanic cavity is reinforced with connective tissue, which in adult primates becomes ossified by outgrowths of the petrosal bone into an auditory bulla.

Three bones in the tympanic cavity (malleus, incus, and stapes) develop initially in loose tissues above the tympanic cavity but this connective tissue later undergoes resorption with the expansion of the tympanic cavity so that the ossicles ultimately come to lie within the tympanic cavity. These bones conduct sound through the tympanic cavity between the tympanic membrane and an opening in the petrosal bone into which fits the foot plate of the stapes (fenestra ovalis or oval window) leading to the auditory sensory apparatus (cochlea) within the inner ear. Several muscles attach to the tympanic membrane and the stapes that serve to stiffen the tympanic membrane (m. tensor

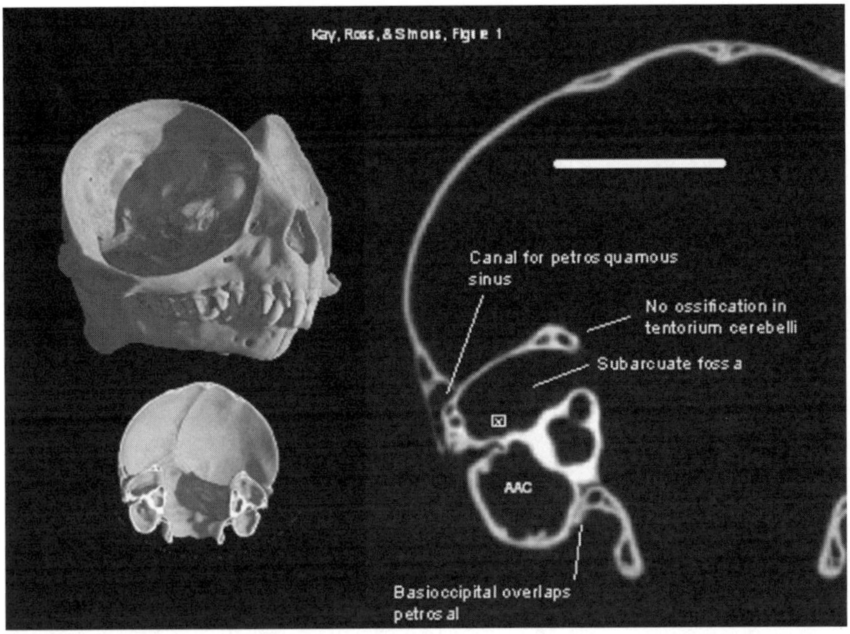

Fig. 1 *Tarsius bancanus* (Tarsiidae) (USNM 488084). Coronal CTs through the cranium at the level of the subarcuate fossa illustrating the position of the anterior accessory cavity (AAC), absence of a tentorium cerebelli, absence of a Cartmill's canal in floor of subarcuate fossa (x) and overlap of the basioccipital on the petrosal. The inset illustrates the level of the coronal section in relation to the rest of the cranium. Scale bar equals 5 mm.CT

tympani) and dampen sound transmitted to the stapes (m. stapedius). Amongst omomyids, and *Tarsius*, the stapedius takes its origins from a funnel shaped opening in the ossified petrosal bulla and its tendon passes through the pyriform aperture adjacent and coalescent with the stylomastoid foramen to reach the stapes. The m. stapedius is wholly encased in bone in strepsirrhines and anthropoids.

The tympanic membrane is reinforced by the tympanic, a c-shaped ribbon of bone. In the early stages of development, the tympanic lies in almost a horizontal position underneath the otic capsule so that the unossified floor of the tympanic cavity is very narrow transversely. With further development, the tympanic cavity expands laterally and the tympanic ring comes to lie more dorsoventrally, increasing the volume of the tympanic cavity. In strepsirrhines and Eocene omomyids, the tympanic cavity extends further ventrally underneath and more lateral to the tympanic carrying with it a membrane of epithelium derived from endoderm and forming a hypotympanic recess. When this membrane is unossified it appears in osteological specimens as though the tympanic lies 'free' within the auditory bulla. Thus, among strepsirrhines,

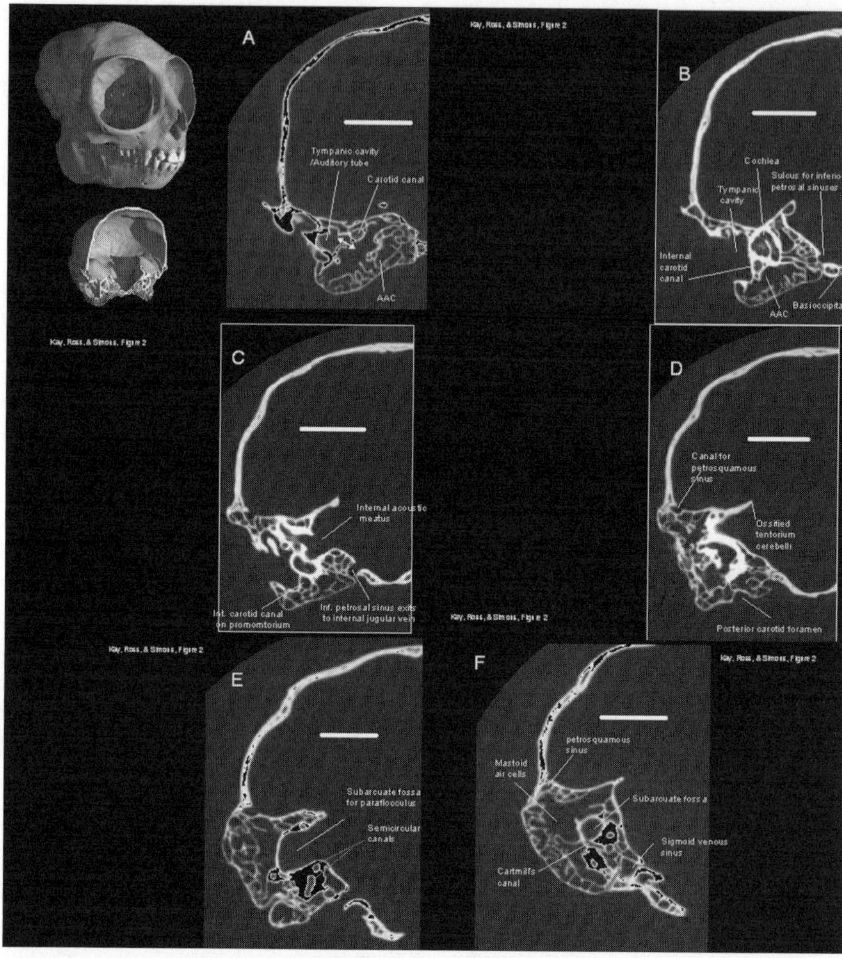

Fig. 2 *Aotus trivirgatus* (Platyrrhini) (USNM 464844). Coronal CTs through the ear region. Sections A through E are successively more caudad. The inset illustrates the position of section A. A. Tympanic cavity and auditory tube connected via an ostium (double-headed arrow) with a trabeculated Anterior Accessory Chamber (AAC). B. Internal carotid canal situated on the promontorium, the relations of the tympanic cavity, and anterior accessory chamber, and the sulcus for inferior petrosal sinus. C. Internal acoustic meatus and sulcus for inferior petrosal sinus exiting through the jugular foramen. D. Posterior carotid foramen marks entrance of internal carotid artery; tentorium cerebelli ossified. E. Subarcuate fossa; semicircular canals. F. Shows the relations of Cartmill's canal with the subarcuate fossa and sigmoid sinus; mastoid air cells and petrosquamous sinus. Scale bar equals 5 mm

the tympanic ring is said to be "intrabullar" or aphaneric. In *Tarsius* and anthropoids the tympanic ring is fused against the outer rim of the petrosal and there is no hypotympanic recess – this is the 'extrabullar,' or phaneric, condition (Figs. 1–4).

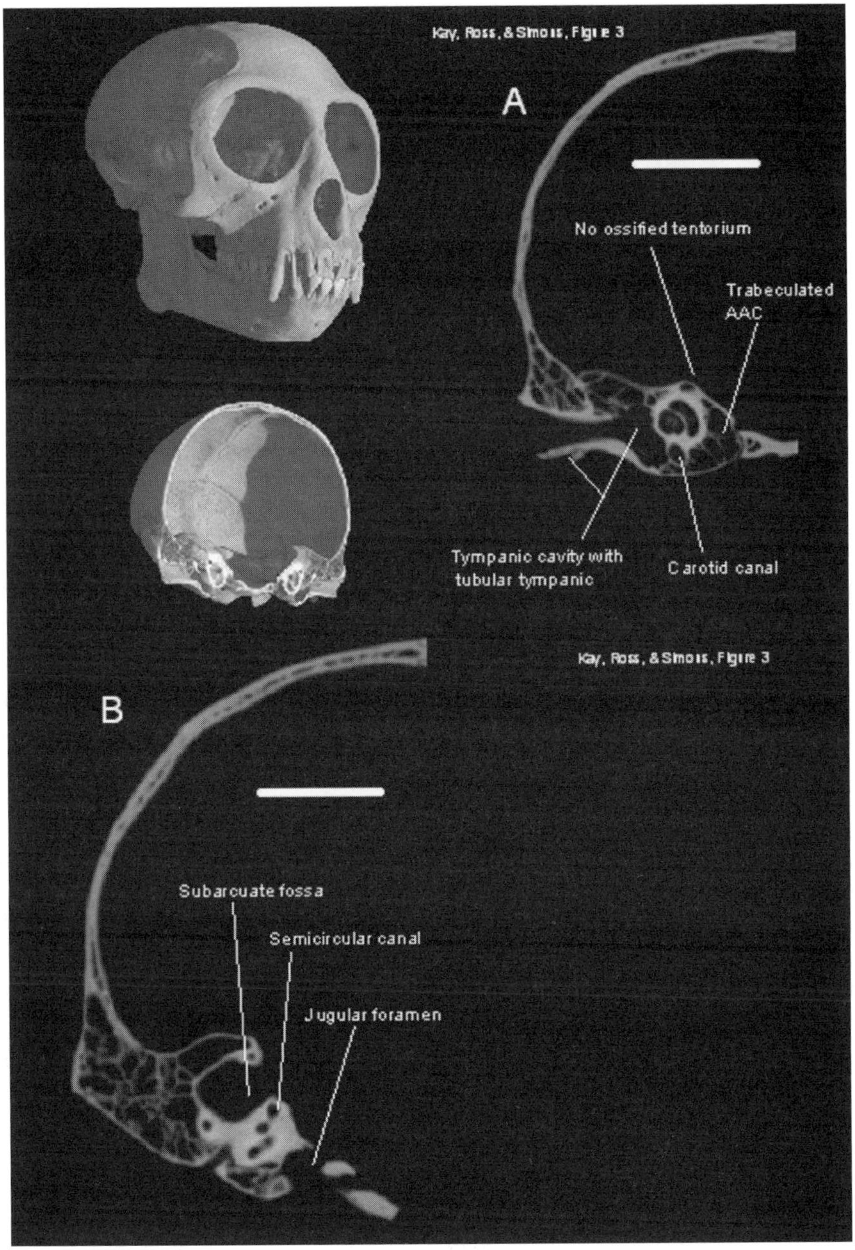

Fig. 3 *Miopithecus talapoin*, an extant cercopithecoid catarrhine (Kay collection). A–B: successively more caudal coronal CT-cross sections illustrating several of the features discussed in the text. The inset illustrates the position of section A. A. Trabeculated AAC separated from tympanic cavity with carotid canal running on the promontorium; tubular tympanic. Note absence of ossified tentorium. B. Note absence of Cartmill's canal in floor of subarcuate fossa; a sulcus for a petrosquamous sinus is also absent. Scale bar equals 5 mm

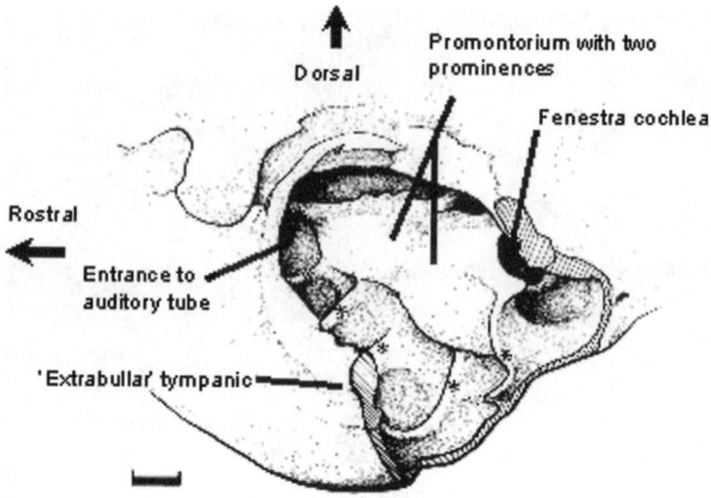

Fig. 4. Left ear region of *Callithrix* sp. (Platyrrhini). From ventral and posterolateral aspect with portions of the bulla and tympanic removed to display the transbullar septa (x), promontorium with two bulges, and the 'extrabullar' position of the tympanic. Scale bar equals 1 mm. Illustration after Cartmill et al. (1981)

Lateral to the tympanic, the external auditory meatus is reinforced by cartilage. In adult *Tarsius* (Fig. 1) and extant catarrhines (Fig. 3) this cartilage partially ossifies as an extension of the tympanic bone, whereas in extant platyrrhines (Figs. 4–5) the cartilage of the meatus remains unossified.

Extant primates vary in the extent of pneumatization of the petrosal bone. Pneumatization is a condition in which sinuses develop in the cranial bones in connection with the nasal cavity or pharynx. In the adult, these sinuses are air-filled. In the case of the ear region of primates, pneumatized spaces are in continuity with the tympanic cavity or auditory tube and, ultimately, to the pharynx. We have already mentioned one such cavity, the hypotympanic recess of strepsirrhines and omomyids, extending beneath, and lateral to, the tympanic membrane and tympanic bone. Another outgrowth found in anthropoids and some strepsirrhines (but not in *Tarsius*) begins in the epitympanic recess of the tympanic cavity and invades the mastoid bone to produce the mastoid air cells (Fig. 2F). Mastoid air cells are absent in *Tarsius* (MacPhee and Cartmill, 1986). In *Tarsius* and anthropoids a third outgrowth occurs at the point where the auditory tube expands outward into the tympanic cavity. This pneumatic cavity is confined to the petrosal bone and forms a large sinus anterior or anteromedial to the tympanic cavity called the anterior accessory chamber (AAC) (Figs. 1, 2A–B). The AAC engulfs a part of the canal for the internal carotid artery (see below) and is separated from the tympanic cavity by a thin bony transbullar septum, or partition. One important difference between the AAC

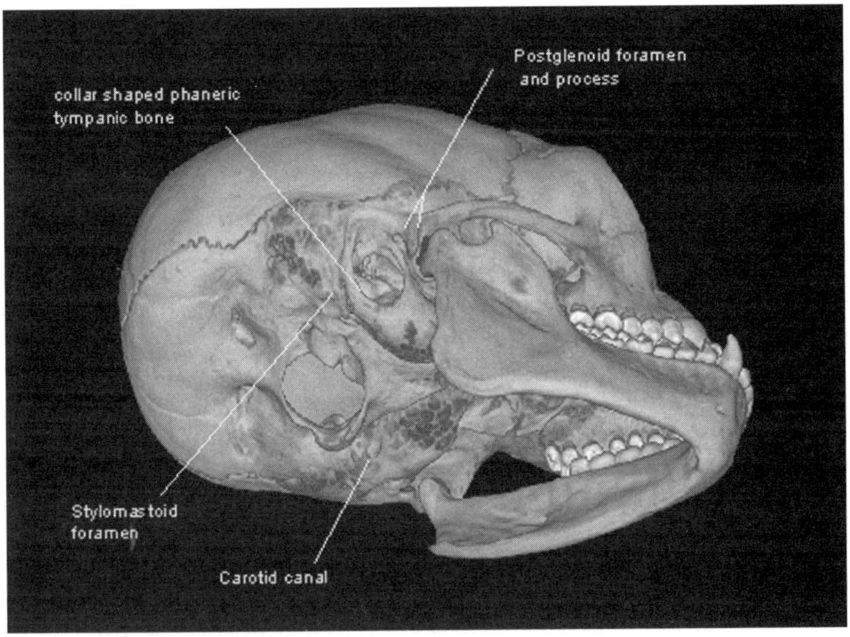

Fig. 5 *Saimiri sciureus* (Platyrrhini) (USNM 518538). Reconstruction from serial CT images. View from ventral and lateral perspective. Scale bar equals 5 mm

of *Tarsius* and that of Anthropoidea is that in the former the chamber is relatively open whereas in the latter it is composed of multiple interconnected chambers partially separated by spicules of bone (i.e., trabecular).

Among primitive eutherians, the brain is supplied by two main arteries, the vertebral and internal carotid. The latter is intimately associated with the ear region. The vertebral artery arises from the subclavian artery in the neck. It passes craniad through paired transverse foramina of cervical vertebrae before entering the braincase with the spinal cord through the foramen magnum. It supplies the brainstem, cerebellum and posterior parts of the cerebrum. The common carotid artery splits in the neck into external and internal branches. The ascending pharyngeal artery is a third branch, often subsidiary to the internal carotid artery. The internal carotid gives off several branches in the neck, then enters the auditory bulla through the posterior carotid foramen in the petrosal bone (Figs. 2D, 5). In fetal life, the artery divides into a promontorial and stapedial branch. The promontory artery is so named because it crosses the promontorium before entering the middle cranial fossa (via the anterior carotid foramen) to supply the orbit and more anterior parts of the cerebrum. The stapedial artery runs through a foramen in the stapes bone and divides into several branches (MacPhee and Cartmill, 1986; Wible, 1987).

The primate internal carotid and its bullar branches are wholly or partially enclosed by bone to form canals as they pass through the tympanic cavity. This makes it possible to reconstruct the course and branches of this artery in extinct primates. In the Eocene primates Omomyidae and Adapoidea, both promontorial and stapedial branches are present.

In *Tarsius*, the proximal part of the stapedial artery is extremely reduced, possibly absent in the adult, although its bony canal persists. *Tarsius* retains a very large promontory artery. Living anthropoids lose the stapedial artery altogether (except in rare cases, Diamond, 1992) and retain only the promontory artery. (In human anatomy the promontory artery is just called the internal carotid artery).

A number of nerves either supply structures of the ear or pass through the ear region on the way to supplying other structures. The list includes branches of the trigeminal and facial nerves, the vestibulocochlear nerve, and the glossopharyngeal nerve. On the endocranial surface of the petrous temporal, the internal acoustic meatus (Fig. 2C) receives the facial and the vestibulocochlear nerves. The vestibulocochlear supplies the organs of balance and hearing within the inner ear. The facial nerve has a complex, but apparently uniform pattern in primates. The most obvious bony manifestation of the facial nerve is the stylomastoid foramen where it emerges behind the bulla.

The endocranial surface of the petrosal also is invaginated by the subarcuate fossa in strepsirrhines, *Tarsius* (Fig. 1), platyrrhines (Fig. 2E) and smaller catarrhines (Fig. 3B). The fossa is situated caudad to the internal acoustic meatus and has a close relationship to the semicircular canals. The utricle and the posterior and horizontal semicircular canals reside ventral to the fossa. The superior semicircular canal delimits its medial border. The subarcuate fossa encloses the parietal paraflocculus, a lobe of the cerebellum (Gannon et al., 1988).

Finally, several grooves and foramina in the petrosal bone are associated with the endocranial venous system. To more fully understand the venous system, it is important to understand the layers of tissue that invest the brain in the adult, and to summarize the pattern of development of these venous channels.

As with other parts of the central nervous system, several protective layers, an innermost pia mater, a weblike arachnoid mater, and an outer dura mater envelop the brain. Additionally, inside the braincase there is an outermost dura, equivalent to the periosteum, lining the inner surface of the braincase. The inner dura projects as a doubled layer between the cerebral hemispheres (as the falx cerebri) and between the cerebrum and cerebellum (as the tentorium cerebelli). Some ossification occurs at the margins of the tentorium in some platyrrhine taxa (Fig. 2D) but in *Tarsius* and catarrhines there is no ossification.(Figs. 1, 3A).

Veins draining the brain connect with a system of more superficial veins called venous sinuses lying between the inner and outer layers of dura mater within the cranium. The venous sinuses take their origin from an embryonic anterior cardinal vein, known as the Primary Head Vein, as well as from three cranial venous plexuses (anterior, middle and posterior) which appear early in embryonic development (Streeter, 1915, 1918). At an early stage in venous development, the anterior and middle venous plexuses join and exit the presumptive middle cranial fossa through the developing postglenoid foramen (Fig. 5). The latter foramen also receives blood draining from the posterior cranial fossa via the petrosquamous sinus.

A petrosquamous sinus exists in all nonhuman primates and occasionally in *Homo*. Saban (1963) illustrates the sinus in cheirogaleids, non-cheirogaleid lemuriforms, and lorisoids. In CT images of *Tarsius*, we observed a large petrosquamous sinus connecting to the transverse sinus caudally and, cranially, partially encased by bone into a tube, running forward to join the postglenoid foramen (Fig. 1) and a second foramen, the suprameatal. In *Cebus* and *Macaca* and, when present, in *Homo*, this sinus connects caudally with the transverse sinus proximal to where it meets with the superior petrosal sinus. In *Cebus* and *Macaca* (again occasionally in *Homo)*, the petrosquamous sinus also transmits blood craniad between the petrous and squamosal parts of the temporal to empty through the postglenoid foramen into the retromandibular vein located in the squamosal bone (Weinstein and Hedges Jr., 1962). Thus, the pattern in strepsirrhines, *Tarsius* and *Cebus* (and *Aotus*, Fig. 2D) is essentially the same, whilst in *Homo*, and presumably among other taxa that have a reduced postglenoid foramen, only the caudad connection persists.

In humans, many other catarrhines, as well as in some platyrrhines, the postglenoid foramen is involuted and the principal drainage of the brain exist more posteriorly through what will become the jugular foramen.

Early in human development, the posterior plexus drains into the caudal part of Primary Head Vein through what is to become the sigmoid sinus through the developing jugular foramen in the posterior cranial fossa into the internal jugular vein (Waltner, 1944). A secondary channel, the transverse sinus, connects the middle to the posterior plexus. Blood draining from the transverse sinus dumps into the sigmoid sinus.

The cavernous sinus, surrounding the stalk of the hypophysis, receives blood directly from vessels draining the brain itself and from veins outside the braincase in the maxillary and orbital regions. In *Homo*, blood drains caudad from the cavernous sinus through the superior petrosal sinus reaching the transverse sinus at a point where its name changes to the sigmoid sinus. In *Macaca*, the superior petrosal sinus reportedly does not have a direct connection with the cavernous sinus. Instead, it begins where the middle cerebral vein enters the dura along the margin of the tentorum cerebelli (Weinstein and Hedges Jr., 1962) and reaches the transverse sinus as in humans. In *Cebus*, the superior petrosal sinus does not reach the transverse sinus and its blood drains craniad into the cavernous sinus (Madeira and Watanabe, 1975). The sinus is not observed in cheirogaleids, noncheoirogaleid lemuriforms, and lorisoids (Saban, 1963).

The inferior petrosal sinus provides a second connection between the cavernous sinus and the sigmoid sinus just where it empties into the internal jugular vein. This second connection is reported in cheirogaleids, non-cheirogaleid lemuriforms, and lorisoids (Saban, 1963) in the platyrrhine *Cebus* (Madeira and Watanabe, 1975), and the catarrhines *Homo* and *Macaca* (Weinstein and Hedges Jr., 1962).

An additional venous sinus and canal not reported in humans or other catarrhines has received some attention. As Cartmill et al. (1981) first noted in platyrrhines, a canal leads from the depths of the subarcuate fossa, caudally and medially to an opening in the channel of the sigmoid sinus. They report the canal in ten genera of extant platyrrhines. They suggested that this canal is a venous channel having noted that a dried vessel was found in this position in a skull of the platyrrhine *Saguinus*. Gannon et al. (1988) confirm this hypothesis through examination of histological sections of a specimen of the platyrrhine *Aotus*. We further examined the canal and its constituents in histologically-sectioned late fetal *Aotus* (Fig. 6). In these sections, we found a subarcuate sinus between two layers of dura inside of the subarcuate space. It commences at the anterior margin of the paraflocculus and runs posteriorly along its lateral edge. On reaching the posterior end of the paraflocculus, this

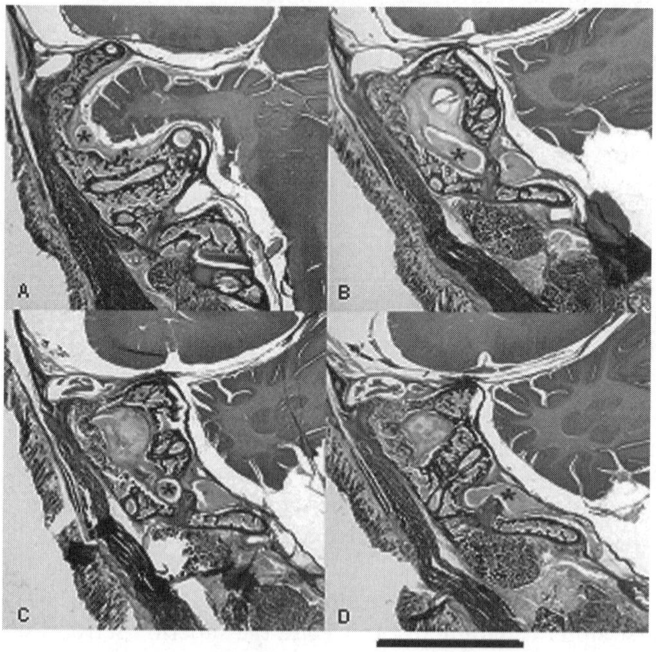

Fig. 6 20μm coronal sections of a fetal specimen of *Aotus trivirgatus* from the Duke University, Department of Biological Anthropology and Anatomy histology collections (Crown to rump length 145 mm, head length, 49 mm); trichrome stain. Sections A–D are successively more craniad. A. Section 83 – 1648 B. Section 86 – 1718 C. 87 – 1728 Section D. Section 87 – 1738. Scale bar equals 5 mm

vessel dips into a groove on the lateral side of the subarcuate fossa and passes caudally and medially to enter the canal leading to the sigmoid sinus.[1]

Review of the Characters of the Auditory Region of Eocene to Recent Haplorhines

In the preceding section, we have laid out the basic structural patterns of the ear region and some of its variations in haplorhines, placing emphasis on the anatomical features that leave bony traces, thus allowing inferences about fossil taxa. Based on a literature survey, supplemented by the examined material listed in Appendix 1 we can now summarize in more detail the distribution of states of the cranial characters of the ear region, basicranium, and temporomandibular joint of extant strepsirrhines, Omomyidae, *Tarsius*, extant platyrrhines, selected extant catarrhines, and African Eocene/Oligocene anthropoids (Tables 1 and 2). Below, we examine the distribution of these characters against a phylogenetic framework of the extant forms, as established from genetic studies.

Characters of Limited Utility for Phylogenetic Inference

A number of characters examined in Tables 1 and 2 exhibit almost as much variation within clades established by molecular data as between the clades. Such variation often limits one's ability to reconstruct the character state of basal nodes for platyrrhines, catarrhines, and anthropoids. Even when that is possible, to use the character state to slot a taxon into a definitive place in the phylogeny is unreliable. Several examples are as follows.

1) The postglenoid process is strong in omomyids, stem anthropoids *Aegyptopithecus* (Fig. 7A) and crown catarrhines. It apparently is secondarily reduced in *Tarsius*. Its size is quite variable in platyrrhines It is reduced size in marmosets and tamarins and, independently in *Saimiri* (Fig. 5) but large in atelids and pitheciids. Its size also is variable in stem platyrrhines—very large in *Homunculus*, but small in *Dolichocebus* (Kay et al., in press).

[1] We also found another vessel in the subarcuate space, which seems to exist unilaterally in the *Aotus trivirgatus* specimen and may or may not have a connection to the subarcuate sinus. Where this vessel commences in the subarcuate fossa is thus uncertain. We first see this vessel in the subarcuate space near the point where the subarcuate sinus makes its turn into the canal leading into the sigmoid. This vessel then runs cranially and laterally and continues in a groove, above the subarcuate space. It finally makes a connection with the petrosquamous sinus. The distribution of this vessel in other primates deserves further investigation.

Table 1 Distribution of ear region characters among haplorhines and extant strepsirrhines. The states given are those reconstructed for the node of the particular clade in question. If the state is uncertain, alternative possible states are listed

Characters	Extant strepsirrhines[1]	Omomyids[1]	*Tarsuis*	Cercopithecoidea	*Hylobates*	*Alouatta, Brachyteles, Lagothrix* & *Ateles*	*Callicebus, Pithecia, Cacajao* & *Chiropotes*	*Aotus*	*Saimiri* & *Cebus*	*Callithrix, Cebuella, Callimico, Saguinus* & *Leontopithecus*
1. Postglenoid foramen (Character 12 in Horovitz, 1999)	Large in *Lemur*, *Cheirogaleus*[2]	Large in *Shoshonius*[5]	large	absent	absent	absent	small or absent	small	small or absent	absent
2. Temporomandibular joint (Character 40 in Ross and Covert, 2000)	Biconcave & transversely wide	Antero-posteriorly oriented trough	Antero-posteriorly oriented trough	Biconcave & transversely wide	Biconcave & transversely wide	Biconcave & transversely wide	Biconcave & transversely wide	Biconcave & transversely wide	Biconcave & transversely wide	Biconcave & transversely wide
3. Postglenoid (entoglenoid) process size[3] (Character Cr41 in Ross and Covert, 2000)	weak	strong	Weak	strong	strong	Strong or very strong	strong	strong	weak to strong	weak or absent
4. Cochlear housing as exposed in middle ear[4] (Character 12 in Horovitz, 1999)	dual	n/a	dual	single	single	single	Single or dual	dual	single or dual	dual
5. Transbullar septa in middle ear (Character 14 in Horovitz, 1999)	Character not applicable	Not observed; ?Character not applicable	septa absent	septa absent	septa absent	septa absent	septa absent	septa absent	septa absent	septa sometimes present
6. Transverse septum arising from the cochlear housing (Character Cr 1 in Ross et al., 1998)[5]	absent	absent	present	present	present	present	present	present	present	present

7. Extent & type of pneumatization in anterior accessory cavity(AAC) (Cr 2 in Ross et al., 1998)	AAC absent	AAC absent	Anterior to tympanic cavity & not trabeculated	Anterior & medial to tympanic cavity & trabeculated	Anterior & medial to tympanic cavity & trabeculated	Anterior & medial to tympanic cavity & trabeculated	Anterior & medial to tympanic cavity & trabeculated	Anterior & medial to tympanic cavity & trabeculated	Anterior & medial to tympanic cavity & trabeculated	Anterior & medial to tympanic cavity & trabeculated
8. Perbullar pathway for the carotid artery (Cr 4 in Ross et al., 1998)	absent	absent	present	present	present	present	present	present	present	present
9. Antero-posterior location of posterior carotid foramen in bulla (Cr 5 in Ross et al., 1998)	posterior	posterior	anterior	posterior	posterior	posterior	posterior	posterior	posterior	posterior
10. Mediolateral position of posterior carotid foramen in bulla (Cr 6 in Ross et al., 1998)	lateral (except medial in lorisoids)	medial (except lateral in *Shoshonius*)	lateral	midline	medial	medial	medial	medial	medial	medial
11. Position of posterior carotid foramen relative to fenestra cochleae (Cr 8 in Ross et al., 1998)	crosses ventral lip of fenestra cochlea	posterior	anterior	anterior	anterior	anterior	anterior	anterior	anterior	anterior
12. Position of the portion of the internal carotid / promontory artery (or its accompanying nerves) on the promontorium anterior to fenestra cochlea (Cr 10 in Ross et al., 1998)	ventrolateral	ventrolateral	contacts cupola only	ventrolateral	ventrolateral	ventrolateral	ventrolateral	ventrolateral	ventrolateral	ventrolateral

(continued)

Table 1 (continued)

Characters	Extant strepsirrhines[1]	Omomyids[1]	*Tarsuis*	Cercopithecoidea	*Hylobates*	*Alouatta, Brachyteles, Lagothrix* & *Ateles*	*Callicebus, Pithecia, Cacajao* & *Chiropotes*	*Aotus*	*Saimiri* & *Cebus*	*Callithrix, Cebuella, Callimico, Saguinus* & *Leontopithecus*
13. Size of stapedial (SC) & promontory (PC) canals (Cr 11 in Ross et al., 1998)	PC & SC absent, (PC reduced, canal absent in non-cheirogaleid lemuroids)	Both SC & PC large, subequal (SC < PC in *Necrolemur*)	SC absent, PC large	SC absent, PC large	SC absent, PC large	SC absent, PC large	SC absent, PC large	SC absent, PC large	SC absent, PC large	SC absent, PC large
14. Morphology of promontory canal on promontorium (Cr 12 in Ross et al., 1998)	absent (trough in non-cheirogaleid lemuroids)	complete	complete	complete	complete	complete	complete	complete	complete	complete
15. Presence or absence of canal for internal carotid artery or nerves (Cr 13 in Ross et al., 1998)	present	present	present	present	present	present	present	present	present	present
16. Position of ventral edge of the tympanic bone (Cr 14 in Ross et al., 1998)	Intrabullar (except extrabullar in *Aloocebus* & lorisoids)	intrabullar	extrabullar	extrabullar	extrabullar	extrabullar	extrabullar	extrabullar	extrabullar	extrabullar
17. Shape of tympanic bone (Cr 15 in Ross et al., 1998)	annular (laterally expanded in lorisoids)	annular in *Shoshonius*[4]	laterally expanded	laterally expanded	laterally expanded	annular	annular	annular	annular	annular

18. Flange of basioccipital overlapping medial bulla wall (Cr 20 in Ross et al., 1998)	absent	extensive	extensive	absent	absent	absent	absent	absent	absent	absent
19. Basioccipital stem (Character 12 in Beard and MacPhee, 1994)	broad	narrow in *Shoshonius*[5]	narrow	broad	broad	broad	broad	broad	broad	broad
20. Suprameatal foramen (Cr 21 in Ross et al., 1998)	absent	present in rostral root of zygomatic (*Shoshonius*); present above external auditory meatus (*Necrolemur*[5])	large, above external auditory meatus	absent	absent	absent or present in rostral root of zygomatic	absent	absent	absent	absent
21. Parotic fissure (Cr 22 in Ross et al., 1998)	closed	open	open	closed	closed	closed	closed	closed	closed	closed
22. Epitympanic crest (Cr 48 in Ross et al., 1998)	absent (except present in non-cheirogaleid lemuroids)	present	absent	absent	absent	absent	absent	absent	absent	absent
23. Pneumatization of mastoid from epitympanic recess (Cr 3 in Ross et al., 1998)	present in lorisoids & *Allocebus*; absent in other cheirogaleids & non-cheirogaleid lemuroids	present (*Necrolemur*); absent (*Shoshonius*)	absent	present	present	present	present	present	present	present
24. Tentorium cerebelli ossification (character 13 in Horovitz, 1999)	not examined	Absent in *Omomys*[6]	absent or minimal	absent or minimal	absent or minimal	Extensively ossified	Extensively ossified	Extensively ossified	absent or minimal	absent or minimal

(continued)

Table 1 (continued)

Characters	Extant strepsirrhines[1]	Omomyids[1]	*Tarsuis*	Cercopithecoidea	*Hylobates*	*Alouatta, Brachyteles, Lagothrix & Ateles*	*Callicebus, Pithecia, Cacajao & Chiropotes*	*Aotus*	*Saimiri & Cebus*	*Callithrix, Cebuella, Callimico, Saguinus & Leontopithecus*
25. Canal connecting sigmoid sinus with subarcuate fossa (Cartmill's canal)	Present in Lemur[7], Hapalemur[8]	Absent in *Omomys*[9]	absent	absent	absent	complete	complete	complete	complete	complete

[1] Except as noted, character states are from Ross and Covert (2000) who give the condition in non-cheirogaleid lemuroids, cheirogaleids, and lorisoids. Omomyids are represented by *Shoshonius*, *Omomys*, *Tetonius*, and *Necrolemur*.
[2] (Saban, 1963)
[3] 100 x (postglenoid process length/ prosthion-inion length)
[4] Figure 2 in Cartmill (1981) shows dual prominences on the promontorium in *Callithrix*.
[5] Ross and Covert (2000) define Cr1 slightly differently, (present or absent). We follow Ross et al. (1998) and use a three-state character to indicate whether the septum is absent or, if present, separates the tympanic cavity from a pneumatized space formed either from the auditory tube (state 1, as in *Tarsius* and anthropoids) or from air cells via the epitympanic recess (state 2, as in lorises).
[6] (Beard and MacPhee, 1994)
[7] (Gannon et al., 1988)
[8] CT scan of *Hapalemur griseus*; Kay, personal collection.
[9] Absent in *Omomys* specimen Ni (Personal communication)

Beard, K. C.,MacPhee, R. D. E., (1994). Cranial anatomy of *Shoshonius* and the antiquity of Anthropoidea. In: Fleagle, J. G.,R. F. Kay (Eds.), *Anthropoid Origins*, pp. 55–98. Plenum Press: New York.

Cartmill, M., Mac Phee, R. D. E.,Simons, E. L., (1981). Anatomy of the temporal bone in early anthropoids with remarks on the problem of anthropoid origins. *Am J. Phys. Anthrop*. 56, 3–22.

Gannon, P. J., Eden, A. R., Laitman, J. T., (1988). The subarcuate fossa and cerebellum of extant primates: Comparative study of a skull-brain interface. *Am J. Phys. Anthrop*. 77, 143–164.

Horovitz, I., (1999). A phylogenetic study of living and fossil platyrrhines. *Am. Mus. Novitates* 3269, 1–40.

Ross, C., Covert, H. H., (2000). The petrosal of *Omomys carteri* and the evolution of the primate basicranium. *J. Hum. Evol*. 39, 225–231.

Ross, C., Williams, B. A., Kay, R. F., (1998). Phylogenetic analysis of anthropoid relationships. *J. Hum. Evol*. 35, 221–306.

Saban, R., (1963). Contribution à l'étude de l'os temporal des Primates. Description chez l'Homme et les Prosimiens. Anatomie comparée et phylogénie. Mémoires du Muséum National D'Histoire Naturelle Nouvelle Série, Série A, Zoologie, 29, 1–378.

Table 2 Distribution of ear region characters among haplorhines and Fayum anthropoids. The states given are those reconstructed for the node of the particular clade in question. If the state is uncertain, alternative possible states are listed

Characters	Omomyids[1]	*Tarsuis*	Crown catarrhines	*Crown Platyrrhines*	*Proteopithecus slyviae*	*Apidium phiomense*	*Parapithecus grangeri*	*Catopithecus browni*	*Aegyptopithecus zeuxis*
1. Postglenoid foramen (Character 15 in Horovitz, 1999)	Large in *Shoshonius*[5]	large	absent	absent	small	n/a	small	small	small
2. Temporomandibular joint (Character 40 in Ross and Covert, 2000)	Antero-posteriorly oriented trough	Antero-posteriorly oriented trough	Biconcave & transversely wide	Biconcave & transversely wide	Biconcave & transversely wide	Biconcave & transversely wide	Biconcave & transversely wide	Biconcave & transversely wide	Biconcave & transversely wide
3. Postglenoid (entoglenoid) process size[1] (Character Cr41 in Ross and Covert, 2000)	strong	weak	strong	strong	strong	n/a	strong	strong	strong
4. Cochlear housing as exposed in middle ear[2] (Character 12 in Horovitz, 1999)	not observed	dual	single	single		dual	single	n/a	single
5. Transbullar septa in middle ear (Character 14 in Horovitz, 1999)	not observed; ?Character not applicable	septa absent	septa absent	septa absent	septa absent	Septa present	septa absent	n/a	septa absent
6. Transverse septum arising from the cochlear housing (Character Cr 1 in Ross et al., 1998)[3]	absent	present	present	present	present	present	present	n/a	present
7. Extent & type of pneumatization in anterior accessory cavity(AAC) (Cr 2 in Ross et al., 1998)	AAC absent	anterior to tympanic cavity & not trabeculated	anterior to tympanic cavity & not trabeculated	anterior to tympanic cavity & not trabeculated	anterior to tympanic cavity & not trabeculated	anterior to tympanic cavity & not trabeculated	anterior to tympanic cavity & not trabeculated	anterior to tympanic cavity & not trabeculated	anterior to tympanic cavity & not trabeculated

(continued)

Table 2 (continued)

Characters	Omomyids[1]	*Tarsuis*	Crown catarrhines	*Crown Platyrrhines*	*Proteopithecus slyviae*	*Apidium phiomense*	*Parapithecus grangeri*	*Catopithecus browni*	*Aegyptopithecus zeuxis*
8. Perbullar pathway for the carotid artery (Cr 4 in Ross et al., 1998)	absent	present	present	present	present	present	present	present	present
9. Antero-posterior location of posterior carotid foramen in bulla (Cr 5 in Ross et al., 1998)	posterior	anterior	posterior	posterior	posterior	n/a	posterior	posterior	posterior
10. Mediolateral position of posterior carotid foramen in bulla (Cr 6 in Ross et al., 1998)	medial (except lateral in *Shoshonius*)	lateral	midline	medial	medial	medial	medial	medial	midline
11. Position of posterior carotid foramen relative to fenestra cochleae (Cr 8 in Ross et al., 1998)	posterior	anterior	anterior	anterior	anterior	anterior	anterior	anterior	anterior
12. Position of the portion of the internal carotid / promontory artery (or its accompanying nerves) on the promontorium anterior to fenestra cochlea (Cr 10 in Ross et al., 1998)	ventrolateral	contacts cupola only	ventrolateral	ventrolateral	ventrolateral	ventrolateral	ventrolateral	ventrolateral	ventrolateral
13. Size of stapedial (SC) & promontory (PC) canals (Cr 11 in Ross et al., 1998)	Both SC & PC large, subequal (SC < PC in *Necrolemur*)	SC absent, PC large	SC absent, PC large	SC absent, PC large	n/a	SC absent, PC large	n/a	n/a	SC absent, PC large
14. Morphology of promontory canal on promontorium (Cr 12 in Ross et al., 1998)	complete	complete	complete	complete	complete	complete	complete	n/a	complete
	present	present	present	present	present	present	present	n/a	present

15. Presence or absence of canal for internal carotid artery or nerves (Cr 13 in Ross et al., 1998)									
16. Position of ventral edge of the tympanic bone (Cr 14 in Ross et al., 1998)	intrabullar	extrabullar	extrabullar	extrabullar	extrabullar	extrabullar	extrabullar	extrabullar	extrabullar
17. Shape of tympanic bone (Cr 15 in Ross et al., 1998)	annular in *Shoshonius*[4]	laterally expanded	laterally expanded	annular	annular	annular	annular	annular	annular
18. Flange of basioccipital overlapping medial bulla wall (Cr 20 in Ross et al., 1998)	extensive	extensive	absent	absent	absent	n/a	extensive	absent	absent
19. Basioccipital stem (Character 12 in Beard and MacPhee, 1994)	narrow in *Shoshonius*[5]	narrow	broad	broad	broad	n/a	broad	broad	broad
20. Suprameatal foramen (Cr 21 in Ross et al., 1998)	present in rostral root of zygomatic (*Shoshonius*); present above external auditory meatus (*Necrolemur*[5])	large, above external auditory meatus	absent	absent	absent	n/a	absent	absent	absent
21. Parotic fissure (Cr 22 in Ross et al., 1998)	open	open	closed	closed	n/a	closed	closed	closed	closed
22. Epitympanic crest (Cr 48 in Ross et al., 1998)	present	absent	absent	absent	n/a	absent	absent	absent	absent
23. Pneumatization of mastoid from epitympanic recess (Cr 3 in Ross et al., 1998)	present (*Necrolemur*); absent (*Shoshonius*)	absent	present	present	present	present	present	present	present

(continued)

Table 2 (continued)

Characters	Omomyids[1]	*Tarsuis*	Crown catarrhines	*Crown Platyrrhines*	*Proteopithecus slyviae*	*Apidium phiomense*	*Parapithecus grangeri*	*Catopithecus browni*	*Aegyptopithecus zeuxis*
24. Tentorium cerebelli ossification (character 13 in Horovitz, 1999)	absent in *Omomys*[5]	absent	absent	equivocal[6]	extensively ossified	n/a	extensively ossified	extensively ossified	absent
25. Canal connecting sigmoid sinus with subarcuate fossa (Cartmill's canal)	absent in *Omomys*[5]	absent	absent	present	present	absent	absent	present	Present only as a diverticulum

[1] 100 x (postglenoid process length/ prosthion-inion length)
[2] Figure 2 in Cartmill (1981) shows dual prominences on the promontorium in *Callithrix*.
[3] Ross and Covert (2000) define Cr1 slightly differently, (present or absent). We follow Ross et al. (1998) and use a three-state character to indicate whether the septum is absent or, if present, separates the tympanic cavity from a pneumatized space formed either from the auditory tube (state 1, as in *Tarsius* and anthropoids) or from air cells via the epitympanic recess (state 2, as in lorises).
[4] (Beard and MacPhee, 1994)
[5] Absent in *Omomys* specimen Ni (Personal communication)
[6] Probably extensive but character distribution equivocal for extant taxa: ossifiaction seen in Pitheciidae and Atelidae, *Aotus*, *Callithrix*. Reduced or absent in other marmosets, tamarins, *Saimiri* and *Cebus*

Beard, K. C., MacPhee, R. D. E., (1994). Cranial anatomy of *Shoshonius* and the antiquity of Anthropoidea. In: Fleagle, J.G., R.F. Kay (Eds.), *Anthropoid Origins*, pp. 55–98. Plenum Press, New York.

Cartmill, M., MacPhee, R. D. E., Simons, E. L., (1981). Anatomy of the temporal bone in early anthropoids with remarks on the problem of anthropoid origins. *Am J. Phys. Anthrop*. 56, 3–22.

Horovitz, I., (1999). A phylogenetic study of living and fossil platyrrhines. *Am. Mus. Novitates* 3269, 1–40.

Ross, C., Covert, H. H., (2000). The petrosal of *Omomys carteri* and the evolution of the primate basicranium. *J. Hum. Evol.* 39, 225–231.

Ross, C., Williams, B. A., Kay, R. F., (1998). Phylogenetic analysis of anthropoid relationships. *J. Hum. Evol.* 35, 221–306.

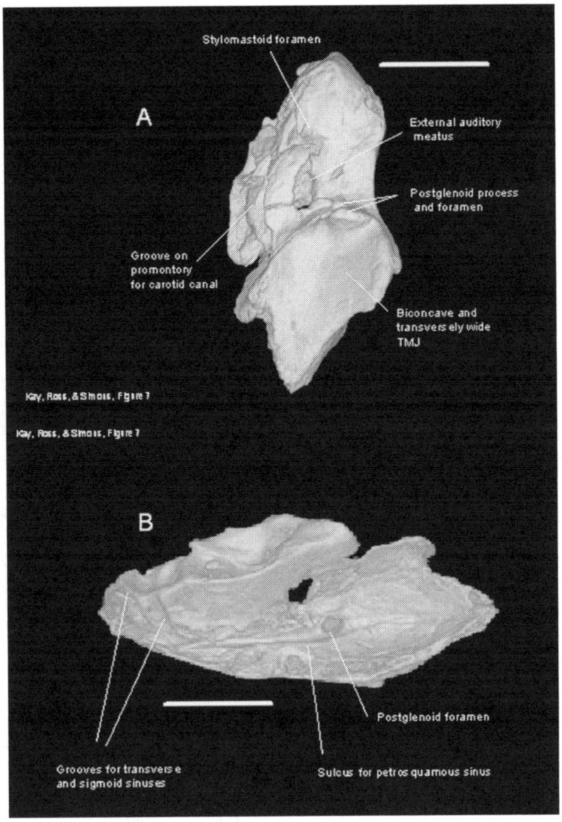

Fig. 7 *Aegyptopithecus zeuxis* (early Oligocene, Egypt; DPC 5401). Right temporal bone from ventral and lateral perspective reconstructed from a series of CT sections. Lateral is to the right and craniad is to the bottom. Scale bar equals 5 mm. The auditory bulla is broken away but an extrabullar tympanic remains. The tympanic cavity is broken away showing part of the groove for the carotid canal on the promontorium. B. Same specimen reconstructed from a series of CT section from dorsal perspective. Lateral is at the top and craniad is to the left. . It shows a sulcus for the petrosquamous sinus leading to the postglenoid foramen. Also shown are sulci for the transverse and sigmoid sinuses. Scale bar equals 5 mm

2) The distribution of the single versus dual expression of the coils of the cochlea (Fig. 4) on the promontorium is difficult to interpret. If, as it appears from the condition in omomyids is for a single turn to be exposed, than a plausible interpretation of the data is for stem anthropoids to preserve this condition and that the condition where two coils are expressed on the promontorium evolved independently in some platyrrhines and *Tarsius*.

Tarsius *autapomorphy*

One characteristic of the ear region of *Tarsius* is absent in omomyids and anthropoids and might be considered as a *Tarsius* autapomorphy. The posterior position of the posterior carotid foramen in all anthropoids is shared with omomyids. Apparently the more anterior position in *Tarsius* is a derived trait.

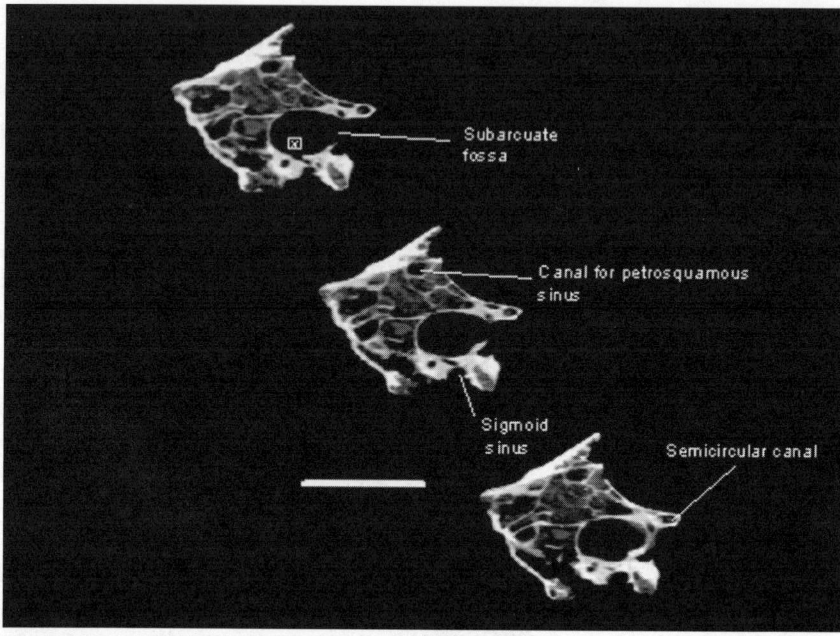

Fig. 8 *Aegyptopithecus zeuxis* (early Oligocene, Egypt; DPC 5401). Successively more caudad sections showing Cartmill's canal (x) exiting the subarcuate fossa but disappearing before it reaches the sigmoid sinus. Scale bar equals 5 mm

Crown Haplorhine Synapomorphies

Several synapomorphies are present in the ear region of extant haplorhines (*Tarsius* and Anthropoidea) as well as the known late Eocene- early Oligocene African anthropoids and early Miocene platyrrhines but are absent in omomyids. These include:

1) In crown haplorhines, the posterior carotid foramen opens into a carotid canal that is located in the primitive medial wall of the auditory bulla (a perbullar pathway) whereas in omomyids, the carotid canal does not run inside the bulla wall (Ross and Covert, 2000).
2) Omomyids do not develop an accessory pneumatic chamber anterior to the tympanic cavity. In *Tarsius* and anthropoids an anterior accessory chamber (AAC) develops by an evagination from the auditory tube. This leaves the carotid canal sandwiched in a transbullar septum between the tympanic cavity and the AAC (Figs. 2C, 3A).
3) In omomyids, the posterior carotid foramen is positioned posterior to the fenestra cochlea whereas in crown haplorhines the posterior carotid foramen is shifted into a more anterior position.

4) Omomyids have a proximal stapedial artery enclosed in a bony tube whereas extant haplorhines experience a reduction of the stapedial artery and its canal, which persists only as a diverticulum in adult *Tarsius* and is absent in crown anthropoids.

5) The tympanic ring of extant haplorhines is phaneric ("extrabullar"), whereas it is aphaneric ("intrabullar') in omomyids. As noted above the tympanic ring is intrabullar only in a topological sense because it supports the tympanic membrane that delimits the tympanic cavity and separates it from the external ear. Rather, a bulging of the tympanic cavity pushes the membranous coverings of the tympanic cavity ventral and lateral to the tympanic ring. On occasion, as in *Necrolemur*, this membrane becomes ossified (Simons, 1961) but the tympanic is still 'intrabullar'. *Tarsius* and crown catarrhines the phaneric tympanic is ossified outward to form a broad collar. This is clearly an independent acquisition because all stem and crown platyrrhines have a ribbon-like tympanic and early Oligocene catarrhine *Aegyptopithecus* retains a ribbon-like tympanic and even the early Miocene catarrhine *Pliopithecus* exhibits only partial formation of the tube.

As has been repeated my many authors in the past going back to Cartmill and Kay, 1978, *Tarsius* and Anthropoidea share a number of features to the exclusion of omomyids suggesting that they are sister taxa. This does not rule out the possibility that the *Tarsius*-anthropoid clade could have arisen from an omomyid-like taxon, always bearing in mind that the Omomyidae would then be a paraphyletic taxon.

Stem Anthropoid Synapomorphies

Several synapomorphies in the ear region of stem and crown anthropoids are absent in *Tarsius* and omomyids.

1) The shape of the temporomandibular joint (TMJ) is biconcave and transversely wide in anthropoids (Figs. 7, 9) whereas it is an anteroposteriorly-oriented trough in *Tarsius* and omomyids. This distribution suggests that the TMJ morphology is a synapomorphy of anthropoids but we would caution that the condition of stem and crown anthropoids also is present in Eocene strepsirrhines. Therefore, it appears equally likely that the anthropoid condition is a primate synapomorphy lost in *Tarsius* and omomyids.

2) The postglenoid foramen is large in omomyids and *Tarsius*, but underwent size reduction in anthropoids. Its size is further reduced in crown catarrhines (but not in *Aegyptopithecus* (Fig. 7) or any other Fayum taxon (e.g., *Parapithecus*, Figs. 9–10) and, independently in some platyrrhines.

3) In all known anthropoids the anterior accessory chamber is trabeculated whereas the chamber is open and untrabeculated in *Tarsius*.

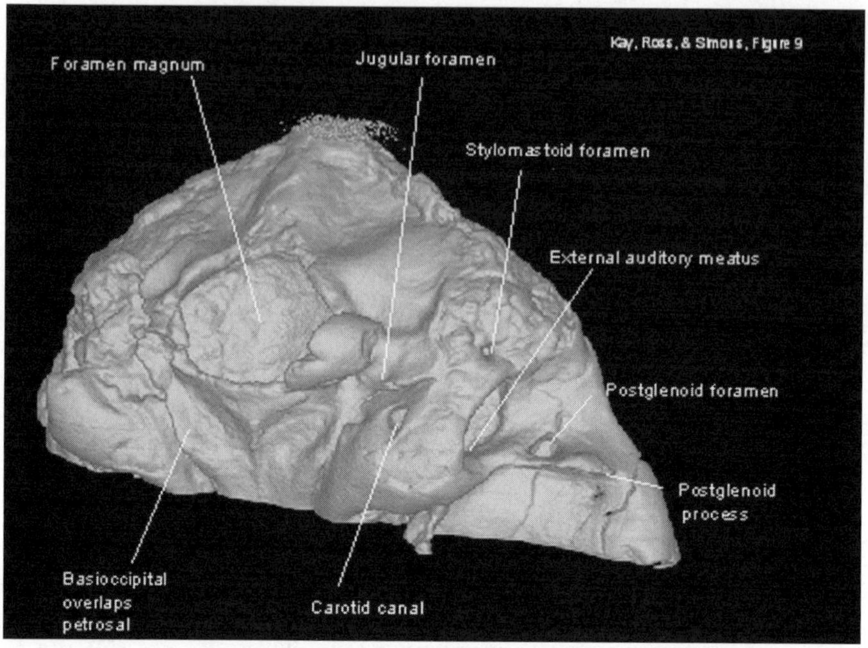

Fig. 9 *Parapithecus grangeri* (early Oligocene, Egypt; DPC 18651) from a ventral and slightly lateral perspective showing the basicranium, glenoid fossa and right ear region. Reconstruction from serial CT images. Cloud of material at inion is artifactual. Scale bar equals 5 mm

4) Stem anthropoids and platyrrhines share the medial placement of the posterior carotid foramen whereas this foramen has migrated laterally in *Aegyptopithecus* and crown catarrhines.

5) Omomyids (at least *Shoshonius* and *Necrolemur*) share with *Tarsius* an extensive overlap of the basioccipital onto the medial surface of the auditory bulla. In contrast, crown Anthropoidea lack a basioccipital bullar flange. Such a flange also is absent in *Proteopithecus* (Fig. 11), *Aegyptopithecus* (Figs. 7–8) and *Catopithecus* (Fig. 12) supporting the conclusion that the flangeless basioccipital is primitive for anthropoids as a whole. Interestingly, *Parapithecus grangeri* has evolved a substantial basioccipital flange.

6) Omomyids and *Tarsius* have a transversely narrow basioccipital whereas in stem and crown anthropoids the basioccipital is broad.

7) Omomyids and *Tarsius* have suprameatal foramina. The suprameatal foramen of *Necrolemur* appears to be functionally homologous with that of *Tarsius*. In both, the foramen is a port for intracranial venous drainage whereas in *Shoshonius* the foramen is located on the root of the zygoma and may have provided only venous drainage for the diploë in that region (Beard and MacPhee, 1994). Stem and crown anthropoids occasionally have

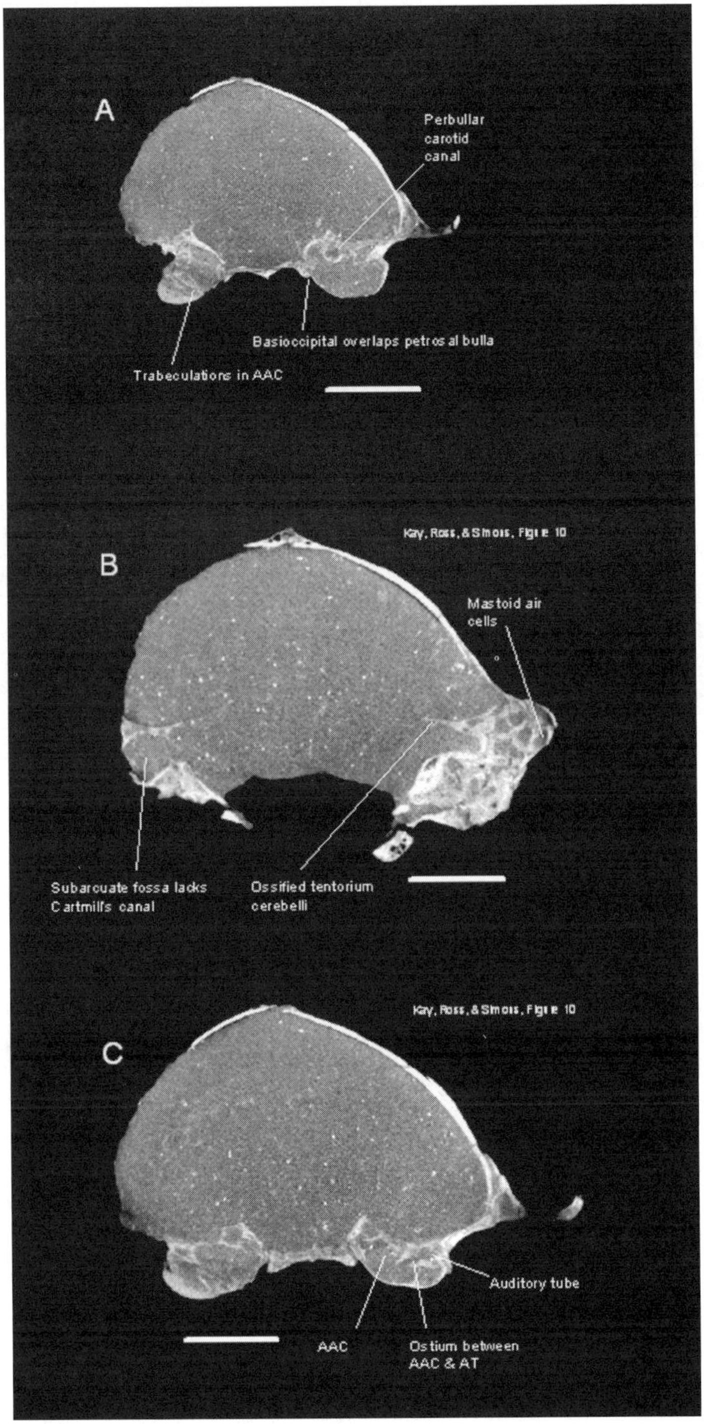

Fig. 10 *Parapithecus grangeri* (early Oligocene, Egypt; DPC 18651). Coronal CT slices through the cranium. Sections A through E are successively more craniad. A. Internal carotid canal on promontorium; basioccipital overlaps petrosal; trabeculated AAC B. Cartmill's canal absent from floor of subarcuate fossa. C. Slice 57. Anterior accessory chamber (AAC connected via an ostium to the auditory tube. Scale bars equals 5 mm

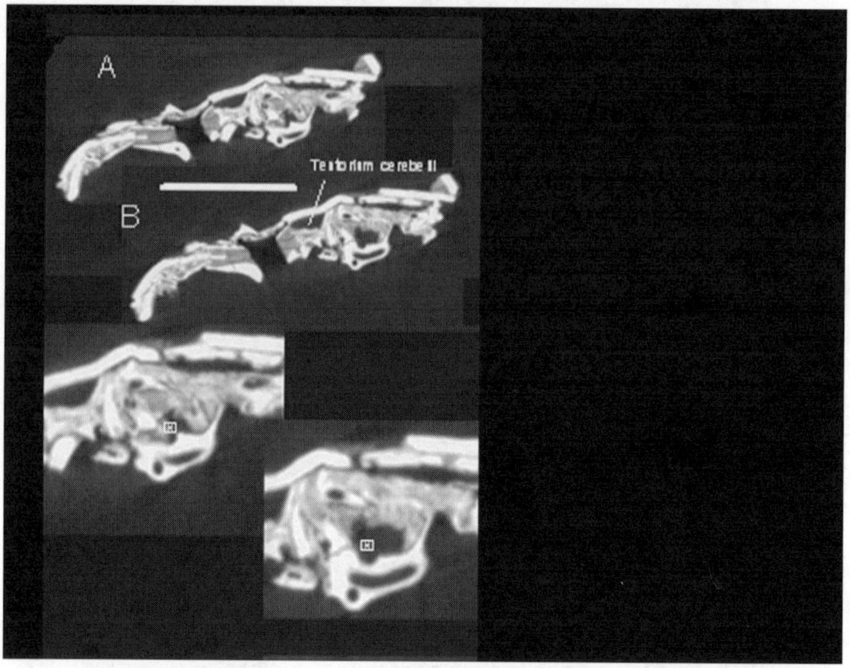

Fig. 11 *Proteopithecus sylviae* (late Eocene, Egypt; DPC 14095). Successively caudad coronal CT slices (A and B) through the cranium at the level of the subarcuate fossa illustrating the broken remnants of a tentorium cerebelli, and the exit of a large Cartmill's canal in floor of subarcuate fossa (x). Images at the bottom are enlargements of images A and B. Further sections, not illustrated, show the canal merging with the sigmoid sinus. Scale bar equals 5 mm

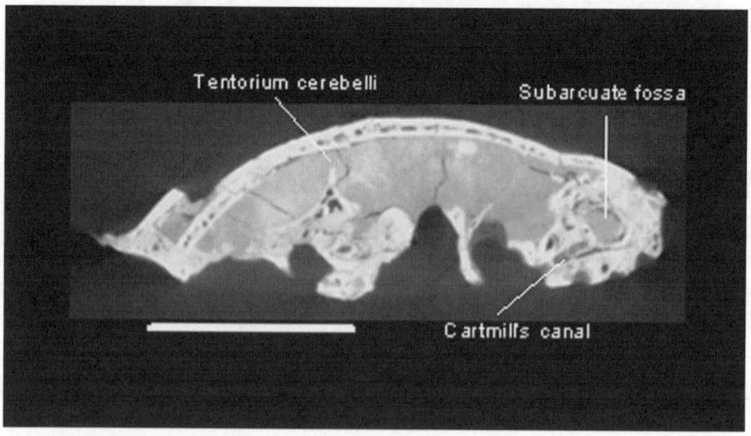

Fig. 12 *Catopithecus browni* (late Eocene, Egypt; DPC 11388. Coronal CT through the cranium at the level of the subarcuate fossa illustrating an ossified tentorium cerebelli, and a large Cartmill's canal. Scale bar equals 5 mm

the diploic foramen but do not have a foramen providing intracranial drainage.

8) *Tarsius* and omomyids possess a parotic fissure. This fissure provides space for tendon of an extrabullar stapedius muscle. Stem and crown anthropoids possess an intrabullar m. stapedius and consequently lack a parotic fissure.

9) Omomyids exhibit an incomplete osseous septum (epitympanic crest) for the stapedial artery, bordering the anterior margin of the epitympanic recess (Ross and Covert, 2000), as do Strepsirrhines (MacPhee and Cartmill, 1986). The crest is invariably absent in *Tarsius* and stem and crown anthropoids.

Are there Synapomorphies in the Ear Region of Crown Platyrrhines or Platyrrhines?

Returning to the questions posed in the introduction, we examine two characters of the ear region that distinguish crown platyrrhines from crown catarrhines, and might, therefore, offer some evidence for the existence of a stem platyrrhine or platyrrhine amongst the known African late Eocene- early Oligocene anthropoids of the Fayum, Egypt. These are ossification of the tentorium cerebelli and the presence of Cartmill's canal.

Tentorium Cerebelli

Early Miocene to Recent stem and crown platyrrhines exhibit ossification in the double-layered sheet of dura mater that forms the tentorium cerebelli (Fig. 2D). The ossification begins proximally along the petrosal and encases the superior petrosal venous sinus, if present. It extends medially although never forming a complete sheet of bone as occurs in some carnivorans. Extensive ossification is present in extant Pitheciidae and Atelidae but is minimal in Cebidae, although the cebids *Aotus* and *Callimico* possess some ossification (Hershkovitz, 1977; Horovitz, 1999). The early Miocene platyrrhines *Dolichocebus* and *Tremacebus* exhibit extensive ossifications (Kay et al., in press). *Aegyptopithecus* and extant catarrhines (Figs. 3, 7, 8) lack such an ossification but ossification is present in *Parapithecus* (Fig. 10B), *Proteopithecus,* and *Catopithecus* (Figs. 11–12). The presence of a tentorium cannot be confirmed in specimens of *Apidium* described by Cartmill et al. (1981). The omomyid *Omomys* lacks ossification (Ni, personal communication) as does *Tarsius*. In spite of the absence of ossification in haplorhine outgroups, our interpretation of these facts is that ossification of the tentorium is a synapomorphy of Anthropoidea preserved in stem anthropoids and many platyrrhines but lost independently in *Aegyptopithecus* and crown catarrhines on one hand and Cebidae on the other. Thus, loss of the tentorium ossification can be viewed as a synapomorphy of at least some stem catarrhines.

Cartmill's Canal

Neither *Tarsius* (Fig. 1) nor catarrhines (Fig. 3) possess a Cartmill's canal in the adult; however, fetal and adult platyrrhines have a well-developed canal (Figs. 2F, 6) containing a vein draining the dura of the subarcuate fossa into the sigmoid venous sinus (Fig. 6; see also Gannon et al, 1987). It is tempting to suggest that the presence of such a canal is a synapomorphy of platyrrhines, a supposition that is reinforced further by the presence of a canal in early Miocene platyrrhines (e.g., *Dolichocebus* and *Tremacebus*; (Kay et al., in press). However, the variable distribution of the canal in Fayum taxa suggests that the story is considerably more complex. Firstly, *Aegyptopithecus* exhibits an interesting intermediate condition where a canal begins in the subarcuate fossa but pinches out before it reaches the sigmoid sinus (Fig. 8). If, as is supposed by many, *Catopithecus* is a primitive catarrhine, it is notable for the presence of an unreduced Cartmill's canal (Fig. 12). This distribution suggests that a canal was present in adult stem catarrhines but lost in crown catarrhines.

Further confusion is introduced by the observation that *Proteopithecus* has a large canal (Fig. 11) but *Apidium* and *Parapithecus* do not (Fig. 12 and Cartmill et al., 1981). If *Proteopithecus* is a parapithecoid (Seiffert et al., 2004) then the canal was present in early parapithecoids but lost in later parapithecoids (*Apidium* and *Parapithecus*). Whatever else this distribution implies, it is certain that the canal appears and disappears independently in several groups and cannot provide an unequivocal signpost to the platyrrhine status of a late Eocene anthropoid.

Conclusions

As already noted by many studies, a very large number of characters of the ear region unite *Tarsius* with Anthropoidea to the exclusion of the Omomyidae. Also, an even larger number of synapomorphies unite the African Eocene/Oligocene anthropoids with crown anthropoids.

Looking at all the similarities among crown platyrrhines, crown catarrhines and stem anthropoids, some of which are derived from a more distant omomyid-like ancestor and others being synapomorphies, makes one point very clearly: the anthropoid ear region was essentially modern in form by at least in late Eocene times by 35 million years ago and has undergone only a few and minor structural changes since. The African Fayum taxa do little if anything to bridge the morphological gap between modern anthropoids, especially platyrrhines and nonanthropoid haplorhines —i.e., *Tarsius* and to an even greater degree omomyids. Overall, the few structural details of the ear region that separate Miocene to Recent platyrrhines from crown catarrhines represent further modifications in the latter. Extant platyrrhines are ideal structural models for late Eocene African anthropoids in the anatomy of the arteries and veins, the design of the tympanic cavity, its accessory pneumatic sinuses, and the structure and relations of the tympanic bone.

On the other hand, several synapomorphies may now be identified that support linkage between *Aegyptopithecus* and crown catarrhines. It appears that loss of partial ossification of the tentorium cerebelli and less certainly, reduction of Cartmill's canal and its constituent vein may be synapomorpies of catarrhines.

However, we have been unable to certainly identify shared derived features uniting the platyrrhines. And, we are still left wanting for a convincing synapomorphy in the ear region linking platyrrhines with any known African Eocene-Oligocene taxon. In its known morphology, *Proteopithecus* remains a possible candidate but only because of shared retentions from a more distant common ancestor.

Acknowledgment The manuscript has benefited greatly from discussion with Matt Cartmill, and several anonymous reviewers. T. Ryan and A. Walker provided the micro-CT scans of the ear regions of a series of Fayum anthropoids. RFK was supported by NSF grants BCS 0090255 and 0087636, from National Geographic Society and from L. S. B. Leakey Foundation. ELS was supported by NSF BCS-0416164 for Fayum field research. ELS thanks the Chairman and staff of the Egyptian Mineral Resources Authority and of the Egyptian Geological Museum, Maadi and the help of members of recent Fayum field crews, Prithijit Chatrath field manager, and the team of local workers from Kom Aushim led by M. Hassen-Taha. This is Duke Lemur Center publication #1026.

References

Beard, K. C., Krishtalka, L., and Stucky, R. (1991). First skulls of the early Eocene primate Shoshonius cooperi and the anthropoid-tarsier dichotomy. *Nature* 349: 64–67.

Beard, K. C., and MacPhee, R. D. E. (1994). Cranial anatomy of Shoshonius and the antiquity of Anthropoidea. In: Fleagle, J. G., and Kay, R. F. (eds.), *Anthropoid Origins*, Plenum Press, New York, pp. 55–98.

Cartmill, M., and Kay, R. F. (1976). Craniodental morphology and development and the problem of tarsier affinities. *VI Int. Cong. Primatology, Abstracts*: 93.

Cartmill, M., and Kay, R. F. (1978). Craniodental morphology, tarsier affinities, and primate suborders. In: Chivers, D. J., and Joysey, K. A. (eds.), *Recent Advances in Primatology: Evolution*, Academic Press, London, pp. 205–214.

Cartmill, M., MacPhee, R. D. E., and Simons, E. L. (1981). Anatomy of the temporal bone in early anthropoids with remarks on the problem of anthropoid origins. *Am J. Phys. Anthropol.* 56: 3–22.

Diamond, M. K. (1992). Homology and evolution of the orbitotemporal venous sinuses of humans. *Am J. Phys. Anthropol.* 88: 211–244.

Fleagle, J. G., and Bown, T. M. (1983). New primate fossils from late Oligocene (Colhuehuapian) localities of Chubut Province, Argentina. *Folia Primatol.* 41: 240–266.

Fleagle, J. G., and Kay, R. F. (1987). The phyletic position of the Parapithecidae. *J. Hum. Evol.* 16: 483–532.

Fleagle, J. G., and Kay, R. F. (1989). The dental morphology of *Dolichocebus gaimanensis*, a fossil monkey from Argentina. *Am J. Phys. Anthropol.* 78: 221.

Gannon, P. J., Eden, A. R., and Laitman, J. T. (1988). The subarcuate fossa and cerebellum of extant primates: comparative study of a skull-brain interface. *Am J. Phys. Anthropol.* 77: 143–164.

Gunnell, G. F., and Miller, E. R. (2001). Origin of Anthropoidea: dental evidence and recognition of early anthropoids in the fossil record, with comments on the Asian anthropoid radiation. *Am J. Phys. Anthropol.* 114: 177–191.

Hershkovitz, P. (1974). A new genus of late Oligocene monkey (Cebidae, Platyrrhini) with notes on postorbital closure and platyrrhine evolution. *Folia Primatol.* 21: 1–35.

Hershkovitz, P. (1977). *Living New World monkeys (Platyrrhini) with an introduction to Primates, Vol. 1.* University of Chicago Press, Chicago.

Hershkovitz, P. (1982). Supposed squirrel monkey affinities of the late Oligocene *Dolichocebus gaimanensis. Nature* 298: 201–202.

Hoffstetter, R. (1977). Phylogenie des primates. *Bulletin et Mémoire de la Societie d-Anthropologie de paris* 4, Série XIII: 327–346.

Horovitz, I., (1997). Platyrrhine systematics and the origin of Greater Antilles monkeys. Ph. D. dissertation, State University of New York.

Horovitz, I. (1999). A phylogenetic study of living and fossil platyrrhines. *Am. Mus. Novit.* 3269: 1–40.

Horovitz, I., and MacPhee, R. D. E. (1999). The quaternary Cuban platyrrhine *Paralouatta varonai* and the origin of Antillean monkeys. *J. Hum. Evol.* 36: 33–68.

Kay, R., F., Vizcaino, S. F., Tauber, A., Bargo, M. S., Williams, B. A., Luna, C., and Colbert, M. W. (2005). Three newly discovered skulls of *Homunculus patagonicus* support its position as a stem platyrrhine and establish its diurnal arboreal folivorous habits. *Am J. Phys. Anthropol.* Supplement 40: 127.

Kay, R. F. (1977). The evolution of molar occlusion in Cercopithecidae and early catarrhines. *Am J. Phys. Anthropol.* 46: 327–352.

Kay, R. F. (1978). Molar structure and diet in extant Cercopithecoidea. In: Butler, P. M., and Joysey, K. (eds.), *Development, Function and Evolution of Teeth*, Academic Press, London, pp. 309–339.

Kay, R. F. (1980). Platyrrhine origins: a reappraisal of the dental evidence. In: Ciochon, R., and Chiarelli, B. (eds.), *Evolutionary Biology of the New World Monkeys and Continental Drift*, Plenum Press, New York, pp. 159–188.

Kay, R. F., and Delson, E., (2000). Oligopithecidae. In: Delson, E., Tattersall, I., VanCouvering, J. A., and Brooks, A. (eds.), *Encyclopedia of Evolution and Prehistory, 2nd Edition*, Garland Pub. Co., New York, pp. 490–493.

Kay, R. F., Fleagle, J. G., Mitchell, T. R. T., Colbert, M. W., Bown, T. M., and Powers, D. W. (in press). The anatomy of *Dolichocebus gaimanensis*, a primitive platyrrhine monkey from Argentina. *J. Hum. Evol.*

Kay, R. F., Fleagle, J. G., and Simons, E. L. (1981). A revision of the African Oligocene Apes of the Fayum Province, Egypt. *Am J. Phys. Anthropol.* 55: 293–322.

Kay, R. F., Ross, C. F., Williams, B. A. (1997). Anthropoid Origins. *Science* 275: 797–804.

Kay, R. F., Williams, B. A., 1994. Dental evidence for anthropoid origins. In: Fleagle, J. G., and Kay, R. F. (eds.), *Anthropoid Origins*, Plenum Press, New York, pp. 361–446.

Kay, R. F., Williams, B. A., Ross, C. F., Takai, M., and Shigehara, N., (2004). Anthropoid origins: a phylogenetic analysis. In: Ross, C. F., and Kay, R. F. (eds.), *Anthropoid Origins: New Visions*, Kluwer/Plenum, New York, pp. 91–135.

MacPhee, R. D. E., Beard, K. C., and Qi, T. (1995). Significance of primate petrosal from middle Eocene fissure-fillings at Shanghuang, Jiangsu Provincve, People's Republic of China. *J. Hum. Evol.* 29: 501–514.

MacPhee, R. D. E., and Cartmill, M., (1986). Basicranial structures and primate systematics. In: Swindler, D. R., and Erwin, J. (eds.), *Comparative Primate Biology, Volume 1: Systematics, Evolution, and Anatomy*, Alan R. Liss, New York, pp. 219–275.

Madeira, M. C., and Watanabe, I.-S. (1975). Anatomical data on the intracranial venous drainage of the tufted capuchin, *Cebus apella* Linnaeus 1758. *Anat. Rec.* 183: 589–598.

Miller, E. R., and Gunnell, G. F., and Martin, R. D. (2005). Deep time and the search for anthropoid origins. *Yearb. Phys. Anthropol.* 48: 60–95.

Patten, B. M., (1968). *Human Embryology, Third Edition*. McGraw Hill Book Company, New York.

Rasmussen, D. T., and Simons, E. L. (1992). Paleobiology of the oligopithecines, the earliest known anthropoid primates. *Int. J. Primatol.* 13: 1–32.

Rosenberger, A. L. (1985). In favor of the *Necrolemur*-tarsier hypothesis. *Folia Primatol.* 45: 179–194.

Rosenberger, A. L., and Szalay, F. S., (1980). On the tarsiiform origins of Anthropoidea. In: Ciochon, R. L., and Chiarelli, A. B. (eds.), *Evolutionary Biology of the New World Monkeys and Continental Drift*, Plenum Press, New York, pp. 139–157.

Ross, C., and Covert, H. H. (2000). The petrosal of *Omomys carteri* and the evolution of the primate basicranium. *J. Hum. Evol.* 39: 225–231.

Ross, C., Williams, B. A., and Kay, R. F. (1998). Phylogenetic analysis of anthropoid relationships. *J. Hum. Evol.* 35: 221–306.

Ross, C. F., 1994. The craniofacial evidence for anthropoid and tarsier relationships. In: Fleagle, J. G., and Kay, R. F. (eds.), *Anthropoid Origins*, Plenum Press, New York, pp. 469–547.

Saban, R. (1963). Contribution à l'étude de l'os temporal des Primates. Description chez l'Homme et les Prosimiens. Anatomie comparée et phylogénie. *Mémoires du Muséum National D'Histoire Naturelle* Nouvelle Série, Série A, Zoologie, 29: 1–378.

Seiffert, E. R., Simons, E. L., Simons, C. V. M. (2004). Phylogenetic, biogeographic, and adaptive implications of new fossil evidence bearing on crown anthropoid origins and early stem catarrhine evolution. In: Ross, C., and Kay, R. F. (eds.), *Anthropoid Origins: New Visions*, Kluwer/Plenum, New York, pp. 157–182.

Simons, E. (1989). Description of two genera and species of Late Eocene Anthropoidea from Egypt. *P. Natl. Acad. Sci. USA* 86: 9956–9960.

Simons, E. (1995). Egyptian Oligocene Primates: A Review. *Yearb. Phys. Anthropol.* 38: 199–238.

Simons, E. L. (1961). Notes on Eocene tarsioids and a revision of some Necrolemurinae. *Bull. Brit. Mus. (Nat. Hist.) Geology* 5: 45–69.

Simons, E. L. (1962). Two new primate species from the African Oligocene. *Postilla, Yale Peabody Mus.* 64: 1–12.

Simons, E. L. (1965). New fossil apes from Egypt and the initial differentiation of the Hominoidea. *Nature* 205: 135–139.

Simons, E. L. (1992). Diversity in the early Tertiary anthropoidean radiation in Africa. *P. Natl. Acad. Sci. USA* 89: 10743–10747.

Simons, E. L., and Rasmussen, D. T. (1989). Cranial anatomy of *Aegyptopithecus* and *Tarsius* and the question of the tarsier-anthropoidean clade. *Am J. Phys. Anthropol.* 79: 1–23.

Streeter, G. L. (1915). The development of the venous sinuses of the dura mater in the human embryo. *Am. J. Anat.* 18: 145–178.

Streeter, G. L. (1918). The developmental alterations in the vascular system of the brain of the human embryo. *Carnegie Institute, Contributions to Embryology* 8 (24): 5–38.

Szalay, F. S. (1976). Systematics of the Omomyidae (Tarsiiformes, Primates): taxonomy, phylogeny and adaptations. *B. Am. Mus. Nat. Hist.* 156: 157–450.

Szalay, F. S., and Delson, E. (1979). *Evolutionary History of the Primates*. Academic Press, New York.

Takai, M., Anaya, F., Shigehara, N., and Setoguchi, T. (2000). New fossil materials of the earliest new world monkey, *Branisella boliviana*, and the problem of platyrrhine origins. *Am J. Phys. Anthropol.* 111: 263–281.

Waltner, J. G. (1944). Anatomic variation of the lateral and sigmoid sinuses. *Archives of Otolaryngology* 39: 307–312.

Weinstein, J. D., and Hedges Jr., T. R. (1962). Studies of intracranial and orbital vasculature of the rhesus monkey (*Macaca mulatta*). *Anat. Rec.* 144: 37–41.

Wible, J. R. (1987). The eutherian stapedial artery: character analysis and implications for superordinal relationships. *Zool. J. Linn. Soc.- Lond.* 91: 107–135.

Appendices

Appendix 1. Material examined for the present study. Many other specimens of extant primates were examined visually in the collections of the Smithsonian Institution, Washington DC. CT images of all extant specimens and all specimens of fossil platyrrhines were made for RFK at the High-Resolution X-ray Computed Tomography (CT) Facility at the University of Texas at Austin, with collaboration and assistance of Dr Matt Colbert. Images of fossils from the Eocene and Oligocene of Africa were made for ELS with collaboration and assistance from Prof. Alan Walker, Pennsylvania State University, State College, PA.

Extant specimens examined visually and by CT imagery: Tarsiidae: *Tarsius bancanus*, USNM 488084. Catarrhini: *Miopithecus talapoin*, Kay collection. Platyrrhini: *Aotus trivirgatus*, USNM 464844; *Saguinus fuscicollis*, USNM 518577; *Callithrix jacchus*, USNM 503885, 503895; *Callithrix argentata*, USNM 239463; *Callimico goeldii*, USNM 303323; *Cebuella pygmaea*, USNM 337324; *Saimiri sciureus*, USNM 518538; *Cebus nigrivittatus*, USNM 338960; *Pithecia monachus*, USNM 518223, Fleagle collection specimen; *Ateles geoffroyi*, USNM 291056; *Alouatta fusca*, USNM 518255; *Callicebus torquatus*, USNM 406411.

African fossil taxa examined visually and by CT imagery: *Catopithecus browni* (late Eocene, Egypt), DPC 11388; *Proteopithecus sylviae* (late Eocene, Egypt), DPC 14095; *Apidium phiomense* (early Oligocene, Egypt); YPM 23968; *Parapithecus grangeri* (early Oligocene, Egypt), DPC 18651; *Aegyptopithecus zeuxis* (early Oligocene, Egypt), DPC 5401, CGM 85785, a juvenile skull

Additional African fossil taxa examined visually: *Aegyptopithecus zeuxis* (early Oligocene, Egypt) DPC 6642, CGM 40237; *Apidium phiomense* (early Oligocene, Egypt) YPM 25972; YPM 25973; YPM 25974.

South American fossil taxa examined visually and by CT imagery: *Tremacebus harringtoni*, (early Miocene, Argentina), Tucuman, TYPE skull; *Dolichocebus gaimanensis* (early Miocene, Argentina): MACN TYPE skull; *Homunculus patagonicus* (late early Miocene, Argentina) PLC RK 04-43.

Paleontological Exploration in Africa

A View from the Rukwa Rift Basin of Tanzania

Nancy J. Stevens, Michael D. Gottfried, Eric M. Roberts, Saidi Kapilima, Sifa Ngasala and Patrick M. O'Connor

Introduction

The Mesozoic – Cenozoic transition was a period of dramatic global change during which time the Earth's continents were in the process of fragmenting from a large, relatively continuous landmass to assume a configuration similar to that seen today. The most significant tectonic activity in the southern hemisphere occurred during the Cretaceous-Paleogene interval, when the large Gondwanan sub-regions of Africa, South America, Australia, Indo-Madagascar and Antarctica became increasingly isolated from one another (Smith et al., 1994; Scotese, 2001). Continental dynamics of this scale are not only geologically significant, they also profoundly influenced the evolution of both terrestrial and marine biotas (Forster, 1999; Krause et al., 1999; Sereno, 1999; Lieberman, 2000; Upchurch et al., 2002; Humphries and Ebach, 2004). Indeed, the Cretaceous-Paleogene transition marks large-scale faunal turnover of major vertebrate and invertebrate taxa (e.g., extinction of nonavian dinosaurs, radiation of "modern" mammals and birds; Cracraft, 2001; Springer et al., 2003, 2004; Archibald and Fastovsky, 2004; Kielan-Jaworowska et al., 2004; Rose and Archibald, 2004; Clarke et al., 2005).

Numerous hypotheses have been proposed to explain the origin, diversification, and extinction of many vertebrate groups living on, or dispersing through, Gondwana during the Cretaceous and Paleogene. For example, molecular studies have postulated a Cretaceous-Paleogene African origin for a number of higher-level amniote clades, including Placentalia (Murphy et al., 2001 and references therein), Afrotheria (Hedges et al., 1996; Springer et al., 1997, 2003,

Nancy J. Stevens
Department of Biomedical Sciences, Ohio University, 228 Irvine Hall, Athens, OH 45701, Ph: 740-597-2785, FAX: 740-597-2778
stevensn@ohio.edu
Patrick M. O'Connor
Department of Biomedical Sciences, Ohio University, 228 Irvine Hall, Athens, OH 45701, Ph: 740-593-2110, FAX: 740-597-2778
oconnorp@ohio.edu

J. G. Fleagle, C. C. Gilbert (eds.), *Elwyn Simons: A Search for Origins.*

2005; Madsen et al., 2001; van Dijk et al., 2001), and neornthine birds (Cracraft, 2001). In particular, an ancient (~Cretaceous/Paleocene) Gondwanan primate origin has been proposed, with a strepsirrhine-haplorhine divergence occurring shortly thereafter (e.g., Tavare et al., 2002). African origins have also been proposed for a number of Malagasy terrestrial and freshwater groups (e.g., etropline cichlids (Vences et al., 2001); lemurs (Yoder et al., 2003, Poux et al., 2005); tenrecs (Poux et al., 2005)). Yet divergence time estimates retrieved by molecular studies for various clades often vastly predate the first occurrences of those groups in the fossil record (e.g., Smith and Peterson, 2002), instigating considerable debate as to the time of origin and path of dispersal for a broad range of taxa (e.g., Martin, 2000; de Wit, 2003; Schrago and Russo, 2003; Rose and Archibald, 2004; de Queiroz, 2005; Masters et al., 2006). This is perhaps not surprising, as Martin and others have demonstrated that by any measure, the vertebrate fossil record (particularly in places like Africa) is dismayingly incomplete, such that dates derived from paleontological data alone are likely to significantly underestimate true divergence times (Martin, 1993, 2000; Paul, 1998; Tavare et al., 2002; Miller et al., 2005). Whereas questions remain regarding the reliability of molecular clocks with respect to calibration and rate heterogeneity (Smith and Peterson, 2002), it is also clear that sustained work is needed to improve sampling of the fossil record and test molecular hypotheses by providing fossil data that can be used to more rigorously calibrate and refine divergence time estimates (Seiffert et al., 2003; Yoder et al., 2003). This is particularly true of undersampled regions where new discoveries can have a profound effect on hypotheses based on presence/absence data (e.g., a Cretaceous gondwanatherian mammal from Tanzania; Krause et al., 2003b; O'Connor et al., 2006). Moreover, recent studies examining the robusticity of biogeographic reconstructions demonstrate that even a single new outgroup or ingroup fossil can powerfully influence area-of-origin interpretations (e.g., Stevens and Heesy, 2004, 2006; Heesy et al., 2006).

Uneven Paleontological Sampling

Much of our knowledge of terrestrial communities during the Cretaceous-Paleogene transition has been based on research conducted on northern continents. These studies have sought to address global-level issues such as species diversity and extinction, in addition to providing information concerning the phylogenetic and biogeographic histories of major terrestrial vertebrate groups. In recent decades, efforts have increased to improve the fossil record from Gondwanan landmasses and rectify the Laurasian sampling bias. Field research in South America, supra-equatorial Africa, India, and Madagascar has significantly improved the Southern Hemisphere vertebrate record over the last thirty years (e.g., Bonaparte and Powell, 1980; Molnar, 1980; Rich et al., 1983; Archer et al., 1985; Flynn et al., 1987; Prasad and Sahni, 1988; Godinot, 1994; Prasad and

Godinot, 1994; Sereno et al., 1996, 2004; Sampson et al., 1998; Sigogneau-Russell et al., 1998; Krause et al., 1999, 2006; Chiappe et al., 2001; Kay et al., 2001; Smith et al., 2001; Flynn et al., 2003; Krause, 2003b; Prasad et al., 2005; Rana et al., 2005), but large portions of the former supercontinent remain relatively unexplored. Especially problematic is the patchiness of terrestrial sequences on continental Africa and its surrounding islands, rendering early Cenozoic vertebrate diversification in many of these locales a virtual mystery.

African Vertebrate Paleontology—A Cretaceous-Paleogene Perspective

Historically, African Mesozoic terrestrial vertebrate assemblages have been recovered from only select portions of the continent, notably the Permo/Triassic-Jurassic Karoo basins in southern Africa, the Upper Jurassic Tendaguru series in southeastern Tanzania, and a number of "middle" Cretaceous locales scattered throughout circum-Saharan Africa (Fig. 1A). Such faunas bear witness to pioneering paleontological exploration during the last century by workers such as Lavocat (1954), Lapparent (1960), and others, for example, in Niger, Algeria, Tunisia, and Morocco. A few Cretaceous-age fossil-bearing sequences are known from sub-equatorial Africa, although only the ~Aptian Dinosaur Beds of Malawi (Dixey, 1928; Jacobs et al., 1990, 1993; Gomani, 2005) and the Berriasian-Valanginian Kirkwood Formation of South Africa (Rich et al., 1983; De Klerk et al., 2000) have revealed diverse vertebrate assemblages. Until recently, Paleogene sampling has largely been restricted to a few isolated localities, mostly concentrated in the northern Saharan arc (Fig. 1B), along with a single locality in the southwestern portion of the continent (Pickford, 1986). Cenozoic faunas recovered from East Africa have traditionally been limited to deposits of Miocene and later age, but recent work has demonstrated that this region also preserves a deeper record of vertebrate evolutionary history (Leakey et al., 1995; Harrison et al., 2001; Gunnell et al., 2002; Kappelman et al., 2003; Stevens et al., 2004). Yet few places in Africa have received the sustained, long-term paleontological effort that has been dedicated to the Eocene-Oligocene deposits of Egypt.

A Window into the Early Cenozoic of Africa

Without doubt, the exemplar for intensive fossil collecting in the African Paleogene has been the work of Elwyn Simons and his collaborators in the Fayum Depression of Egypt. Prior to these expeditions, a modest number of fossils from the Fayum had been collected and described (e.g., Andrews, 1901, 1902; Beadnell, 1902; Osborn, 1908; Matsumoto, 1921). However, since 1961,

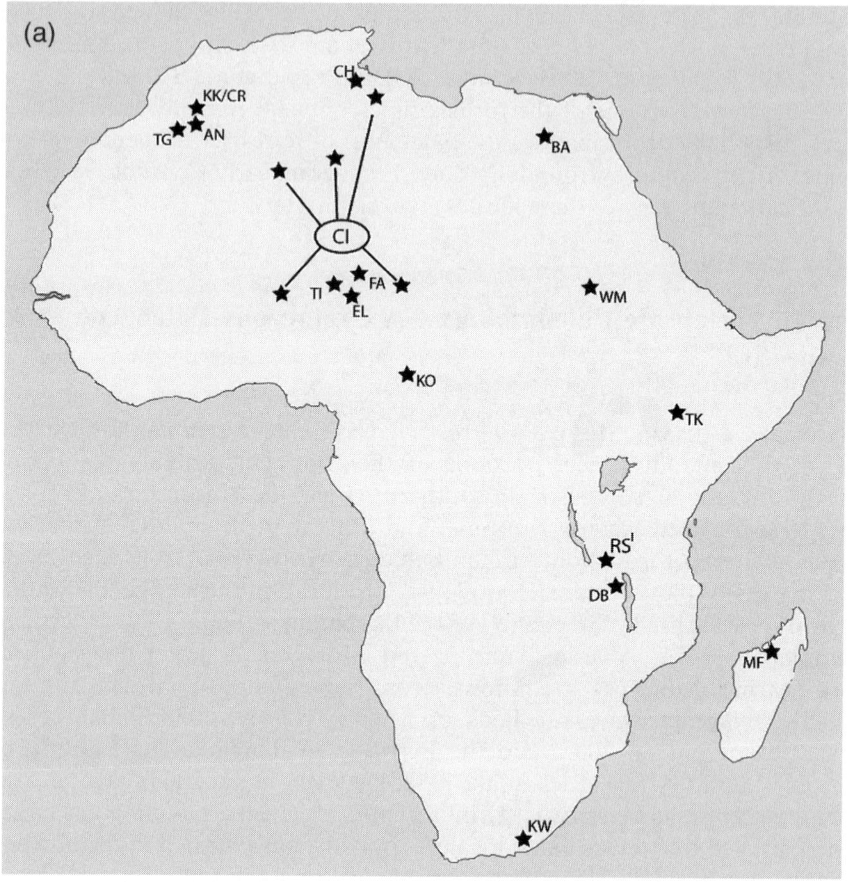

Fig. 1 Map of Africa and its surrounding islands illustrating the principal Cretaceous (**A**) and Paleogene (**B**) sequences preserving terrestrial vertebrates. Note: only locales preserving relatively diverse terrestrial vertebrate assemblages are indicatedAbbreviations for **Fig. 1A** as follows: **AN**, Anoual, Lower Cretaceous, Morocco;**BA**, Baharija Fm., Upper Cretaceous, Egypt; **CH**, Chenini Fm., Lower Cretaceous, Tunisia; **CI**, "Continental Intercalaire," Lower Cretaceous-lowermost Upper Cretaceous, multiple locations throughout circum-Saharan Africa; **CR**, Continental Red Beds, Upper Cretaceous, Morocco; **DB**, Dinosaur Beds, Lower Cretaceous, Malawi; **EL**, Elrhaz Fm., Lower Cretaceous, Niger; **FA**, Farak Fm., Lower Cretaceous, Niger; **KK**, Kem Kem Beds, Upper Cretaceous, Morocco; **KO**, Koum Fm., Lower Cretaceous, Cameroon; **KW**, Kirkwood Fm., Lower Cretaceous, South Africa; **MV**, Maevarano Fm., Upper Cretaceous, Madagascar; **RS^{I}**, Red Sandstone Group (Unit I), "mid" Cretaceous, Tanzania; **TI**, Tiourarén Fm., Lower Cretaceous, Niger; **TG**, Tegana Fm., Lower Cretaceous, Morocco; **TK**, Turkana Grits, Upper Cretaceous, Kenya; **WM**, Wadi Milk Fm. Upper Cretaceous, Sudan.
Abbreviations for **Fig. 1B** as follows: **AM**, Adrar Mgorn, Paleocene, Morocco; **BZ**, Buel Haderait/Zella, Dor et Talha, Eocene-Oligocene, Libya; **CG**, Chambi/Gebel Bou Gobrine, Eocene-Oligocene, Tunisia; **CH**, Chilga, Oligocene, Ethiopia; **FD**, Fayum Depression, Eocene-Oligocene, Egypt; **GN**, Glib Zegdou/El Kohol/Nementcha, Eocene, Algeria; **MH**, Mahenge, Eocene, Tanzania; **ML**, Malembe, Oligocene, Angola; **RS^{II}**, Red Sandstone Group (Unit II), Oligocene, Tanzania

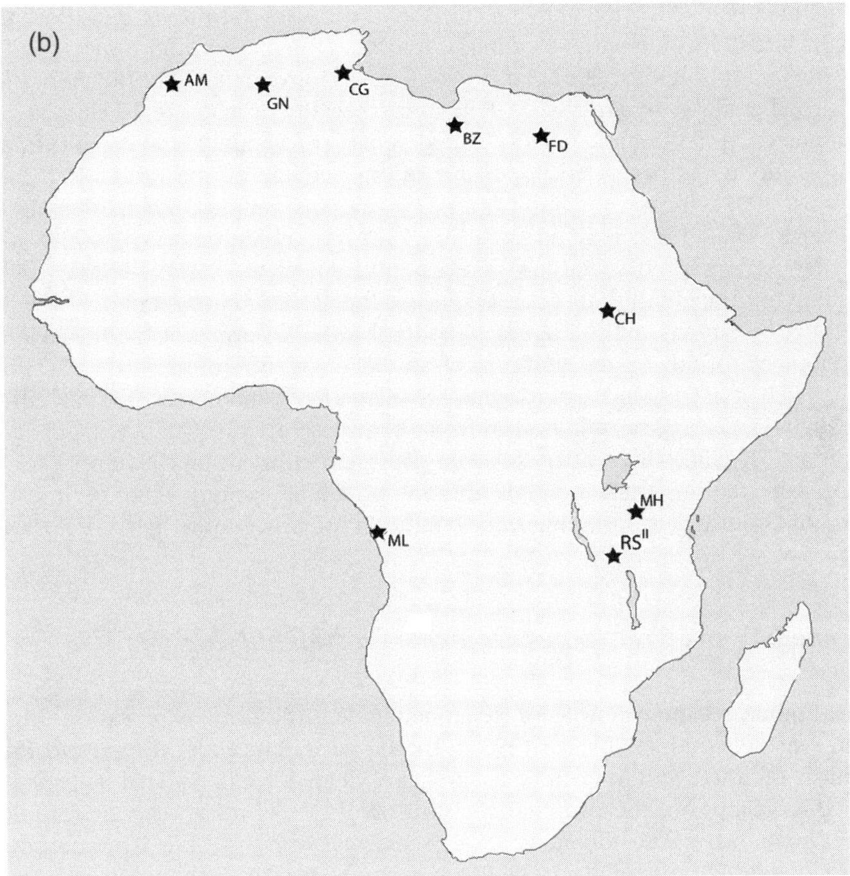

Fig. 1 (continued)

each of the 38 (∼7-week) Fayum expeditions organized by Simons has assembled a crew of approximately 17 members, resulting in over 207,000 '*person-hours of collecting effort*' on the ground searching for fossils. This effort has yielded tens of thousands of vertebrate fossil specimens, which have in turn formed the basis for hundreds of publications. From a new order of mammals (Ptolemaiida Simons and Bown, 1995), to primitive elephant shrews (Simons et al., 1991) and abundant rodents (Wood, 1968; Holroyd, 1994), from the minute *Wadilemur* (Simons, 1997; Seiffert et al., 2005) and *Widanelfarasia* (Seiffert and Simons, 2000) to *Megalohyrax* (e.g., Thewissen and Simons, 2001), the Fayum Depression has provided a tantalizing glimpse into Africa's past. This collection has inspired a generation of researchers to muse over aspects of the paleoecological setting (Bown, 1982; Bown et al., 1982; Simons et al., 1994), incorporating studies of plants, fish (Murray and Attia, 2004), tortoises (Holroyd and Parham, 2003), birds (Rasmussen et al., 2001), hyraxes (Rasmussen and Simons, 2000; De Blieux and Simons, 2002), anthracotheres (Holroyd et al., 1996), creodonts (Holroyd et al., 1996; Holroyd, 1999), and especially primates (e.g., Simons and Rasmussen, 1995; Simons, 1998). The diversity and abundance of the primate sample alone has permitted important reconstructions of diets (e.g., Kirk and Simons, 2001), locomotor habits (e.g., Fleagle and Simons, 1982, 1983, 1995; Hamrick et al., 1995; Rasmussen et al., 1998, Ankel-Simons et al., 1998), and activity patterns and social systems (e.g., Fleagle et al., 1980; Fleagle, 1999), fleshing out the inhabitants of an Oligocene forest in striking detail.

Work by Simons and colleagues in the Jebel Qatrani Formation has fostered the development of other significant projects in the region, for example in Wadi Hitan (Gingerich et al., 1990); Wadi Moghara (Miller, 1999), and Birket Qarun (Seiffert et al., 2003). Perhaps most importantly, the continued discovery of new species and additional localities even after so much dedicated effort in the Fayum serves to underscore the importance of this area for providing exciting windows into the past, and offering encouragement for long-term projects in other parts of Africa.

Expanding the Cretaceous Record on a Neighboring Isle

As another recent example of sustained paleontological exploration, eight field seasons conducted since 1993 in the Upper Cretaceous Maevarano Formation of northwestern Madagascar have yielded similarly striking results. Since its inception, the Mahajanga Basin Project has more than quadrupled the known vertebrate diversity (>40 species) from the Late Cretaceous of Madagascar (Krause et al., 1999, 2006; Krause, 2003b). Although fossils from the Maevarano Formation were first discovered and published in the late 1800s (Depéret, 1896), intensive field research with large crews was not undertaken until the early 1990s. Several ~6 week expeditions, with crew sizes ranging between 15 and 20 individuals, transformed a virtually unknown terrestrial fauna into one of the most comprehensive Late Cretaceous assemblages in all of Gondwana (e.g., Sampson et al., 1998; Krause et al., 1999). This diversity encompasses fishes (Gottfried et al., 2001), frogs (Asher and Krause, 1998), turtles (Gaffney and Forster, 2003), lizards (Krause et al., 2003a), crocodyliforms (Buckley and Brochu, 1999; Buckley et al., 2000), nonavian dinosaurs (Sampson et al., 1998; Curry Rogers and Forster, 2001; Carrano et al., 2002; O'Connor, 2007), birds (Forster et al., 1996, 1998), and mammals (Krause et al., 1997; Krause, 2001, 2003a). Hand in hand with announcements of species new to science, contributions have detailed the geological and paleoenvironmental context of the Maevarano Formation (Rogers et al., 2000; Rogers, 2005), exploring paleobiological aspects of the fauna (e.g., Buckley et al., 2000; Rogers et al., 2003; O'Connor and Claessens, 2005). Krause et al. (1999, 2006) provide a more comprehensive overview of the project's history.

The abovementioned projects have set a standard for how to conduct and sustain successful field projects in historically undersampled regions. Over the years both projects have included numerous student participants, graduates and undergraduates alike. Perhaps most significant is the fact that so many of these former student participants have pursued paleontological projects in other locales (e.g., Fleagle and Bown, 1983; Covert and Hamrick, 1993; Wing et al., 1993; Kay et al., 2001; Kappelman et al., 2003; Rasmussen, 2008). One such project still in its early stages is discussed below.

New Views into Vertebrate Evolutionary History: The Rukwa Rift Basin Project, Southwestern Tanzania

Developed in the spirit of the aforementioned endeavors, the Rukwa Rift Basin Project (RRBP) is seeking to expand the record of Cretaceous and Paleogene terrestrial vertebrates in sub-equatorial Africa. Situated in southwestern Tanzania between Lakes Nyasa (Malawi) and Tanganyika, the Rukwa Rift Basin (Fig. 2A) forms part of the western branch of the East African Rift System (EARS), and contains some of the thickest sedimentary deposits in East Africa (Wescott et al., 1991; Kilembe and Rosendahl, 1992; Morley et al., 1999). At least three major sedimentary episodes are recorded in the RRB, resulting from intermittent Permian to Recent tectonic events (Quenell et al., 1956; Wheeler and Karson, 1994). The middle sedimentary sequence, commonly referred to as the Red Sandstone Group (RSG), has been poorly understood in terms of depositional timing, with age assessments ranging from the Middle Jurassic through the late Cenozoic (e.g., Damblon et al., 1998; Morley et al., 1999). Since 2002, work by our group has revised the geological context of the RSG, uncovering nearly 40 fossil-bearing localities that clearly document the presence of both Cretaceous (Unit I) and Oligocene (Unit II) sequences (O'Connor et al., 2003, 2006; Gottfried et al., 2004; Roberts et al., 2004; Stevens et al., 2004, 2005 [a, b]; 2006). Fossils from both units are typically recovered from laterally extensive sandstone bodies, with occasional lenticular mudstones (Figs 2B–C; see Roberts et al., 2004 for a discussion and historical overview of the RSG). Table 1 provides a faunal list for Units I and II.

Cretaceous Finds from the Rukwa Rift Basin

Cretaceous (Unit I) localities are characterized by specimens ranging from small, isolated elements such as teeth and jaws to large, articulated dinosaur skeletons (Fig. 3; O'Connor et al., 2006). The close of the 2005 field season witnessed the discovery of the first multi-taxon bonebeds; initial analyses suggest that these densely packed accumulations contain specimens representing multiple dinosaurian taxa. Notable finds from other localities include

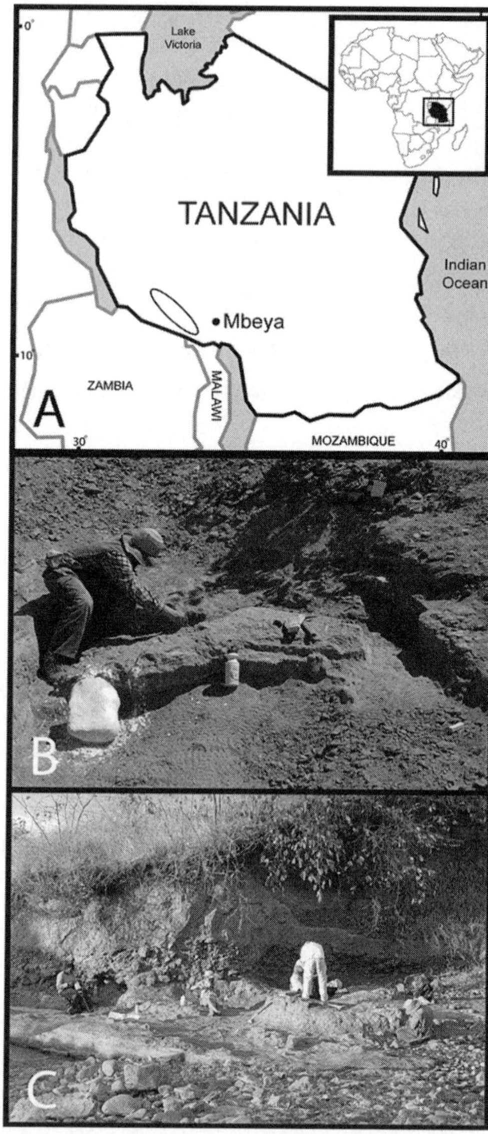

Fig. 2 Map of Tanzania (inset of Africa depicting position of Tanzania) illustrating general location and orientation of the Rukwa Rift Basin (oval outline) between Lake Nyasa (Malawi) to the southeast and Lake Tanganyika to the northwest (**A**), and representative photographs of Unit I—Cretaceous (**B**) and Unit II—Oligocene (**C**) exposures of the Red Sandstone Group (Note: informal Unit designation follows that outlined in Roberts et al., 2004; the definition of formal type sections is currently underway; Roberts et al., *In Prep*)

articulated theropod and sauropod dinosaurs (O'Connor et al., 2003, 2006), megaloolithid dinosaur eggshell (Gottfried et al., 2004), one of the most complete mammal specimens yet recovered from the Cretaceous of continental Africa (Krause et al., 2003a), as well as crocodyliform, testudine, ceratodontid lungfish, and osteoglossomorph teleost representatives (e.g., Gottfried et al., 2005).

Saurischian dinosaurs include a theropod (Fig. 3A) and at least two titanosaurian sauropod taxa distinguishable on the basis of dental morphology. Two

Table 1 Faunal list of vertebrate and invertebrate groups recovered from Unit I (Cretaceous) and Unit II (Oligocene) of the Red Sandstone Group in the Rukwa Rift Basin, southwestern Tanzania

Unit I:	Unit II
Mammalia	Mammalia
Gondwanatheria	Primates
Mammalia *incertae sedis*	Macroscelideans (x2)
	Phiomorph Rodents (x4)
Dinosauria	
Theropoda (x2)	Avialae
Sauropoda (x3)	
Crocodyliformes	Crocodyliformes
Testudines	Testudines
Osteichthyes	Osteichthyes
Sarcopterygii	Sarcopterygii
Ceratodontidae	Ceratodontidae
Actinopterygii	Actinopterygii
Osteoglossomorpha	Teleostei *incertae sedisi*
Mollusca	Mollusca
Gastropoda	Gastropoda
Bivalvia	Bivalvia
	Arthropoda
	Decapoda

tooth morphs are represented, one exhibiting the thin, cylindrical crowns with high-angle wear facets typical of many titanosaurians (Fig. 3) and a second characterized by a crown exhibiting moderate apical-flattening (i.e., a crown that is less convex lingually than labially) reminiscent of *Malawisaurus dixeyi* from the Aptian Dinosaur Beds of Malawi (Gomani, 2005). Multiple vertebral morphologies are also present, some preserving features (e.g., procoelous middle caudal vertebrae; Fig. 3C) consistent with a titanosaurian assignment (see O'Connor et al., 2006 for additional information on the RRBP dinosaurian fauna).

During the 2002 field season, the team also recovered a small (~2 cm) left dentary (TNM 02067; Fig. 3D) tentatively referred to sudamericid gondwanatherians, a poorly-known group of Cretaceous-Paleogene mammals restricted to Madagascar, India, South America, and Antarctica (Krause et al., 2003). Although moderately abraded, it is clear that the specimen possessed a single, laterally compressed, procumbent central incisor and at least four (likely five) columnar, extremely hypsodont, single-rooted cheek teeth (Fig. 3D). If the gondwanatherian assignment of TNM 02067 stands with the recovery of additional specimens, it represents the first occurrence of the group on the African continent. Moreover, it would extend the clade into the 'mid' Cretaceous (~Aptian-Cenomanian; see O'Connor et al., 2006 for a

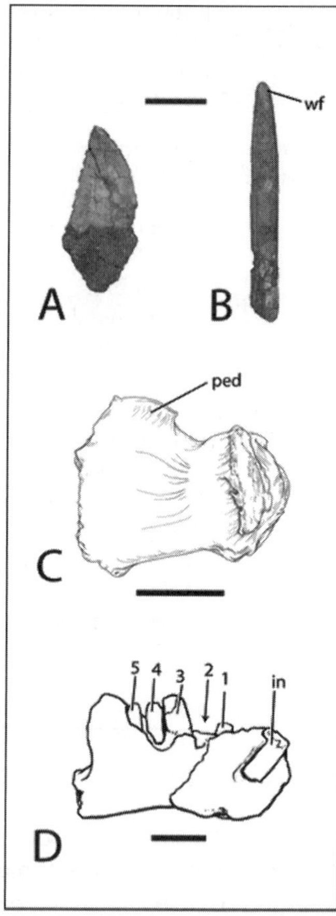

Fig. 3 Fossils recovered from Unit I (Cretaceous) of the Red Sandstone Group, including: (**A**) Nonavian theropod dinosaur tooth (TNM 02088); (**B**) sauropod dinosaur tooth (TNM 02093); (**C**) titanosaurian sauropod caudal vertebra (TNM 02072), left lateral view (modified from O'Connor et al., 2006); (**D**) left gondwanatherian(?) dentary (TNM 02067), lingual view (modified from Krause et al., 2003). Abbreviations as follows: in, incisor; ped, pedicle; wf, wear facet; 1–5, cheek teeth. Scale equals 1 cm in **A** and **B**, 5 cm in **C**, and 5 mm in **D**

discussion of the likely temporal affinities of Unit I taxa), a notable range extension for the group as all other gondwanatherians have been recovered from Campanian to Eocene deposits. With considerable debate surrounding the timing and location of the origin of 'modern' placental groups (e.g., Hedges et al., 1996; Stanhope et al., 1998; Foote et al., 1999a, b; Rich et al., 1999; Ji et al., 2002; Springer et al., 2003, 2004), recovery of mammalian fossils (e.g., TNM 2067) from Unit I may provide critical new data for evaluating and/or refining competing hypotheses.

Paleogene (Oligocene) Finds from the Rukwa Rift Basin

Oligocene (Unit II) localities are dominated by small (<4 cm) vertebrate and invertebrate remains (Fig. 4). These sites have produced articulated crustacean specimens, in addition to dozens of well-preserved gastropods, ranging in size

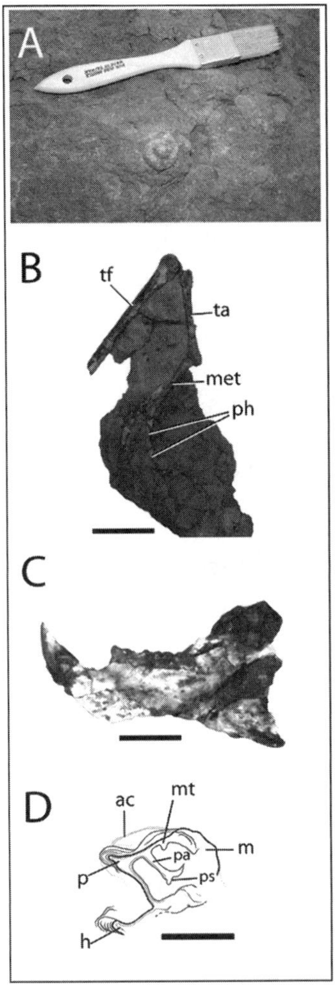

Fig. 4 Fossils recovered from Unit II (Oligocene) of the Red Sandstone Group, including (**A**) *in situ* gastropod (TNM 04008); (**B**) anuran hind limb (TNM 03014); (**C**) right phiomorph rodent mandible (TNM 04150), lingual view; (**D**) partial lower left molar of *Metaphiomys beadnelli* (TNM 03111), occlusal view (drawing modified from Stevens et al., 2006). Abbreviations as follows: ac, anterior cingulum; h, hypoconid; m, metaconid; met, metatarsus; mt, metalophid; p, protoconid; pa, posterior arm of protoconid; ph, phalanges; ps, protospur; ta, tarsoastragalus; tf, tibiofibula. Scale equals 1 cm in **B**, 5 mm in **C**, and 1 mm in **D**

between 1 mm and 3 cm (Fig. 4A). Numerous vertebrate gnathic elements and postcranial specimens have also been recovered. For example, Unit II localities have revealed an abundance of well-preserved anuran cranial and postcranial elements (Holman et al., 2004). Some of these materials are found still in articulation, with limbs found in postures ranging from flexed to extended (e.g., Fig. 4B). A number of isolated teeth representing small crocodyliforms have also been found.

Micromammals are well-represented in the fauna, including a number of phiomorph rodents previously restricted to deposits in North Africa and the Arabian Peninsula (Fig. 4C, Stevens et al., 2006). *Metaphiomys* preserves hystricognathous mandibular morphology and low-crowned lower molars exhibiting acute buccal cusp margins and a posterior arm of the protoconid that ends in a weak protospur (Fig. 4D). Upper molars of *Metaphiomys* similar to specimens recovered from Zalla, Libya (Fejfar, 1987) and the Fayum Depression of Egypt (Holroyd, 1994) have also been found. Multiple species of *Phiomys* are present, one with lower molar morphology featuring a small cuspulid on the anterior cingulum, an anteriorly-positioned metaconid and broad shallow basins that open lingually. A diminutive humerus (TNM 03100; Stevens et al., 2005b) bears strong resemblance to early anthropoid specimens from the Fayum (e.g., Seiffert et al., 2000). This specimen preserves an entepicondylar foramen and a dorso-epitrochlear fossa (Stevens et al., 2005b), features commonly observed in parapithecids and platyrrhines (Fleagle and Simons, 1982). Finally, Unit II localities have produced dental and postcranial elements representing multiple novel macroscelidean taxa, offering rare documentation of the diversification of the sengis, or elephant shrews (Stevens et al., 2005a).

To date, specimens recovered from the RRPB study area are limited to taxa characteristic of Afro-Arabia prior to the faunal exchange with Eurasia that began around 24 Ma with no evidence of the small-bodied immigrant taxa such as cricetodontids, sciurids, murids, or leporids that began to appear in the Miocene (e.g., Winkler, 1994). All Oligocene finds in the first two field seasons were restricted to a single locality, yet three additional highly fossiliferous Unit II sites discovered during the 2004 field season indicate promise for future discoveries.

Outcrops

Like much of sub-equatorial Africa, the Rukwa Rift Basin consists of a mixed vegetation-rock surface cover; hence outcrops of the RSG are unlike the badlands typically explored for vertebrate fossils in other parts of the world. For this reason, we have employed remote-sensing technology (computer-assisted satellite photographic analysis) to assist with the identification of outcrops throughout the region. To do so, a ratio operation is performed to emphasize contrast between outcrops and vegetated surfaces, and an image classification procedure is performed on the ratio data to produce a land cover map of the study area. Known outcrops and ground cover information recorded by GPS during previous field expeditions are overlaid on base maps, resulting in a refined plan for prospecting regions for new outcrops. A pilot study utilizing these methods conducted prior to the 2004 field season successfully identified new RSG exposures in the heavily vegetated southern end of the study area, clearly demonstrating that such approaches are useful not only to help refine search areas, but also to identify access points into more remote regions. We wholeheartedly agree with Miller et al. (2005), regarding the need for paleontological exploration of "non-traditional" deposits.

Potential for Future Discoveries

In short, four brief (2–4 week) field seasons in southwestern Tanzania have significantly expanded the Cretaceous and Oligocene vertebrate record in sub-equatorial Africa. This increased sampling is providing the requisite comparative data for evaluating regional, continental, and supercontinent-level biogeographic hypotheses related to the distribution of Gondwanan vertebrates during the Cretaceous and Paleogene (O'Connor et al., 2006; Stevens et al., 2006). A comparison of the RRBP's discoveries and '*person hours of collecting effort*' with estimates from projects such as the Fayum suggests that the developing RSG fauna still occupies a basal position on the discovery asymptote (Meehl, 1983), underscoring the role of the Rukwa Rift Basin for producing many more fossils in years to come.

Prospectus

Clearly, continental Africa still holds many secrets about long extinct biotas, with great promise for discovery of novel faunas and floras, as demonstrated by sustained efforts in locales such as the Fayum Depression. The Rukwa Rift Basin in southwestern Tanzania preserves a situation found in few other places in Africa, the presence of co-localized Cretaceous and Paleogene fossil-bearing sequences. In addition to expanding our knowledge of the relatively patchy African vertebrate fossil record, paleontological efforts can:

- Continue developing technological approaches (e.g., satellite-assisted photographic survey) to expand search areas into *less typical* environments (e.g., forested and bush environments with minimal sedimentary exposures).
- Serve a role in training not only US students, but also African geoscientists and paleobiologists to facilitate the development of natural resource maintenance infrastructure in host-countries. This is consistent with efforts in other fields (e.g., http://africaarray.psu.edu/) to support scientific infrastructure throughout Africa[1].
- Provide unique opportunities for student training, not only in aspects of paleontology, but ideally, in how to develop and maintain strong international research collaborations that will benefit the field for years to come.

Conducted in collaboration with faculty and students from the University of Dar es Salaam, the RRBP has resulted in the development of multiple research projects for both Tanzanian and US participants. During the field season we

---

[1] Editorial. (2005). Africa 2005. Nature 433:669.

liaise with a representative from the Antiquities Division of the Tanzanian National Museums, who accompanies the expedition and facilitates interactions with the local communities in order to explain the project goals and outcomes. In this regard, we have spent 1–2 days in each of the last two field seasons conducting workshops and lectures on biology, geology, paleontology and biogeography at local primary and secondary schools, reaching dozens of teachers and hundreds of students. US faculty are active in the African Studies programs at their respective universities, taking advantage of language courses and mentoring African students studying in the US. These activities maintain an interactive learning environment, not only with regard to fossil vertebrates, but also in promotion of international scientific exchange.

Members of the Rukwa Rift Basin Project participated in field work in the Fayum (NJS) and/or the Mahajanga Basin Projects (PMO, EMR, NJS, MDG), inspiring them to develop a project in East Africa. Realizing the importance of such involvement early in an academic career, the RRBP routinely includes US and Tanzanian graduate and undergraduate students as part of the field crew and in laboratory based research, in hopes that they too will begin projects of their own.

Acknowledgment The authors wish to thank F. Oakley, J. Oakley, and J. Fleagle for organizing this symposium, J. Fleagle for the invitation to participate in this volume, and most importantly, mentors E. Simons and D. Krause for sharing their passion for exploration and many practical insights of how to conduct field paleontology in different parts of Africa and Madagascar. Information on the Fayum project was kindly provided by P. Chatrath and F. Ankel-Simons. Research in Tanzania would not have been possible without input and assistance from the following individuals: Tanzanian colleagues at the University of Dar es Salaam (E. Mbede, A. Mruma), Division of Antiquities (D. Kamamba, R. Chami, C. Msuya), and COSTECH [Tanzanian Commission for Science and Technology] (H. Gideon); A. Njao, E. and K. Johansen, J. Garcia-Massini, T. Hieronymus, E. Rasmusson, V. Simons & Y. Tulu for assistance in the field; D. Krause, M. O'Leary, J. Groenke, V. Heisey and J.P. Cavigelli for fossil preparation; J. A. Holman, M. Carrano, M. Dawson, R. Feldmann, T. Fiorillo, J. Fleagle, G. Gunnell, T. Harrison, P. Holroyd, B. Jacobs, L. Jacobs, J. Salton, D. Krause, Z-X. Luo, P. Rose, E. Schrank, E. Seiffert, D. Unwin, and A. Winkler for invaluable discussions and P. Chatrath, S. Bell, B. Burger, S. McLaren, J. Wible, C. Beard, and H. Kafka for providing access to comparative materials in their care. NJS and PMO particularly thank S. Howard and A. Nikoi in the Ohio University African Studies Program and J. Rota in International Studies. Funding was provided by the following sources: National Geographic Society—CRE (2003, 2004, 2005), LSB Leakey Foundation (2004, 2005), Office of the Vice-President for Research and Graduate Studies at MSU, Paleobiological Fund, Jurassic Foundation, Ohio University Office of Research and Sponsored Programs, Ohio University Research Council, and the OU-COM Research and Scholarly Affairs Committee.

References

Andrews, C. W. (1901). Preliminary note on some recently discovered extinct vertebrates from Egypt. Part I. *Geol. Mag.* 4(8):400–409.

Andrews, C. W. (1902). Preliminary note on some recently discovered extinct vertebrates from Egypt. Part III. *Geol. Mag.* 4(9):291–295.

Ankel-Simons, F., Fleagle, J. G., and Chatrath, P. S. (1998). Femoral anatomy of *Aegyptopithecus zeuxis*, an early Oligocene anthropoid. *Am. J. Phys. Anthropol.* 106:413–424.
Archer, M., Flannery, T. F., Ritchie, A., and Molnar, R. E. (1985). First Mesozoic mammal from Australia – an early Cretaceous monotreme. *Nature* 318:363–366.
Archibald, J. D., and Fastovsky, D. E. (2004). Dinosaur extinction. In: Weishampel, D. B., Dodson, P., and Osmólska, H. (eds.), *The Dinosauria* (Second Edition). University of California Press, Berkeley, pp. 672–684.
Arnason, U., Adegoke, J. A., Bodin, K., Born, E. W., Esa, Y. B., Gullberg, A., Nilsson, M., Short, R. V., Xu, X., and Janke, A. (2002). Mammalian mitogenomic relationships and the root of the eutherian tree. *P. Natl. Acad. Sci. USA* 99: 8151–8156.
Asher, R. J., and Krause, D. W. (1998). The first pre-Holocene (Cretaceous) record of Anura from Madagascar. *J. Vert. Paleon.* 18:696–699.
Beadnell, H. J. L. (1902). A preliminary note on *Arsinoitherium zitteli* Beadnell from the upper Eocene strata of Egypt. *Egypt Surv. Dept., Pub. Works Min.* 1902:1–9.
Bonaparte, J. F., and Powell, J. W. (1980). A continental assemblage of tetrapods from the Upper Cretaceous beds of El Brete, northwestern Argentina (Sauropoda Coelurosauria-Carnosauria-Aves). *Mem. Soc. Geol. France* 139:19–28.
Bown, T. M. (1982). Ichnofossils and rhizoliths of the nearshore fluvial Jebel Qatrani Formation (Oligocene), Fayum Province, Egypt. *Palaeogeogr. Palaeoclimatol. Palaeoecol.* 40:255–309.
Bown, T. M., Kraus, M. J., Wing, S. L., Fleagle, J. G., Tiffany, B., Simons, E. L., and Vondra, C. F. (1982). The Fayum forest revisited. *J. Hum. Evol.* 11(7):603–632.
Buckley, G. A., and Brochu, C. A. (1999). An enigmatic new crocodile from the Upper Cretaceous of Madagascar. In: Unwin, D. (ed.), *Cretaceous Fossil Vertebrates: Special Papers in Palaeontology*, No. 60. The Palaeontological Association, London, pp. 149–175.
Buckley, G. A., Brochu, C. A., Krause, D. W., and Pol, D. (2000). A pug-nosed crocodyliform from the Late Cretaceous of Madagascar. *Nature* 405:941–944.
Carrano, M. T., Sampson, S. D., and Forster, C. A. (2002). The osteology of *Masiakasaurus knopfleri*, a small abelisauroid (Dinosauria: Theropoda) from the Late Cretaceous of Madagascar. *J. Vertebr. Paleontol.* 22:510–534.
Chiappe, L. M., Salgado, L., and Coria, R. A. (2001). Embryonic skulls of titanosaur sauropod dinosaurs. *Science* 293:2444–2446.
Clarke, J. A., Tambussi, C. P., Noriega, J. I., Erickson, G. M., and Ketcham, R. A. (2005). Definitive fossil evidence for the extant avian radiation in the Cretaceous. *Nature* 433:305–308.
Covert, H. H., and Hamrick, M. W. (1993). Description of new skeletal remains of the early Eocene anaptomorphine primate *Absarokius* (Omomyidae) and a discussion about its adaptive profile. *J. Hum. Evol.* 25:351–362.
Cracraft, J. (2001). Avian evolution, Gondwanan biogeography, and the Cretaceous-Tertiary mass extinction event. *P. Roy. Soc. Lond. B. Biol.* 268:459–469.
Curry Rogers, K. A., and Forster, C. A. (2001). The last of the dinosaur titans: A new sauropod from Madagascar. *Nature* 412:530–534.
Damblon, F., Gerrienne, P., D'Outrelepont, H., Delvaux, D., Beeckman, H., and Back, S. (1998). Identification of a fossil wood specimen in the Red Sandstone Group of southwestern Tanzania: stratigraphical and tectonic implications. *J. Afr. Earth Sci.* 26:387–396.
De Blieux, D. D., and Simons, E. L. (2002). Cranial and dental anatomy of *Antilohyrax pectidens*: A late Eocene hyracoid (Mammalia) from the Fayum, Egypt. *J. Vertebr. Paleontol.* 22(1):122–136.
De Klerk, W. J., Forster, C. A., Sampson, S. D., Chinsamy, A., and Ross, C. F. (2000). A new coelurosaurian dinosaur from the Early Cretaceous of South Africa. *J. Vertebr. Paleontol.* 20(2):324–332.
Depéret, C. (1896). Note sur les dinosauriens sauropodes et théropodes du Crétacé supérieur de Madagascar. *Bul. Soc. Géol. Fr.* 21:176–194.

de Queiroz, A. (2005). The resurrection of oceanic dispersal in historical biogeography. *Trends Ecol. Evol.* 20:68–73.

de Wit, M. J. (2003). Madagascar: Heads it's a continent, tails it's an island. *Ann. Rev. Earth Planet. Sci.* 31:213–248.

Dixey, F. (1928). The Dinosaur Beds of Lake Nyasa. *Trans. Geol. Soc. S. Afr.* 16:55–66.

Fejfar, O. (1987). Oligocene rodents from Zallah Oasis, Libya. *Munchner Geowiss. Abh. A* 10:265–268.

Fleagle, J. G. (1999). *Primate adaptation and evolution*. Academic Press, London.

Fleagle, J. G., and Bown, T. M. (1983). New primate fossils from late Oligocene (Colhuehuapian) localities of Chubut Province, Argentina. *Folia Primatol.* 41(3–4):240–266.

Fleagle, J. G., and Simons, E. L. (1982). The humerus of *Aegyptopithecus zeuxis*: A primitive anthropoid. *Am. J. Phys. Anthropol.* 59:175–193.

Fleagle, J. G., and Simons, E. L. (1983). The tibio-fibular articulation in *Apidium phiomense*, an Oligocene anthropoid. *Nature* 301:238–239.

Fleagle, J. G., and Simons, E. L. (1995). Limb skeleton and locomotor adaptations of *Apidium phiomense*, an Oligocene anthropoid from Egypt. *Am. J. Phys. Anthropol.* 97:235–289.

Fleagle, J. G., Kay, R. F., and Simons, E. L. (1980). Sexual dimorphism in early anthropoids. *Nature* 287:328–330.

Flynn, L. J., Brillanceau, A., Brunet, M., Coppens, Y., Dejax, J., Duperon-Laudoueneix, M., Ekodeck, G., Flanagan, K. M., Heintz, E., Hell, J., Jacobs, L. L., Pilbeam, D. R., Sen, S., and Djallo, S. (1987). Vertebrate fossils from Cameroon, West Africa. *J. Vertebr. Paleontol.* 7(4):469–471.

Flynn, J. J., Wyss, A. R., Croft, D. A., and Charrier, R. (2003). The Tinguiririca Fauna, Chile: Biochronology, paleoecology, biogeography, and a new earliest Oligocene South American Land Mammal Age. *Palaeogeogr. Palaeoclimatol. Palaeoecol.* 195(3/4):229–259.

Foote, M., Hunter, J. P., Janis, C. M., and Sepkoski, J. J., Jr. (1999a). Evolutionary and preservational constraints on origins of biologic groups: Divergence times of eutherian mammals. *Science* 283:1310–1314.

Foote, M., Hunter, J. P., Janis, C. M., and Sepkoski, J. J., Jr. (1999b). Divergence times of eutherian mammals. *Science* 285:2031a.

Forster, C. A. (1999). Gondwanan dinosaur evolution and biogeographic analysis. *J. Afr. Earth Sci.* 28:169–185.

Forster, C. A., Chiappe, L. M., Krause, D. W., and Sampson, S. D. (1996). The first Cretaceous bird from Madagascar. *Nature* 382:532–534.

Forster, C. A., Sampson, S. D., Chiappe, L. M., and Krause, D. W. (1998). The theropod ancestry of birds: New evidence from the Late Cretaceous of Madagascar. *Science* 279(5358):1915–1919.

Gaffney, E. S., and Forster, C. A. (2003). Side-necked turtle lower jaws (Podocnemididae, Bothremydidae) from the Late Cretaceous Maevarano Formation of Madagascar. *Am. Mus. Nov.* 3397:1–13.

Gingerich, P. D., Smith, B. H., and Simons, E. L. (1990). Hind limbs of Eocene *Basilosaurus*: evidence of feet in whales. *Science* 249:154–157.

Godinot, M. (1994). Early North African primates and their significance for the origin of Simiformes (=Anthropoidea). In: Fleagle, J. G. and Kay, R. F. (eds.), *Anthropoid Origins*. Plenum Press, New York, pp. 235–295.

Gomani, E. M. (2005). Sauropod dinosaurs from the Early Cretaceous of Malawi, Africa. *Palaeo. Electronica* 8:1–37.

Gottfried, M. D., Rabarison, J. A., and Randriamiaramanana, L. L. (2001). Late Cretaceouselasmobranchs from the Mahajanga Basin of Madagascar. *Cret. Res.* 22:491–496.

Gottfried, M. D., O'Connor, P. M., Jackson, F. D., Roberts, E. M., and Chami, R. (2004). Dinosaur eggshell from the Red Sandstone Group of Tanzania. *J. Vertebr. Paleontol.* 24(2):494–497.

Gottfried, M. D., O'Connor, P. M., Roberts, E. M., Stevens, N. J., and Chami, R. (2005). First Cretaceous lungfish from Tanzania: Ceratodontid from the Rukwa Rift Basin. *J. Vertebr. Paleontol.* 25(suppl., 3):65A.

Gunnell, G. F., Jacobs, B. F., Herendeen, P. S., Head, J. J., Kowalski, E., Msuya, C., Mizambwa, F. A., Harrison, T., Habersetzer, J., and Storch, G. (2002). The earliest placental mammal from sub-Saharan Africa: Eocene microbat from Tanzania—Evidence for early evolution of sophisticated echolocation. *Palaeo. Electronica* 5(3):1–10.

Hamrick, M. W., Meldrum, D. J., and Simons, E. L. (1995). Anthropoid phalanges from the Oligocene of Egypt. *J. Hum. Evol.* 28:121–145.

Harrison, T., Msuya, C. P., Murray, A. M., Jacobs, B. F., Baez, A. M., Mundil, R., and Ludwig, K. R. (2001). Paleontological investigations at the Eocene locality of Mahenge in north-central Tanzania, East Africa. In: Gunnell, G. F. (ed.), *Eocene Biodiversity: Unusual Occurrences and Rarely Sampled Habitats.* Kluwer Academic/Plenum, New York, pp. 39–74.

Hedges, S. B., Parker, P., Sibley, C., and Kumar, S. (1996). Continental breakup and the ordinal diversification of birds and mammals. *Nature* 381:226–229.

Heesy, C. P., Stevens, N. J., and Samonds, K. E. (2006). Biogeographic origins of primate higher taxa. In: Fleagle, J. G., and Lehman, S. (eds.), *Primate Biogeography.* Springer, New York, pp. 419–437.

Holman, J. A., Stevens, N. J., O'Connor, P. M., Gottfried, M. D., and Roberts, E. M. (2004). An enigmatic neobatrachian frog from the Paleogene of Southwestern Tanzania. *J. Vertebr. Paleontol.* 24(suppl., 3):71A.

Holroyd, P. A. (1994). An examination of dispersal origins for Fayum Mammalia. Ph.D. dissertation, Duke University, Durham, North Carolina, 328 pp.

Holroyd, P. A. (1999). New Pterodontinae (Hyaenodontidae: Creodonta) from the late Eocene - early Oligocene Jebel Qatrani Formation, Fayum, Province, Egypt. *PaleoBios* 19: 1–18.

Holroyd, P. A., and Parham, J. F. (2003). The antiquity of African tortoises. *J. Vertebr. Paleontol.* 23(3):688–690.

Holroyd, P. A., Simons, E. L., Bown, T. M., Polly, P. D., and Kraus, M. J. (1996). New records of terrestrial mammals from the Upper Eocene Qasr el Sagha Formation, Fayum Depression, Egypt. *Palaeovertebrata* 25(2–4):175–192.

Humphries, C. J., and Ebach, M. C. (2004). Biogeography on a dynamic earth. In: Lomolino, M. V., and Heaney, L. R. (eds.), *Frontiers of Biogeography.* Sinauer Associates, Sunderland, pp. 67–86.

Jacobs, L. L., Kaufulu, Z. M., and Downs, W. R. (1990). The Dinosaur Beds of northern Malawi, Africa. *Nat. Geo. Res.* 6:196–204.

Jacobs, L. L., Winkler, D. A., Downs, W. R., and Gomani, E. M. (1993). New material of an Early Cretaceous titanosaurid sauropod dinosaur from Malawi. *Palaeontology* 36(3):523–534.

Ji, Q., Luo, Z.-X., Yuan, C.-X., Wible, J. R., Zhang, J.-P., and Georgi, J. A. (2002). The earliest known eutherian mammal. *Nature* 416:816–822.

Kappelman, J., Rasmussen, D. T., Sanders, W. J., Feseha, M., Bown, T., Copeland, P., Crabaugh, J., Fleagle, J., Glantz, M., Gordon, A., Jacobs, B., Maga, M., Muldoon, K., Pan, A., Pyne, L., Richmond, B., Ryan, T., Seiffert, E., Sen, S., Todd, L., Wiemann, M., and Winkler, A. (2003). Oligocene mammals from Ethiopia and faunal exchange between Afro-Arabia and Eurasia. *Nature* 426:549–552.

Kay, R. F., Williams, B. A., and Anaya, F. (2001). The adaptations of *Branisella boliviana*, the earliest South American Monkey. In: Plavcan, J. M., Kay, R. F., Jungers, W. L., and van Schaik, C. (eds.), *Reconstructing Behavior in the Primate Fossil Record.* Plenum Publishing Corporation, New York, pp 339–370.

Kielan-Jaworowska, Z., R. L. Cifelli, and Z. -X. Luo. (2004). *Mammals from the age of dinosaurs: origins, evolution, and structure.* Columbia University Press, New York.

Kilembe, E. A., and Rosendahl, B. R. (1992). Structure and stratigraphy of the Rukwa rift. *Tectonophysics* 209:143–158.

Kirk, E. C., and Simons, E. L. (2001). Diets of fossil primates from the Fayum Depression of Egypt: a quantitative analysis of molar shearing. *J. Hum. Evol.* 40:203–229.
Krause, D. W. (2001). Fossil molar from a Madagascan marsupial. *Nature* 412:497–498.
Krause, D. W. (2003a). Discovery of a relatively complete mammalian specimen from the Late Cretaceous of Madagascar. *J. Vertebr. Paleontol.* 23(suppl., 3):69A.
Krause, D. W. (2003b). Late Cretaceous vertebrates from Madagascar: A window into Gondwanan biogeography at the end of the Age of Dinosaurs. In: Goodman, S. M., and Benstead, J. P. (eds.), *The Natural History of Madagascar*. University of Chicago Press, Chicago, pp. 40–47.
Krause, D. W., Prasad, G. V. R., von Koenigswald, W., Sahni, A., and Grine, F. E. (1997). Cosmopolitanism among Late Cretaceous Gondwanan mammals. *Nature* 390:504–507.
Krause, D. W., Rogers, R. R., Forster, C. A., Hartman, J. H., Buckley, G. A., and Sampson, S. D. (1999). The Late Cretaceous vertebrate fauna of Madagascar: Implications for Gondwanan paleobiogeography. *GSA Today* 9:1–7.
Krause, D. W., Evans, S. E., and Gao, K. -Q. (2003a). First definitive record of Mesozoic lizards from Madagascar. *J. Vertebr. Paleontol.* 23:842–856.
Krause, D. W., Gottfried, M. D., O'Connor, P. M., and Roberts, E. M. (2003b). A Cretaceous mammal from Tanzania. *Acta Palaeont. Pol.* 48(3):321–330.
Krause, D. W., O'Connor, P. M., Curry Rogers, K., Sampson, S. D., Buckley, G. A., and Rogers, R. R. (2006). Late Cretaceous terrestrial vertebrates from Madagascar: implications for Latin American Biogeography. *Annals of the Missouri Botanical Garden* 93:178–208.
Lapparent, A. F. (1960). De les dinosariens du 'Contienental intercalaire' du Sahara central. *Mem. Soc. Geol. France* 88A:1–57.
Lavocat, R. J. M. (1954). Sur les dinosauriens du continental intercalaire des Kem-Kem de la Doura. *C.R. 19th Cong. Géol. Intern.* 15:65–68.
Leakey, M. G., Ungar, P. S., and Walker, A. (1995). A new genus of large primate from the Late Oligocene of Lothidok, Turkana District, Kenya. *J. Hum. Evol.* 28:519–531.
Lieberman, B. S. (2000). *Paleobiogeography: Using fossils to study global change, plate tectonics, and evolution.* Kluwer Academic, New York.
Madsen, O., Scally, M., Douady, C. J., Kao, D. J., DeBry, R. W., Adkins, R., Amrine, H. M., Stanhope, M. J., de Jong, W. W., and Springer, M. A. (2001). Parallel adaptive radiations in two major clades of placental mammals. *Nature* 409:610–614.
Martin, R. D. (1993). Primate origins, plugging the gaps. *Nature* 363:223–234.
Martin, R. D. (2000). Origins, diversity and relationships of lemurs. *Int. J. Primatol.* 21(6):1021–1049.
Masters, J. C., Stankiewicz, J., Thiart, C., and de Wit, M. J. (2006). How good is Simpson's "sweepstakes" model in explaining the mammalian colonization of Madagascar? *Am. J. Phys. Anthropol. Suppl.*129(S42):127.
Matsumoto, H. (1921). *Megalohyrax* Andrews and *Titanohyrax* g.n.—A revision of the genera of hyracoids from the Fayum, Egypt. *P. Zool. Soc. Lond.* 1921:839–850.
Meehl, P. E. (1983). Consistency tests in estimating the completeness of the fossil record: A neo-Popperian approach to statistical paleontology. *Minn. Studies Phil. Sci.* 40:135–139.
Miller, E. R. (1999). Faunal correlation of Wadi Moghara, Egypt: implications for the age of *Prohylobates tandyi*. *J. Hum. Evol.* 36:519–533.
Miller, E. R., Gunnell, G. F., and Martin, R. D. (2005). Deep time and the search for anthropoid origins. *Yrbk. Phys. Anthropol.* 48:60–95.
Molnar, R. E. (1980). Australian late Mesozoic terrestrial tetrapods: some implications. *Mem. Soc. Geol. France* 139:131–143.
Morley, C. K., Cunningham, S. M., Harper, R. M., and Westcott, W. A. (1999). Geology and geophysics of the Rukwa Rift, East Africa. *Tectonics* 11:69–81.
Murphy, W. J., Eizirik, E., O'Brien, S. J., Madsen, O., Scally, M., Douady, C. J., Teeling, E., Ryder, O. A., Stanhope, M. J., de Jong, W. W., and Springer, M. S. (2001). Resolution

of the early placental mammal radiation using Bayesian phylogenetics. *Science* 294:2348–2351.

Murray, A. M., and Attia, Y. S. (2004). A new species of *Lates* (Teleostei: Perciformes) from the Lower Oligocene of Egypt. *J. Vertebr. Paleontol.* 24(2):299–308.

O'Connor, P. M. (2007). The postcranial axial skeleton of *Majungatholus atopus* (Theropoda: Abelisauridae) from the Late Cretaceous of Madagascar. *J. Vert. Paleo.* 27:127–162.

O'Connor, P. M., and Claessens, L. P. A. M. (2005). Basic avian pulmonary design and flow through ventilation in non-avian theropod dinosaurs. *Nature* 436:253–256.

O'Connor, P. M., Gottfried, M. D., Roberts, E. M., Stevens, N. J., Jackson, F. D., and Rasmusson, E. L. (2003). Closing the 'African Gap' – a new Cretaceous vertebrate fauna from Tanzania. *J. Vertebr. Paleontol.* 23(suppl., 3):84A.

O'Connor, P. M., Gottfried, M. D., Stevens, N. J., Roberts, E. M., Ngasala, S., Kapilima, S., and Chami, R. (2006). A new vertebrate fauna from the Cretaceous Red Sandstone Group, Rukwa Rift Basin, southwestern Tanzania. *J. Afr. Earth Sci.* 44:277–288.

Osborn, H. F. (1908). New fossil mammals from the Fayum Oligocene, Egypt, *B. Am. Mus. Nat. Hist.* 24:265–272.

Paul, C. R. C. (1998). Adequacy, completeness and the fossil record. In: Donovan, S. K., and Paul, C. R. C. (eds.), *The adequacy of the fossil record.* John Wiley and Sons, New York, pp 1–22.

Pickford, M. (1986). Première découverte d'une faune mammalienne terrestre paléogène d'Afrique sub-saharienne. *C. R. Acad. Sci. II* 302(19):1205–1210.

Poux, C., Madsen, O, Marquard, E., Vieites, D. R., de Jong, W. W., and Vences, M. (2005). Asynchronous colonization of Madagascar by the four endemic clades of primates, tenrecs, carnivores, and rodents. *Syst. Biol.* 54(5):719–730.

Prasad, G. V. R. and M. Godinot. (1994). Eutherian tarsal bones from the Late Cretaceous of India. *J. Palaeontol.* 68:892–902.

Prasad, G. V. R., and A. Sahni. (1988). First Cretaceous mammal from India. *Nature* 332: 638–640.

Prasad, V., C. A. E. Stromberg, H. Alimohammadian, and A. Sahni. (2005). Dinosaur coprolites and the early evolution of grasses and grazers. *Science* 310:1177–1180.

Quenell, A. M., Mcinley, A. C. M., and Aitken, W. G. (1956). Summary of the geology of Tanganyika: *Geol. Surv. Tanganyika Mem.* 1956:1–264.

Rana, R. S., Singh, H., Sahni, A., Rose, K. D., and Saraswati, P. K. (2005). Early Eocene chiropterans from a new mammalian assemblage (Vastan Lignite Mine, Gujarat, Western Peninsular Margin): oldest known bats from Asia. *J. Palaeontol. Soc. India* 50(1):93–100.

Rasmussen, D. T. and Simons, E. L. (2000). Ecomorphological diversity among Paleogene hyracoids (Mammalia): a new cursorial browser from the Fayum, Egypt. *J. Vertebr. Paleontol.* 20(1):167–176.

Rasmussen, D. T., Conroy, G. C., and Simons, E. L. (1998). Tarsier-like locomotor specializations in the Oligocene primate *Afrotarsius*. *Proc. Nat. Acad. Sci. USA* 95:14848–14850.

Rasmussen, D. T., Simons, E. L., Hertel, F., and Judd, A. (2001). Hindlimb of a giant terrestrial bird from the upper Eocene, Fayum, Egypt. *Palaeontology* 44:325–337.

Rasmussen, D. T., Tshakreen, S. O., Abugares, M., and Smith, J. B. (2008). Return to Dor al-Talha: Paleontological Reconaissance of the Early Tertiary of Libya. In: Fleagle, J. G. and Gilbert, C. C. (eds.), Elywn Simons: *A Search for Origins*. Springer, New York, pp 181–196.

Rich, T. H., Molnar, R. E., and Vickers-Rich, P. (1983). Fossil vertebrates from the Late Jurassic or Early Cretaceous Kirkwood Formation, Algoa Basin, Southern Africa. *Trans. Geol. Soc. S. Afr.* 86:281–291.

Rich, T. H., Vickers-Rich, P., and Flannery, T. F. (1999). Divergence times of eutherian mammals. *Science* 285:2031A.

Roberts, E. M., O'Connor, P. M., Gottfried, M. D., Stevens, N. J., Kapilima, S., and Ngasala, S. (2004). Revised stratigraphy and age of the Red Sandstone Group in the Rukwa Rift Basin, Tanzania. *Cret. Res.* 25:749–759.

Rogers, R. R. (2005). Fine-grained debris flows and extraordinary vertebrate burials in the Late Cretaceous of Madagascar. *Geology* 33:297–300.

Rogers, R. R., Hartman, J. H., and Krause, D. W. (2000). Stratigraphic analysis of Upper Cretaceous rocks in the Mahajanga Basin, northwestern Madagascar: Implications for ancient and modern faunas. *J. Geol.* 108:275–301.

Rogers, R. R., Krause, D. W., and Curry Rogers, K. (2003). Cannibalism in the Madagascan dinosaur *Majungatholus atopus*. *Nature* 422:515–518.

Rose, K. D., and Archibald, J. D. (2004). *The rise of placental mammals*. The Johns Hopkins, University Press, Baltimore.

Sampson, S. D., Witmer, L. M., Forster, C. A., Krause, D. W., O'Connor, P. M., Dodson, P., and Ravoavy, F. (1998). Predatory dinosaur remains from Madagascar: Implications for the Cretaceous biogeography of Gondwana. *Science* 280:1048–1051.

Schrago, C. G., and Russo, C. A. M. (2003). Timing the origin of New World monkeys. *Mol. Biol. Evol.* 20(10):1620–1625.

Scotese, C. R. (2001). *Atlas of Earth History*. PALEOMAP project, Arlington, Texas.

Seiffert, E. R., and Simons, E. L. (2000). *Widanelfarasia*, a diminutive placental from the late Eocene of Egypt. *P. Natl. Acad. Sci. USA* 97(6):2646–2651.

Seiffert, E. R., Simons, E. L., and Attia, Y. (2003). Fossil evidence for an ancient divergence of lorises and galagos. *Nature* 422:421–424.

Seiffert, E. R., Simons, E. L., and Fleagle, J. G. (2000). Anthropoid humeri from the late Eocene of Egypt. *P. Natl. Acad. Sci. USA* 97:10062–10067.

Seiffert, E. R., Simons, E. L., Ryan, T. M., and Attia, Y. (2005). Additional remains of *Wadilemur elegans*, a primitive stem galagid from the late Eocene of Egypt. *P. Natl. Acad. Sci. USA* 102:11396–11401.

Sereno, P. C. (1999). The evolution of dinosaurs. *Science* 284:2137–2147.

Sereno, P. C., Dutheil, D. B., Larochene, M., Larsson, H. C. E., Lyon, G. H., Magwene, P. M., Sidor, C. A., Varricchio, D. J., and Wilson, J. A. (1996). Predatory dinosaurs from the Sahara and Late Cretaceous faunal differentiation. *Science* 272: 986–991.

Sereno, P. C., Wilson, J. A., and Conrad, J. L. (2004). New dinosaurs link southern landmasses in the mid-Cretaceous. *P. Roy. Soc. Lond. B. Biol.* 271:1325–1330.

Sigogneau-Russell, D., Evans, S. E., Levine, J. F., and Russell, D. A. (1998). The Early Cretaceous microvertebrate locality of Anoual, Morocco: a glimpse at the small vertebrate assemblage of Africa. In: Lucas, S. J., Kirkland, J. I., and Estep, J. W. (eds.), Lower and Middle Cretaceous Terrestrial Ecosystems. *Bull. New Mexico Mus. Nat. Hist.*, Bulletin 14:pp. 177–182.

Simons, E. L. (1997). Discovery of the smallest Fayum Egyptian primates (Anchomomyini, Adapidae). *P. Natl. Acad. Sci. USA* 94:180–184.

Simons, E. L. (1998). The prosimian fauna of the Fayum Eocene/Oligocene deposits of Egypt. *Folia Primatol.* 69:286–294.

Simons, E. L., and Bown, T. M. (1984). A new species of *Peratherium* (Didelphidae; Polyprotodonta): the first African marsupial. *J. Mammal.* 65(4):539–548.

Simons, E. L., and Bown, T. M. (1995). Ptolemaiida, a new order of Mammalia—with description of the cranium of *Ptolemaia grangeri*. *P. Natl. Acad. Sci. USA* 92:3269–3273.

Simons, E. L., and Rasmussen, D. T. (1995). A whole new world of ancestors: Eocene anthropoideans from Africa. *Evol. Anthropol.* 3:128–139.

Simons, E. L., Holroyd, P. A., and Bown, T. M. (1991). Early tertiary elephant-shrews from Egypt and the origin of the Macroscelidea. *P. Natl. Acad. Sci. USA* 88:9734–9737.

Simons, E. L., Rasmussen, D. T., Bown, T. M., and Chatrath, P. S. (1994). The Eocene origin of anthropoid primates: adaptation, evolution, and diversity. In: Fleagle, J. G., and Kay, R. F., (eds.), *Anthropoid Origins*. Plenum Press: New York, pp. 179–201.

Smith, A. B., and Peterson, K. J. (2002). Dating the time of origin of major clades: molecular-clocks and the fossil record. *Annu. Rev. Earth. Pl. Sc.* 30:65–88.

Smith, A. G., Smith, D. G., and Funnell, B. M. (1994). *Atlas of Mesozoic and Cenozoic Coastlines*. Cambridge University Press, Cambridge, England.

Smith, J. B., Lamanna, M. C., Lacovara, K. J., Dodson, P. D., Smith, J. R., Poole, J. C., Giegengack, R., and Attia, Y. (2001). A giant sauropod dinosaur from an Upper Cretaceous mangrove deposit in Egypt. *Science* 292:1704–1706.

Springer, M. S. (1997). Molecular clocks and the timing of the placental and marsupial radiations in relation to the Cretaceous-Tertiary boundary. *J. Mammal. Evol.* 4: 285–302.

Springer, M. S., Cleven, G. C., Madsen, O., de Jong, W. W., Waddell, V. G., Amrine, H. M., and Stanhope, M. J. (1997). Endemic African mammals shake the phylogenetic tree. *Nature* 388:61–64.

Springer, M. S., Murphy, W. J., Eizirik, E., and O'Brien, S. J. (2003). Placental mammal diversification and the Cretceous-Tertiary boundary. *P. Natl. Acad. Sci. USA* 100(3):1056–1061.

Springer, M. S., Stanhope, M. J., Madsen, O., and De Jong, W. W. (2004). Molecules consolidate the placental mammal tree. *Trends Ecol. Evol.* 19(8):430–438.

Springer, M. S., Murphy, W. J., Eizirik, E., and O'Brien, S. J. (2005). Molecular evidence for major placental clades. In: Rose, K. D. and Archibald, J. D.(eds.), *The rise of placental mammals*. Johns Hopkins University Press, Baltimore, pp. 37–49.

Stanhope, M. J., Waddell, V. G., Madsen, O., de Jong, W. W., Hedges, S. B., Cleven, G. C., Kao, D., and Springer, M. S. (1998). Molecular evidence for multiple origins of Insectivora and for a new order of endemic African insectivore mammals. *P. Natl. Acad. Sci. USA* 95:9967–9972.

Stevens, N. J., and Heesy, C. P. (2004). Reliability of biogeographic reconstructions for determining the area of origin of primate higher taxa. *Intl. Primate Cong.* (suppl., 1):118.

Stevens, N. J., and Heesy, C. P. (2006). Malagasy primate origins: Fossils, phylogenies and biogeographic reconstructions. *Folia Primatol* 77:419–433.

Stevens, N. J., Gottfried, M. D., O'Connor, P. M., Roberts, E. M., and Kapilima, S. D. (2004). A new Paleogene fauna from the East African Rift, southwestern Tanzania. *J. Vertebr. Paleontol.* 24(suppl., 3):118A.

Stevens, N. J., O'Connor, P. M., Gottfried, M. D., Roberts, E. M., Ngasala, S., and Kapilima, S. (2005a). New Paleogene mammals and other vertebrates from the Rukwa Rift Basin, southwestern Tanzania. *J. Vertebr. Paleontol.* 25(suppl., 3):118A.

Stevens, N. J., O'Connor, P. M., Gottfried, M. D., Roberts E. M., and Ngasala, S. (2005b). An anthropoid primate humerus from the Rukwa Rift Basin, Paleogene of southwestern Tanzania. *J. Vertebr. Paleontol.* 25(4):986–989.

Stevens, N. J., O'Connor, P. M., Gottfried, M. D., Roberts, E. M., Ngasala, S., and Dawson, M. (2006). *Metaphiomys* (Rodentia: Phiomyidae) from the Paleogene of southwestern Tanzania. *J. Paleontol.* 80(2):407–410.

Tavare, S., Marshall, C. R., Will, O., Soligo, C., and Martin, R. D. (2002). Using the fossil record to estimate the age of the last common ancestor of extant primates. *Nature* 416:726–729.

Thewissen, J. G. M., and Simons, E. L. (2001). Skull of *Megalohyrax eocaenus* (Hyracoidea, Mammalia) from the Oligocene of Egypt. *J. Vertebr. Paleontol.* 21(1):98–106.

Upchurch, P., Hunn, C. A., and Norman, D. B. (2002). An analysis of dinosaurian biogeography: Evidence for the existence of vicariance and dispersal patterns caused by geological events. *Proc. Roy. Soc. Lond., Ser. B* 269:613–621.

van Dijk, M. A. M., Madsen, O., Catzefils, F., Stanhope, M. J., de Jong, W. W., and Pagel, M. (2001). Protein sequence signatures support the African clade of mammals. *Proc. Nat. Acad. Sci. USA* 98(1): 188–193.

Vences, M., Freyhof, J., Sonnenberg, R., Kosuch, J., and Veith, M. (2001). Reconciling fossils and molecules: Cenozoic divergence of cichlid fishes and the biogeography of Madagascar. *J. Biogeogr.* 28:1091–1099.

Wescott, W. A., Krebs, W. N., Engelhardt, D. W., and Cunningham, S. M. (1991). New biostratigraphic age dates from the Lake Rukwa rift basin in western Tanzania. *AAPG Bull.* 75(7):1255–1263.

Wheeler, W. H., and Karson, J. A. (1994). Extension and subsidence along a "weak" continental transform: An example from the Rukwa Rift, East Africa. *Geology* 22:625–628.

Wing, S. L., Hickey, L. J., and Swisher, C. C. (1993). Implications of an exceptional fossil flora for Late Cretaceous vegetation. *Nature* 363, 342–344.

Winkler, A. J. (1994). The middle/upper Miocene dispersal of major rodent groups between southern Asia and Africa. In: Tomida, Y., Li, C. K., Setoguchi, T. (eds.), *Rodent and Lagomorph Families of Asian Origins and Diversification*, No. 8. National Science Museum Monographs, Tokyo, pp. 173–184.

Wood, A. E. (1968). The African Oligocene Rodentia. *Bull. PMNH* 28:23–105.

Yoder, A. D., Burns, M. M., Zehr, S., Delfosse, T., Veron, G., Goodman, S. M., and Flynn, J. J. (2003). Single origin of Malagasy Carnivora from an African ancestor. *Nature* 421: 734–737.

Return to Dor al-Talha

Paleontological Reconnaissance of the Early Tertiary of Libya

D. Tab Rasmussen, Sefau O. Tshakreen, Miloud M. Abugares and Joshua B. Smith

Introduction

No individual has contributed more to the growth of knowledge about mammalian evolution in the Early Tertiary of Africa than Elwyn Simons. For 45 years he has steadfastly worked the Fayum sites, and the efforts of his field teams have yielded an incredible richness of paleontological treasures. Referring to the Egyptian sites by the monolithic term "The Fayum" does not do justice to the diversity of fossil localities contained within the research area. Simons and his colleagues have developed quarries at multiple stratigraphic levels within the early Oligocene, the late Eocene, and in recent years, the middle Eocene (Simons, 1968; El-Kashab et al., 1983; Simons and Rasmussen, 1995; Seiffert et al., 2003, 2005, 2008; Seiffert, 2006). These localities include several famous vertebrate-bearing sites, but also ones that yield fossil terrestrial plants, ichnofossils, and invertebrates (Wing and Tiffney, 1982; Bown et al., 1982; Bown and Kraus, 1988). His forays into neighboring marine Eocene and early Miocene sites have fostered research in those areas as well (e.g., Gingerich, 1992, this volume; Miller, 1999). In sum, the Fayum project represents the most important natural laboratory on the African Early Tertiary in terms of temporal span covered, taxonomic diversity represented, completeness of skeletal remains, and the richness of accompanying paleobiological and paleoecological research.

The research led by Simons also has been important in illuminating paleontological work at other, non-Egyptian Early Tertiary sites of the Arabian-African continent. In the early 20th Century, the Fayum provided the first glimpse of the archaic, endemic mammal faunas of Africa (Andrews, 1906; Osborn, 1908; Schlosser, 1911; El-Kashab et al., 1983). Now, many decades later, similar Early Tertiary mammals have been found at a handful of sites in Mali, Algeria, Tunisia, and Oman (e.g., Mahboubi et al., 1986; Rasmussen et al., 1992; Thomas et al., 1999; Gheerbrant et al., 2002). The Fayum's magneto- and biostratigraphy has proven to be a central axis by which to compare other African sites

D. Tab Rasmussen
Department of Anthropology, Washington University, St. Louis, MO 63130, USA
dtrasmus@wustl.edu

J. G. Fleagle, C. C. Gilbert (eds.), *Elwyn Simons: A Search for Origins.*

(Kappelman et al., 1992; Seiffert, 2006). The Simons influence on the Early Tertiary is also manifested by the enlarging ripples of research activity by his students and colleagues at the ever-growing number of other localities found on the continent. Several veterans of the Fayum field crews have gone on to make notable contributions to the early Tertiary of African nations other than Egypt (e.g., Court and Hartenberger, 1992; Godinot, 1994; Kappelman et al., 2003; Stevens et al., 2005, 2006; 2008).

While Simons himself has not strayed from Egypt (if one considers only the Early Tertiary of Africa), at one point he had planned to do so. He organized an expedition to Libya in the late 1960's when the smooth continuation of his Egypt project was disrupted by war in the region. While Simons' Libya project never came to fruition, stories of the aborted effort told around the Fayum field camp in the 1980's intrigued graduate students at the time. This paper describes a new field project in Libya that represents the continuation, after a 36 year hiatus, of paleontological exploration at one of the most important early Tertiary sites of Libya: a long escarpment deep in the Sahara called Dor al-Talha.

History

While Simons was initiating his museum work on Fayum mammals in the 1950s and long before he took to the field in the early 1960s, French geologists exploring for petroleum resources in the deserts of Libya reported the occurrence of Tertiary vertebrates at a few locations (Bellair et al., 1954). The French paleontologist Camille Arambourg provided a preliminary description of these geological outcrops and the fossil mammals they yielded (Arambourg and Magnier, 1961; Arambourg, 1963). Most importantly, Arambourg showed that a few archaic proboscideans and hyracoids recovered from Libya were nearly identical to ones from the Fayum that had been discovered in the early part of the century (Andrews, 1906; Osborn, 1908; Schlosser, 1911). These mammals could be found at two Libyan localities. One of these was near the oasis of Zella, the potential of which was limited, Arambourg complained, by a lack of appropriate exposure (Arambourg and Magnier, 1961; Savage, 1971; Fejfar, 1987). The other locality was a much more extensive escarpment in the desert to the south called Dor al-Talha, "home of the acacias".*

* The name Dor al-Talha was first applied to the escarpment in a geological context by Bellair et al. (1954). For some reason, Arambourg coined his own term, "Gebel Coquin" for the same escarpment. Savage (1971) rejected that term as "singularly inappropriate" in favor of the original Dor al-Talha. Savage's student A. W. R. Wight used the spelling Dur at-Talha. Savage reports that the term means "long escarpment" but, in Arabic, it actually means "home or place of the acacia trees." There are no acacia trees evident there now, and it is unknown whether the name refers to real acacia stands now gone, or was an allusion to the numerous fossil tree trunks evident in the area.

Shortly thereafter, the British paleontologist Robert J. G. Savage of the University of Bristol initiated a series of field projects in Libya. Savage and his students packed their equipment into vehicles and took a drive-and-ferry marathon from Great Britain through France, Spain, Morocco, Algeria, Tunisia, Tripoli, and finally out to eastern Libya. Savage's primary focus was on the rich Miocene deposits of Jebel Zelten (Arambourg, 1961; Savage, 1971; Gaziry, 1987). But Savage and his student A. W. R. Wight also spent several weeks in 1968 and 1969 working from a camp set up at the escarpment of Dor al-Talha to characterize the geology of the scarp and to collect fossil mammals (Savage, 1969, 1971; Wight, 1980). In the few publications resulting from these trips, Savage and Wight presented taxonomic lists of fossils recovered, and a very good outline of the local stratigraphy. No supporting photos of the fossils or of the geological outcrops were published at the time. Years later, two of the important mammal specimens from Dor al-Talha were illustrated as part of subsequent museum research projects (Tassy, 1981; Court, 1995).

In Egypt, where Simons had established his Fayum field effort in 1961, the Six-Day War of the summer of 1967 led to restrictions being placed on desert travel in the country by foreigners and Egyptians alike. Simons tried to run an Egyptian field project in 1968, but when that faltered, he looked to Libya as a potential new research area. He dispatched a small crew to Libya in the late summer of 1969 that consisted of Prithijit Chatrath, Thomas Bown, and Tony Gaston. The trio arrived in Libya after a drive from India to Italy and a ferry crossing from there to Tripoli. They were staying in a Tripoli hotel and preparing to take to the field when the Libyan Revolution occurred on September 1st, 1969, which led to the abandonment of the project (T. M. Bown, P. S. Chatrath, personal communication). The Savage expeditions of 1968 and 1969 thus constituted the end of major research activity on Early Tertiary vertebrates of Libya for many years.

The most important new addition to the study of Libyan early Tertiary mammals since that time has been the description of beautifully articulated Oligocene hyracoid skeletons found at Jebel al-Hasawnah by a private collector (Thomas et al., 2004). This site is far to the west of Dor al-Talha and the Sirt Basin in an area where the Oligocene beds are associated with older exposures of Early Tertiary and Cretaceous rocks.

In the meantime, the geological sciences in Libya developed and grew, with an obvious focus being the country's rich petroleum resources and their associated sedimentary contexts (e.g., Wennekers et al., 1996; Ahlbrandt, 2001). Knowledge of Early Tertiary sediments, their distribution, precise age and sedimentological characteristics have been well elucidated for a number of areas, primarily in the Sirt Basin (Selley, 1969, 1997; Gumati and Kanes, 1985; Salem et al., 1996). Ongoing basic research has led to great gains in understanding Libyan stratigraphy, micropaleontology, and other areas related to the Early Tertiary history of the country.

Following political changes that developed in 2004, the four authors of this paper began collaborating on plans to launch new work on vertebrate fossils in the Early Tertiary of Libya. The four included two American vertebrate paleontologists and two Libyan geologists who shared interests in the geological history of Libya and Africa. A trip was planned that consisted of two parts: (1) a look at some Cretaceous exposures from which dinosaurs had been reported (Smith et al., in press), and (2) a look at the early Tertiary sites worked by Arambourg and Savage. This preliminary reconnaissance trip under the auspices of the Petroleum Research Centre (PRC) of Tripoli was completed Aug. 7th–28th, 2005, and the preliminary results of this trip are presented here. Funding was by the PRC and by Washington University, St. Louis.

The Early Tertiary of Libya

Sedimentary deposits of early Tertiary age are exposed primarily in the northeastern part of the country. Arambourg (1963) had included the Libyan sites and those of the Fayum together under a term "The Nilotic-Saharian Zone", reflecting that they accumulated under in a common geological context. The northern margin of Africa during the early Tertiary was tectonically fairly quiet (with a few exceptions outlined below). River systems flowing to the Tethys Sea generated aggrading fluvial deposits as the rivers and the load of sediment they carried leveled off on a broad African coastal plane (e.g., Bown and Kraus, 1988).

Kortlandt (1980) and others have erroneously attributed the Fayum deposits, only 60 km from the Nile River, to the action of a "proto-Nile," an idea criticized by geologists working in the Fayum (e.g., Bown et al., 1982). The extension of fluvial deposits similar to those of the Fayum westwards into Libya is one of several lines of evidence that the depositional context was not generated by one massive river originating deep in the African interior, but rather by a series of streams draining what was, at the time, tropical forests under a high rainfall regime. Suggestions that evaporites found in some early Tertiary sediments of North Africa indicate an arid climate at the time of deposition ignore the fact that gypsum and other evaporites can develop diagenetically (Bain, 1990). All inferences about the North African climate during the early Tertiary that are based on the communities of fossil plants, birds, and mammals, indicate lush tropical conditions (Bown et al., 1982; Olson and Rasmussen, 1986; Wing, personal communication). The watershed patterns that developed along the Tethyan floodplain of early Tertiary Africa are not analogous or ancestral to the Nile.

The shoreline of Africa during the early Tertiary was geographically positioned further to the south than it is today. Most of the Eocene beds of northeastern Africa are marine, but several deposits represent nearshore or terrestrial accumulations, particularly westwards towards the Saharan Atlas (e.g., Mahboubi et al., 1986). The Eocene was a time of marine transgression and regression across the region, with marine conditions predominating in

Egypt and, apparently, in Libya. Terrestrial lenses of Eocene age (representing temporary or local regressions) embedded within broader marine deposits have been found in Egypt and Oman (e.g., Seiffert et al., 2005; Thomas et al., 1999). The Eocene-Oligocene transition was marked by a worldwide regressive event that should be identifiable in terrestrial sequences by an erosional unconformity (Bown and Kraus, 1987; Kappelman et al., 1992; Seiffert, 2006), and in theory at least, by a transition in other places from nearshore marine to terrestrial conditions. A relatively uninterrupted stratigraphic column at any one geographic spot should show, from the Paleocene through the Oligocene, a transition from offshore marine, to nearshore marine and estuarine conditions, and finally to terrestrial environments. The precise pattern of marine, nearshore and terrestrial sediments of early Tertiary age across northern Egypt and Libya were presumably influenced by local permutations in river flow, surface topography, and subtle tectonic action. Analysis of drill cores from the rich petroleum fields of northeastern Libya provides just such a test of these general geological models. Cores from the Sirt Basin there confirm a general marine to terrestrial transition as one progresses through the Early Tertiary, and provide quite a bit of detail on the local stratigraphy, lithology and dating (e.g., Salem, 1996; Ahlbrandt, 2001).

A distinctive feature of Libyan geology, when compared to that of Egypt, is that there was small-scale but significant rifting from the Cretaceous into the Paleocene and Eocene (Gumati and Kanes, 1985). There were three primary grabens of this age range formed in what is now terrestrial Libya. The two eastern ones are covered by Quaternary deposits, but the westernmost one, the Hun Graben, is subareally exposed. Field study by Libyan geologists (Geological Map of Libya, 1985) identified the rock contained within the graben as primarily Eocene marine calcareous muds, but with a variety of other sedimentary types, while the graben walls are Paleocene and Cretaceous marine sediments uncapped by younger sediment. We visited the Hun Graben to examine some of the local facies; the sedimentary rock we saw in our brief stop there was consistent with the previous, more detailed mapping work in revealing no evidence of terrestrial or nearshore conditions.

The problem of finding local or extensive terrestrial lenses within the Libyan Tertiary series ultimately comes down to surface prospecting for fossils. Much of the Early Tertiary deposits are covered by aeolian sand, or are exposed in very low relief in the form of great expansive desert flats crusted by serir, or desert pavement. Prospecting for fossil vertebrates in these conditions ranges from impossible to difficult. Savage reported a spot in which numerous fossil sirenians were weathering on the surface in the middle of these expansive Tertiary deposits (Savage, 1971). Local, low-relief escarpments exist that expose shallow stratigraphic sections; the outcrops that yielded the vertebrates of Zella represent such an occurrence (Arambourg and Magnier, 1961; Fejfar, 1987). We spent a single day surface prospecting outcrops in the Zella region, without being able to pinpoint Arambourg's precise locality. While we encountered no terrestrial sediments or vertebrates, we were able to confirm that the local stratigraphic column was fairly complex, comprising a variety of marine rocks, including limestones, calcareous muds, sandstones, and mudstones, including beds lacking fossils and others with a mash of marine invertebrates. Prospecting the flat Paleogene exposures of Libya in the future will presumably rely on intensive survey of outcrops of this kind.

For the most part, the extensive early Tertiary of Libya has yielded little to paleontologists interested in terrestrial vertebrates, with two impressive exceptions. The first is the articulated hyracoids and frogs of Jebel al Hasawnah (Thomas et al., 2004), the second is Dor al-Talha. The Paleogene beds of Libya have a gentle dip to the north. The southernmost rim of this vast tilted bank of Early Tertiary sediment lies at the south-facing escarpment of Dor al-Talha, which stretches along an east-west axis for over 100 km. This is where Bellair et al. (1954), Arambourg and Magnier (1961), Savage (1969) and Wight (1980) found fossil vertebrates of estuarine and terrestrial origin.

Dor al-Talha

Accessing the escarpment involved a 400 km drive on sandy desert tracks from the Zella oasis, curving around the eastern margin of the al-Haruj Basalts, and arriving on the top of the scarp. The al-Haruj basalts represent six or seven flows ranging in age from late Pliocene to late Pleistocene or early Holocene (Klitsch, 1967, Busrewil et al., 1996). Much of the drive looping around to the east of al-Haruj is over Quaternary wind-blown sand and early Tertiary sediment, but the latter is usually poorly exposed.

The broken escarpment stands north of a great depression filled with *feshfesh*, extremely fine-grained aeolian accumulations. The feshfesh basin represents a sea of dry loess-like powder, covering at least 8500 km^2, which Savage (1971) described as being "very difficult to traverse by car" The escarpment is shown in Figs. 1–3. The scarp is generated by erosive action on a resistant sandstone deposit that forms the escarpment rim, rich in fossil wood, which overlies finer-grained and softer mudstones, siltstones, and sandstone lenses. The vertebrates come from the finer-grained rocks which, by virtue of the erosional resistance of the rimming sandstones, are exposed as cliffs and bluffs (Fig. 1). In our trip, we contacted the scarp at a point not well sampled by Savage and Wight, and so our work represents an independent assessment of the geology and fossils. The 13 localities we found were new, but the results we obtained were consistent with those of Wight (1980).

Wight divided the sediments of Dor al-Talha into three members. The lowest of these members—which at its base is overlapped by the feshfesh to the south—is called the Evaporite Unit (Fig. 2). Wight interpreted this as a marine deposit based on the limited mammalian fauna of moeritheres and barytheres. The preservation of vertebrate bones at this level is, in our limited experience, very

Fig. 1 View from the top of the Dor al-Talha escarpment looking south. The yellow-brown sandstones of the foreground erode as dark, broken cliffs (visible at background right), with broad wadis washing through softer, underlying sediment towards the feshfesh basin at the distant horizon

poor (Fig. 4). The bones are soft, fractured, and chalky. The abundant evaporites seem to be diagenetic, and not related to depositional conditions.

The next oldest stratigraphic unit of Wight is the Idam Unit (Fig. 3). This unit consists of fine to medium-grained clastics. While not uniformly fossiliferous, local spots produce vertebrate fossils in notable concentrations. The Idam Unit was the primary producer of the fossil mammals listed by Savage. We found a wealth of vertebrates there. Our finds suggest that environments of deposition were varied, and contained at least some terrestrial contexts. Some of the Idam Unit localities consisted of broken and unassociated fragments that represent a mixed assemblage of aquatic and terrestrial vertebrates. For example, locality DT-01 (Fig. 3) includes broken and dissociated specimens of sawfish, catfish, gar, turtles, crocodilians (including *Tomistoma*), and the mammals *Moeritherium* and indeterminate members of Hyracoidea. In contrast, another PRC locality, DT-13 (found in the last hour of dusk on our last day in the field by S. T. and M. M. A.), was characterized by extremely well-mineralized fossils layed out as loosely associated skeletons in fine-grained clasts. The vertebrate taxa included silurid catfish, the crocodilian *Tomistoma* (Fig. 5), and the mammal *Moeritherium*. These two localities attest to the

Fig. 2 Sedimentary rock of the Evaporite Unit eroding as rounded bluffs and soft slopes (kneeling J. B. S. as a scale bar in center left). A butte of the overlying Idam Unit is seen rising in the center background

potential diversity of environmental and taphonomic conditions among sites at Dor al-Talha.

The uppermost stratigraphic unit of Wight was reported to be devoid of vertebrates but it did contain fossil wood. We spent no time prospecting for vertebrates here, given our limited schedule, and our paleontological observations are basically limited to the fact that, in local places, we had to drive carefully to avoid fossil tree trunks distributed on the landscape. No vertebrates were seen in this unit.

Paleoenvironmental reconstructions of the Dor al-Talha stratigraphic units are particularly important. The Evaporite Unit seems to be primarily marine, albeit nearshore marine, given the prevalence of *Moeritherium*, *Barytherium*, and some freshwater or estuarine fish and crocodilian groups (Wight, 1980). The uppermost unit is full of fossil trees, some quite large, indicating stable terrestrial conditions. The Idam Unit contains non-aquatic vertebrates like *Arsinoitherium*, *Palaeomastodon* and Hyracoidea, suggesting that at certain times and places, the accumulations are derived from terrestrial conditions. However, the Idam Unit also contained many aquatic, and in some cases, marine vertebrates as well; for example, we collected in this unit a vertebra of *Gigantophis*, a huge snake believed to be marine (Andrews, 1906). The interpretation that local terrestrial conditions did exist is supported by massive tree trunks in the Idam Unit (Fig. 6); further analysis of the Idam Unit's fossil forest

Fig. 3 Typical appearance of the Idam Unit, a highly fossiliferous unit containing a diversity of rock types and fossils. Locality PRC DT-01 consists of the slope in the left foreground below the bench-forming light sandstone (M. M. A. prospects for fossils on the right end of the outcrop)

Fig. 4 A complete but badly deteriorated humerus of *Barytherium* from locality PRC DT-11, in the Evaporite Unit. The proximal end is to the left, the deltoid crest is visible spiraling around the shaft at center, and the broad distal end is on the right. The bone is chalky, and badly fissured from weathering

Fig. 5 Fossil remains of the long-nosed crocodilian *Tomistoma*, a genus currently extant in Southeast Asia, from the Idam Unit locality PRC DT-13. These fossils are heavily mineralized and well preserved, in contrast to those from the Evaporite Unit. The individual pieces illustrated here represent bones of a single individual (including cranial fragments, scutes, and the upper and lower jaw seen in the foreground) that were scattered on the surface over a few square meters (they have been arranged close together for the photo). Remains of several other individuals of *Tomistoma*, large fish, and *Moeritherium*, were also found at this site

localities will be required to confirm that these are *in situ* forests, rather than transported, floating wood (cf. Bown et al., 1982).

Fossil Mammals

We concluded the fieldwork three weeks before the Simons conference, and have not analyzed the mammals and other vertebrates we collected in any detail. We can report that the fauna we encountered in a couple days of intensive field collection matched well taxonomically with the fauna reported by Savage (1971) and Wight (1980). The most common mammals found in the old pioneering trips to Dor al-Talha were quickly duplicated by us: we found remains of the Early Tertiary proboscideans *Moeritherium, Barytherium* and *Palaeomastodon*, of the embrithopod *Arsinoitherium*, and bone fragments of smaller, indeterminate mammals, most likely hyracoids. The most abundant fossil vertebrates were those of fish, particularly silurid catfish, lungfish, sawfish and gars. Among reptiles, remains of the long-nosed crocodilian *Tomistoma* were particulary common.

Fig. 6 Large fossil tree trunks exposed at locality DT-04 of the Idam Unit (J. B. S. serves as scale). Vertebrates were not found at this locality, but it occurs at nearly the same stratigraphic level as DT-13 and other vertebrate-bearing sites. This view looking south provides another view of landscapes within the Idam Unit

Moeritherium is a primitive proboscidean with bunodont teeth and short, robust chisel-like tusks (Tassy, 1981). It has been recovered at several Early Tertiary sites in Africa, and the depositional contexts of these finds suggest it was amphibious. Based on a specimen recovered by the early Simons expeditions to the Fayum, the body shape also supports this interpretation: *Moeritherium* was a heavy, long-bodied, short-legged form. We recovered several postcranial and dental specimens of *Moeritherium* from the Idam Unit, including a well-preserved and lightly worn maxillary dentition that will contribute to the investigation of species-level taxonomy in this genus. In addition, we found an eroded cranium from the Evaporite Unit that is comparable in preservation to the important Arambourg specimen described by Tassy (1981).

Barytherium graves is a very large proboscidean with specialized bilophodont teeth, closely resembling much later deinotheres in that one respect, and differing dramatically from bunodont *Moeritherium* and *Palaeomastodon*. *B. graves* was a common component of the early collections from Dor al-Talha, including a partial postcranial skeleton, but the material recovered by Arambourg and Savage has never been fully described. *Barytherium* has not been found outside Dor al-Talha except for a few, poor specimens from Eocene levels of the Fayum (Andrews, 1906; Simons, 1968); this genus remains a Libyan specialty item. We found several dental specimens of *Barytherium* during our reconnaissance trip.

Wight (1980) suggested that a smaller species of *Barytherium* was present at Dor al-Talha but did not describe it formally. Later, a lower jaw representing this taxon was described as a new species of the very primitive proboscidean *Numidotherium* (Mahboubi et al., 1984; Court, 1995). Whether this really represents a species of *Barytherium*, *Numidotherium*, or something new, is a question that deserves further study, and it will probably require the recovery of new material. We found one molar specimen of the smaller bilophodont proboscidean. The primitive proboscideans of Africa are taxa of great interest to evolutionary biologists (e.g., Gheerbrant et al., 2002).

Palaeomastodon is a proboscidean well known from the Early Tertiary of Africa, and is best characterized to date from the Fayum, Egypt, and Chilga, Ethiopia (Tobien, 1971; El-Kashab, 1979; Sanders et al., 2004). Unlike *Moeritherium, Barytherium,* and *Numidotherium*—which represent distant outgroups to the lineages that gave rise to Late Tertiary mastodons, deinotheres, elephants, and their kin—*Palaeomastodon* is considered to be an early member of the elephantiform stock (Tassy, 1994). It was an animal with bunodont but linearly seriated tooth cusps that provide an obvious ancestral condition for gomphotheres and deinotheres (Sanders et al., 2004). The anterior teeth were elongated into tusks, spatulate in shape, and appressed at the midline, unlike those of modern elephants. We recovered a few dental fragments of *Palaeomastodon.*

Arsinoitherium is a massive, horned mammal with bizarre high-crowned teeth that is known only from the Early Tertiary of a few African sites (Court, 1992a, 1992b, 1993). It is best known from latest Eocene and early Oligocene levels of the Fayum (Andrews, 1906; El-Kashab, 1977; Tanner, 1978), and from the Late Oligocene of Chilga, Ethiopia (Sanders et al., 2004). Postcranial remains of *Arsinoitherium* have been reported from the Fayum (Andrews, 1906; Court, 1993). At Dor al-Talha, we recovered one specimen that consists of an associated complete tibia and fibula, very well preserved, from the Idam Unit. The specimen matches very well with a tibia of *Arsinoitherium* described by Andrews (1906). The tibia and fibula of *Barytherium*, a similarly sized animal, remain undescribed, and so comparisons to that taxon are impossible. The discovery of well-preserved, complete leg bones of *Arsinoitherium* in the Idam Unit is of interest because *Arsinoitherium*, in contrast to *Moeritherium* and possibly *Barytherium*, was indisputably an animal of terrestrial environments (Court, 1993; Kappelman et al., 2003). Wight reported a very few number of specimens of *Arsinoitherium* from Dor al-Talha.

In addition to the large probosideans and embrithopods, we recovered several fragments representing smaller mammals, probably hyracoids, pending further study. Hyracoids were a dominant order of terrestrial mammals in the Early Tertiary of Africa, both in terms of abundance and diversity (Rasmussen, 1989; Rasmussen and Gutierrez, in press). We collected an occipital bone with its condyles, an innominate with the acetabulum, and several mammalian vertebral fragments that all fall within the size range of hyracoids ranging in body mass between *Thyrohyrax* and *Pachyhyrax*. These specimens, collected in only a couple of days on the outcrops at Dor al-Talha, suggest that there is potential to find additional mammalian taxa. The obvious similarities to the Fayum fauna suggest the possibility of finding primates (Simons, 1987, 1990, 1995, 1997, 2001).

Conclusions

The stratigraphic work conducted by A. W. R. Wight in the late 1960s, during short trips to Dor al-Talha, was very useful to us, and it proved to be reliable and accurate in a general sense (Wight, 1980). The abundance of fossil vertebrates at the escarpment is confirmed. The Evaporite Unit appears to be primarily marine. The poor preservation of fossils there and the abundant evaporites are probably diagnetic in nature, not necessarily reflecting an arid environment. The Idam Unit is particularly fossiliferous and holds much promise for further work on the Libyan mammal communities of the Early Tertiary. The fossils are well preserved, and consist of both terrestrial and semi-aquatic taxa. The Dor al-Talha faunal community, as currently construed based on very preliminary sampling, may represent one of the best assemblages of basal proboscideans and embrithopods known from Africa. The diversity among Dor al-Talha sites and the great geographic extent of the exposures suggests that future work on fossils, dating, and ecological context will continue to illuminate evolutionary and paleoenvironmental events in the northern part of the African-Arabian continent during the Early Tertiary.

References

Ahlbrandt, T. S. (2001). The Sirte Basin Province of Libya – Sirte-Zelten total petroleum system. *U. S. Geological Survey Bulletin* 2202-F: 1–29.

Andrews, C. W. (1906). *A Descriptive Catalogue of the Tertiary Vertebrata of the Fayum, Egypt*. British Museum (Natural History), London, 324 pages, 26 plates.

Arambourg, C. (1961). Note préliminaire sur quelques Vertébrés nouveaux du Burdigalien de Libye. *Comptes Rendus Sommaire des Seances de la Société Géologique de France* 4: 107–109.

Arambourg, C. (1963). Continental vertebrate faunas of the Tertiary of North Africa. In: Howell, F. C., and Bourliere, F. (eds.), *African Ecology and Human Evolution*. Aldine, Chicago, pp. 55–60.

Arambourg, C., and Magnier, P. (1961). Gisements de vertébrés dans le bassin tertaire de Syrte (Libie). *Comptes Rendus Hebdomadaires des Séances de l'Academie de Sciences* 252: 1181–1183.

Bain, R. J. (1990). Diagenetic, nonevaporative origin for gypsum. *Geology* 18, 447–450.

Bellair, P., Freulon, J., and Lefranc, J. (1954). Découverte d'une formation á vertébrés et végétaux d'âge tertiaire au bord occidental du désert libyque (Sahara occidental). *C. R. Acad. Sci. Paris* 239: 1822–1824.

Bown, T. M., and Kraus M. J. (1988). Geology and paleoenvironment of the Oligocene Jebel Qatrani Formation and adjacent rocks, Fayum Depression, Egypt. *U. S. Geol. Surv. Prof. Paper* 1452: 1–60.

Bown, T. M., Kraus, M. J., Wing, S. L., Fleagle, J. G., Tiffney, B. H., Simons, E. L., and Vondra, C. F. (1982). The Fayum primate forest revisited. *J. Hum. Evol.* 11: 603–632.

Busrewil, M. T., Mriheel, I. Y., and Al-Fasatwi, T. A. (1996). Volcanism, tectonism, and hydrocarbon potential of parts of Al Haruj area, SW Sirt Basin, Libya. In: Salem, M. J., Mouzughi, A. J., and Hammuda, O. S. (eds.), *The Geology of Sirt Basin, Volume 3*. Elsevier, Amsterdam, pp. 317–329.

Court, N. (1992a). The skull of *Arsinoitherium* (Mammalia, Embrithopoda) and the higher order interrelationships of ungulates. *Palaeovertebrata* 22, 1–43.

Court, N. (1992b). A unique form of dental bilophodonty and a functional interpretation of peculiarities in the masticatory system of *Arsinoitherium* (Mammalia, Embrithopoda). *Hist. Biol.* 6: 91–111.

Court, N. (1993). Morphology and functional anatomy of the postcranial skeleton in *Arsinoitherium* (Mammalia, Embrithopoda). *Palaeontolographica Abhandlungen* A 226: 125–169.

Court, N. (1995). A new species of *Numidotherium* (Mammalia: Proboscidea) from the Eocene of Libya and the early phylogey of the Proboscidea. *J. Vertebr. Paleontol.* 15: 650–671.

Court, N., and Hartenberger, J.-L. (1992). A new species of the hyracoid mammal *Titanohyrax* from the Eocene of Tunisia. *Palaeontology* 35: 309–317.

El-Kashab, B. (1977). Approach on genus *Arsinoitherium* and the diagnosis of its closely related sympatric fossil species, in Lower Oligocene of Fayum Province Egypt. *The Geological Survey of Egypt and Mining Authority*, Cairo, Paper No. 62: 29–39.

El-Kashab, B. (1979). A brief account of Egyptian Paleogene Proboscidea. *Annals of the Geological Survey of Egypt* 9: 245–260.

El-Kashab, B., Simons, E. L., and Fleagle, J. G. (1983). Annotated bibliography of Egyptian vertebrate fossils up to the end of 1980. *Geological Survey of Egypt and Mining Authority*, Cairo, Paper No. 65: 1–111.

Fejfar, O. (1987). Oligocene rodents from Zallah Oasis, Libya. *Münchner Geowissenschaften Abhandlungen* A 10: 265–268.

Gaziry, A. W. (1987). New mammals from the Jabal Zaltan site, Libya. *Senckenbergiana lethaea* 68: 69–89.

Geological Map of Libya (1985). First Edition. Geological Researches and Mining Department, Industrial Research Center, Tripoli.

Gheerbrant, E., Sudre, J., Cappetta, H., Iarochene, M., Amaghzaz, M., and Bouya, B. (2002). A new large mammal from the Ypresian of Morocco: Evidence of surprising diversity of early proboscideans. *Acta Palaeontol. Pol.* 47: 493–506.

Gingerich, P. D. (1992). Marine mammals (Cetacea and Sirenia) from the Eocene of Gebel Mokattam and Fayum, Egypt: stratigraphy, age, and paleoenvironments. *University of Michigan, Papers on Paleontology* 30, 1–84.

Gingerich, P. D. (this volume). Early evolution of whales: A century of research in Egypt. In: Fleagle, J. G. (ed.), *Elwyn L. Simons: Mapping the Progress of Primate Evolution*. Plenum Press, New York.

Godinot, M. (1994). Early North African primates and their significance for the origin of Simiiformes (= Anthropoidea). In: Fleagle, J. G., and Kay, R. F. (eds.), *Anthropoid Origins*. Plenum Press, New York, pp 235–295.

Gumati, Y. D., and Kanes, W. H. (1985). Early Tertiary subsidence and sedimentary facies – northern Sirte Basin, Libya. *Am. Assoc. Petr. Geol. B.* 69: 39–52.

Kappelman, J., Simons, E. L., and Swisher, C. C., III (1992). New age determinations for the Eocene-Oligocene boundary sediments in the Fayum depression, northern Egypt. *J. Geol.* 100: 647–667.

Kappelman, J., Rasmussen, T., Sanders, W., Feseha, M., Bown, T. M., Copeland, P., Crabaugh, J., Fleagle, J., Glantz, M., Gordon. A., Jacobs, B., Maga, M., Muldoon, K., Pan, A., Pyne, L., Richmond, B., Ryan, T., Seiffert, E., Sen, S., Todd, L., Wiemann, M. C.,

and Winkler, A. (2003). Oligocene mammals from Ethiopia and faunal exchange between Afro-Arabia and Eurasia. *Nature* 426: 549–552.
Klitsch, E. (1967). Der Basaltvulkanismus des Jabal Haruj, Ostfezzan/ Libyen. *Geol. Rdsch.* 57: 585–601.
Kortlandt, A. (1980). The Fayum primate forest: Did it exist? *J. Hum. Evol.* 9: 277–297.
Mahboubi, M., Ameur, R., Crochet, J. Y., and Jaeger, J.-J. (1984). Earliest known proboscidean from the early Eocene of north-west Africa. *Nature* 308: 543–544.
Mahboubi, M., Ameur, R., Chrochet, J. Y., and Jaeger, J.-J. (1986). El Kohol (Saharan Atlas, Algeria), a new Eocene mammal locality in northwestern Africa: stratigraphic, phylogenetic and paleobiogeographical data. *Palaeontographica Abteilung A*, 192, 15–49.
Miller, E. R. (1999). Faunal correlation of Wadi Moghara, Egypt: Implications for the age of *Prohylobates tandyi*. *J. Hum. Evol.* 36: 519–533.
Olson, S. L., and Rasmussen, D. T. (1986). The paleoenvironment of the earliest anthropoids: new evidence from the Oligocene avifauna of Egypt. *Science* 233: 1202–1204.
Osborn, H. F. (1908). New fossil mammals from the Fayum Oligocene, Egypt. *B. Am. Mus. Nat. Hist.* 24: 265–272.
Rasmussen, D. T. (1989). The evolution of the Hyracoidea: a review of the fossil evidence. In: Prothero, D. R., and Schoch, R. M. (eds.), *The Evolution of Perissodactyls*. Oxford University Press, New York, pp. 57–78.
Rasmussen, D. T., Bown, T. M., Simons, E. L. (1992). The Eocene-Oligocene transition in continental Africa. In: Prothero, D. R., and Berggren, W. A. (eds), *Eocene-Oligocene Climatic and Biotic Evolution*. Princeton Univ., Princeton, New Jersey, pp. 548–566.
Salem, M. J., El-Hawat, A. S., and Sbeta, A. M. (1996). *The Geology of the Sirt Basin, Volume 1*. Elsevier, Amsterdam.
Sanders, W. J., Kappelman, J., Rasmussen, D. T. (2004). New large-bodied mammals from the late Oligocene site of Chilga, Ethiopia. *Acta Palaeontol. Pol.* 49: 365–392.
Savage, R. J. G. (1969). Early Tertiary mammal locality in southern Libya. *P. Geol. Soc. London* 1657: 167–171.
Savage, R. J. G. (1971). Review of the fossil mammals of Libya. In: Gray, C. (ed.), *Symposium on the Geology of Libya*. University of Libya, Tripoli, pp. 215–226.
Schlosser, M. (1911). Beitrage zur kentniss der Oligozanen Landsaugetiere aus dem Fayum, Egypt. *Beitr. Palaeont. Geol. Ost. Ung., Wien* 24: 51–167.
Seiffert, E. R. (2006). Revised age estimates for the later Paleogene mammal faunas of Egypt and Oman. *P. Natl. Acad. Sci. USA* 103: 5000–5005.
Seiffert, E. R., Simons, E. L., and Attia, Y. (2003). Fossil evidence for an ancient divergence of lorises and galagos. *Nature* 422: 421–424.
Seiffert, E. R., Simons, E. L., Clyde, W. C., Rossie, J. B., Attia, Y., Bown, T. M., Chatrath, P. S., and Mathison, M. E. (2005). Basal anthropoids from Egypt and the antiquity of Africa's higher primate radiations. *Science* 310: 300–304.
Seiffert, E. R. (2008). Geology, paleoenvironment, and age of Birket Qarun Locality 2 (BQ-2), Fayum Depression, Egypt. In: Fleagle, J. G. and Gilbert, C.C. (eds.), *Elwyn Simons: A Search for Origins*. Springer, New York, pp.71–86.
Selley, R. C. (1969). Nearshore marine and continental sediments of the Sirte Basin, Libya. *Quarterly Journal of the Geological Society of London* 124: 419–460.
Selley, R. C. (1997). The Sirte Basin of Libya. In: Selley, R. C. (ed.), *The Sedimentary Basins of the World, Volume 3*. Elsevier, Amsterdam, pp. 27–37.
Simons, E. L. (1968). Early Cenozoic mammalian faunas, Fayum Province, Egypt, Part I, African Oligocene mammals: introduction, history of study, and faunal succession. *Peabody Museum of Natural History, Yale University, Bulletin* 28:1–21.
Simons, E. L. (1987). New faces of *Aegyptopithecus* from the Oligocene of Egypt. *J. Hum. Evol.* 16: 273–289.
Simons, E. L. (1990). Discovery of the oldest known anthropoidean skull from the Paleogene of Egypt. *Science* 247: 1521–1612.

Simons, E. L. (1995). Crania of *Apidium*: primitive anthropoidean (Primates, Parapithecidae) from the Egyptian Oligocene. *Am. Mus. Nov.* 3124: 1–10.
Simons, E. L. (1997). Preliminary description of the cranium of *Proteopithecus sylviae*, an Egyptian Late Eocene anthropoidea primate. *P. Natl. Acad. Sci. USA* 94: 14970–14975.
Simons, E. L. (2001). The cranium of *Parapithecus grangeri*, an Egyptian Oligocene anthropoidean primate. *P. Natl. Acad. Sci. USA*, 98: 7892–7897.
Simons, E. L., and Rasmussen, D. T. (1995). A whole new world of ancestors: Eocene anthropoideans from Africa. *Evol. Anthropol.* 3: 128–139.
Smith, J. B., Askar, A. S., Bergig, K. A., Tshakreen, S. O., Abugares, M. M., Sliman, O., Lamanna, M. C., and Rasmussen, D. T. (in press). An abelisauroid theropod dinosaur from the Early Cretaceous of Libya. *Naturwissenschaften* 00, 000–000.
Stevens, N., O'Conner, P. M., Gottfried, M. D., Roberts, E. M., and Ngasala, S. (2005). An anthropoid primate humerus from the Rukwa Rift Basin, Paleogene of southwestern Tanzania. *J. Vertebr. Paleontol.* 25: 986–989.
Stevens, N., O'Conner, P. M., Gottfried, M. D., Roberts, E. M., Ngasala, S., and Dawson, M. R. (2006). *Metaphiomys* (Rodentia: Phiomyidae) from the Paleogene of southwestern Tanzania. *J. Paleontol.* 80: 407–410.
Stevens, N. J., Gottfried, M. D., Roberts, Saidi Kapilima, E. M., Ngasala, and O'Conner, P. M. (2008). Paleontological Exploration in Africa: A View from the Rukwa Rift Basin of Tanzania. In: Fleagle, J. G. and Gilbert, C.C. (eds.), *Elwyn Simons: A Search for Origins.* Springer, New York, pp.159–180.
Tanner, L. G. (1978). Embrithopoda. In: Maglio, V. J., and Cooke, H. B. S. (eds.), *Evolution of African Mammals.* Harvard University Press, Cambridge, Massachusetts, pp. 279–283.
Tassy, P. (1981). Le crâne de *Moeritherium* (Proboscidea, Mammalia) de l'Éocène de Dor el Talha (Libye) et le problème de la classification phylogénétique du genre dans les Tethytheria McKenna, 1975. *Bull. Mus. Natn. Hist. Nat., Paris, 4e sér., 3, section C.*, no 1, 87–147.
Tassy, P. (1994). Origin and differentiation of the Elephantiformes (Proboscidea, Mammalia). *Verhandlungen Naturwissenschaftlichen Vereins in Hamburg 34*, 73–94.
Thomas, H., Roger, R., Sen, S., Pickford, M., Gheerbrant, E., Al-Sulaimani, Z., and Al-Busaidi, S. (1999). Oligocene and Miocene terrestrial vertebrates in the southern Arabian Peninsula (Sultanate of Oman) and their geodynamic and palaeogeographic settings. In: Whybrow, P. J., and Hill, A. (eds.) *Fossil Vertebrates of Arabia.* Yale Univ., New Haven, pp. 430–442.
Thomas, E., Gheerbrant, E., and Pacaud, J. M. (2004). Découverte de squelettes subcomplets de mammifères dans le Paléogène d'Afrique (Libye). *C. R. Palevol* 3: 209–218.
Tobien, H. (1971). *Moeritherium, Palaeomastodon, Phiomia* aus dem Paläogen Nordafricas und die Abstammung der Mastodonten (Proboscidea, Mammalia). *Mitteilungen aus dem Geolischen Institut der Technischen Universität Hannover* 10, 141–163.
Wennekers, J. H. N., Wallace, F. K., and Abugares, Y. I. (1996). The geology and hydrocarbons of the Sirt Basin: A synopsis. In: Salem, M. J., El-Hawat, A. S., and Sbeta, A. M. (eds.), *The Geology of the Sirt Basin, Volume 1.* Elsevier, Amsterdam, pp. 3–56.
Wight, A. W. R. (1980). Palaeogene vertebrate fauna and regressive sediments of Dur at Talha, southern Sirt Basin, Libya. In: Salem, M. J., and Busrewil, M. T. (eds.), *The Geology of Libya, Volume 1.* Academic Press, London, pp. 309–325.
Wing, S. L., and Tiffney, B. H. (1982). A paleotropical flora from the Oligocene Jebel Qatrani Formation of northern Egypt: A preliminary report. *Miscellaneous Series of the Botanical Society of America* 162, 67.

Revisiting Haritalyangar, the Late Miocene Ape Locality of India

Rajeev Patnaik

Introduction

Around four decades ago between 1967 and 1969, Elwyn Simons organized a large paleoanthropological project at the site of Haritalyangar, Himachel Pradesh, India (Fig. 1) in collaboration with the Department of Anthropology, Panjab University. This project used the then novel approach of collecting small fossil vertebrates through large scale dry and wet screening of mammal bearing horizons (Fig. 2). Simons' project was primarily inspired by the finds at the same locality by G. E. Lewis (Fig. 3), who in 1934 named *Ramapithecus brevirostris,* Rama's short-faced ape, along with contributing other revisions of fossil apes from the Siwaliks (e.g., Lewis, 1937; Gregory et al., 1938). Most importantly, Lewis placed *Ramapithecus* in Hominidae, a position Simons and Pilbeam later advocated in the sixties and seventies (Simons, 1961, 1963, 1964, 1972, 1976a & b; Simons and Pilbeam, 1972; Simons et al., 1971; Pilbeam, 1966, 1969a & b), including in their landmark revision of fossil apes (Simons and Pilbeam, 1965). Later, in the early eighties, the discovery of facial remains of *Sivapithecus* (Pilbeam, 1982) disproved the notion of *Ramapithecus* as a hominid (Pilbeam, 1983). In addition to the facial remains recovered by Simons and Pilbeam, Andrews and Tekkya (1980) and Andrews and Cronin (1982) also showed that the *Sivapithecus* facial part they had recovered from Turkey was very similar to the face of an orangutan and argued that *Ramapithecus* and *Sivapithecus* were indistinguishable dentally.

The Yale-Panjab University collaboration, though short lived, produced one of the most remarkable fossil primate finds from this area (Fig. 4), a mandible (CYP359/68) of *Gigantopithecus bilaspurensis* (Simons and Chopra, 1969; Simons and Ettel, 1970; Simons and Pilbeam, 1973). Earlier, Von Koenigswald (1950) had referred a molar (GSI D-175) of *Dryopithecus giganteus* (Pilgrim,

Rajeev Patnaik
Center of Advanced Study in Geology, Panjab University, Chandigarh-160014, India
rajeevpatnaik@gmail.com

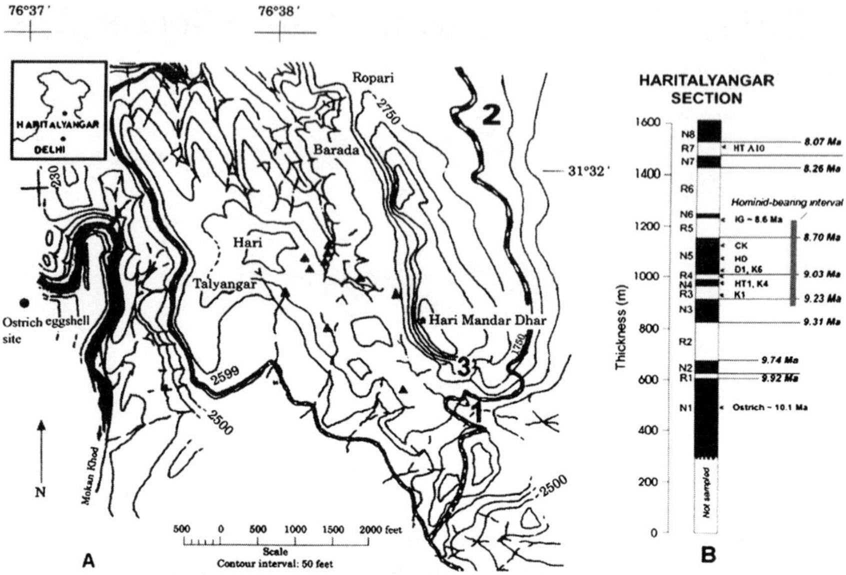

Fig. 1 Geographical and chronological placement of the some of the important fossil vertebrates at Haritalyangar (modified after Patnaik et al., 2005). (A) (1) the Dangar I locality; (2) the *Gigantopithecus* (*Indopithecus*) mandible locality; (3) *Gigantopithecus* (*Indopithecus*) molar site Hari Devi I; triangles denote other major hominoid localities. (B) Magnetic polarity stratigraphy near Haritalyangar showing the levels of important fossil findings (modified after Pillans et al, 2005): HTA10-*S.sivalensis* (Sankhyan, 1985), IG-*Indopithecus giganteus* (*Gigantopithecus* mandible site), CK-Chob Ka-nala *Svalhippus* site HD-Hari Devi-*Gigantopithecus* site, D1-Dangar *Sivapithecus* site HT-1-*Sivaladapis* site

1910 a,b) from Salt Range, Pakistan to a new genus *Indopithecus giganteus*, because it showed resemblance to that of *Gigantopithecus blacki*, the large ape best known from Pleistocene of China. Recently, the mandible has been reassigned to *Indopithecus* von Koenigswald (1950) based on both the similaries between the Indian and Pakistan specimens, and distinct differences in mandibular and dental morphologies between the Indian and Chinese *Gigantopithecus* (Cameron, 2001, 2003).

Since the late seventies, researchers have continued work on the palaeontology and stratigraphy of this locality, in some cases describing fossils originally collected by Simons' expeditions, and in other cases reporting the results of new fieldwork. In addition to large apes, *Pliopithecus, Indraloris, Sivaladapis* and *Palaeotupaia* have also been reported from Haritalyangar (Gingerich and Sahni, 1979, 1984; Chopra and Kaul, 1979; Chopra and Vasishat, 1979). Sahni and Khare (1976) reported an insectivore from Ladhyani (Haritalyangar), which also yielded a rich collection of fish remains (Sahni and Khare,1977). Vasishat (1985), provided a comprehensive account

Fig. 2 Scenes from the Yale-Chandigarh Expedition of 1968–1969. Above: John Fleagle (left) and Samvit Kaul (right) dry screening sediment at Quarry D. Below: Team members wet screening sediment from Quarry D in the Sutleg River

Fig. 3 Dr. Ed Lewis at Haritalyanger in 1932

of additional primates, rodents, suids, tragulids, proboscideans and equids from the Hari Scarp Haritalyangar). Sankhyan (1985) recovered a right second lower molar of *Sivapithecus sivalensis* from ~5.5 Ma (based on dates by Johnson et al., 1983) deposits near Haritalyangar (see Fig. 1B). Flynn et al., (1990) described murid and rhizomyid rodents from Haritalyangar (Fig. 4). Additional fauna from Ladhyani fish horizon include murids and tree shrews (Tiwari, 1996).

More recently, fossil ape material coming from Haritalyangar includes a second upper molar (Fig. 4b–d) of *Gigantopithecus* ('*Indopithecus*'), a first upper incisor (Fig. 4e–f) and a third upper premolar (Fig. 4g–i) of *Sivapithecus sivalensis* (Patnaik et al., 2005; Pillans et al., 2005; Patnaik and Cameron,1997). Patnaik and Cameron (1997) assigned the P^3 from Dangar I

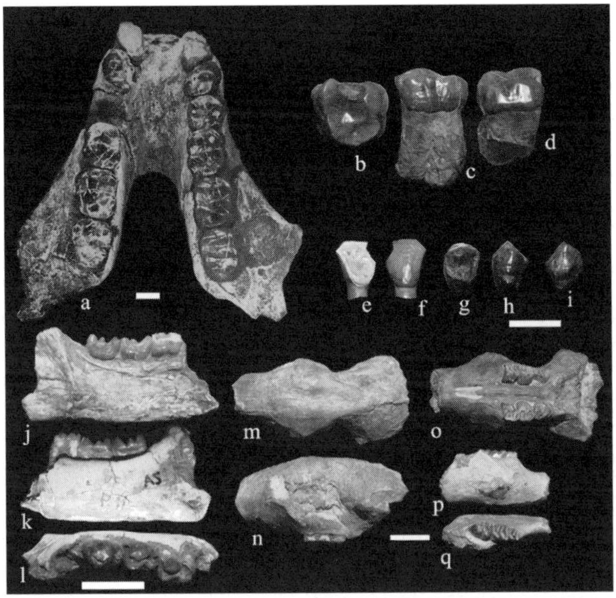

Fig. 4 Select fossil mammals from the hominid interval at Haritalyangar. a, occlusal view of *Gigantopithecus* (*Indopithecus*) mandible (CYP359/68 -Simons and Chopra, 1969); b-d, occlusal, labial and lingual views of *Gigantopithecus* (*Indopithecus*) M^2 (VPL/HD I-1, Patnaik et al, 2005); e and f, lingual and labial views of *Sivapithecus* I^1(VPL/HD I-2); g-i, occlusal, lingual and labial views of P^3 *Sivapithecus* (VPL/RP-H1, Patnaik and Cameron, 1997); j-l, lingual, labial and occlusal view of *Sivaladapis* mandible (P11AS, Gingerich and Sahni, 1979); m-q, *Brachyrhizomys choristos*, m-o, dorsal, lateral and ventral views of the skull (VPL/BS/PU 101), p and q labial and occlusal view of a mandible (VPL/BS/PU 102, Flynn et al., 1990). Bar scales represent 1 cm

locality to *S. sivalensis* based on a detailed comparison of its metric values with those of *S. parvada*, *S. indicus*, and *S. sivalensis*. The M^2 from the Hari Devi site was found to be similar in size to that of a modern gorilla (Pillans et al., 2005). Its allocation to either *Sivapithecus* or *Indopithecus* was a bit problematic because no upper molars of the Siwalik *Indopithecus* have previously been found. In order to assign the M^2 to *Sivapithecus* or *Indopithecus*, a comparative analysis of upper and lower second molars of extant and fossil hominids, was undertaken, as there is a strong correlation between the breadths of upper and lower M2s. The M^2 clearly fell outside the range of *S. sivalensis* and to a lesser degree, *S. indicus*, but well within the expected range of both *Indopithecus* and *S. parvada*. Also the occlusal morphology of the M^2 is more similar to *Indopithecus* than *Sivapithecus* (Patnaik et al., 2005; Pillans et al., 2005).The isolated upper incisor shows close resemblance to those of *Sivapithecus*. A site called Dharamsala (dated to 10.1 Ma) near Haritalyangar has yielded ostrich eggshell fragments showing pore patterns similar to those of *Struthio* (Patnaik et al., in review).

Chronology of the Siwalik Fossil Apes

Three species of *Sivapithecus* are now recognized in the Siwaliks: *Sivapithecus indicus*, *S. sivalensis* and *S. parvada*. Based on early paleomagnetic studies, *Sivapithecus indicus* ranges from 12.5 to 9.3 and *S. sivalensis* ranges from 8.8 to 7.4. *S. parvada* comes from a single site dated to 9 Ma (Berggren et al., 1985; Kappelman et al., 1991; Flynn et al., 1995). Because of revisions in the Geomagnetic Polarity Timescale (GPTS), these dates have now become ~1 Ma older following Cande and Kent (1995). Therefore, the combined data from the Pakistan and Indian Siwaliks suggest that the Siwalik apes appeared at ~13.5 Ma and suddenly disappeared by ~7.5 Ma. Johnson et al., (1983) carried out a detailed palaeomagnetic stratigraphy of the Haritalyangar region and found that the *Sivapithecus* fossils occur between 850 and 1100 m above the base of the Haritalyangar section. Johnson et al. (1983) placed *Gigantopithecus bilaspurensis* (*G. giganteus*/*Indopithecus giganteus*) at ~1200 m level (Fig. 1B). Sankhyan (1985) recorded *Sivapithecus* from ~1500 m level. Therefore, these fossil apes ranged from ~7.5 to 5.5 Ma. It was previously assumed that sivapithecines disappeared from the Siwaliks of Pakistan by 7.4 Ma and they survived in the Indian part until the Latest Miocene. Based on new correlation of the Haritalyangar section to the GPTS, the interval has been placed now between 9.23 and 8.85 Ma (Pillans et al., 2005). The absolute age and provenance of *Gigantopithecus* (*Indopithecus*) has always been quite uncertain because neither the mandible nor the M_3 (from Salt Range, Pakistan) were found *in situ*. However, Johnson et al. (1983) proposed a tentative age of 6.3 Ma for the Haritalyangar mandible (IG in Fig. 1B). Pillans et al., (2005) have provided a date of 8.6 Ma for the *Gigantopithecus* (*Indopithecus*) mandible. The M^2 from Hari Devi site provided the first accurate stratigraphic occurrence of *Gigantopithecus* (*Indopithecus*) at Haritalyangar (Pillans et al., 2005). This occurrence of *Gigantopithecus* (*Indopithecus*) has been dated to 8.85 Ma (Pillans et al., 2005). The *Sivapithecus* incisor (also came from the same level. The *Sivapithecus* specimen can now be assigned to 8.08 Ma following Pillans et al. (2005). With new dates, the hominid material from Haritalyangar can now be placed between 9.23 and 8.10 Ma (Fig. 1B). This provides a better correlation between the Pakistan and Indian hominid bearing deposits.

Ecology and Climate in the Late Miocene

The hominid interval at Haritalyangar ranging between ~9.23 and ~8.10 Ma has yielded a diverse faunal assemblage besides the fossil apes. The characteristic fauna include: apes (*Sivapithecus*, *Gigantopithecus* (*Indopithecus*)), primitive catarrhines (*Pliopithecus*), adapid primates (*Sivaladapis*), tree shrews (*Palaeotupaia*), horses (*Sivalhippus*), mouse deer (*Dorcatherium*), suids

(*Propotamochoerus*), bamboo rats (*Rhizomyides, Brachyrhizomys*) and other rodents (e.g., *Sayimys*). The large dental caries observed in the occlusal surface of the *Gigantopithecus* (*Indopithecus*) molar specimen are rare for Miocene hominids (Fig. 4b). Pleistocene *Gigantopithecus* from China shows extensive development of dental caries (Ciochon et al., 1990; Han and Zhao, 2002). Opal phytoliths found in the teeth enamel of *Gigantopithecus* indicates a varied diet comprising of bamboos, grasses and fruits (Ciochon et al., 1990). This probably indicates presence of grasses, bamboos and forests at around 9 Ma when *Gigantopithecus* (*Indopithecus*) and *Sivapithecus* lived in close proximity to each other. The overall fauna, sedimentary structures and palaeosols are indicative of a mosaic environment comprised of streams, gallery forests, and a floodplain with thick forest cover containing patches of grasslands. Arid conditions probably set in a little earlier, as indicated by the presence of ostrich eggshells at around 10 Ma.

By the Late Miocene (~7.5–8 Ma ago) both *Sivapithecus* and *Sivaladapis* disappeared from the Siwalik sediments. Their disappearance has been found to coincide fairly well with major tectonic and climatic events in South Asia. Tibeto-Himalayan uplift at around 12–9 Ma (Amano and Taira, 1992; Harrison et al., 1993) probably changed the heat budget leading to the intensification of the monsoon system in South Asia (Ruddiman and Kutzbach, 1989; Raymo and Ruddiman, 1992; Kutzbach et al., 1993; Hay, 1996; Ramstein et al., 1997). Prell and Kutzbach (1992) have suggested that a rapid uplift started at ~10 Ma and stopped at ~5 Ma (after the present elevation of ~5 km was attained). They went on to suggest that monsoons similar to present day might have started at around 8–7 Ma, when the elevation was half (2.5 km) of the present elevation. An upwelling of endemic foraminifers and radiolarians fauna took place at ~8 Ma. (Prell et al.,1992; see Fig. 5). A very similar process occurs in the Indian Ocean today, where strong summer monsoon winds induce upwelling in the upper water column that controls oceanic primary productivity and promotes the blooming of distinct fauna and flora in different parts of the northern Indian Ocean (Gupta and Melice, 2003). The Indian Ocean diatom record is also indicative of an intensification of monsoons between 11 and 7 Ma (Schrader, 1974; Burkle, 1989). Siwalik soil carbonate isotopic studies indicate a C3 (mainly bushes and trees) dominant vegetation prior to 8 Ma, but by 7 Ma the C4 grasses dominated the Siwalik floodplain biomass (Quade et al., 1989, 1995). Because of the monsoonal condition, C4 grasses dominate riverine habitats at low elevations all across the Indian subcontinent. It has been found that C4 plants are more efficient than C3 plants at high temperatures (Ehleringer, 1978; Quade and Cerling, 1995). However, C3-C4 shift has been observed on a global scale at ~8–6 Ma ago (Cerling et al., 1993; Quade and Cerling, 1995) and at the Miocene/Pliocene boundary (Cerling et al., 1997). Raymo and Ruddiman (1992) and Quade et al. (1995) have emphasized that regional climates could affect global CO_2 levels (see also Molnar et al., 1993; Filippelli, 1997). Hoorn et al. (2000) are of the opinion that the Himalayan foothills and the Gangetic

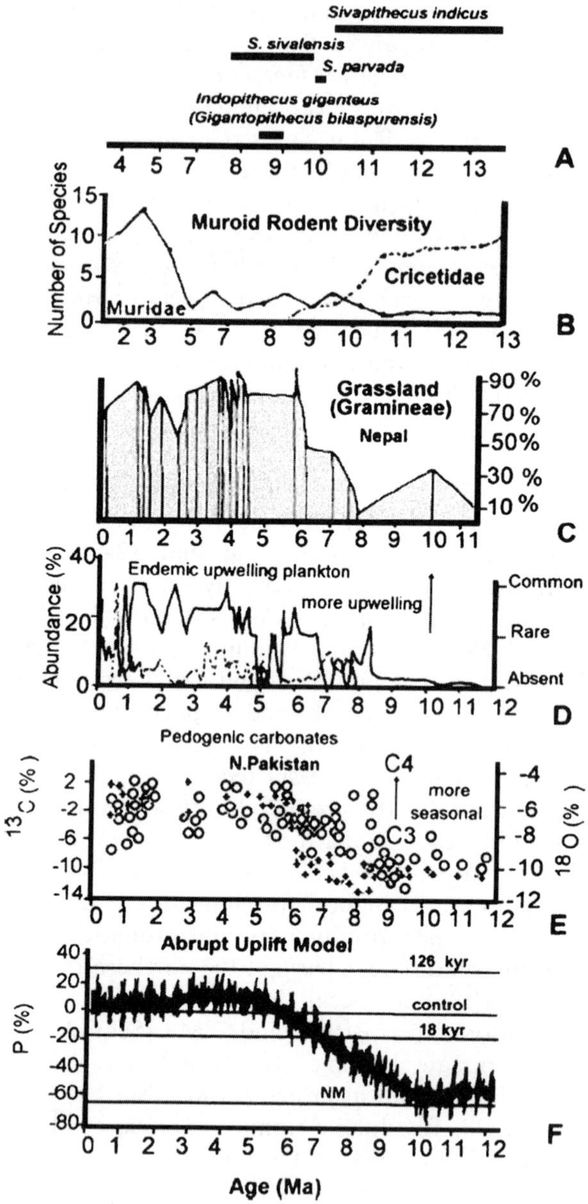

Fig. 5 Hominoid occurrences and paleoclimatological and paleoenvironmental indicators of the Siwalik sequence (modified after Patnaik et al., 2005). (A) Age ranges of Siwalik apes, data from this study and Flynn et al. (1995), adjusted following Cande and Kent (1992, 1995). (B) Muroid rodent diversity, from Patnaik (2003). (C) Percentage of grassland, redrawn from Hoorn et al. (2000). (D) Time series of endemic upwelling species of plankton: relative abundance of foraminifera, *Globigerina bulloides* (dotted line), and radiolarian *Actinomma* spp. (solid line), as indices of upwelling induced by monsoon winds, redrawn from Prell and Kutzbach (1992) Fig. 4c; arrow indicates direction of increased upwelling. (E) Time series of carbon (crosses) and oxygen (circles) isotopes in soil carbonates in Pakistan as an index of a shift in regional climate and C4 vegetation, redrawn from Prell and Kutzbach (1992) Fig. 4d, after Quade et al. (1989); arrow indicates direction of a more arid, highly seasonal environment. (F) Change in precipitation (P) using the abrupt uplift model, redrawnfrom Prell and Kutzbach (1992)

floodplain in Nepal were mainly forested with subtropical to temperate broad-leafed and tropical forest taxa during the late Middle Miocene to early Late Miocene (~11.5–8 Ma), respectively. Between the early to late Late Miocene (~8–6.5 Ma), grassland replaced subtropical and temperate broad-leafed forest in the same regions. Intensification of monsoon and disturbance of the vegetation on the slopes due to uplift might have facilitated such a change (Hoorn et al., 2000). Palaeosols from various Siwalik sequences also indicate marked seasonality in rainfall (Retallack, 1991, 1995). Recently, Barry et al. (2002) observed three very brief periods of high Siwalik faunal turnover at 10.3, 7.8, and 7.37–7.04 Ma. Latest Miocene faunal turnover at 7.37 and 7.04 Ma, particularly, has been correlated with expansion of C4 grasses, the oxygen isotope and sedimentological evidence indicating an increasingly drier and more seasonal climate (Barry et al., 2002). A dramatic change in the diversity of muroid rodents (from cricetid-dominated to murid-dominated) at ~9–8 Ma has also been noted (Patnaik, 2003, see Fig. 5). Patnaik (2003) has attributed this replacement of cricetids by murids in the Late Miocene to an intensification of the monsoons, as most of the present-day murids are found in the monsoon-affected region of the world and their reproduction-oriented life history strategy is better suited to unpredictable and seasonal climatic conditions. Therefore, a change in climate in the Late Miocene time probably led to expansion of grasslands at the expense of rainforests, which was probably the preferred habitat of *Sivapithecus* and *Sivaladapis* (Nelson, 2003a,b).

Fig. 6 Elwyn Simons and colleagues at Haritalynger in 1968. From left to right: Lakmir Singh, Prithijit Chatrath, Steve Boyer, and Elwyn Simons

Fig. 7 A view of terraced fields in Haritalyangar today

Conclusions

Elwyn Simons' efforts in the late sixties and early seventies led to an upsurge in research around Haritalyangar, resulting in some outstanding discoveries of fossil primates (Fig. 6). In recent years, Haritalyangar has yielded additional new evidence of fossil apes, information about their habitat and the main causes of their extinction. But, there remain many key questions yet to be answered. We still know virtually nothing about *Gigantopithecus* (*Indopithecus*) except the size and shape of its teeth and lower jaw. Nothing is known about the morphology of the cranium or about its locomotor habits. Nor do we have any published information about the cranial or postcranial anatomy of *Sivaladapis*. The fossiliferous deposits at Haritalyangar still have still a lot to offer, although there is increasingly less nonagricultural area available each year (Fig. 7). The vision and direction shown by Elwyn and his field crews need to be taken up more vigorously and systematically by the new generation of paleoprimatologists to further our understanding of primate evolution in India.

Acknowledgment I would like to thank Prof. John Fleagle for inviting me to contribute to this volume in honour of Prof. Elwyn Simons. Prof. Ashok Sahni has generously permitted the rodent and sivaladapid specimens from Haritalyangar to be photographed. Financial support from DST, New Delhi (No SR/S4ES-171/2005) and the Wenner-Gren Foundation is thankfully acknowledged.

References

Amano, K., and Taira, A. (1992). Two-phase uplift of higher Himalayas since 17 Ma. *Geology* 20: 391–394.
Andrews, P., and Cronin, J. (1982). The relationships of *Sivapithecus* and *Ramapithecus* and the evolution of the orangutan. *Nature* 297: 541–546.
Andrews, P., and Tekkaya, I. (1980). A revision of the Turkish Miocene hominoid *Sivapithecus meteai*. *Palaeontology* 23:83–95.
Barry, J. C., Morgan, M. E., Flynn, L. J., Pilbeam, D., Behrensmeyer, A. K., Raza, S. M., Khan I. A., Badgley C., Hicks J., and Kelley, J. (2002). Faunal and environmental change in the latest Miocene Siwaliks of northern Pakistan. *Paleobiol. Mem.* 28(2): 1–72.
Berggren, W. A., Kent, D. V., and van Couvering, J. A. (1985). Neogene geochronology and chronostratigraphy. In: Snelling, N. J.(ed.), *The Chronology of the Geological Record*. Blackwell Scientific, Oxford, pp. 211–260.
Burkle, L. H. (1989). Distribution of diatoms in sediments of the northern Indian ocean: Relationship to physical oceanography. *Mar. Micropaleontol.*, 15: 53–65.
Cameron, D. W. (2001). The taxonomic status of the Siwalik Late Miocene hominid *Gigantopithecus* ('*Indopithecus*'). *Himalayan Geol.* 22: 29–34.
Cameron, D. W. (2003). A functional and phylogenetic interpretation of the late Miocene Siwalik hominid *Indopithecus* and the Chinese Pleistocene hominid *Gigantopithecus*. *Himalayan Geol.* 24: 19–28.
Cande, S. C., and Kent, D. V. (1992). A new geomagnetic polarity time scale for the Late Cretaceous and Cenozoic. *J. Geophys. Res.* 97:13917–13951.
Cande, S. C., and Kent, D. V. J. (1995). Revised calibration of geomagnetic polarity time scale for the Late Cretaceous and Cenozoic. *J. Geophys. Res.* 100(B): 6093–6095.
Cerling, T. E., Wang, Y., and Quade, J. (1993). Expansion of C4 ecosystems as an indicator of global ecological change in the Late Miocene. *Nature* 361: 344–345.
Cerling, T. E., Harris, J. M., MacFadden, M. G., Leakey, M. G., and Quade, J. (1997). Global vegetation change through the Miocene/Pliocene boundary. *Nature* 389: 153–158.
Chopra, S. R. K., and Kaul, S. (1979). A new species of *Pliopithecus* from the Indian Siwaliks. *J. Hum. Evol.* 8:475–477.
Chopra, S. R. K., and Vasishat, R. N. (1979). Siwalik fossil tree shrew from Haritalyangar, India. *Nature* 281:214–215.
Ciochon, R. L., Piperno, D. R., and Thompson, R. G. (1990). Opal phytoliths found on the teeth of the extinct ape *Gigantopithecus blacki*: Implications for paleodietary studies. *P. Natl Acad. Sci. USA*, 87: 8120–8124.
Ehleringer, J. R. (1978). Carbon Isotope ratios and physiological processes in aridland plants. In: Rundel, P. W., Ehlinger, J. R., and Nagy, K. R. (eds.), *Stable Isotopes in Ecological Research*. Springer, New York, pp. 41–44.
Filippelli, G. M. (1997). Intensification of the Asian Monsoon and a chemical weathering event in the late Miocene–early Pliocene: Implications for late Neogene climate change. *Geology* 25: 27–30.
Flynn, L. J., Barry, J. C., Morgan, M. E., Pilbeam, D. R., Jacobs, L. L., and Lindsay, E. H. (1995). Neogene Siwalik mammalian lineages: Species longevities, rates of change, and modes of speciation. *Palaeogeogr. Palaeocl.* 115: 249–264.
Flynn, L. J., Sahni, A., Jaeger, J. J., Singh, B., Bhatia, S. B. (1990). Additional fossil rodents from the Siwalik Beds of India. *Proc. Koninklijke Nederlandse Akademie van. Wetenschappen* 93: 7–20.

Gingerich, P. D. and Sahni, A. (1979). *Indraloris* and *Sivaladapis*: Miocene adapid primates from the Siwaliks of India and Pakistan. *Nature* 279:415–416.
Gingerich, P. D. and Sahni, A. (1984). Dentition of *Sivaladapis nagrii* (Adapidae) from the Late Miocene of India. *Int. J.Primatol.* 5(1):63–79.
Gregory, W. K., Hellman, M., and Lewis, G. E. (1938). Fossil anthropoids of the Yale Cambridge India Expedition of 1935. Carnegie Ins. Washington Pub. No. 495.
Gupta, A. K. and Melice, J.-L. (2003). Orbital forcing of the Plio–Pleistocene Indian monsoons: Benthic foraminiferal proxies from ODP Site 758. *Curr. Sci. India* 85: 179–184.
Han, K. Z., and Zhao, L. X. (2002). Dental caries of *Gigantopithecus blacki* from Hubei Province of China. *Acta Anthropologica Sinica* 21: 191–198.
Harrison, T. M., Copeland, P., Hall, S. A., Quade, J., Burner, S., Ojha, T. P., and Kidd W. S. F. (1993). Isotopic preservation of Himalaya-Tibetan uplift, denudation, and climatic histories in two molasse deposits. *J Geol.* 101: 157–175.
Hay, W. W. (1996). Tectonics and climate. *Geologische Rundschau* 85: 409–437.
Hoorn, C., Ojha, T., and Quade, J. (2000). Palynological evidence for vegetation development and climatic change in the sub-Himalayan zone (Neogene, Central Nepal). *Palaeogeogr. Palaeocl.* 163: 133–161.
Johnson G. D., Opdyke N. D., Tandon S. K., and Nanda A. C. (1983) The magnetic polarity stratigraphy of the Siwaliks group at Haritalyangar (India) and a new last appearance datum for *Ramapithecus* and *Sivapithecus* in Asia. *Palaeogeogr. Palaeocl.* 44: 223–249.
Kappelman, J., Kelley, J., Pilbeam, D. R., Sheikh, K. A., Ward, S., Anwar, M., Barry, J. C., Brown, B., Hake, P., Johnson, N. M., Raza, S. M., and Imbrahim Shah, S. M. (1991). The earliest occurrence of *Sivapithecus* from the middle Miocene Chinji Formation of Pakistan. *J. Hum. Evol.* 21: 61–73.
Kutzbach, J. E., Prell, W. L., and Ruddiman, W. F. (1993). Sensitivity of Eurasian climate to surface uplift of the Tibetan Plateau. *J. Geol.* 101: 177–190.
Lewis, G. E. (1934). Preliminary notice of man-like apes from India. *Am. J. Sci.* 27:161–179.
Lewis, G. E. (1937). Taxonomic syllabus of Siwalik fossil anthropoids. *Am. J. Sci.* 34:139–147.
Molnar, P., England, P., and Martinod, J. (1993). Mantle dynamics, uplift of the Tibetan plateau, and the Indian monsoon. *Rev. Geophys.* 31: 357–396.
Nelson, S. (2003a). The preferred habitats of *Sivapithecus* and paleoenvironmental changes leading to its extinction in the Siwaliks of Pakistan. *Asian Paleoprimatol.* 3: 36.
Nelson, S. V. (2003b). *The Extinction of Sivapithecus: Faunal and Environmental Changes Surrounding the Disappearance of a Miocene Hominoid in the Siwaliks of Pakistan.* American School of Prehistoric Research Monograph Series. Brill, Boston.
Patnaik, R., Cameron, D., Sharma, J. C., and Hogarth, J. (2005). Extinction of Siwalik fossil apes: A review based on a new fossil tooth and on palaeoecological and palaeoclimatological evidence *Anthropol. Sci.* 113: 65–72.
Patnaik, R. (2003). Reconstruction of Upper Siwalik palaeoecology and palaeoclimatology using microfossil palaeocommunities. *Palaeogeogr. Palaeocl.* 197: 133–150.
Patnaik, R., and Cameron, D. W. (1997). New Miocene fossil ape locality, Dangar, Hari-Talyangar region, Siwaliks, northern India. *J. Hum. Evol.* 32: 93–97.
Patnaik, R., Cameron, D., Sahni, A., Pillans, B., Chatrath, P., Simons, E., Williams, M., and Bibi, F. (in review). Ostrich eggshells of a new oospecies (*Struthiolithus sivalensis* sp. nov) from 10.1 Ma old Middle Siwalik sediments of India. *Curr. Sci. India.*
Pilbeam, D. R. (1966). Notes on *Ramapithecus*, the earliest known hominid, and *Dryopithecus*. *Am. J. Phys. Anthropol.* 25:1–6.
Pilbeam, D. R. (1969a). Early hominidae and cranial capacity. *Nature* 224 (5217):386
Pilbeam, D. R. (1969b). Newly recognized mandible of *Ramapithecus*. *Nature* 222:1093–1094.
Pilbeam, D. R. (1982). New hominoid skull material from the Miocene of Pakistan. *Nature* 295: 232–234.

Pilbeam, D. R. (1983). *Ramapithecus* disowned. *Science* 23(1):24–25.
Pilgrim, G. E. (1910a). Notices of new mammalian genera and species from the Tertiaries freshwater deposits of India. *Records Geological Survey of India* 40:63–71.
Pilgrim, G. E. (1910b). New Siwalik Primates and their bearing on the question of the evolution of Man and the Anthropoids. *Records Geological Survey of India* 45:1–74.
Pillans, B., Williams, M., Cameron, D., Patnaik, R., Hogarth, J., Sahni, A., Sharma, J. C., Williams, F., and Bernor, R. (2005). Revised correlation of the Haritalyangar magnetostratigraphy, Indian Siwaliks: implications for the age of the Miocene hominids *Indopithecus* and *Sivapithecus*, with a note on a new hominid tooth. *J. Hum. Evol.* 48: 507–515
Prell, W. L., and Kutzbach, J. E. (1992). Sensitivity of the Indian monsoon to forcing parameters and implications for its evolution. *Nature* 360: 647–652.
Prell, W. L., Murray, D. W., Clemens, S. C., and Anderson D. M. (1992). Evolution and variability of the Indian Ocean summer monsoon: Evidence from the western Arabian Sea drilling program. In: Duncan, R. A. (ed.), *The Indian Ocean: A Synthesis of Results from the Ocean Drilling Program.* Geophysical Monograph, No. 70, American Geophysical Union, Washington DC, pp. 447–469.
Quade, J. T., and Cerling, T. E. (1995). Expansion of C4 grasses in the Late Miocene of northern Pakistan: evidence from stable isotopes in paleosols. *Palaeogeogr. Palaeocl.* 115: 91–116.
Quade, J. T., Cerling, T. E., and Bowman, J. R. (1989). Development of the Asian monsoon revealed by marked ecological shift in the latest Miocene of northern Pakistan. *Nature* 342: 163–166.
Quade, J., Cater, J. M., Ojha, T. P., Adam, J., and Harrison, T. M. (1995). Late Miocene environmental change in Nepal and the northern Indian subcontinent: stable isotopic evidence from paleosols. *Geol. Soc. Am. Bull.* 107:1381–1397.
Ramstein, G., Fluteau, F., Besse, J., and Joussaume, S. (1997). Effect of orogeny, plate motion and land-sea distribution on Eurasian climate change over the past 30 million years. *Nature* 386: 788–795.
Raymo, M. E., and Ruddiman, W. F. (1992). Tectonic forcing of late Cenozoic climate. *Nature* 359: 117–122.
Retallack, G. J. (1991). *Miocene Paleosols and Ape Habitats of Pakistan and Kenya.* Oxford University Press, New York.
Retallack, G. J. (1995). Palaeosols of the Siwalik Group as a 15 Mys record of South Asian palaeoclimate. *Mem. Geol. Soc. India,* 32: 36–51.
Ruddiman, W. F., and Kutzbach, J. E. (1989). Forcing of late Cenozoic northern hemisphere climate by plateau uplift in Southern Asia and the American West. *J. Geophys. Res.* 94: 409–418.
Sahni, A., and Khare, S. K. (1976). Siwalik insectivore. *J. Geol. Soc. India* 17:114–116.
Sahni, A., and Khare, S. K. (1977). A Middle Siwalik fish fauna from Ladhyanai (Haritalyangar) Himachal Pradesh. *Biol. Mem. Vertebr. Palaeontol.* 2:187–221.
Sankhyan, A. R. (1985). Late Occcurrence of *Sivapithecus* in Indian Siwaliks. *J. Hum. Evol.* 14:573–578.
Simons, E. L. (1961). The phyletic position of *Ramapithecus. Peabody Mus. Postilla* 57:1–9.
Simons, E. L. (1963). Some fallacies in the study of hominid phylogeny. *Science* 141: 879–889.
Simons, E. L. (1964). On the mandible of *Ramapithecus. P. Natl. Acad. Sci. USA* 51: 528–535.
Simons, E. L. (1972). *Primate Evolution: An introduction to Man's place in Nature.* Macmillan, New York, p. 322.
Simons, E. L. (1976a). The nature of transition in the dental mechanism from Pongids and hominids. *J. Hum. Evol.* 5:511–528.
Simons, E. L. (1976b). Relationship between *Dryopithecus, Sivapithecus* and *Ramapithecus* and their bearing on hominid origins, In: Tobias, P. V., and Coppens, Y. (eds), *Les Plus*

Anciens Hominides (Colloque 6,9 Union Internationale des Sciences Prehistorique et Protohistorique, Nice). CNRS, Paris, pp. 60–67.
Simons, E. L., and Chopra, S. R. K. (1969). *Gigantopithecus* (Pongidae, Hominoidea): A new species from Northern India. *Postilla* 138: 1–8.
Simons, E. L., and Ettel, P. C. (1970). *Gigantopithecus*. *Sci. Am.* 222: 77–85.
Simons, E. L., and Pilbeam, D. (1965). Preliminary revision of Dryopithecinae (Pongidae, Anthropoidae). *Folia. Primatol.* 3:1–152.
Simons, E. L., and Pilbeam, D. (1972). Hominid Paleoprimatology, In: Tuttle, R. (ed.), *The functional and evolutionary Biology of Primates*. Aldine, Atherton, pp. 36–62
Simons, E. L., and Pilbeam, D. (1973). A gorilla-sized ape from the Miocene of India. *Science* 173:23–27.
Simons, E. L., Pilbeam, D., and Boyer, S. (1971). Appearance of *Hipparion* in the Tertiary of the Siwalik Hills of North India, Kashmir and West Pakistan. *Nature* 229: 408–409.
Schrader, H. J. (1974). Cenozoic marine planktonic diatom stratigraphy of the tropical Indian Ocean. *Initial Reports of the Deep Sea Drilling Project* 24: 887–968.
Tiwari, B. N. (1996). Middle Siwalik murid rodents from Ladhyani, Bilaspur, Himachal Pradesh. In: Dehra, D., Pandey, J., Azmi, R. J., Bhandari, A., and Dave, A. (Eds.), *Ind. Coll. Micropalaeo. Strat.* 15: 513–518.
Von Koenigswald, G. H. R. (1950). Bemerkungen zu *Dryopithecus giganteus* Pilgrim. *Eclogae Geologicae Helvetiae* 42: 515–519.
Vasishat, R. N. (1985). *Antecedents of early man in northwestern India: Paleontological and Paleoecological evidences*. Inter India Publications, New Delhi.

Revisiting Primate Postcrania from the Pondaung Formation of Myanmar

The Purported Anthropoid Astragalus

Gregg F. Gunnell and Russell L. Ciochon

Introduction

Late middle Eocene primates from the Pondaung Formation in Myanmar (Burma) have been known since the early 20th century with *Pondaungia* being the first taxon formally described (Pilgrim, 1927), followed ten years later by *Amphipithecus* (Colbert, 1937). The remains originally described for both of these taxa were fragmentary, each consisting of a single individual represented by jaw fragments. Since that time, an additional 20 specimens or so (single individuals are often represented by several numbers) of *Pondaungia* and *Amphipithecus* have been recovered but only four represent anything other than lower or upper dentitions. Of these four, two represent possible cranial fragments (Gunnell et al., 2002) although these have recently been questioned both as to their taxonomic assignment and anatomical identification (Beard et al., 2005).

The other two non-dental specimens that have been associated with *Pondaungia* or *Amphipithecus* are NMMP (National Museum of Myanmar Primate) 20 and NMMP 39. NMMP 20 consists of a complete left humerus, fragments of a right humerus, left and right ulnar fragments, and a left calcaneum (Ciochon et al., 2001). NMMP 39 is an isolated left astragalus (Marivaux et al., 2003). NMMP 20 is from Sabapondaung Kyitchaung in the Bahin area, while NMMP 39 was found at Segyauk Kyitchaung in the Mogaung area, approximately 12 km northwest of the NMMP 20 locality (Fig. 1).

Amphipithecidae (including *Amphipithecus, Pondaungia, Myanmarpithecus,* and *Siamopithecus*) have had a colorful taxonomic history. Depending on one's preferred interpretation of dental evidence, these animals have been viewed alternatively as anthropoids, as a sister taxon to anthropoids (Colbert, 1938; Simons, 1971; Ba Maw et al., 1979; Ciochon et al., 1985; Godinot, 1994; Jaeger et al., 1998; Chaimanee et al., 2000; Takai et al., 2000; Marivaux et al., 2003)

Gregg F. Gunnell
Museum of Paleontology, University of Michigan, 1109 Geddes Avenue, Ann Arbor, Michigan 48109-1079, USA, Telephone: – (734) 936-1385, FAX: – (734) 936-1380
ggunnell@umich.edu

J. G. Fleagle, C. C. Gilbert (eds.), *Elwyn Simons: A Search for Origins.*

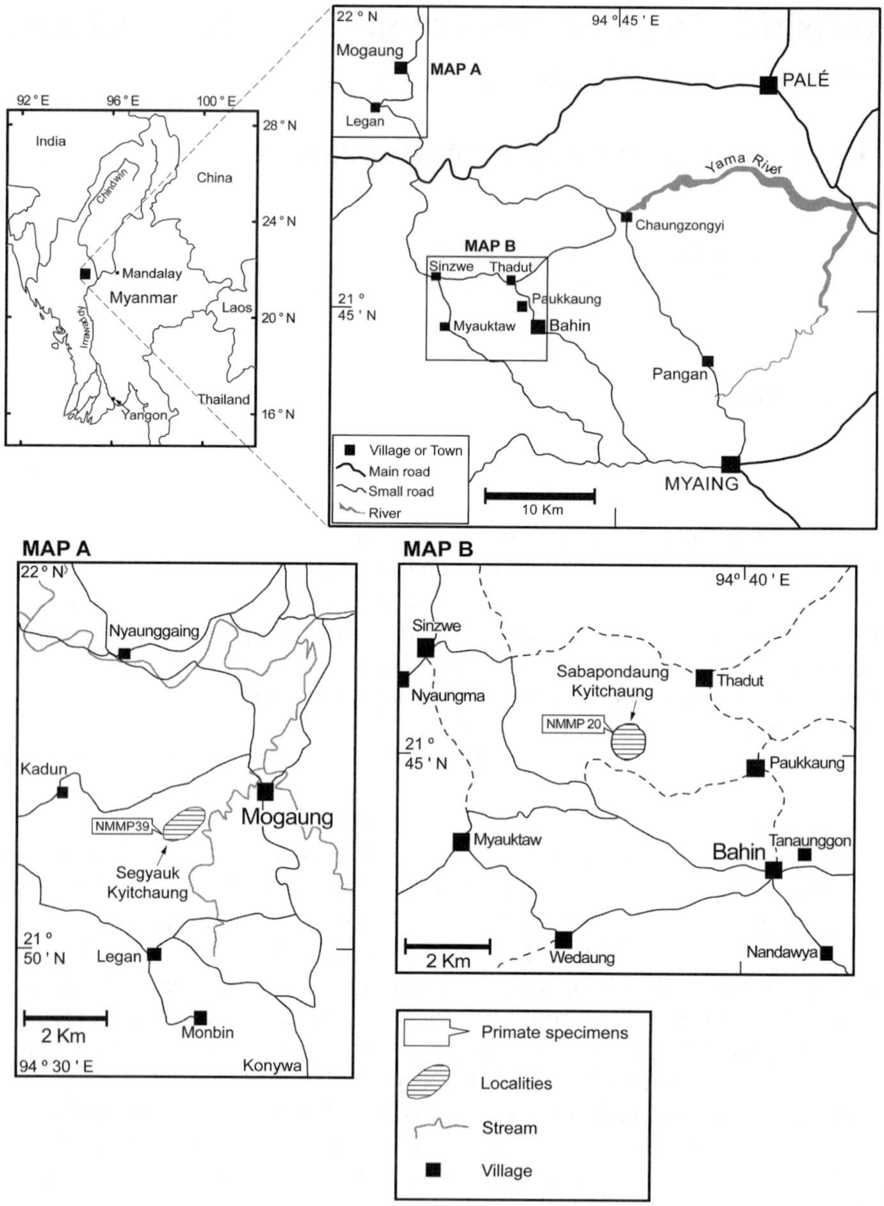

Fig. 1 Maps of Myanmar (Burma) showing localities where postcranial remains of amphipithecids have been found. Map A depicts the Mogaung area showing the position of Segyauk Kyitchaung, the locality of NMMP 39. Map B depicts the Bahin area showing the location of Sabapondaung Kyitchaung, the locality of NMMP 20

as nonanthropoid euprimates, or even not representing primates at all (Von Koenigswald, 1965; Simons, 1972; Szalay, 1970, 1972; Ciochon and Holroyd, 1994; Ciochon et al., 2001; Ciochon and Gunnell, 2002; Gunnell et al., 2002). Plausible cases can be made interpreting dental characters as shared and derived with those of some Fayum anthropoids (Jaeger et al., 1998) or as convergent on Fayum anthropoids because of hard object feeding (Ciochon and Gunnell, 2004; Kay et al., 2004).

The cranial material is equally enigmatic. Two, nearly identical partial bone fragments (NMMP 19 and NMMP 27) have been interpreted as possibly representing frontal bones of *Amphipithecus* (Gunnell et al., 2002; Takai et al., 2003). NMMP 19 was found in apparent association with NMMP 18, a maxilla of *Amphipithecus*, bolstering the interpretation that both elements do belong to an amphipithecid (Takai and Shigehara, 2004). If these two frontal fragments are correctly identified they indicate that whatever animal they belonged to lacked postorbital closure, had large olfactory bulbs, and the frontal bones were unfused. These features are not those expected of a crown or advanced stem, anthropoid.

Beard et al. (2005) have questioned the identification of these cranial fragments as those of an amphipithecid or even as cranial fragments of any mammal. They note the unusual thickness of the dermal bone, the inflated nature of the frontal trigon, and the robust nature of the descending process of the frontal bone as features arguing against these fragments being representative of a primate. The authors do not offer their own taxonomic interpretation of these elements only noting that previous interpretations are incorrect and that there are similarities to turtles.

We would observe that the morphology presented by these frontals, while unusual in some respects, is not as dramatically different from primates as Beard et al. (2005) have suggested. One of the main differences cited by Beard et al. (2005) as being distinct from primates is in the relatively great thickness of these elements. However, this is also true of the archaic, small-brained, primate-like mammal *Plesiadapis* (Gingerich and Gunnell, 2005) and suggests that relatively thick frontal bones may well be primitive for primates. We suggest that the thick dermal bone is probably the result of a combination of a relatively large, primitive primate and a relatively small brain. Further, if interpretations of hard-object feeding are correct for amphipithecids (Ciochon and Gunnell, 2004; Kay et al., 2004), then it would not be surprising to find reinforced and thickened cranial bones at the precise point where the main chewing forces (above the molars) are directed through the skull. Nevertheless, until more complete specimens are found, it will not be possible to determine if these cranial fragments are those of an amphipithecid or not.

We believe the two postcranial specimens offer further evidence of the non-anthropoid nature of amphipithecids. Both represent primates of an appropriate size to be either *Amphipithecus* or *Pondaungia* (Egi et al., 2004). Ciochon et al. (2001) described NMMP 20 and noted several features of the distal humerus that are shared between it and North American notharctine adapiforms. These features include (Fig. 2): (1) a rounded capitulum; (2) the presence of a distinct trochlear gutter; (3) a straight medial epicondyle; (4) absence of a dorsoepitrochlear pit; and (5) an expanded lateral epicondyle. In contrast, primitive anthropoids (such as *Aegyptopithecus, Catopithecus,* and *Proteopithecus*) have a cylindrical capitulum, lack a trochlear gutter, have a posteriorly angled medial epicondyle, possess a large dorsoepitrochlear pit, and lack an expanded lateral epicondyle (Fleagle and Simons, 1978; Seiffert et al., 2000). The proximal humerus of NMMP 20 also resembles North American notharctines in sharing a broad, round, and proximally extended humeral head and a shallow bicipital groove.

The calcaneal fragment of *Pondaungia* (Fig. 3) preserves the proximal calcaneal facet and the distal tuber. The distal tuber is relatively elongate, the peroneal tuberosity is relatively robust and located proximally, the proximal calcaneal facet is relatively broad and angled plantomedially, the distal

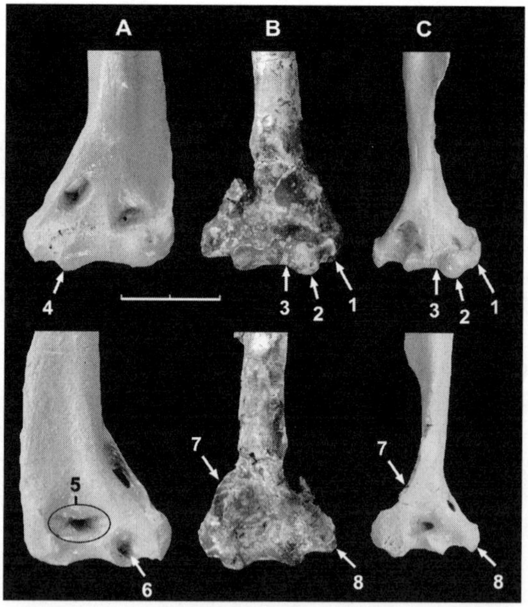

Fig. 2 Comparison of the distal humerus of NMMP 20 assigned to *Pondaungia* (B) with the early Oligocene anthropoid *Aegyptopithecus* (A) from Egypt and the middle Eocene adapiform *Smilodectes* (C) from western North America (Wyoming). Features shared between *Pondaungia* and *Smilodectes* include: 1) expanded lateral epicondyle and capitular tail; 2) rounded capitulum; 3) distinct trochlear gutter; 7) distinct supinator crest; and 8) straight, medially extended medial epicondyle. Features distinguishing *Aegyptopithecus* (and other anthropoids) from *Pondaungia* and *Smilodectes* include: 4) distally extended medial trochlear lip; 5) large and deep olecranon fossa; 6) presence of a distinct dorsoepitrochlear pit. Also note the lack of an expanded lateral epicondyle and supinator crest, the lack of a trochlear gutter, and the more cylindrical capitulum in *Aegyptopithecus*. Scale in centimeters

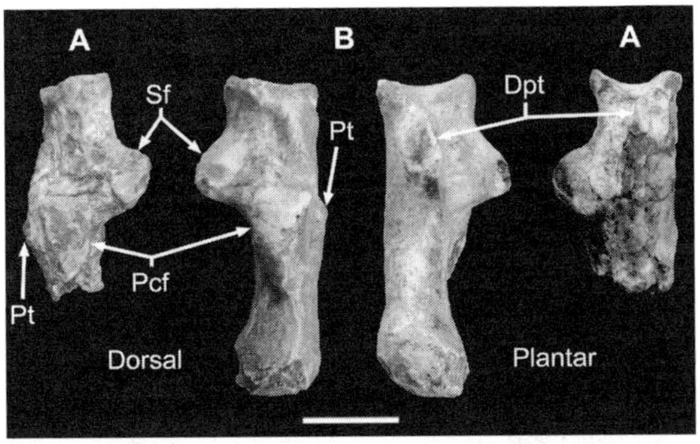

Fig. 3 Comparison of left distal calcaneum of *Pondaungia* (A) with right calcaneum of the European adapiform *Leptadapis* (B) in dorsal and plantar views. Arrows and abbreviations indicate the following: Dpt, distal plantar tubercle; Pcf, posterior calcaneal facet; Pt, peroneal tubercle; Sf, sustentacular facet. Scale = 1 cm

calcaneal facet (sustentacular facet) is mediolaterally narrow and extends nearly to the distal margin, and the distal plantar tubercle is robust and positioned at the distal margin. The calcaneocuboid facet is oriented nearly dorsoplantarly, is relatively narrow mediolaterally, and has a very deep pit that is notched medially. The distal calcaneum of *Pondaungia* most closely resembles that of European adapines (Fig. 3) and North American notharctines (Ciochon et al., 2001; Ciochon & Gunnell, 2004).

Many of the features of the humerus and calcaneum shared in common between NMMP 20 and adapiforms have traditionally been viewed as shared, primitive characters of primates (Ciochon and Gunnell, 2004). However, given the apparent antiquity of the anthropoid lineage and the number of additional possible sister taxa that now are brought into consideration because of this (Ford, 1994; Beard, 2004; Seiffert et al., 2005; Miller et al., 2005), it has now become more problematic to determine character polarities, especially given the fact that it is not possible to directly determine primate ancestry as yet. It is now distinctly possible that features of the postcranial skeleton of strepsirhines, many adapiforms, some omomyiforms, and amphipithecids may, in fact, be derived and distinct from those of anthropoids, regardless of whether they are shared or not.

We note for the purposes of this paper that we consider Strepsirhini to include only Lemuriformes and Lorisiformes that share a tooth-comb and strepsirhine nasal characteristics and Haplorhini to include only Anthropoidea and living and extinct Tarsiidae. Fossil strepsirhines may exist in North Africa (Seiffert et al., 2003, 2005) but await definitive proof of the presence of a tooth-comb. We consider Adapiformes (including the family Amphipithecidae) and Omomyiformes to be distinct groups whose relationships to living primates have yet to be adequately demonstrated. We consider Eosimiidae to possibly represent a basal simian group but contend that this has yet to be convincingly demonstrated, nor has its inclusion in Haplorhini been unequivocally confirmed.

Analysis of NMMP 39

The astragalus (Fig. 4) that has been attributed to an amphipithecid was interpreted to be anthropoid-like by Marivaux et al. (2003). This specimen (NMMP 39) is a left astragalus from Segyauk Kyitchaung in the Mogaung Area of west-central Myanmar (Fig. 1, Map A). It is the only primate element found at this locality but appears to be of the right size to be associated with either *Pondaungia* or *Amphipithecus* (Egi et al., 2004). Among the features of this astragalus cited as anthropoid-like by Marivaux et al. (2003) are steep medial and lateral trochlear facets, medial tibial facet flat, limited in extent, and lacking a cotylar fossa, and lacking a prominent trochlear shelf. Additionally, Marivaux et al. (2003) noted the broad and oval-shaped astragalar head that is medially extended, stating that this feature is also seen in the Fayum anthropoid *Apidium phiomense*.

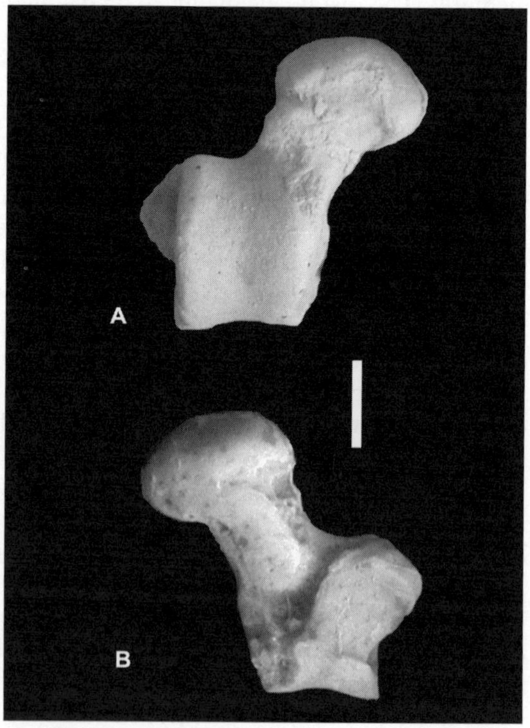

Fig. 4 NMMP 39, left astragalus of an amphipithecid from Segyauk Kyitchaung, Mogaung, Pondaung Hills, Myanmar in A) dorsal view and B) plantar view. Scale = 5 mm

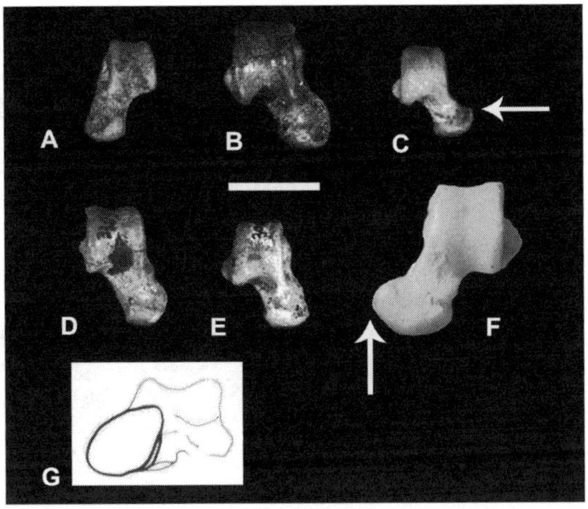

Fig. 5 NMMP 39 (F) compared with five *Apidium phiomense* (A–G) astragali from the upper sequence, Jebel Qatrani Formation, Fayum, Egypt. A) DPC 2450; B) DPC 1130; C) DPC 1001; D) DPC 3105; E) DPC 3054. Note the expanded medial astragalar head (arrows) in NMMP 39 and DPC 1001 and lack of medially expanded head in all other specimens of *Apidium*. Also note dorsally angled head of *Apidium* (image at bottom) (G). Scale = 1 cm. *Apidium* photographs from Fleagle and Simons (1995)

Taking the latter comment first, it is true that NMMP 39 does have a medially extended astragalar head and that one published specimen of *Apidium phiomense* (DPC 1001) shares a similar morphology (Fig. 5). However, four other published specimens of *Apidium phiomense* (Fleagle and Simons, 1995) lack any medial expansion of the astragalar head – in fact if anything the medial astragalar head is compressed rather than expanded (see Fig. 5A, B, D, E). One could just as easily say that the medially expanded astragalar head of NMMP 39 is also seen in some ischyromyid rodents, a statement that is both equally as true and equally as meaningless phylogenetically as noting that NMMP 39 looks like DPC 1001 in this feature. Additionally, the shape and orientation of the astragalar head in all known specimens of *Apidium* (e.g. Fig. 5G) is strikingly different from that seen in NMMP 39.

NMMP 39 does have relatively steep medial and lateral trochlear facets with the lateral facet flaring plantarly. A comparison of these features among a series of extant and fossil primates (Fig. 6) reveals that the lateral facet is similar to that seen in *Cebus* but less similar to the more primitive *Dolichocebus* and unlike the primitive anthropoids *Catopithecus* and *Proteopithecus*, the former having a gentle sloping lateral facet and the latter having a steep-sided lateral facet lacking any apparent plantar flaring.

As noted by Marivaux et al. (2003), the medial tibial facet of NMMP 39 is flat, limited in extent, and lacks a cotylar fossa. However, these features hardly suggest anthropoid affinities, in fact they appear much more similar to living strepsirhines and fossil adapiforms (Fig. 7). The most primitive known

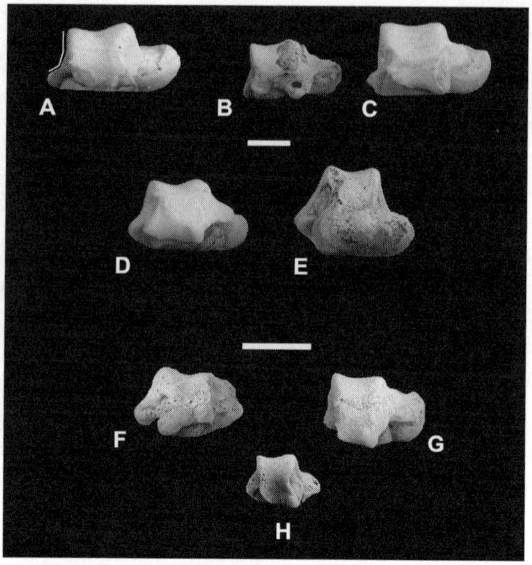

Fig. 6 Series of primate astragali in proximal view showing disposition of lateral trochlear facet. A) NMMP 39, amphipithecid; B) MACN-CH 362, *Dolichocebus gaimanensis*; C) UMMZ 126129, *Cebus apella*; D) UMMZ 160910, *Lemur catta*; E) USNM 425518, *Notharctus robustior*; F) DPC 10037, *Catopithecus browni* (reversed); G) DPC 15417, *Proteopithecus sylviae;* H) UM 98604, *Omomys carteri*. Scales = 5 mm

anthropoid astragali of *Catopithecus* and *Proteopithecus* (Seiffert and Simons, 2001) both have medial tibial facets that extend onto the medial neck of their astragali and there is a small yet distinct cotylar facet developed in each as well (less distinct in *Proteopithecus* but clearly present). Figure 7 shows the extent of the medial tibial facet in NMMP 39 compared with *Cebus*, *Lemur*, and the fossil adapiform *Smilodectes*. In general shape, the medial facet of NMMP 39 is not unlike *Cebus* but it also is not terribly different from that in *Smilodectes*. Unlike *Cebus*, NMMP 39 lacks a cotylar fossa (indicated as the rounded distal extent

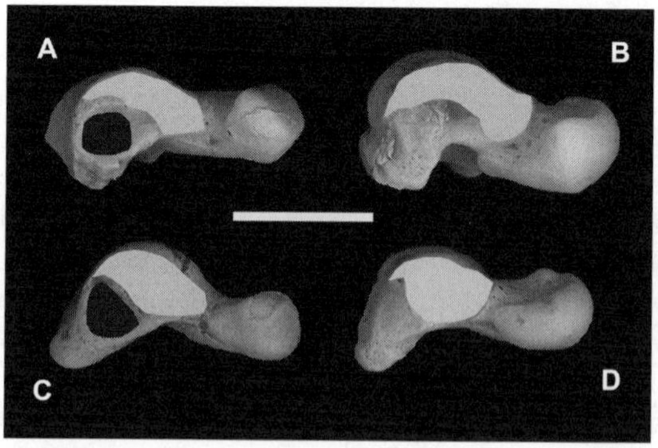

Fig. 7 Comparison of medial tibial facet in a series of primates. A) NMMP 39, amphipithecid; B) UMMZ 126129, *Cebus apella*; C) UM 95526, *Smilodectes mcgrewi*; D) UMMZ 160910, *Lemur catta*. Light shading indicates extent of medial tibial facet – note rounded area at distal extent of facet in *Cebus* (B) indicating presence of cotylar fossa. Dark shading indicate distinct concavity shared by amphipithecid and notharctine adapiform *Smilodectes*. Scale = 1 cm

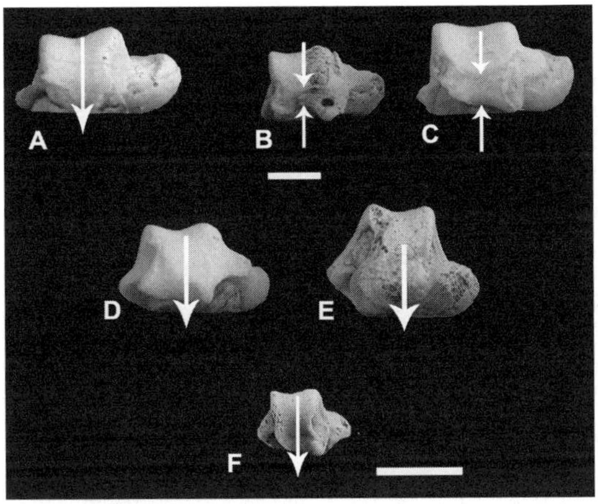

Fig. 8 Series of primate astragali in proximal view showing disposition of trochlear shelf. A) NMMP 39, amphipithecid; B) MACN-CH 362, *Dolichocebus gaimanensis*; C) UMMZ 126129, *Cebus apella*; D) UMMZ 160910, *Lemur catta*; E) USNM 425518, *Notharctus robustior*; F) UM 98604, *Omomys carteri*. Scales = 5 mm. Arrows in A, D-F indicate continuous nature of trochlear articular surface, top arrows in B-C show proximal extent of trochlear articular surface, bottom arrows indicate trochlear shelf

of the medial tibial facet of *Cebus* in Fig. 7B) as do *Lemur* and *Smilodectes*. Additionally, like *Smilodectes*, NMMP 39 possesses a distinctive concavity placed proximal and plantar to the medial tibial facet (indicated by the dark circular area in Fig. 7A and 7C). This concavity appears to only be present in notharctine adapiforms and amphipithecines and may well represent a shared, derived character between these two groups, although further comparisons are necessary to confirm this hypothesis.

NMMP 39 does appear to lack a prominent trochlear shelf (Fig. 8), however it is difficult to be certain of the disposition of this shelf because of damage to the bone. Even if this is the case, lack of a trochlear shelf is apparently a primitive feature of primates (Dagosto and Gebo, 1994) and thus of limited phylogenetic value. A closer examination of the posterior aspect of the astragalar trochlea reveals another subtle difference as well. The trochlear surface in NMMP 39 is continuous, lacking any break between the trochlea and the posterior shelf area. This is found in adapiforms, omomyiforms, and living strepsirhines and could represent either a derived feature for these animals or more likely a functional convergence involved with extreme plantar flexion of the foot. In anthropoids (including the most primitive known African forms) there is a distinct break between the trochlear surface and the trochlear shelf area represented by a non-articular surface. This is similar to other archontans and suggests that this feature may be primitive for primates.

An additional possible shared, derived feature of NMMP 39 and adapiforms concerns the disposition of the sustentacular facet. In both the sustentacular

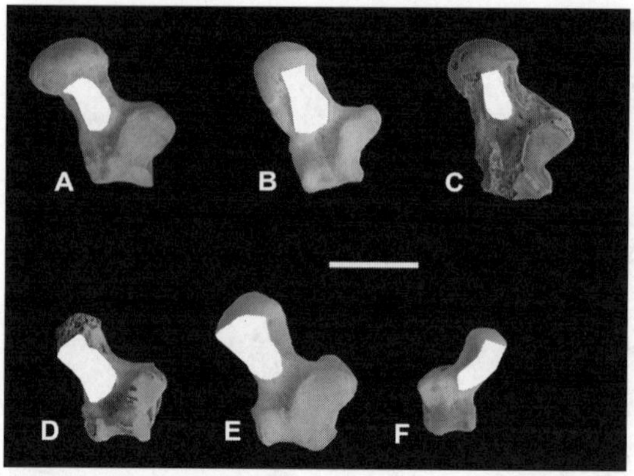

Fig. 9 Series of primate astragali in plantar view showing disposition of sustentacular facet indicated by light shading. A) NMMP 39, amphipithecid; B) UMMZ 160910, *Lemur catta*; C) UM 100886, *Notharctus robinsoni*; D) MACN-CH 362, *Dolichocebus gaimanensis*; E) UMMZ 126129, *Cebus apella*; F) UMMZ 46415, *Saimiri sciureus*. Scale = 1 cm. Note limited extent of sustentacular facet in NMMP 39, the strepsirhine *Lemur*, and the adapiform *Notharctus* and the larger sustentacular facet that extends at least to the medial border in the anthropoids (D-F)

facet is positioned along the midline of the plantar aspect of the astragalar neck and does not extend to the medial margin of the neck (Fig. 9). In most anthropoids the sustentacular facet extends at least to the medial margin of the neck and may even extend onto the medial surface.

Taking all of the morphological features of NMMP 39 in combination does not suggest any close relationship between amphipithecids and anthropoids, but instead indicates that amphipithecids probably are more closely related to adapiforms. This interpretation is also supported by other known postcrania of amphipithecids (Ciochon et al., 2001) and is not contradicted by any dental or cranial evidence now known (Ciochon and Gunnell, 2004).

To further illustrate our contention that NMMP 39 can best be interpreted as representing a non-anthropoid we undertook a morphometric analysis. We took standard measurements (Gebo, 1988) on a series of haplorhine (13 extant and extinct taxa), omomyiform (2 taxa), strepsirhine (3 extant taxa), and notharctine adapiform (2 taxa) astragali and compared them with those of NMMP 39 and with the same measurements taken from the literature for eosimiids (Tables 1 and 2; Gebo et al., 2000). While doing so, we also compared our measurements of NMMP 39 with the same measurements taken by Marivaux et al. (2003) finding in all cases but one that our measurements were comparable. The one notable difference between our measurements and those of Marivaux et al. (2003) was in lateral body height (LHT). It appears from their Fig. 2 (Marivaux et al. 2003, p. 13174) that they, in fact, measured medial body height instead of LHT and this may explain the discrepancy

Table 1 Astragalar Measurements (in millimeters) and Indices (see Gebo, 1988 for descriptions of indices). Index number corresponds to numbers shown in Fig. 10

Measurement	This Paper	Marivaux et al. (2003)
Astragalar Length (AL)	16.33	16.21
Neck Length (NL)	8.66	9.25
Trochlear Length (TRL)	9.12	8.73
Midtrochlear Width (MTRW)	7.87	7.90
Astragalar Width (AW)	10.70	10.52
Lateral Body Height (LHT)	7.37	8.81
Head Width (HW)	9.08	8.46
Head Height (HHT)	6.58	6.03
Neck Angle	30°	32°
Index	This Paper	Marivaux et al. (2003)
1) NL/AL	53	57
2) NL/TRL	95	106
3) NL/MTRW	110	117
4) LHT/MTRW	94	111
5) LHT/TRL	81	101
6) MTRW/TRL	86	90
7) HW/HHT	138	140
8) HW/MTRW	115	107
9) AW/AL	66	65

between our measurements. Comparisons of the astragalar indices (Table 1) presented by Marivaux et al. (2003) and those calculated by us are also comparable except for those involving LHT as one of the factors.

Figure 10 is a graphic representation of the nine astragalar indices used by Marivaux et al. (2003) and recalculated by us. The first graph (10A) is based on our measurements, the second graph (10B) includes the measurements for NMMP 39 given by Marivaux et al. (2003). In both graphs, haplorhines, omomyiforms, strepsirhines, and adapiforms are presented as means for each index. There are some interesting features of these graphs. First, there is almost no difference in index means for haplorhines vs. strepsirhines calling into question the usefulness of any of these indices for phylogenetic analysis (included in the data matrix of Marivaux et al., 2003). NMMP 39 nearly matches the distributions for all primates compared except in two indices – Index 3, neck length vs. mid-trochlear width and Index 7, head width vs. head height. NMMP 39 has a shorter neck relative to trochlear width than does any of the comparative sample. Interestingly, eosimiids differ from the comparative sample in the opposite manner, having a relatively longer neck relative to trochlear width, a feature shared only with omomyiforms in our sample. NMMP 39 also has a broader astragalar head relative to height, a feature similar to that found in notharctine adapiforms and approached by eosimiids as well. According to the measurements of Marivaux et al. (2003), NMMP 39 also differs from other primates in Index 5, lateral trochlear height

Table 2 Astragalar measurements for extant and fossil primates

Specimen	Genus	Species	AL	NL	TRL	MTRW	AW	LHT	HW	HHT	Angle
NMMP-39	Amphipithecid	sp.	16.33	8.66	9.12	7.87	10.7	7.37	9.08	6.58	30
UMMZ 63503	*Alouatta*	sp.	25.02	10.35	13.46	11.16	16.29	10.82	10.39	8.37	49
UMMZ 126129	*Cebus*	*apella*	17.37	9.66	9.23	8.77	12.34	8.97	7.97	7.03	49
UMMZ 46415	*Saimiri*	*sciureus*	11.72	6.92	6.1	5.39	7.07	5.43	4.29	4.58	52
UMMP No #	*Saguinus*	sp.	8.5	4.77	4.65	4.15	5.59	3.79	3.85	3.18	52
Unnumbered	*Cebuella*	sp.	5.35	2.92	2.83	2.33	3.14	1.98	2.12	1.7	42
UMMZ 58927	*Aotus*	*trivirgatus*	11.94	6.02	7.09	5.17	7.74	5.22	5.99	4.63	35
UMMZ 125576	*Callicebus*	sp.	10.92	6.21	6.3	5.07	7.03	5.53	4.75	3.6	35
MACN-CH-362	*Dolichocebus*	*gaimanensis*	14.53	7.97	8.47	6.57	10.07	7.11	6.13	5.02	45
DPC 10037	*Catopithecus*	*browni*	10.32	5.78	5.07	3.88	6.27	4.79	4.17	4.06	29
DPC15417	*Proteopithecus*	*sylviae*	9.04	4.98	5.16	4.03	5.63	4.47	3.97	3.88	37
DPC 3052	*Aegyptopithecus*	*zeuxis*	20.21	8.85	11.76	8.9	19.26	9.47	8.62	6.5	46
DPC Mean	*Apidium*	*phiomense*	13	7.72	8.15	6.4	11.04	6	6.11	5.24	41
UMMZ 95741	*Tarsius*	sp.	5.78	4.33	4.05	3.39	4.13	2.64	2.67	2.18	39
UM 98604	*Omomys*	*carteri*	7.81	4.8	4.24	3.57	4.88	3.85	3.51	2.88	40
UM 98885	*Washakius*	*insignis*	6.89	4.18	3.88	3.04	4.19	3.03	2.64	2.17	28
UMMZ 160910	*Lemur*	*catta*	17.55	8.48	10.04	7.67	10.75	7.33	6.95	5.8	25
UMMZ 172669	*Varecia*	sp.	20.23	10.48	11.18	8.4	12.81	9.24	8.68	6.98	35
UMMZ 113349	*Galago*	*demidovii*	8.95	5.54	4.68	4.02	5.88	3.82	3.59	3.05	27
UM 95526	*Smilodectes*	*mcgrewi*	18.18	8.85	9.86	6.91	9.78	8.02	7.4	5.08	35
USNM 425518	*Notharctus*	*robustior*	23.36	10.49	11.43	9.22	12.71	10.4	9.6	8.05	38
IVPP Mean	Eosimiid	sp.	6.05	2.98	2.93	2.17	2.92	2.5	2.29	1.82	27

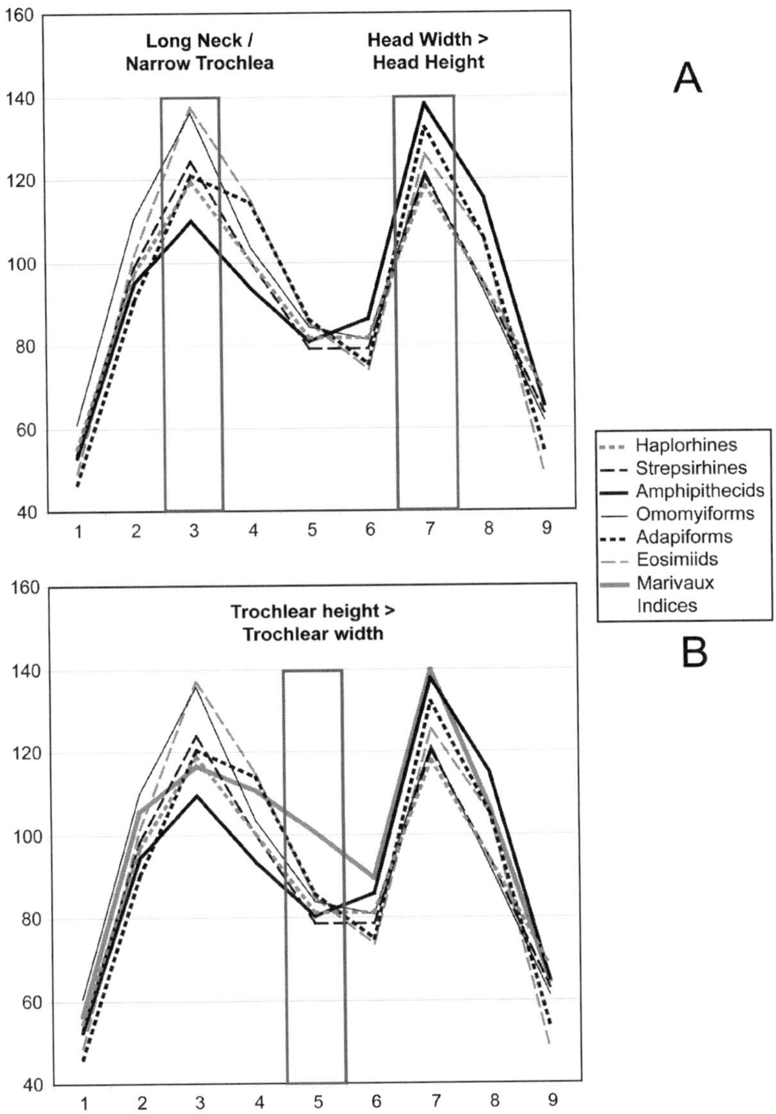

Fig. 10 Graphic representation of astragalar indices (see Table 1) for a series of haplorhines, omomyiforms, strepsirhines, and adapiforms compared with amphipithecids (NMMP 39) and eosimiids. Top graph (A) is based on data from this paper, bottom graph (B) includes data for NMMP 39 taken from Marivaux et al. (2003) (indicated by light dashed line). Note in A: 1) haplorhine and strepsirhine means for each index differ very little from each other; 2) amphipithecids differ from all other primates sampled in having a relatively shorter astragalar neck compared to trochlear width and are similar to adapiforms (and to a slightly lesser extent eosimiids) in having a relatively broader head compared to head height; 3) eosimiids are nearly identical to omomyiforms in having a longer neck relative to trochlear width. Note in bottom graph that measurements provided by Marivaux et al. (2003) also suggest that NMMP 39 differs from other groups in having a relatively higher trochlea compared to trochlear width. However, this may be due to measurement of medial trochlear height instead of the standard lateral trochlear height (see text for further discussion)

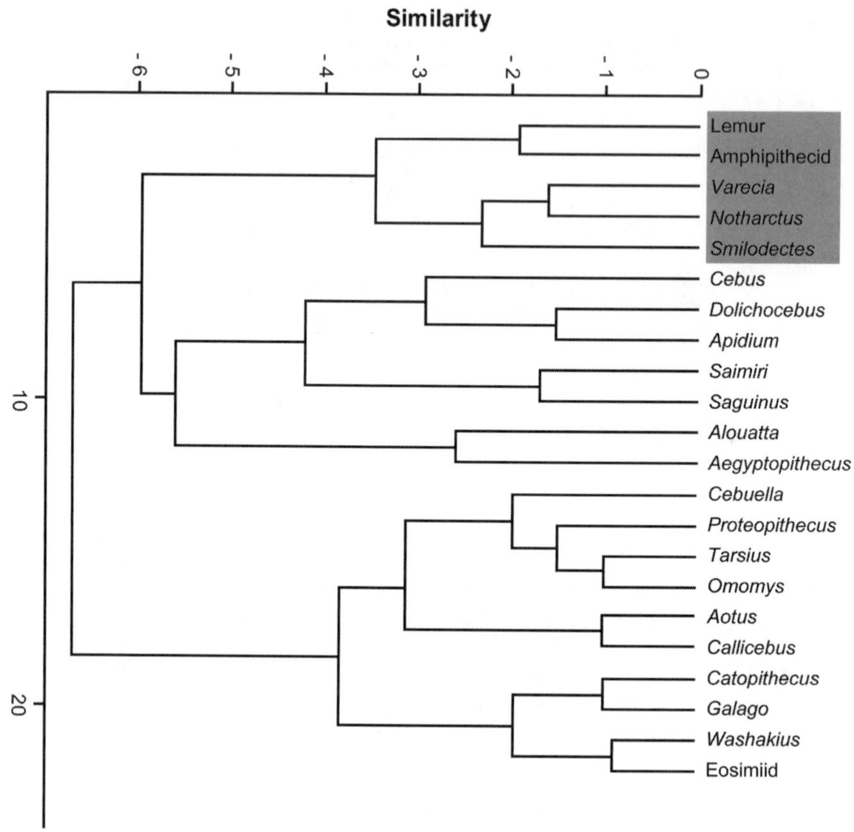

Fig. 11 Cluster analysis of astragalar measurement data utilizing Euclidian Distance Measure. Note that NMMP 39 groups with strepsirhines and notharctine adapiforms, not with anthropoids

vs. trochlear length. However, this is probably due to measuring the medial trochlear height instead of the lateral trochlear height as discussed above. Otherwise our index distribution and that of Marivaux et al. (2003) is quite similar.

We then subjected our astragalar metrical data to a Euclidean Cluster Analysis of Similarity, the results of which are shown in Fig. 11. NMMP 39 grouped with *Lemur* and this group with another group consisting of *Varecia* and two notharctine adapiforms to the exclusion of all other primate taxa used in the analysis. We tested our results by using the data of Marivaux et al. (2003) in a similar Cluster Analysis and found that the position of NMMP 39 changed slightly to group more closely with the adapiforms instead of *Lemur* but in neither analysis did NMMP 39 ever group with anthropoids.

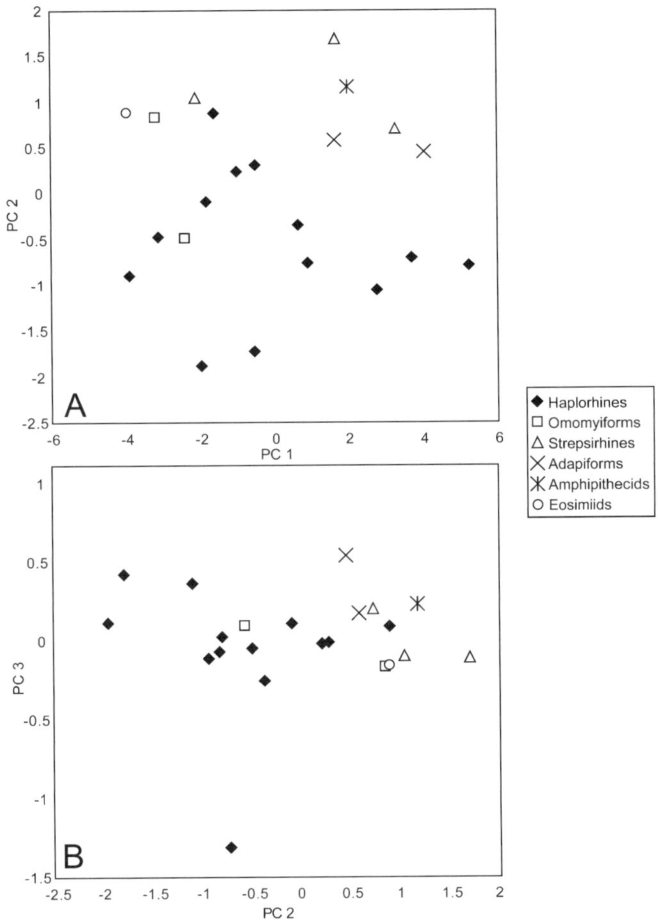

Fig. 12 Graphic representation of component loadings resulting from Principle Component Analysis of astragalar measurement data utilizing normalized variance-covariance to account for differences in measurement units (millimeters and degrees). Top graph (A) includes loadings for PC I and PC II, bottom graph (B) PC II and PC III. PC I accounts for 86.5% of variance and is interpreted as a size component. PC II accounts for 10.7 % of the variance and essentially separates haplorhines from all other primates – note that amphipithecid and eosimiid astragali are included within non-haplorhine grouping. PC III accounts for only 1.4% of the variance and separates *Aegyptopithecus* from all other taxa

In a final analysis we examined the distribution of variance through Principle Components analysis of our astragalar metrical data. We conducted a normalized variance-covariance analysis because the data was measured in two different units (centimeters and degrees). The results are presented in Fig. 12. Figure 12A presents Principle Components I and II in graphic form. PC I represents the variance in astragalar size of the entire sample and accounts for almost 87% of the variance identified in the analysis. It can be seen that NMMP 39 falls in with the larger included taxa while the eosimiid astragalus falls in with the small taxa. PC II accounts for about 11% of the variance, and essentially separates all other primates from haplorhines with the exception of *Catopithecus* which falls among the non-haplorhines and *Omomys* (an omomyiform) which falls with haplorhines. Both the eosimiid and NMMP 39 fall in with non-haplorhines on PC II. Figure 12B shows Principle Components II and III in graphic form. Again it can be seen that NMMP 39 and the eosimiid group with non-haplorhines on PC II. PC III accounts for less than 2% of the variance and separates *Aegyptopithecus* from all of the other taxa included in the analysis.

Conclusions

We have presented a re-analysis of NMMP 39, a left astragalus of a presumed amphipithecid from the Pondaung Formation in Myanmar. When this astragalus was originally described it was claimed that it provided strong evidence for a close relationship between amphipithecids and anthropoid primates (Marivaux et al., 2003). Our reanalysis fails to support these previous conclusions. Instead we believe that gross comparative morphology and morphometric analysis both support the conclusion that NMMP 39 is more closely related to adapiform primates than to any other known group. This is not especially surprising given that other known postcranial remains of amphipithecids are undeniably adapiform in morphology (Ciochon et al. 2001; Ciochon and Gunnell, 2004).

Acknowledgment We thank Erik Seiffert for providing casts of *Proteopithecus* and *Catopithecus*. We thank Erik Seiffert and Daniel Gebo for comments on a previous version of the manuscript. We thank Freddie and John Oakley for all of their hard work in organizing the symposium honoring Elwyn Simons and John Fleagle for inviting us to participate in both the symposium and this book. We thank J. Rogers, Univ. of Iowa, for his work on the figures. Laboratory support was provided by the Human Evolution Research Fund of the University of Iowa Foundation.

References

Ba Maw, Ciochon, R. L., and Savage, D. E. (1979). Late Eocene of Burma yields earliest anthropoid primate, *Pondaungia cotteri*. *Nature* 282:65–67.

Beard, K. C. (2004). *The Hunt for the Dawn Monkey – Unearthing the Origins of Monkeys, Apes, and Humans*. University of California Press, Berkeley.

Beard, K. C., Jaeger, J.-J., Chaimanee, Y., Rossie, J. B., Aung Naing Soe, Soe Thura Tun, Marivaux, L., and Marandat, B. (2005). Taxonomic status of purported primate frontal bones from the Eocene Pondaung Formation of Myanmar. *J. Hum. Evol.* 49:468–481.

Chaimanee, Y., Tin Thein, Ducrocq, S., Aung Naing Soe, Benammi, M., Than Tun, Thit Lwin San Wai, and Jaeger, J. J. (2000). A lower jaw of *Pondaungia cotteri* from the late middle Eocene Pondaung Formation (Myanmar) confirms its anthropoid status. *P. Natl. Acad. Sci. USA* 97: 4102–4105.

Ciochon, R. L., Gingerich, P. D., Gunnell, G. F., and Simons, E. L. (2001). Primate postcrania from the late middle Eocene of Myanmar. *P. Natl. Acad. Sci. USA* 98:7672–7677.

Ciochon, R. L., and Gunnell, G. F. (2002). Eocene primates from Myanmar: Historical perspectives on the origin of Anthropoidea. *Evol. Anthropol.* 11:156–68.

Ciochon, R. L., and Gunnell, G. F. (2004). Eocene large-bodied primates of Myanmar and Thailand: Morphological considerations and phylogenetic affinities. In: Ross, C. F., and Kay, R. F. (eds.), *Anthropoid Origins: New Visions*. Kluwer Academic/Plenum Publishers, New York, pp. 249–282.

Ciochon, R. L., and Holroyd, P. A. (1994). The Asian origin of Anthropoidea revisited. In: Fleagle, J. G., and Kay, R. F. (eds.), *Anthropoid Origins*. Plenum Press, New York, pp. 143–162.

Ciochon, R. L., Savage, D. E., Thaw Tint, and Ba Maw. (1985). Anthropoid origins in Asia? New discovery of *Amphipithecus* from the Eocene of Burma. *Science* 229:756–59.

Colbert, E. H. (1937). A new primate from the upper Eocene Pondaung Formation of Burma. *Am. Mus. Nov.* 951:1–18.

Colbert, E. H. (1938). Fossil mammals from Burma in the American Museum of Natural History. *B. Am. Mus. Nat. Hist.* 74:255–436.

Dagosto, M., and Gebo, D. L. (1994). Postcranial anatomy and the origin of the Anthropoidea. In: Fleagle, J. G., and Kay, R. F. (eds.), *Anthropoid Origins*. Plenum Press, New York, pp. 567–593.

Egi, N., Takai, M., Shigehara, N., and Tsubamoto, T. (2004). Body mass estimates for Eocene eosimiid and amphipithecid primates using prosimian and anthropoid scaling models. *Int. J. Primatol.* 25:211–236.

Fleagle, J. G., and Simons, E. L. (1978). Humeral morphology of the earliest apes. *Nature* 276:705–707.

Fleagle, J. G., and Simons, E. L. (1995). Limb skeleton and locomotor adaptations of *Apidium phiomense*, an Oligocene anthropoid from Egypt. *Am. J. Phys. Anthropol.* 97:235–289.

Ford, S. M. (1994). Primitive platyrrhines? Perspectives on anthropoid origins from platyrrhine, parapithecid, and preanthropoid postcrania. In: Fleagle, J. G., and Kay, R. F. (eds.), *Anthropoid Origins*, Plenum Press, New York, pp. 595–673.

Gebo, D. L. (1988). Foot morphology and locomotor adaptation in Eocene primates. *Folia Primatol.* 50:3–41.

Gebo, D. L., Dagosto, M., Beard, K. C., and Qi, T. (2000). The smallest primates. *J. Hum. Evol.* 38:585–94.

Gingerich, P. D., and Gunnell, G. F. (2005). Brain of *Plesiadapis cookei* (Mammalia, Proprimates): Surface morphology and encephalization compared to those of Primates and Dermoptera. *Contrib. Mus. Paleontol., Univ. Mich.* 31:185–195.

Godinot, M. (1994). Early North African primates and their significance for the origin of Simiiformes (= Anthropoidea). In: Fleagle, J. G., and Kay, R. F. (eds.), *Anthropoid Origins*. Plenum Press, New York, pp. 235–295.

Gunnell, G. F., Ciochon, R. L., Gingerich, P. D., and Holroyd, P. A. (2002). New assessment of *Pondaungia* and *Amphipithecus* (Primates) from the late middle Eocene of Myanmar, with a comment on 'Amphipithecidae'. *Contrib. Mus. Paleontol., Univ. Mich.* 30:337–372.

Jaeger, J.-J., Aung Naing Soe, Aye Ko Aung, Benammi, M., Chaimanee, Y., Ducrocq, R. M., Than Tun, Tin Thein, and Ducrocq, S. (1998). New Myanmar middle Eocene anthropoids. An Asian origin for catarrhines? *C. R. Acad. Sci. Paris* 321:953–959.

Kay, R. F., Schmitt, D., Vinyard, C. J., Perry, J. M., Shigehara, N., Takai, M., and Egi, N. (2004). The paleobiology of Amphipithecidae, South Asian late Eocene primates. *J. Hum. Evol.* 41:3–25.

Koenigswald, G. H. R. von. (1965). Critical observations upon the so-called higher primates from the upper Eocene of Burma. *Proc. Koninkl. Nederl. Akad. Weten. Amsterdam* 68:165–167.

Marivaux, L., Chaimanee, Y., Ducrocq, S., Sudre, J., Aung Naing Soe, Soe Thura Tun, Wanna Htoon, and Jaeger, J. J. (2003). The anthropoid status of a primate from the late middle Eocene Pondaung Formation (Central Myanmar): Tarsal evidence. *P. Natl. Acad. Sci. USA* 100:13173–13178.

Miller, E. R., Gunnell, G. F., and Martin, R. D. (2005). Deep time and the search for anthropoid origins. *Yrbk. Phys. Anthropol.* 28:60–95

Pilgrim, G. E. (1927). A *Sivapithecus* palate and other primate fossils from India. *Palaeontol. Indica* 14:1–26.

Seiffert, E. R., and Simons, E. L. (2001). Astragalar morphology of late Eocene anthropoids from the Fayum Depression (Egypt) and the origin of catarrhine primates. *J. Hum. Evol.* 41:577–606.

Seiffert, E. R., Simons, E. L., and Attia, Y. (2003). Fossil evidence for an ancient divergence of lorises and galagos. *Nature* 422:421–424.

Seiffert, E. R., Simons, E. L., Clyde, W. C., Rossie, J. B., Attia, Y., Bown, T. M., Chatrath, P., and Mathison, M. E. (2005). Basal anthropoids from Egypt and the antiquity of Africa's higher primate radiation. *Science* 310:300–304.

Seiffert, E. R., Simons, E. L., and Fleagle, J. G. (2000). Anthropoid humeri from the late Eocene of Egypt. *P. Natl. Acad. Sci. USA* 97:10062–10067.

Seiffert, E. R., Simons, E. L., Ryan, T. M., and Attia, Y. (2005). Additional remains of *Wadilemur elegans*, a primitive stem galagid from the late Eocene of Egypt. *P. Natl. Acad. Sci. USA* 102:11396–11401.

Simons, E. L. (1971). Relationships of *Amphipithecus* and *Oligopithecus*. *Nature* 232:489–491.

Simons, E. L. (1972). *Primate Evolution – An Introduction to Man's Place in Nature*. The Macmillan Company, New York.

Szalay, F. S. (1970). Late Eocene *Amphipithecus* and the origins of catarrhine primates. *Nature* 227:355–57.

Szalay, F. S. (1972). *Amphipithecus* Revisited. *Nature* 236:179–180.

Takai, M., and Shigehara, N. (2004). The Pondaung primates, enigmatic "possible anthropoids" from the latest middle Eocene , central Myanmar. In: Ross, C. F., and Kay, R. F. (eds.), *Anthropoid Origins New Visions*. Kluwer Academic/Plenum Publishers, New York, pp. 283–321.

Takai, M., Shigehara, N., Egi, N., and Tsubamoto, T. (2003). Endocranial cast and morphology of the olfactory bulb of *Amphipithecus mogaungensis* (latest middle Eocene of Myanmar). *Primates* 44:137–144.

Takai, M., Shigehara, N., Tsubamoto, T., Egi, N., Aye Ko Aung, Tin Thein, Aung Naing Soe, and Soe Thura Tun. (2000). The latest middle Eocene primate fauna in Pondaung area, Myanmar. *Asian Paleoprimatol.* 1:7–28.

A Haplorhine First Metatarsal from the Middle Eocene of China

Daniel L. Gebo, Marian Dagosto, K. Christopher Beard, Xijun Ni and Tao Qi

Introduction

We are honored to celebrate the career and scholarship of Professor Elwyn L. Simons. His body of work covering the field of paleoprimatology is staggering in quantity and quality. Elwyn's record of recovery, description, and analysis of the Fayum fossil primates from Egypt is undoubtedly his greatest contribution in a phenomenally long and productive career. We thank him for his insights, hard work, and patience, having trained most of us studying primate evolution. Our contribution concerns the description of a new fossil – something Elwyn has done so many times before. We describe a primate first metatarsal from the Eocene of Asia.

A newly discovered first metatarsal from the middle Eocene (~45 mya) Shanghuang fissures (fissure D) in southern Jiangsu Province, China, provides the first evidence for the anatomy of the grasping mechanism of the Shanghuang primates. Since the form of the grasping foot is a fundamental character complex that distinguishes primitive ("prosimian") and more derived (crown-group anthropoid) primates, this bone also allows us to assess the taxonomic status of Shanghuang primates and to consider the changing role of primate grasping through time and its relevance to higher primate evolution.

Description

V13018 is a tiny right first metatarsal (Fig. 1). It is only 6.9 mm in length (there is a slight amount of erosion and destruction at both the proximal and distal ends, but total length would not have been more than 7.0 mm). Among living primates, the most similar in overall metric dimensions is *Microcebus berthae*, the pygmy mouse lemur (body mass 24.6 to 38 g, with a mean value of 30.6 g;

Daniel L. Gebo
Department of Anthropology, Northern Illinois University, DeKalb, IL 60115
dgebo@niu.edu

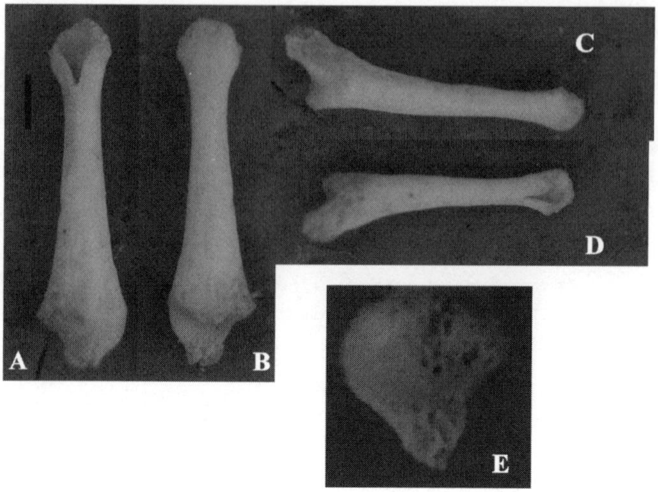

Fig. 1 Views of V13018: plantar (a), dorsal (b), medial (c), lateral (d), and proximal (e). Bar equals 1 mm

Rasoloarison et al., 2000). One exception to the metric similarity of the fossil to *M. berthae* is its length, which in V 13018 is longer than that of *Microcebus berthae* and compares better with metatarsal length of the larger *Microcebus murinus*. Other taxa such as *Microcebus murinus* (54–69 g; Rasoloarison et al., 2000), *Galagoides demidoff* (44–97 g; Nash et al., 1989), *Cebuella pygmaea* (110–132 g; Groves, 2001), and *Tarsius syrichta* (83–182 g (Dagosto et al., 2003) have first metatarsal lengths of 6.9–7.4 mm, 9.2 mm, 7.4–7.9 mm, and 11.8–12.5 mm, respectively. V13018 is smaller in all metatarsal dimensions than *Shoshonius cooperi*, a small fossil primate estimated to weigh between 60 and 90 g (Dagosto et al., 1999). Given its morphological similarity to primitive primates and the body sizes of the taxa examined, we would estimate that the V13018 specimen belongs to a primate body in the size range of 30 g.

The V13018 shaft is relatively straight with some torsion at the distal end, as seen in other primates. In contrast to most primates where the shaft begins to narrow at mid-length, the narrowing does not begin until near the metatarsal head in V13018. In this feature, V13018 is more similar to *Necrolemur* than to tarsiids, platyrrhines, or adapiforms. The proximal joint surface has a wide arc of curvature like that of adapiforms (Szalay and Dagosto, 1988), tarsiers, or anthropoids in comparison to the narrower arc observed in omomyids and microchoerids. V13018 has a saddle-shaped proximal joint surface similar to adapiform (includes Notharctidae, Adapidae, and Sivaladapidae), lemuriform, (includes Lemuridae, Indriidae, Lepilemuridae, Cheirogaleidae, and Daubentoniidae), lorisiform (includes Lorisidae and Galagidae), and tarsiiform (includes Omomyidae, Microchoeridae, and Tarsiidae) primates, not the broad, circular, flattened shape of crown-group anthropoids. Nor does it have the peroneal tubercle offset to the postaxial side of the proximal joint surface as in crown-group anthropoids, relative to the mid-joint position in adapiforms, lemuriforms, lorisiforms, and tarsiiforms. This joint surface lacks the oblique part of this facet as observed in omomyids (Dagosto et al., 1999). The peroneal tubercle is long, rough and irregular in shape mediolaterally, being very broad, and most similar in overall shape to that of adapiform and microchoerid primates. It is not like the shortened but broad tubercles found among crown-group anthropoids.

Taxonomic Allocation

At Shanghuang, we have been able to identify four different small haplorhine taxa (<125 g) based on tarsal morphology with several size differentiated paleospecies within each group. Only one postcranial specimen from Shanghuang is attributed to a large haplorhine primate (~350 g; Gebo et al., 2001) while another *Macrotarsius* (900–1,221 g; MacPhee et al., 1995) is represented only by teeth at this site. Both of these haplorhine primates, like both of the adapiform primates at Shanghuang, are much larger than the smaller haplorhines, and the V13018 first metatarsal is too small to belong to any of them. It can only be allocated to one of the four small-sized haplorhine groups: the eosimiid or new haplorhine basal anthropoids, the "omomyid-like" haplorhines, or the tarsiids. Each of these groups have tarsals representing individuals with a body mass near 30 g (Gebo et al., 2001). Therefore, the size of V13018 does not help us allocate it to a specific haplorhine group at Shanghuang.

In terms of commonality, eosimiids and adapiforms predominate among the dental samples at Shanghuang, suggesting that V13018 most probably represents a first metatarsal from a small eosimiid. Associated specimens will be needed to confirm or reject this assumption. However, if this assumption is correct, it implies that basal anthropoids at Shanghuang possessed a primitive "prosimian" type of grasping big toe, unlike that of crown-group anthropoids. Ultimately, the taxonomic allocation of the V 13018 first metatarsal is indeterminate. It could belong to any of the four small-sized haplorhine groups at Shanghuang. We believe that the tarsiids (due to several morphological similarities) or the eosimiids (due to commonality) at Shanghuang represent the two most likely taxonomic allocations for this specimen.

Measurements and Indices

Eight measurements (Table 1; Fig. 2) were taken on this specimen following Gebo et al. (1999). Corresponding ratios were calculated from these measures (Table 2) and compared with the published data in Gebo et al. (1999) and Dagosto et al. (1999). A prominent, long peroneal tubercle is seen in adapiform,

Table 1 First metatarsal measurements (mm; see Fig. 2)

Measurements	V13018	*Shoshonius cooperi* n = 1	*Microcebus berthae* n = 1	*Microcebus murinus* n = 2	*Galagoides demidoff* n = 1	*Tarsius syrichta* n = 6	*Cebuella pygmaea* n = 5
A	0.65	0.86	0.71	1.19	1.84	1.02	1.17
B	1.75	2.59	1.73	2.10	2.51	2.72	1.72
C	0.9	2.03	1.2	1.34	1.77	1.44	0.98
D	1.8	3.30	1.87	2.45	2.82	2.85	1.94
E	6.15	—	5.27	5.78	7.8	10.75	7.13
F	6.9	—	6.32	7.12	9.18	12.18	7.69
G	0.9	2.36	0.94	1.26	1.64	1.20	1.07
H	0.85	2.23	1.09	1.57	1.14	1.37	0.89

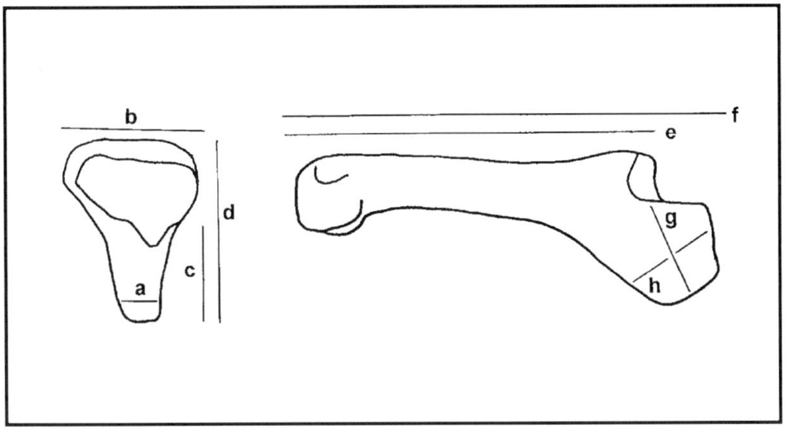

Fig. 2 First metatarsal measurements (a–h; see Table 1)

lemuriform, lorisiform, and tarsiiform primates but crown-group anthropoids have short tubercles (Fig. 3). However relative to first metatarsal length, tarsiers and anthropoids both possess relatively short tubercles, as does V 13018 (Tables 2 and 3). The G/F and F-E/F ratios (Table 2) are two ways to assess peroneal tubercle length relative to total metatarsal length. First metatarsal length is a reasonable choice for the denominator since in a comparison of first metatarsal length and body mass anthropoids and strespsirhines do not differ in the slope of the regression line ($p = 0.668$) and there are no significant shifts in elevation ($p = 0.102$). The common slope of 0.384 indicates positive allometry, but 95% confidence intervals around this slope include isometry. The G/F ratio

Table 2 Mean ratio values for the first metatarsal measurements

Taxa	n	A/B	C/D	E/F	H/G	H/F	D-C/B	G/F	F-E/F
V13018	1	0.37	0.50	0.89	0.94	0.12	0.51	0.13	0.11
Tarsius syrichta	6	0.37	0.51	0.88	1.14	0.11	0.51	0.10	0.12
Cebuella pygmaea	5	0.68	0.51	0.93	0.83	0.12	0.56	0.14	0.07
Saguinus mystax	6	0.53	0.45	0.96	0.88	0.10	0.55	0.11	0.04
Saguinus fuscicollis	4	0.58	0.48	0.96	0.79	0.10	0.57	0.13	0.04
Saguinus geoffroyi	5	0.56	0.46	0.95	0.84	0.11	0.60	0.13	0.05
Saguinus labiatus	4	0.56	0.44	0.98	0.93	0.10	0.55	0.11	0.02
Saimiri sciureus	13	0.55	0.46	0.95	0.92	0.11	0.61	0.12	0.05
Hemiacodon gracilis	1	0.28	0.68	0.89	0.98	0.28	0.49	0.28	0.11
Shoshonius cooperi	1	0.33	0.62	—	0.95	—	0.49	—	—
Necrolemur zitteli	1	0.44	0.55	0.88	0.96	0.18	0.57	0.18	0.12
Cantius abditus	1	0.46	0.55	0.96	0.82	—	0.52	—	—
Notharctus tenebrosus	1	0.44	0.41	0.92	0.82	0.18	0.75	0.22	0.08

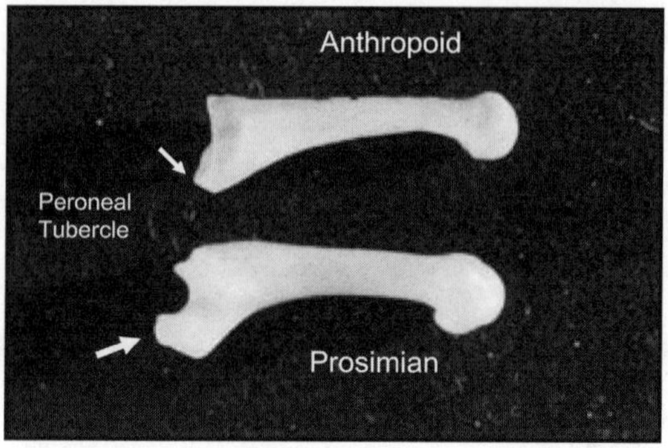

Fig. 3 Peroneal tubercle (arrows) size in anthropoids and prosimians. "Prosimians" equals strepsirhines (adapiforms, lemuriforms, and lorisiforms) plus tarsiiforms (omomyids, microchoerids, and *Tarsius*)

Table 3 Primate comparisons of peroneal tubercle morphology relative to leaping frequency, leg specializations for leaping, and peroneus longus musculature. Locomotor data from: Fleagle and Mittermeier (1980); Crompton (1984); Gebo (1987b); Dagosto and Yamashita (1998); Dagosto et al., 2001; Garber and Preutz (1995); Garber and Leigh (2001). * = captive study; 7 = most specialized or most robust while a 1 = the least specialized or the least robust

Taxa	Leaping Frequency	Rank by leg specializations for leaping	Peroneal tubercle length/first metatarsal length (G/F)	Peroneal tubercle robusticity rank	Peroneus longus muscle weight (gms)/body weight (gms) × 1000
Eulemur fulvus	68%	4	0.16	4	1.4
Propithecus coquereli	89%	5	0.15	5	1.81
Galago moholi	53%	6	0.21	7	2.06
Tarsius syrichta	57%	7	0.10	3	0.63
Nycticebus coucang	0%	1	0.16	2	0.65
Microcebus murinus	38%*	3	0.16	6	0.79
Saguinus labiatus	38%	2	0.11	1	0.38
Saguinus fuscicollis	33%	2	0.13	1	0.63
Saguinus mystax	31%	2	0.11	1	0.30
Saguinus geoffroyi	42%	2	0.13	1	0.54
Saimiri sciureus	42%	2	0.12	1	0.53

shows low values (relatively short tubercle) for V13018, *Tarsius*, and platyrrhines, and higher values (longer tubercle) for *Hemiacodon, Necrolemur*, and *Notharctus*. The ratio F-E/F, on the other hand, indicates that the peroneal tubercle is relatively long in V13018 (higher index), similar in length to that of *Tarsius*, *Hemiacodon*, and *Necrolemur*. Shorter tubercles (low index) are seen in platyrrhines. In terms of relative tubercle length, V13018 does not have the very long tubercle characteristic of omomyids and it does not have a tubercle as short as is typical for crown-group anthropoids. It is most similar to *Tarsius*. Additionally, the tubercle of V13018 is not narrow and thin (ratio A/B) like omomyid primates, but has the same value for this ratio as *Tarsius*. *Necrolemur* (a microchoerid) and adapiforms (*Cantius* and *Notharctus*) possess even higher values (wider tubercles) than V13018, while platyrrhines have the widest tubercles.

The ratio C/D compares the length of the tubercle relative to the height of the proximal end of the first metatarsal. Omomyids have the highest ratios, and *Notharctus* the lowest. V13018 is again most similar to *Tarsius* and anthropoids which have intermediate values. The H/F ratio (tubercle height to shaft length; Table 2) likewise shows V13018 to be similar to the tarsier and platyrrhines. Higher values (tall tubercles relative to metatarsal length) are found among omomyids, microchoerids, and adapiforms.

The ratio D-C/B is a rough measure of joint height divided by width of the proximal end. V13018 has the same value as *Tarsius* (0.51) but is not very distant from platyrrhines (0.55–0.61) or *Hemiacodon* (0.49). *Necrolemur* (0.57) has a higher value for this ratio (taller joint). In contrast to all other taxa *Notharctus* has a very high value for this ratio showing that its proximal joint surface makes up most of its proximal end. Other indices (E/F, H/G) do not distinguish among taxa, with the exception of *T. syrichta* which possesses a higher H/G value than the others.

Overall, the V13018 ratio values in Table 2 show many similarities to haplorhine primates and thus the best taxonomic allocation based on the metric evidence fits well with the allocation based on size. The closest overall match is with *Tarsius*, thus an allocation to *T. eoceanus*, or an allied species is reasonable. There are several significant morphological differences from omomyids (e.g., relatively short tubercle length; indices G/F and C/D, thicker tubercle, index A/B and H/G), and thus it is unlikely that this metatarsal belongs to a member of the "unnamed haplorhine group" which in terms of tarsal morphology is most similar to omomyids. Likewise, it shows some differences from the first metatarsal of a crown-group anthropoid (i.e., a relatively long peroneal tubercle (index F-E/F) and one that is not wide (index A/B). However, the tarsus of *Eosimias* is primitive for anthropoids, and there is no reason to expect the first metatarsal to be as derived as in crown-group anthropoids. Thus, this metatarsal fits best with the tarsiids or the basal anthropoids at Shanghuang.

Functionally, the V13018 first metatarsal from Shanghuang would have worked like that of more primitive haplorhine primates, with a forceful grasping big toe. If this specimen does indeed belong to a basal anthropoid, like *Eosimias*, it means that the transition to a reduced peroneal tubercle characteristic of crown-group anthropoids had not yet occurred. The long peroneal tubercle in a basal anthropoid is a retained primitive character suggesting a greater emphasis on forceful grasping and more frequent vertical climbing. It would also imply that anthropoids evolved from more primitive "prosimian-like" ancestors. If the V13018 specimen does not belong to a basal anthropoid and in fact belongs to a tarsiid or to another primitive haplorhine group present at Shanghuang, this simply implies a "prosimian-like" grasping mechanism, as would be expected for those groups.

Grasping Mechanics and Big Toe Torphology

The modified hallucial-first metatarsal joint of primates, which facilitates hindfoot grasping, is a key innovation in primate evolution (Morton, 1924; Le Gros Clark, 1959; Cartmill, 1972; Szalay and Dagosto, 1988). The long, robust and twisted shaft of the first metatarsal allows primates to swing their first digit along the saddle-shaped joint of the entocuneiform to oppose the four other lateral digits. The grasping ability of the big toe is largely a function of the shape of the entocuneiform- first metatarsal joint and its associated musculature. Robust and long peroneal tubercles in adapiforms, lemuriforms, lorisiforms and tarsiiform primates provide a long lever arm for > the peroneus longus muscle, an extrinsic adductor of the first metatarsal. Other things being equal, muscles that insert farther away from the joint fulcrum increase force and thus primates with long peroneal tubercles have been interpreted to be able to execute a more forceful grasp (Morton, 1924; Gebo, 1987a; Szalay and Dagosto, 1988; but see Boyer et al., 2007). In contrast, the shortened tubercles found in crown-group anthropoid primates should result in a less forceful grasp, or a grasp less dependent upon the action of peroneus longus. A recent paper by Boyer and colleaques (2007) has suggested that peroneus longus is not an active grasping muscle, but primarily a foot evertor. No matter the final outcome of this functional debate, the morphological situation among primates remains the same, with primitive primates possessing a different condition from crown-group anthropoids.

Szalay and Dagosto (1980) suggested that, in addition to providing leverage for peroneus longus, the thick buttressing of the peroneal tubercle, along with the deep, enlarged, sellar shaped joint surface that extends onto the tubercle, functions to resist the compressive loads of both grasping and landing after a leap. They note the association of larger tubercles in frequently leaping strepsirhines and tarsiiforms, and shorter, less robust tubercles in less frequently leaping strepsirhines and other mammals (e.g., lorises, *Adapis*, and marsupials). This association does not, however, hold up well within anthropoids, or in a comparison between anthropoids and "prosimians". For example, *Colobus guereza*, *Saguinus sp.*, or *Saimiri sciureus* leap frequently but none display obviously larger tubercles than anthropoids that leap less frequently, nor are tubercles as long or as robust as in strepsirhines and tarsiers that leap at similar frequencies.

To investigate this further, we measured several aspects of first metatarsal morphology, associated musculature, and locomotor behavior in a selection of strepsirhine, tarsier, and platyrrhine primates (Table 3), including leaping frequency, a rank estimate of anatomical hindlimb specializations for leaping, length of the peroneal tubercle, a rank estimate of peroneal tubercle robusticity, and peroneus longus muscle mass.

These data (admittedly sparse) show that across primates, as suggested by Szalay and Dagosto (1980), the relative length of the peroneal tubercle (as a proportion of MT1 length) and its robusticity are significantly correlated with the relative size of the peroneus longus muscle (tubercle length, $r = 0.75$, $p < 0.000$; robusticity rank, Spearman's rho $= 0.855$, $p < 0.000$). Relative peroneal tubercle length, however, is not correlated with leaping frequency ($r = -0.067$, $p = 0.83$), and tubercle robusticity rank only moderately so (Spearman's rho $= 0.456$, $p = 0.05$). In addition, we cannot support the suggestion of Szalay and Dagosto (1988) that marsupials have a large peroneus longus muscle in conjunction with a short tubercle, since the marsupials we dissected have relatively small muscles (*Caluromys lanatus*, *Phalanger orientalis*, *Petaurus breviceps*, and *Didelphis virginiana*; Table 4). The same results apply within strepsirhines. The length of the tubercle is not correlated with leaping frequency, but its robusticity is (*rho* $= 0.673$, $p = 0.017$).

Rather than a simple relationship between leaping and first metatarsal morphology, the data reflect a major difference in the anatomy and function of the grasping mechanism between strepsirhines and tarsiers on the one hand, and anthropoids on the other. Living strepsirhines and tarsiers have relatively larger peroneus longus muscles than similarly sized anthropoids (Fig. 4; Tables 3–4). This is associated with and likely explains the difference in peroneal tubercle robusticity which is also larger in strepsirhines and tarsiers (with the exception of the lorises) (Fig. 5) than anthropoids. Although most strepsirhines and tarsiers leap more frequently than most anthropoids, these morphological differences cannot be explained by leaping frequency alone, but must reflect the influence of additional factors. There may be differences in leaping "style" (e.g., differences in limb contact on takeoff or landing), size and/or compliance of take-off and landing supports, or different support type use.

For example, Demes et al. (1995) and Demes et al. (1999) show that when leaping, strepsirhines (*Galago*, *Otolemur*, *Eulemur*, *Hapalemur*, and *Propithecus*) experience higher hindlimb peak substrate forces relative to anthropoids (*Pithecia monachus* and *Macaca fuscata*). These experimental data support the hypothesis of a need for additional buttressing of the MT1-entocuneiform joint in strepsirhines and tarsiers advocated by Szalay and Dagosto (1980). *Lemur catta,* however, is an exception showing forces very close to the two anthropoids, even though there is very little difference in peroneal tubercle size or robusticity between it and the other strepsirhines tested.

From these data, we presume that the relatively large peroneal tubercles present in notharctines and omomyids means that they also had a relatively large peroneus longus muscle and that this condition is primitive for euprimates (Szalay and

Table 4 Relative muscle weights from Gebo (1986) and by Gebo (Field Museum of Natural History dissections)

Taxa (n = 1)	Peroneus longus muscle weight (g)/Body weight (g) × 1000
Lemur catta	1.73
Eulemur mongoz	1.24
Eulemur macaco	1.62
Eulemur coronatus	1.69
Hapalemur griseus	1.41
Varecia variegata	1.87
Cheirogaleus medius	0.78
Mirza coquereli	0.66
Perodicticus potto	0.87
Otolemur crassicaudatus	1.84
Otolemur garnettii	2.07
Callithrix jacchus	0.32
Callimico goeldii	0.65
Cebuella pygmaea	0.31
Marsupials	
Caluromys lanatus	0.28
Phalanger orientalis	0.69
Petaurus breviceps	0.58
Didelphis virginiana	0.34

Dagosto, 1980). However, unlike Szalay and Dagosto, we would not use this evidence to predict leaping frequency in these fossil primates. The reduced tubercles exhibited by adapines, lorisines, *Tarsius*, and anthropoids represent (convergently) derived conditions. In the extant primates above, the reduction in the length and the robusticity of the tubercle is associated with a relatively smaller peroneus longus muscle, and we assume the same was true for the fossil species.

Both tarsiers and lorises are (differently) specialized in their emphasis of intrinsic or extrinsic foot musculature to accomplish grasping (Gebo, 1987a, 1989). The same may be true of anthropoids, as well as didelphid and phalangerid marsupials, but data are currently lacking to test this.

Discussion

Most strepsirhine primates have both a large peroneus longus muscle with a longer lever arm and a robust tubercle, yielding increased mechanical advantage for first metatarsal adduction (and opposition) for a forceful grasp. Based

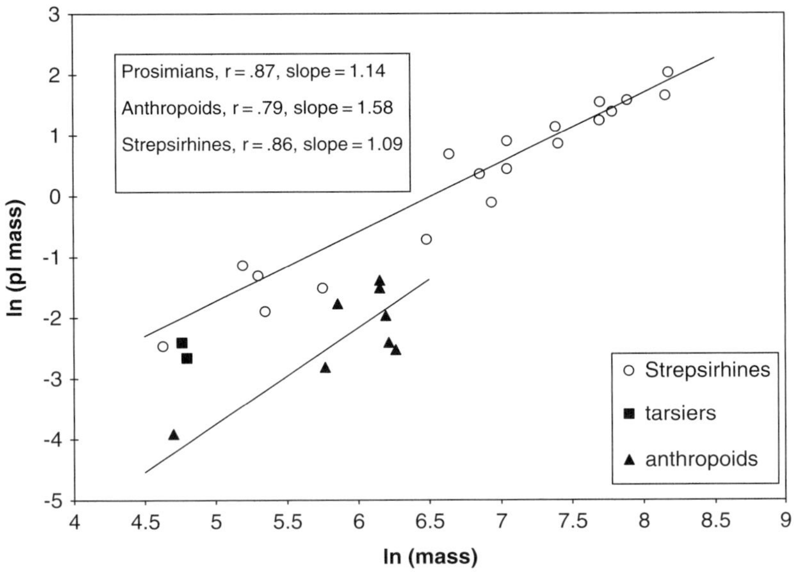

Fig. 4 The relationship between body mass and peroneus longus mass in primates. "Prosimians" equals extant strepsirhines plus *Tarsius*. Both the prosimian line and the strepsirhine RMA lines are significantly elevated over the anthropoid line ($p < 0.000$). The results are based on reduced major axis regression; slopes are compared using SMATR (Falster et al., 2003)

on the distribution of the bony traits among fossil primates, this is likely the ancestral condition for euprimates. There are significant differences in the shape of the peroneal tubercle between notharctid, adapid, sivaladapid, and microchoerid primates relative to omomyid primates. Omomyids have very thin and tall tubercles in contrast to the rugged and mountain-shaped tubercles of the others. From present evidence we cannot decipher the functional meaning of this difference, but would argue that omomyids possess a derived condition relative to euprimates.

Szalay and Dagosto (1980) proposed that the reason for the evolution of this type of grasping mechanism was, at least in part, to buttress the entocuneiform-first metatarsal joint against compressive loads experienced during leaping takeoffs and landings. If there was a primitive relationship between this morphology and the frequency of leaping, it is not strongly evident in extant taxa. Instead, differences among primates may be related to other aspects of leaping style, or may simply reflect different grasping strategies unrelated to leaping. For example, we suspect that grasping force may be most critical when climbing vertical supports (Cartmill, 1974, 1985; Gebo, 1986a,b). Therefore, one strong possibility is that this morphology is related to support use.

In various primate lineages, this ancestral morphology was altered, most dramatically in the anthropoids. Crown-group anthropoids either 1) have a less forceful grasp than strepsirhines and tarsiers; 2) rely on intrinsic rather than

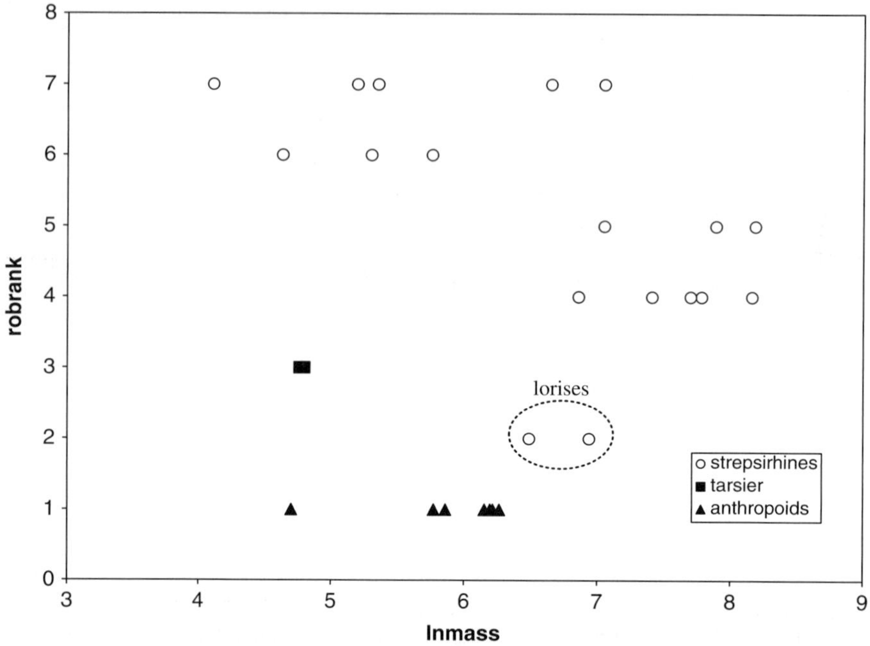

Fig. 5 Peroneal robusticity rank in primates. There is no relationship with body mass, but, extant strepsirhines with the exception of lorises, have more robust tubercles than anthropoids or tarsiers of similar body mass

extrinsic muscles to power the grasp; 3) use horizontal branches more often; 4) climb vertical supports less often; 5) use differently sized supports than strepsirhines and tarsiers; or 6) some combination of the above factors. This may be true for marsupials as well, which also do not have enlarged peroneus longus muscles or long and robust peroneal tubercles. Lemurids, indriids, and galagos may have even further enlarged the peroneus longus and its tubercle. Although the ratio of PL mass to body mass is highest in these taxa, this muscle, like others in the "prosimian" hindlimb (Demes et al., 1998), scales at exponents less than expected for functional similarity.

Conclusion

The first metatarsal from Shanghuang, V13018, exhibits a long peroneal tubercle. If the V 13018 first metatarsal belongs to a tarsiid or to the other non-anthropoid haplorhines from Shanghuang, this morphology is simply primitive and expected for euprimates. However, if an eosimiid allocation can be confirmed for the V13018 first metatarsal, this specimen suggests that the grasping mechanics of basal anthropoids was unlike that of crown-group anthropoids. This observation allows us to begin to decipher the sequence of locomotor changes that had to occur within the evolutionary transition to higher primates.

Acknowledgment We thank the staffs at the FMNH (Chicago), especially Dr. Bill Stanley, as well as those at the USNM (D.C.), and at the IVPP (Beijing). We thank Steve Goodman at the FMNH for access to his *Microcebus* collections, and Scott Williams (NIU) and Amanda Zika (NIU) for their efforts. We also thank several reviewers for their comments. Financial support for this project was provided by the Leakey Foundation, the National Science Foundation, and the Chinese NSF.

References

Boyer, D.M., Patel, B.A., Larson, S.G., and Stern, J.T. (2007). Electromyography of peroneus longus in *Varecia variegata* and *Eulemur rubriventer* helps in grasping primate origins. *J. Hum. Evol.* 53:119–134.

Cartmill, M.C. (1972). Arboreal adaptations and the origin of the order Primates. In: Tuttle, R.H. (ed.), *The Functional and Evolutionary Biology of Primates*. Aldine-Atherton, Chicago, pp. 97–122.

Cartmill, M.C. (1974). Pads and claws in arboreal locomotion. In: Jenkins, F.A. (ed.), *Primate Locomotion*. Academic Press, New York, pp. 45–84.

Cartmill, M.C. (1985). Climbing. In: Hildebrand, M., Bramble, D.M., Liem, K.F., and Wake, D.B. (eds.), *Functional Vertebrate Morphology*. Belknap Press, Cambridge, MA, pp. 73–88.

Crompton, R.H. (1984). Foraging, habitat structure, and locomotion in two species of *Galago*. In: Rodman, P.S. and Cant, J.G.H. (eds.), *Adaptations for Foraging in Nonhuman Primates*. Columbia University Press, New York, pp. 73–111.

Dagosto, M., and Yamashita, N. (1998). Effect of habitat structure on positional behavior and support use in three species of lemur. *Primates* 39:459–472.

Dagosto, M., Gebo, D.L., and Beard, K.C. (1999). Revision of the Wind River Faunas, Early Eocene of Central Wyoming. Part 14. Postcranium of *Shoshonius cooperi* (Mammalia: Primates). *Annals of Carnegie Museum* 68(3):175–211.

Dagosto, M., Gebo, D.L., and Dolino, C. (2001). Positional behavior and social organization of the Philippine Tarsier (*Tarsius syrichta*). *Primates* 42:233–243.

Dagosto, M., Gebo, D.L., and Dolino, C. (2003). The natural history of the Philippine tarsier (*Tarsius syrichta*). In: Wright, P.C., Simons, E.L., and Gursky, S. (eds.), *Tarsiers – Past, Present, and Future*. Rutgers University Press, New Brunswick, pp. 237–259.

Demes, B., Fleagle, J.G., and Jungers, W.L. (1999). Takeoff and landing forces of leaping strepsirhine primates. *J. Hum. Evol.* 37:279–292.

Demes, B., Jungers, W.L., Gross, T.S., and Fleagle, J.G. (1995). Kinetics of leaping primates: Influence of substrate orientation and compliance. *Am. J. Phys. Anthropol.* 96:419–429.

Demes, B., Fleagle, J.G., and Lemelin, P. (1998). Myological correlates of vertical clinging and leaping. *J. Hum Evol.* 34:385–400.

Falster, D.S., Warton, D.I., and Wright, I.J. (2003). (S)MATR: Standardised major axis tests and routines. http://www.bio.mg.au/ecology/SMATR.

Fleagle, J.G., and Mittermeier, R.A. (1980). Locomotor behavior, body size, and comparative ecology of seven Surinam monkeys. *Am. J. Phys. Anthropol.* 52:301–314.

Garber, P.A., and Preutz, J.D. (1995). Positional behavior in moustached tamarin monkeys: Effects of habitat on locomotor variability and locomotor stability. *J. Hum. Evol.* 28:411–426.

Garber, P.A., and Leigh, S.R. (2001). Patterns of positional behavior in mixed-species troops of *Callimico goeldi*, *Saguinus labiatus*, and *Saguinus fuscicollis* in Northwestern Brazil. *Am. J. Primatol.* 54:17–31.

Gebo, D.L. (1986a). Anthropoid origins – the foot evidence. *J. Hum. Evol.* 15:421–430.

Gebo, D.L. (1986b). *The Anatomy of the Prosimian Foot and its Application to the Primate Fossil Record.* Duke University, Durham.

Gebo, D.L. (1987a). Functional anatomy of the tarsier foot. *Am. J. Phys. Anthropol.* 73:9–31.

Gebo, D.L. (1987b). Locomotor diversity in prosimian primates. *Am. J. Primatol.* 13:271–281.

Gebo, D.L. (1989). Postcranial adaptation and evolution in Lorisidae. *Primates* 30(3):347–367.

Gebo, D.L., Dagosto, M., Beard, K.C., and Wang, J. (1999). A first metatarsal of *Hoanghonius stehlini* from the Late Middle Eocene of Shanxi Province, China. *J. Hum. Evol.* 37:801–806.

Gebo, D.L., Dagosto, M., Beard, K.C., and Qi, T. (2001). Middle Eocene primate tarsals from China: implications for haplorhine evolution. *Am .J. Phys. Anthropol.* 116:83–117.

Groves, C. (2001). *Primate Taxonomy*. Smithsonian Institution Press, Washington, DC.

LeGros Clark, W.E. (1959). *The Antecedents of Man: An introduction to the evolution of primates*. British Museum of Natural History, London.

MacPhee, R.D.E., Beard, K.C., and Qi, T. (1995). Significance of primate petrosal from middle Eocene fissure-fillings at Shanghuang, Jiangsu Province, People's Republic of China. *J. Hum. Evol.* 29:501–514.

Morton, D.J. (1924). Evolution of the human foot (II). *Am. J. Phys. Anthropol.* 7:1–52.

Nash, L.T., Bearder, S.K., and Olson, T.R. (1989). Synopsis of *Galago* species characteristics. *Int. J. Primatol.* 10:57–80.

Rasoloarison, R.M., Goodman, S.M., and Ganzhorn, J.U. (2000). Taxonomic revision of mouse lemurs (*Microcebus*) in the western portions of Madagascar. *Int. J. Primatol.* 21:961–1019.

Szalay, F.S., and Dagosto, M. (1980). Locomotor adaptations as reflected on the humerus of Paleogene primates. *Folia Primatol.* 34:1–45.

Szalay, F.S., and Dagosto, M. (1988). Evolution of hallucial grasping in the primates. *J. Hum. Evol.* 17:1–33.

New Data on *Loveina* (Primates: Omomyidae) from the early Eocene Wasatch Formation and Implications for Washakiin Relationships

Patricia A. Holroyd and Suzanne G. Strait

Introduction

Omomyine omomyid primates are comparatively rare in earliest Eocene faunas of North America (Wasatchian North American Land Mammal Age or NALMA), both in species diversity and abundance (Gunnell, 1997), although they are the dominant primate group by the middle Eocene. The origins of the subfamily and relationships among the several named tribes are unclear (e.g., Gunnell, 1995; Gunnell and Rose, 2002) and it has been suggested that Omomyinae are not a holophyletic grouping and that the North American omomyine radiation is the sister taxon of anthropoids+*Tarsius* (Kay and Williams, 1994; Kay et al., 1997; Ross et al., 1998; Kay et al., 2004).

Gunnell (1997) has explicitly suggested that the diversification of omomyines took place during the late early Eocene (Lostcabinian biozone of the Wasatchian), and phylogenetic scenarios that nest anthropoids within omomyines necessitate a minimum divergence of the anthropoid lineage within the early Eocene. As a consequence, early Eocene omomyines are disproportionately important to understanding later omomyid diversification (e.g., Honey, 1990; Gunnell, 1995, 1997) and for understanding character polarities of the earliest primates (e.g., Rose and Bown, 1991; Rose et al., 1994; Ross et al., 1998). Unfortunately, they are comparatively rare in this time period.

A critical taxon for understanding the differentiation of omomyines is the early Eocene genus *Loveina*, comprising four species *Loveina zephyri*, *L. minuta*, *L. wapitiensis*, and *L. sheai*. In his analysis of relationships among washakiinins, Honey (1990) portrayed *Loveina* as a paraphyletic stem taxon at the base of the tribe, with *Loveina minuta* branching off before *L. zephyri (L. sheai and L. wapitiensis* were not included in that study). A better understanding of

Patricia A. Holroyd,
1101 Valley Life Sciences Building Mail Code 4780,University of California, Berkeley, CA 94720, Phone: 510-642-3733; Fax: 570-642-1822
pholroyd@berkeley.edu

J. G. Fleagle, C. C. Gilbert (eds.), *Elwyn Simons: A Search for Origins.*

Loveina's place within the Omomyinae and character transformations in the early differentiation of omomyines has been hampered by the lack of data on the premolar dentition of *L. minuta* or *L. sheai*, as the morphology of p3 and p4 is significant in establishing the relationships of Wasatchian omomyines (e.g., Honey, 1990; Beard et al., 1992). The rare taxon *L. sheai* from the Lostcabinian Big Piney local fauna in the northern Green River Basin, southwestern Wyoming, is particularly important to this discussion as it is intermediate in age between the sole specimen of *L. minuta* and younger *L. zephyri* and later washakiins.

Originally described as *Omomys sheai* by Gazin (1962), the taxon was based on two specimens (one m2–m3, the other m3 only). Szalay (1976) synonymized the taxon with *Utahia kayi* (in the tribe Utahiini), although as noted by Gunnell and Rose (2002) this attribution is not consistent with several features of the molar morphology. In a footnote, Stucky (1984: Table 1, footnote 21) reassigned it and "*Omomys*" *minutus* to the genus *Loveina* based on his own observations as well as a personal communication from Peter Robinson. Gunnell and Rose (2002) concurred with Stucky's observations, but did not think that the specimens differed from *Loveina zephyri* and synonymized it with that species. Recent fieldwork in the Green River Basin, Sublette County, Wyoming has yielded new specimens of *Loveina* species, including the first evidence of the lower premolars and first molar of *L. sheai* as well as additional specimens referable to *L. wapitiensis*. Analysis of these specimens allows us to revalidate *L. sheai*, revise the genus *Loveina*, and provide insights into primitive washakiinin morphology.

Table 1 Measurements of all described specimens of *Loveina* species

	P_3l	P_3w	P_4l	P_4w	M_1l	M_1w	M_2l	M_2w	M_3l	M_3w
L. sheai										
USNM 22384							2.1	1.8	2.6	1.5
UCMP 156831	1.6	1.2	1.7	1.5	2.0	1.6	2.1	1.7		
L. zephyri										
AMNH 32517*	1.6	1.4	1.8	1.6	2.2	1.75				
AMNH 32517†	1.70	1.45	1.9	1.6	2.3	1.75				
AMNH 17554♠			2.0	1.6	2.1	1.9	2.1	1.8		
AMNH 17554†			1.8	1.55	2.2	1.7	2.35	1.85		
AMNH 55219♠			1.9	1.7						
L. wapitiensis										
YPM PU 17317	1.7	1.3	1.9	1.6	2.2	1.6	2.2	1.9	2.8	1.8
UCMP 151125			1.8	1.5	2.2	1.7	2.2	1.8		
UCMP 151127							2.1	1.8	2.8	1.6
L. minutus										
AC 3325†					1.85	1.5	1.75	1.6	2.2	1.3

* as reported by Simpson (1940)
† as reported by Bown and Rose (1984)
♠ as reported by Robinson (1960)

The Age and Faunal Context of Loveina

All *Loveina* species occur in sediments generally regarded as late early Eocene or Lostcabinian in age in the intermontane basins of southern and western Wyoming and northeastern Colorado. The type species *Loveina zephyri* occurs in the Wind River Formation, Lost Cabin Member, Wyoming (Simpson, 1940; Stucky, 1984), and referred specimens were described by Robinson (1960) from both Lostcabinian and slightly younger Gardnerbuttean localities in the Huerfano Basin of Colorado. *L. sheai* is one of several omomyid taxa known from the late Wasatchian Big Piney fauna, originally described by Gazin (1952, 1962). The age of the fauna is generally regarded as late early Eocene (Lostcabinian subage, or Wa7, of the Wasatchian NALMA) (Robinson et al., 2004), although some taxa more characteristic of younger, Bridgerian faunas have also been recovered (Smith and Holroyd, 2003). *L. minuta* was originally described from slightly older Lysitean aged deposits of the Wasatch Formation in the Red Desert area of southwestern Wyoming and has subsequently been noted but not formally described from the Lostcabinian Niland Tongue of the Wasatch Formation in the Washakie Basin (Covert and Hamrick, 1993). *L. wapitiensis* is described from a single specimen from the Lostcabinian Wapiti Valley fauna (Gunnell et al., 1992).

Abbreviations Used

Dental nomenclature: upper case letters with subscripts and superscripts are used to indicate dental position, e.g., P_3 for the third lower premolar.

Other abbreviations: AC, Amherst College, Amherst, MA; AMNH, American Museum of Natural History, New York, NY; LACM, Los Angeles County Natural History Museum, Los Angeles, CA; NALMA, North American Land Mammal Age; UCMP, University of California Museum of Paleontology, Berkeley, CA; USNM, United States National Museum, Smithsonian Institution, Washington, DC; YPM PU, former Princeton University collections now part of the Yale Peabody Museum, New Haven, CT.

Systematic Paleontology

Family Omomyidae Trouessart, 1879
Subfamily Omomyinae Trouessart, 1879
Loveina Simpson, 1940

Synonymies: *Notharctus* Loomis, 1906; *Loveina* Simpson, 1940 (in part), Gazin, 1962 (part), Russell et al., 1967 (part), Stucky, 1984; Bown and Rose,

1984; Gunnell et al., 1992; Gunnell and Rose, 2002; *Tetonius* Seton 1940; *Absarokius* Gazin, 1958 (part); *Omomys* Matthew, 1915 (part), Simpson, 1940 (part) Gazin, 1958 (part) 1962 (part); *Anemorhysis* McKenna, 1960 (part); non *Loveina* Bown, 1979; *Utahia* Szalay 1976 (part).

Type species: *Loveina zephyri* Simpson, 1940.

Included species: *L. minutus* (Loomis, 1906); *L. sheai* (Gazin, 1962), *L. wapitiensis* Gunnell, Bartels, Gingerich, and Torres, 1992.

Emended Diagnosis: Differs from other Washakiini in having narrower P_3–P_4, more weakly-developed P_3 paracristid and P_3 lingual postprotocristid lacking, distinct P_4 paraconid lacking and metaconid slightly smaller, an absence of enamel crenulation, lack of an M_1 postmetacristid; continuous buccal cingulids on molars. Differs from primitive omomyinins (*Steinius* and *Omomys*) in having relatively wider P_3 and P_4, $M_1 > M_2 > M_3$ in length.

Discussion: Species now assigned to *Loveina* have suffered a tortuous taxonomic history. *Loveina* was originally described based on the species *L. zephyri* Simpson 1940. *Tetonius barbeyi*, named by Seton (1940) was synonymized with *Loveina zephyri* by Gazin (1958), Szalay (1976) and Bown and Rose (1984). In describing *Loveina*, Simpson also included Matthew's (1915) *Omomys vespertinus*. This species was also included in *Loveina* by Gazin (1962) and Russell et al. (1967). Szalay (1976) removed it to *Uintanius*, and it was finally reassigned to *Steinius vespertinus* by Bown and Rose (1984).

Loveina minutus was first named by Loomis (1906) as *Notharctus minutus*. It was reassigned to *Omomys minutus* by Matthew (1915), Simpson (1940), Gazin (1958), Guthrie (1967), and Szalay (1976); it was reassigned to *Anemorhysis minutus* by McKenna (1960); it was reassigned to *Loveina minuta* by Bown and Rose (1984) and Stucky (1984). Bown (1979) erroneously assigned UCMP 46705, a dentary of *Chlororhysis knightensis* described by McKenna (1960) and Szalay (1976), to *Loveina minuta*. UCMP 46705 is clearly an anaptomorphine and is here revalidated as *Chlororhysis knightensis*.

Loveina sheai (Gazin, 1962)
Figs.1, 2b, 2d

Synonymies: *Absarokius* Gazin, 1958 (part); *Omomys sheai* Gazin, 1962; *Loveina zephyri* Gunnell and Rose, 2002

Other Illustrations: Gazin, 1962, plate 2, Figs. 3–4.

Holotype: USNM 22384 from 12 miles north of Big Piney, Sublette Co., Wyoming.

Newly referred specimens: UCMP 156831 from UCMP locality V77067, a right dentary with P_3–M_2 (Fig.1b, 2d); UCMP 148981, a P_4 talonid from UCMP locality V96281.

Revised diagnosis: P_3 differs from *L. wapitiensis* and *L. zephyri* in that it is smaller, base less buccolingually inflated, lacks metacristid (slight in *L. wapitiensis* and developed in *L. zephyri*), and crenulation entirely absent on posterior wall of protoconid. Differs from *L. wapitiensis* in that the P_3 lacks a metaconid and lacks

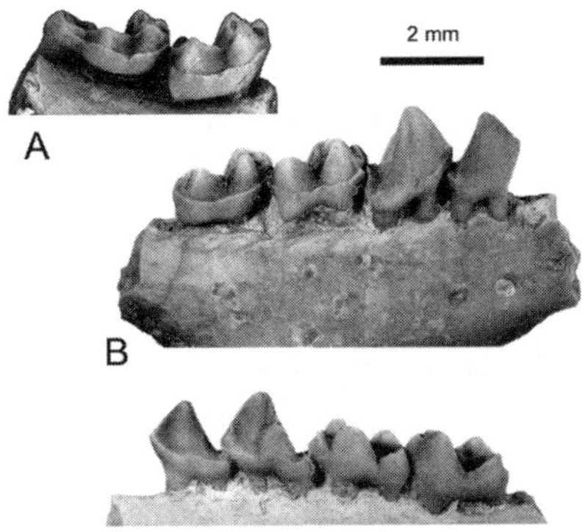

Fig. 1 *Loveina sheai.* A. USNM 22384, holotype, in lateral view; B. UCMP 156831 in lateral and lingual view

metastylids on lower molars. Differs from *L. zephyri* in a more developed P_3 lingual cingulum (like *L. wapitiensis*). Differs from *L. minuta* in larger tooth size, more buccal placement of hypoconulid on M_1–M_2, more anterior placement of hypoconid relative to the entoconid on M_3 (like *L. wapitiensis*). Differs from *L. zephyri* and resembles *L. wapitiensis* in lacking a P_3 metacristid and weaker P_3 postcristids, having no ventral extension of the P_4 buccal surface

Occurrence: The two new UCMP localities occur near the type locality in the main body of the Wasatch Formation, Green River Basin, Sublette Co., Wyoming. Based on description of landmarks in Gazin's field notes (on file at USNM) UCMP locality V77064 (see Smith and Holroyd, 2003) may be the same locality where the holotype was found, and lies approximately 10 meters stratigraphically below the new localities (Holroyd and J.H. Hutchison, unpublished field data). Precise locality data are on file at UCMP and the USNM. All these localities are associated with *Lambdotherium* and so are generally regarded as Lostcabinian (late Wasatchian NALMA) in age. West (1973) also

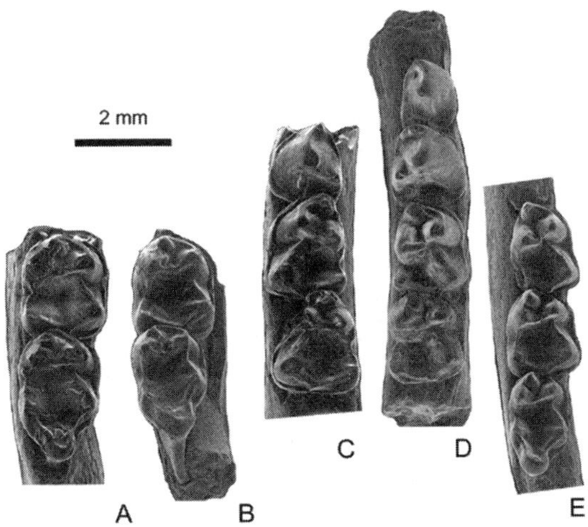

Fig. 2 *Loveina* species in occlusal view. A. *L. wapitiensis*, UCMP 151127; B. USNM 22384, *L. sheai* holotype; C. *L. wapitiensis*, UCMP 151125; D. *L. sheai*, UCMP 156831; E. ACM 3365, *L. minuta* holotype

listed an occurrence of *Omomys* cf. *O. sheai* from locality NF-1 in the stratigraphically higher "New Fork Tongue" (now Alkali Creek Tongue) of the Wasatch Formation, Sublette Co., Wyoming. We have not had the opportunity to view this specimen, and so cannot confirm this occurrence.

Description: The P_3 is known from UCMP 156831. It is a fairly simple, two rooted tooth, lacking both metaconid and metacristid. Unlike other *Loveina* species for which P_3s are known (*L. zephyri* and *L. wapitiensis*), the P_3 is distinctively narrow with less basal buccolingual inflation. A sharp paracristid descends from the apex of the prominent protoconid; half-way down its course it curves slightly lingually until terminating on the anterior margin of the tooth. A faint postcristid descends the posterior wall of the protoconid, joining it to the anterobuccal corner of the talonid. The enamel on the posterior surface of the protoconid is smooth. In this lightly worn tooth, there is no indication of a paraconid. It lacks a buccal cingulid, but there is a shallow lingual cingulid from the base of the protoconid to the base of the paracristid. The talonid is short and is divided by a small, but distinct anteroposteriorly oriented cristid obliqua , just buccal of midline. The buccal side of the talonid is a small enclosed crescent-shaped basin. Lingually the talonid is expanded anteriorly until the base of the lingual cingulid and is not deeply excavated but is open lingually.

The P_4 is also now known from UCMP 156831. This tooth is taller, broader, and more complex than P_3. A small metaconid is developed posterolingual of the dominant protoconid. The paraconid is smaller and lower than the metaconid and is present at the termination of the lingually curved paracristid. On the trigonid, narrow buccal and lingual cingulids are present. The talonids are more developed than on P_3 and the posterior wall is slightly crenulated. The cristid obliqua divides the talonid into a smaller, raised buccal basin, and lingually into a wider shelf-like slope.

UCMP 156831 also includes the first M_1 of this species. The paraconid is smaller and more internal than metaconid. The paracristid is well developed, closing the trigonid basin buccally, while the trigonid remains open posteriorly and lingually. There is a narrow buccal cingulum. The talonid basin and posterior wall show slight crenulation. The hypoconulid is not centrally placed, as in *L. minuta*, but positioned closer to the hypoconid as in *L. wapitiensis*. All molar positions lack metastylids.

The M_2 is the only tooth position shared by the type and the newly referred material. As in the type (USNM 22384) the paraconid and metaconid are close, with the paraconid being more internal and slightly lingual to midline. The cristid obliqua is buccally positioned, meeting the posterior wall at the base of the protoconid. The talonid basin is wide and bounded by a low preentocristid. A distinct and continuous cingulid is present from the base of the paracristid, along the buccal side of the tooth to the base of the entoconid. The M_2 differs from M_1 in that the buccal cingulum is better developed, paraconid closer to metaconid, paraconid is higher and less anteriorly projecting, cristid obliqua is more curved, and the tooth is more square in occlusal outline.

The M_3 is still only known from the type specimen. On the trigonid, the paraconid is more posteriorly placed than on M_2, resulting in a relatively longer paracristid. The bases of the paraconid and protoconid are joined so that the trigonid basin is bisected. Even though this specimen is slightly worn, the talonid basin shows distinct crenulations. The hypoconid is more anterior than the entoconid.

Discussion: Gazin erected this species in 1962 and included no diagnosis and little description. Although Gunnell and Rose (2002) synonymized *L. sheai* with *L. zephyri*, this new material supports the view that it is a distinctive species, or at least that UCMP 156831 is distinct from other *Loveina* species. Referral of UCMP 156831 is based in part on its recovery from the type area and similarity in M_2 morphology. Both UCMP localities V77067 and V96281, as well as the type locality, occur in a complex of fossiliferous horizons within an approximately 15 meter thick interval of the Wasatch Formation occurring 20 meters below the contact with the overlying Green River Formation in the area 12 miles north of Big Piney, Wyoming. The complex of coeval *Loveina* species is difficult to differentiate solely based on their M_2 morphologies, and we acknowledge this difficulty. However, we consider it most parsimonious to refer UCMP 156831 and 148981 to a revalidated *L. sheai*, rather than erecting a new species of *Loveina*.

Loveina wapitiensis Gunnell et al., 1992
Fig.2a, 2c

Other Illustrations: Gunnell et al., 1992, figure 5

Holotype: YPM-PU 17317

Newly referred specimens: UCMP 151125, a left dentary with P_4–M_2 (Fig. 2c); 151126, a right dentary with the posterior portion of m1 and anterior portion of M_2; and 151127, a right dentary with M_2–M_3 (Fig.2a). All from UCMP locality V99511, Sassy Flats, Wasatch Formation, Sublette Co., Wyoming.

Emended Diagnosis: Differs from other *Loveina* species in possessing small but distinct metastylids on M_1–M_3. Further differs from *L. minuta* in larger size and more anterior placement of the M_3 hypoconid relative to the entoconid. Differs from *L. sheai* in having better developed molar metacristids. Differs from *L. zephyri* in having less well developed molar metacristids. Gunnell et al. (1992:261) noted that *L. wapitiensis* differed from *L. zephyri* in that the P_3 had a "lower, relatively smaller paraconid and less bulging metaconid". However, a comparison of the types of these two species reveals that the position and size of the paraconids are similar. Further, *L. zephyri* has no trace of a metaconid, while *L. wapitiensis* has a very small metaconid.

Discussion: These are the first referred specimens of *L. wapitiensis*. Although they duplicate known tooth positions, they are less worn than the type and permit documentation of intraspecific metric variation in this taxon and some additional morphologic features. UCMP 151125 includes the first unworn P_4 of this species. The posterior wall of the P_4 trigonid is slightly crenulated, with two distinct ridges running parallel to one another descending from the protoconid (as on the type of *L. zephyri*). The lingual of these two ridges is longer and is continuous with the cristid obliqua. The cristid obliqua divides the talonid into a smaller, raised buccal basin, and lingually into a shelf which is wider than in *L. sheai*. UCMP 151125 exhibits a small metastylid on m1 (roughly half the size as on M_2). The moderately worn type specimen (YMP-PU 17317) was described as lacking a m1 metastylid (Gunnell et al., 1992); however, the type has a slight expansion of the crest in the same position as a metastylid might represent a worn metastylid.

Discussion

Previous Interpretations of the Phylogenetic Relationships of Loveina

At the time of its description, Gazin (1962) did not reflect at length on the possible relationships of *Omomys sheai*. He considered it close to *"Omomys"* [= *Steinius*] *vespertinus* and "*Omomys minutus*" [= *Loveina minuta*], but did not suggest anything more specific. Studying these three species, Bown (1976) suggested that the more medial position of the M_3 paraconid in these may have been derived from an anaptomorphine condition such as that seen in *Anemorhysis*. Based on these comparisons, he hypothesized that a hypothetical omomyid ancestor had a premolariform-semimolariform P_4, an unreduced M_3/M_3, and lingually positioned paraconids on M_2–M_3. Szalay (1976) did not recognize *Omomys sheai* as distinct, and subsumed it within *Utahia*, and subsequent discussions of washakiin relationships (e.g., Honey, 1990; Beard et al. 1992) have not dealt with the species.

Despite its poor record, the genus *Loveina* has long been noted as a critical taxon for understanding omomyine relationships (Simpson, 1940; Szalay, 1976; Bown and Rose, 1984), because its generally primitive morphology has led to basal placements. Szalay (1976) considered it one of the most primitive omomyines, and placed *Loveina zephyri* at the base of the tribe Washakiini (comprising *Loveina*, *Shoshonius*, *Washakius*, and *Dyseolemur*) and placed "*Omomys*" (= *Loveina*) *minuta* at the base of a clade comprising *Omomys*, *Ourayia*, and *Macrotarsius*. More recently *Steinius* has been recognized as a very primitive omomyine (Bown and Rose, 1991; Rose et al., 1994), and *Loveina* has been placed within the Washakiini as its most primitive member (Honey, 1990; Rose et al. 1994; Gunnell and Rose, 2002). *Loveina*'s origins are hypothesized to be from a species of *Teilhardina* (Bown and Rose, 1984; Rose et al., 1994; Gunnell and Rose, 2002)

Loveina species and other washakiins have been treated in several phylogenetic analyses with differing results (Fig. 3). Honey (1990) considered *L. minuta*

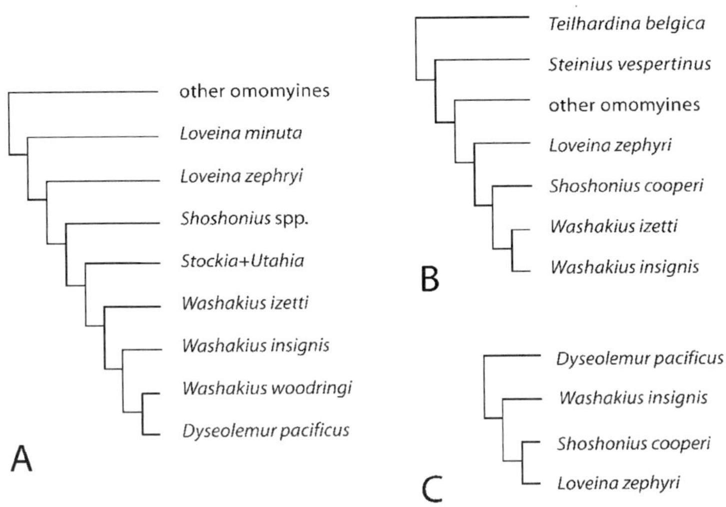

Fig.17.3 Differing hypotheses for the phylogenetic position of *Loveina* species. A. paraphyletic stem to washakiins and utahiins, Honey (1990); B. Basal washakiin, Muldoon and Gunnell (2002); C. Sister taxon to *Shoshonius*, Bloch et al. (1997), Kay et al. (2004) and Ni et al. (2004)

and *L. zephyri* to be the paraphyletic stem genus leading to *Shoshonius*, Utahiini (*Stockia* and *Utahia*), and a paraphyletic *Washakius+Dyseolemur* (Fig.3a). Kay and Williams (1994), considering only dental data and *Washakius* as the only Washakiinae, found it to be sister taxon of *Nannopithex* (the sole Microchoerinae) and that these two taxa were a sister taxon to Anthropoidea+*Tarsius*. Considering only *L. zephyri*, Muldoon and Gunnell (2002) considered it basal to other washakiins. Bloch et al. (1997), Kay et al. (2004) and Ni et al. (2004) all recovered a holophyletic Washakiini comprised of (*Dyseolemur* [*Washakius* {*Shoshonius+Loveina*}]) (Fig. 3c), but differed in placing them as a sister taxon to *Tarsius* (Bloch et al., 1997), as the sister clade to Anthropoidea+*Tarsius* (Kay et al., 2004), as part of a clade of North American omomyines nested within Holarctic omomyids exclusive of *Rooneyia* (Ni et al., 2004), or as a clade whose relationships are unresolved among euprimates (Ross et al., 1998). The consistency of results for washakiin interrelationships across these studies is not surprising, as each builds upon the same base matrices (Kay and Williams, 1997; Kay et al., 1997; Ross et al., 1998), and differences in the broader relationships of washakiins can be attributed to slight differences in taxon sampling and methods of data analysis across these studies.

Using a stratocladistic approach and a different set of taxa and characters, Muldoon and Gunnell (2002) also found washakiins to be holophyletic, but differed in recognizing *Loveina zephyri* as the most basal branch in the clade and *Shoshonius cooperi* as ancestral to *Washakius* species (Fig.3b). Most recently, Seiffert et al. (2005) did not find washakiins to be monophyletic, with

Dyseolemur unresolved among omomyids and *Loveina zephyri*, *Shoshonius*, and *Washakius insignis* only united, without internal resolution, in their Adams consensus tree.

An Alternative Phylogenetic Analysis

In order to examine the impact of our new morphological observations upon the holophyly of Washakiini, and the position of *Loveina* within Washakiini, and of washakiins to other primates, we scored the morphology of *L. minuta*, *L. sheai*, and *L. wapitiensis* within Ni et al.'s (2004) matrix, for a total of 55 included taxa. This matrix, expanded from ones originally published by Kay and Williams (1994) and Ross et al. (1998), has been criticized by some authors (e.g., Seiffert et al., 2005) for the way in which some characters are coded. We selected it for this analysis because among published matrices it has the best taxonomic sampling among omomyids.

Specimens used for scoring the three additional *Loveina* species are those described above as well as previously described specimens. In the hopes of providing better resolution among omomyids, characters of the upper premolars (P_3–P_4) and molars were added for *Anemorhysis savagei* based on UCMP 146720 and 171182 from the type locality. Scorings for *Purgatorius* were also added based on new data for the anterior dentition from Clemens (2004), and we updated the taxon name from *P. unio* to *P. janisae*, based on Van Valen's (1994) referral of UCMP 107406 and LACM 28128 to *P. janisae*. These specimens were used for the original character scorings of *P. unio* in Kay and Williams (1994), and the use of the outdated name has propagated in matrices built upon the Kay and Williams (1994) matrix. These changes to the previously-published matrix are given in Table 2.

A heuristic search with 1000 repetitions was performed with PAUP 4b10 (Swofford, 1998) with the following options: MSTAXA = VARIABLE, ADDSEQ = RANDOM; MAXTREES = AUTOINC. The tree-bisection-reconnection algorithm was used for branch swapping. One hundred nineteen characters were treated as ordered and 184 as unordered, using the typeset of Ni et al. (2004). Of the 303 total characters scored, 9 are constant and 12 are parsimony-uninformative, for a total of 282 parsimony-informative characters. Twenty-one equally most-parsimonious trees of 2106 steps were obtained, and a strict consensus tree computed. This strict consensus tree was rooted using Scandentia, *Purgatorius janisae*, and the plesiadapiform taxa.

Figure 4 shows a simplified version of the strict consensus tree with some terminal branchings collapsed. Decay indices (Bremer, 1994), calculated via a heuristic search of trees with an increasing number of steps, are indicated at the base of each node. As found by other recent studies, the Washakiini form a holophyletic group, but with a different set of internal relationships than found by most of these studies. These taxa are united by a series of changes,

Table 2 Additional or changed scorings to the Ni et al. (2004) matrix

Purgatorius janisae

011-???1?? 2?01?0??1? ????001001 0022000011 1011010000 0200001025 1000001122 2000100001 2000032110 1222333222 2001012221 300??????? ?????????? ??0?1?00?1 1?02000?0? ?0?0210001 1000020010 0211?10??1 ?????????? ?????????? ?????????? ?????????? 0?????0??? ?????????? ?????????? ?????????? ?????????? ?????????? ?????????? ?000031011 0?0

Anemorhysis savagei

11?124-??? 2?0????20? ????100012 111?0?0110 201101(01)0?1 012000012(12) 1000001123 2000(12)01001 2000032330 -21(01)300222 2001122111 (12)10??????? ?????????? ?00?1?1000 0?02010000 1001220002 2-00010010 1200110101 01???????? ?????????? ?????????? ?????????? 0?????0??? ?????????? ?????????? ?????????? ?????????? ?????????? ?????????? ?0-2010121 2?0

Loveina sheai

?????????? ?????????? ?????0?011 020?1?0110 00210100?1 1121011121 2000001121 3000101001 2000032220 2110000222 1001122202 (12)10??????? ?????????? ?????????? ?????????? ?????????? ?????????? ?????????? ?????????? ?????????? ?????????? ?????????? ?????????? ?????????? ?????????? ?????????? ?????????? ?????????? ?????????? ?????20011 2??

Loveina minuta

?????????? ?????????? ?????????? ?????????? ?????????? ?????????? ??00001121 3000101001 2020032220 2110000222 00010022?? ?00??????? ?????????? ?????????? ?????????? ?????????? ?????????? ?????????? ?????????? ?????????? ?????????? ?????????? ?????????? ?????????? ?????????? ?????????? ?????????? ?????????? ?????????? ???????01? 2??

Loveina wapitiensis

?????????? ?????????? ???????011 021?1?0110 00210100?1 1121011122 2000001121 3010101001 2000032220 2110000222 1001112202 200??????? ?????????? ?????????? ?????????? ?????????? ?????????? ?????????? ?????????? ?????????? ?????????? ?????????? ?????????? ?????????? ?????????? ?????????? ?????????? ?????????? ?????????? ?????2001? 2??

primarily in the premolar dentition (character numbers are those of Ni et al., 2004): P_3 paraconid changing from absent to small (char. 30); P_3 lateral protocristid reduced (char. 43); increase in P_4 paraconid size (char. 31) and height (char. 57) and shifting to a more mesial position relative to the metaconid (char. 32); P_4 metaconid more widely spaced from the protoconid (char. 35); P_4 entoconid absent (char. 41); increase in M_1 area (char. 110); and an increase in the length of M_3 relative to M_2 (char. 109). Interestingly, all but two of these character changes (characters 35 and 57 plot out in the tree as reversals to the putative primitive condition, suggesting that basal washakiins may possess some dental character states that more closely approximate the omomyid or even euprimate morphotype. However, the large number of anaptomorphines in the matrix, as well as the more basal position of adapids with respect to omomyids, may explain the appearance of primitive morphologies as reversals, as these taxa generally have more well-developed premolars with stronger cusps and crests. Washakiini are also united in this analysis by a number of characters of the upper dentition (characters 127, 130, 136, 141, 147, 167) which may or may not be present at this node when *Loveina* upper dentitions are known, as well as two cranial characters (characters 203 and 205) that are only known in *Shoshonius* and may ultimately be better placed at an internal node within Washakiini.

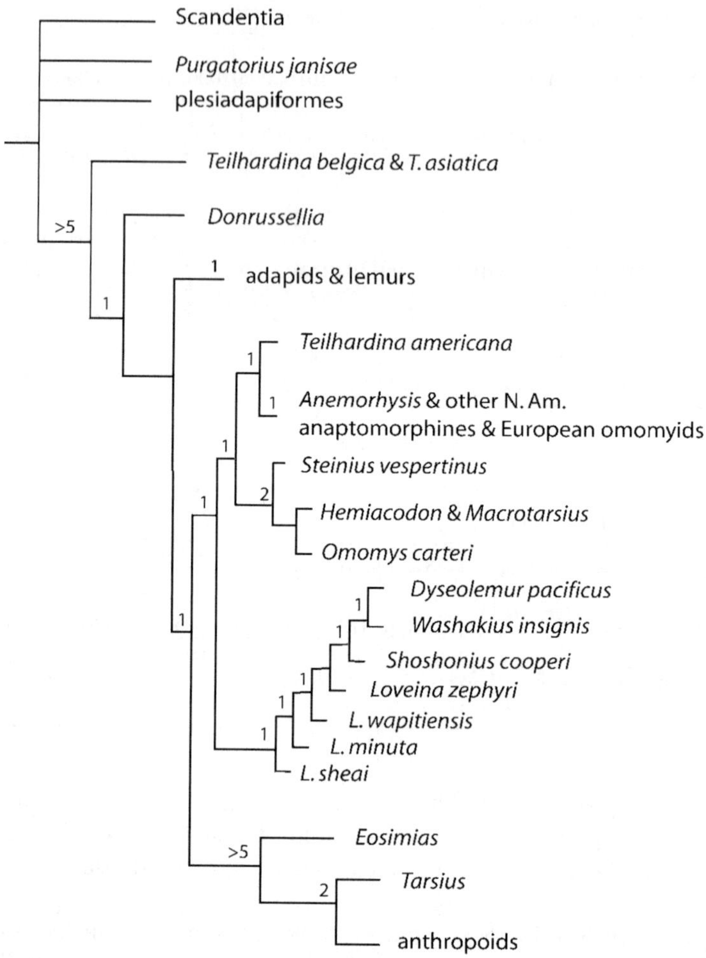

Fig. 4 Strict consensus tree of 21 equally most parsimonious trees of 2106 steps. Decay indices indicated for each node

Within Washakiini, *Loveina* species form a paraphyletic "comb" along the stem leading to middle Eocene taxa, formed by successive increases in size, P_3 cusps, and metastylids. The generally low decay indices for nodes within Washakiini are not surprising, given that most *Loveina* species are known from only a few teeth of the lower dentition, and it is likely that if we had larger samples that allowed us to better understand variation in lower molar and premolar characters or if we had any upper dentition at all, better resolution could be achieved. Alternatively, the presence of generally low decay indices throughout the basal parts of the tree may also reflect adequate sampling of the stem branches of this radiation. Generally speaking, decay indices will be higher for clades lacking good taxonomic representation along their stems, as they reflect not only the robusticity of a set of synapomorphies uniting a clade, but also how great a morphological gap we have in the transition between early forms and their descendents. To put it more simply, if we had sampled all the possible ancestors and descendents in the evolution of a lineage, the decay indices for each node would be one, as each taxon would only be separated from its closest relative by the one character we use to differentiate them. While these indices are often used as an indication of how robust our characters and analysis are, they also reflect how well we have sampled a set of taxa and provide an indication of the relative completeness of the fossil record of its evolutionary history. The fact that the anthropoid/tarsier clade possesses a high decay index and is united by many hypothesized synapomorphies reflects the fact that we lack fossils showing the morphological transitions along that stem that would "break up" the stem and produce lower decay indices along its length.

Resolving omomyid relationships is important for understanding anthropoid origins: As prior studies have made abundantly clear, the character polarities of dental characters, especially those of the premolars and anterior dentition, are key to understanding how omomyids relate to one another and to anthropoids. Establishing these polarities is pivotal to understanding whether omomyids are a holophyletic grouping or one that is paraphyletic with respect to tarsiids and/or anthropoids. This analysis demonstrates how even the inclusion of a handful of additional taxa, particularly those along the stem leading to the better known members of a key clade, can alter relationships and propagate changes elsewhere in the tree. While our analysis does not attempt to fully resolve or even address the polarities of dental characters in early omomyid evolution, it illustrates the importance of further work directed toward these goals.

Acknowledgment Elwyn Simons introduced the first author to fieldwork in Wyoming and first sent her to the Greater Green River Basin in his jeep to see what could be found, inspiring a life-long love of all things Wasatchian. J. H. Hutchison subsequently suggested work in the Big Piney area and has lent his invaluable skills to our field studies, and P.A. Holman ably assisted in the second author's work in the Pinedale area. Fieldwork was conducted under Bureau of Land Management permits, and we are grateful for the assistance of the staffs of the Wyoming State and Pinedale District offices. T.M. Bown, G.F. Gunnell, P. Robinson, and K.D. Rose kindly provided casts for comparison, and R. B. Irmis provided thoughtful discussion of decay indices. This is UCMP contribution #1917.

References

Beard, K. C., Krishtalka, L., and Stucky, R. K. (1992). Revision of the Wind River faunas, early Eocene of central Wyoming. Part 12. New species of omomyid primates (Mammalia: Primates: Omomyidae) and omomyid taxonomic composition across the early-middle Eocene boundary. *Ann. Carnegie Mus.* 61:39–62.

Bloch, J.I., Fisher, D.C., Gingerich, P.D., Gunnell, G.F., Simons, E.L., and Uhen, M.D. (1997). Cladistic analysis and anthropoid origins. *Science* 278: 2134–2135.

Bown, T. M. (1976). Affinities of *Teilhardina* (Primates, Omomyidae) with description of a new species from North America. *Folia Primatol.* 25:62–72.

Bown, T. M. (1979). New omomyid primates (Haplorhini, Tarsiiformes), from middle Eocene rocks of west-central Hot Springs County, Wyoming. *Folia Primatol.*31, 48–73.

Bown, T. M., and Rose, K. D. (1984). Reassessment of some early Eocene Omomyidae, with description of a new genus and three new species. *Folia Primatol.* 43:97–112.

Bown, T. M., and Rose, K. D. (1991). Evolutionary relationships of a new genus and three new species of omomyid primates (Willwood Formation, Lower Eocene, Bighorn Basin, Wyoming). *J. Hum. Evol.* 20:465–480.

Bremer, K. (1994). Branch support and tree stability. *Cladistics* 10:295–304.

Covert, H. H., and Hamrick, M. W. (1993). Description of new skeletal remains of the early Eocene anaptomorphine primate *Absarokius* (Omomyidae) and a discussion about its adaptive profile. *J. Hum. Evol.* 25:351–362.

Clemens, W. A. (2004). *Purgatorius* (Plesiadapiformes, Primates?, Mammalia), a Paleocene immigrant into northeastern Montana: Stratigraphic occurrences and incisor proportions. *Bull. Carnegie Mus. Nat. Hist.* 3:3–13.

Gazin, C. L. (1952). The lower Eocene Knight Formation of western Wyoming, and its mammalian faunas. *Smithsonian Misc. Coll.* 117:1–82.

Gazin, C. L. (1958). A review of middle and upper Eocene primates of North America. *Smithsonian Misc. Coll.* 136:1–112.

Gazin, C.L. (1962). A further study of the lower Eocene mammalian faunas of southwestern Wyoming. *Smithsonian Misc. Coll.* 144:1–98.

Gunnell, G. F. (1995). Omomyid primates (Tarsiiformes) from the Bridger Formation, middle Eocene, southern Green River Basin, Wyoming. *J. Hum. Evol.* 28:147–187.

Gunnell, G. F. (1997). Wasatchian-Bridgerian (Eocene) paleoecology of the Western Interior of North America: Changing paleoenvironments and taxonomic composition of omomyid (Tarsiiformes) primates. *J. Hum. Evol.* 32:105–132.

Gunnell, G. F., Bartels, W. S., Gingerich, P. D., and Torres, V. (1992). Wapiti Valley faunas: Early to middle Eocene fossil vertebrates from the North Fork of the Shoshone River, Park County, Wyoming. *Contr. Mus. Paleontol. Univ. Mich.* 28:247–287.

Gunnell, G.F., and Rose, K.D. (2002). Tarsiiformes: Evolutionary history and adaptation. In: Hartwig, W.C. (ed.), *The Primate Fossil Record.* Cambridge University Press, Cambridge, pp. 45–82.

Guthrie, D. A. (1967). The mammalian fauna of the Lysite Member, Wind River Formation, (early Eocene) of Wyoming. *Mem. Southern Calif. Acad. Sci.* 5:1–53.

Honey, J. G. (1990). New washakiin primates (Omomyidae) from the Eocene of Wyoming and Colorado, and comments on the evolution of the Washakiini. *J. Vertebr. Paleontol.* 10:206–221.

Kay, R.F. and Williams, B.A. (1994). Dental evidence for anthropoid origins. In Fleagle, J. G., and Kay, R. F. (eds.), *Anthropoid Origins.* Plenum Press, New York, pp. 361–446.

Kay, R. F. Ross, C. F., and Williams, B. A. (1997). Anthropoid Origins. *Science* 275:797–804.

Kay, R. F. Williams, B. A., Ross, C. F., Takai, M., and Shigehara, N. (2004). Anthropoid Origins: A phylogenetic analysis. In: Ross, C. F., and Kay, R. F. (eds.), *Anthropoid Origins: New Visions.* Plenum Press, New York, pp. 91–135.

Loomis, F. B. (1906). Wasatch and Wind River primates. *Am. J. Sci.* 21:277–284.

Matthew, W. D. (1915). A revision of the lower Eocene Wasatch and Wind River faunas. IV. Entelonychia, Primates, Insectivora (part). *B. Am. Mus. Nat. Hist.* 34:429–483.

McKenna, M. C. (1960). Fossil Mammalia from the early Wasatchian Four Mile fauna, Eocene of northwest Colorado. *Univ. Calif. Publ. Geol. Sci.* 37(1):1–130.

Muldoon, K. M., and Gunnell, G. F. (2002). Omomyid primates (Tarsiiformes) from the early middle Eocene at South Pass, Greater Green River Basin, Wyoming. *J. Hum. Evol.* 43:479–511.
Ni, X., Wang, Y., Hu, Y., and Li, C. (2004). A euprimate skull from the early Eocene of China. *Nature* 427:65–68.
Robinson, P. (1960). Fossil Mammalia of the Huerfano Formation, Eocene, of Colorado. *Bull. Peabody Museum of Natural History* 21:1–95.
Robinson, P., Gunnell, G. F., Walsh, S. L., Clyde, W. C., Storer, J. E., Stucky, R. K., Froehlich, D. J., Ferrusquia-Villafranca, I., and McKenna, M. C. (2004). Wasatchian through Duchesnean biochronology. In: Woodburne, M. O. (ed.), *Late Cretaceous and Cenozoic Mammals of North America: Biostratigraphy and Geochronology*. Columbia University Press, New York, pp. 106–155.
Rose, K. D., and Bown, T. M. (1991). Additional fossil evidence on the differentiation of the earliest euprimates. *P. Natl. Acad. Sci. USA* 88:98–101.
Rose, K. D., Godinot, M., and Bown, T. M. (1994). The early radiation of euprimates and the initial diversification of Omomyidae. In: Fleagle, J. G., and Kay, R. F. (eds.), *Anthropoid Origins*. Plenum Press, New York, pp. 1–28.
Ross, C. F., Williams, B. A., and Kay, R. F. (1998) Phylogenetic analysis of anthropoid relationships. *J. Hum. Evol.* 35:221–306.
Russell, D. E., Louis, P., and Savage, D. E. (1967). Primates of the French early Eocene. *Univ. Calif. Publ. Geol. Sci.* 73:1–46.
Seiffert, E. R., Simons, E. L., Clyde, W. C., Rossie, J. B., Attia, Y., Bown, T. M., Chatrath, P., and Mathison, M. E. (2005). Basal anthropoids from Egypt and the antiquity of Africa's higher primate radiation. *Science* 310:300–304.
Seton, H. (1940). Two new primates from the lower Eocene of Wyoming. *Proc. New England Zoological Club* 18:39–42.
Simpson, G. G. (1940). Studies on the earliest primates. *B. Am. Mus. Nat. Hist.* 77:185–212.
Stucky, R. K. (1984). The Wasatchian-Bridgerian land mammal age boundary (early to middle Eocene) in western North America. *Ann. Carnegie Mus.* 53:347–382.
Smith, K.T. and Holroyd, P.A. (2003). Rare taxa, biostratigraphy, and the Wasatchian-Bridgerian boundary in North America. *Geol. Soc. Am. Spec. Paper* 369:501–511.
Swofford, D. L. 1998. PAUP*. Phylogenetic Analysis Using Parsimony (*and Other Methods). Version 4. Sinauer Associates, Sunderland, Massachusetts.
Szalay, F. S. (1976). Systematics of the Omomyidae (Tarsiiformes, Primates)—Taxonomy, Phylogeny, and Adaptations. *B. Am. Mus. Nat. Hist.* 156:159–449.
Van Valen, L. M. (1994). Origin of the plesiadapid primates and the nature of *Purgatorius*. *Evol. Monographs* 15:1–79.

The Behavioral Ecology of our Earliest Hominid Ancestors

R. W. Sussman and Donna Hart

Introduction

> Unfortunately not all reconstructions in our science are based on solid evidence from skull and skeleton.... studies of primate and human evolution will persist in mythologizing as long as poorly founded or unsubstantiated claims continue to be taken as proven or presented as fact to the public (Simons, 2000:443).

Although there have been many theoretical reconstructions of the behavioral ecology of our earliest ancestors, the most common theory and one that is widely accepted today is the "Man the Hunter" hypothesis. Cultural anthropologist Laura Klein (2004:10) expresses the current situation well: "While anthropologists argue in scientific meetings and journals, the general public receives its information from more popular sources . . . In many of these forums, the lesson of Man the Hunter has become gospel." This theory of early hominid behavior is still widely debated even within the anthropological community and, as we will show, the evidence to support it remains controversial.

The small skulled australopithecine discovered in 1924 by Raymond Dart was considered by most of his contemporaries a mere ape (Fig. 1). While Piltdown supporters were busy explaining the intellectual endowments of our large-brained ancestors, Dart was convinced his small-brained creature was the first ape-man. At first, Dart theorized that australopithecines were scavengers barely eking out an existence in the harsh savannah environment, scavenging small animals in order to live (Dart 1926). It was not until a quarter of a century later, with the unearthing of many more australopithecines and the discovery in 1953 that Piltdown was a fraud, that students of human evolution realized our earliest ancestors indeed were more ape-like than they were like modern humans. This led to a great interest in using primates to understand human evolution and the evolutionary basis of human nature (Sussman, 2000). With

R.W. Sussman
Department of Anthropology, Washington University, St. Louis, MO 63130,
Phone: 314-935-5264 Fax: 314-935-8535
rwsussma@artsci.wustl.edu

J. G. Fleagle, C. C. Gilbert (eds.), *Elwyn Simons: A Search for Origins.*

Fig. 1 Raymond Dart with the Taung child, the type specimen of *Australopithecus*

these discoveries, began a long list of theories attempting to recreate the behavior, and often the basic morality, of the earliest hominids.

By 1950 Dart developed a new view of hominid behavior based on the fragmented and damaged bones found with the australopithecines, together with dents and holes in these early hominid skulls. From these Dart eventually concluded that this species had used bone, tooth and antler tools to kill, butcher, and consume their prey, as well as to murder one another. Rather than leaving the trees to search out a meager existence in the savannah, Dart espoused that hunting, and a carnivorous lust for blood, drew the man-apes out of the forest and was a main force in human evolution (Dart and Craig, 1959).

This view of the depravity of human nature is related to the idea of man's fall from grace and to the Christian notion of original sin (see Cartmill, 1993; 1997).

Thomas Hobbes, in his 1651 *Leviathan*, "regarded human beings as inherently selfish, competitive, and aggressive –'man is a wolf to man,' in the famous phrase (Pope 2004:315)....Government...is established by social contract to bring some degree of security and safety in the midst of pervasive human evil and unremitting threats to survival presented by the 'state of nature'" (Pope, 2004:316).

Pope (2004) believes Darwin was partially influenced by Hobbesian thinking but was moderated by the natural "moral sense" theories of David Hume and Adam Smith, especially their view of "sympathy." On the other hand, Thomas Huxley could be considered a strict Hobbesian. To Huxley [1893 (1993):36], "intense and unceasing competition of the struggle for existence" permeates the natural world. Human beings are essentially selfish, proud, quick to take offense, and relentless in seeking revenge. Huxley describes the world in explicitly Hobbesian terms (Pope, 2004). There are other views within Western philosophy, such as that of "natural law" professed by Aristotle and Thomas Aquinas, who considered humans as "rational" and "social" animals. To these philosophers humans are political animals and are incomplete without one another: sociality is essential to human well-being, rooted in biology as well as intelligence. However, currently it appears to be a Hobbesian philosophy that pervades many modern views of human nature and the Man the Hunter hypotheses (see Kelly, 2000; Pope, 2004; and, especially, Fry, 2006:244–245).

Between 1961–1976, Dart's view was drawn upon and extensively popularized in the books of the playwright Robert Ardrey (*The Territorial Imperative, African Genesis, The Social Contract, The Hunting Hypothesis*). Although more spectacular than the claims of Raymond Dart, Ardrey's views of human nature did not differ greatly from his, nor from the ancient Christian beliefs of man's fall from grace and original sin.

Dart's evidence for Man the Hunter was not substantive and his particular vision of the human hunter/killer hypothesis did not have much staying power. Upon examination of the evidence, C.K. Brain (1981) noted that the bones associated with the man-apes were exactly like fragments left by leopards and hyenas. It seems that Dart's australopithecines were likely the hunted and not the hunters.

Since the 1950s, it seems that the Man the Hunter hypothesis has been revitalized every decade. The next widely-accepted version of this recurring Man the Hunter theme was presented in the late 1960s by Sherwood Washburn and his colleagues. They claimed that many of the features which define men as hunters again separated the earliest humans from their primate relatives.

> To assert the biological unity of mankind is to affirm the importance of the hunting way of life....The biology, psychology, and customs that separate us from the apes – all these we owe to the hunters of time past. And, for those who would understand the origin and nature of human behavior there is no choice but to try to understand "Man the Hunter" (Washburn and Lancaster, 1968:303).

Like Dart, Washburn related human hunting to human morality, both of which had their biological basis in our evolutionary past.

The next major scientific statement on the importance of hunting in the formulation of human nature was introduced in the mid-1970s by E.O. Wilson and the proponents of sociobiology. Wilson (1975) describes a number of behavioral traits (e.g., territoriality, aggressive dominance hierarchies, male dominance, male-female bonds) that he claims are found in humans generally and that are genetically-based human universals. These traits are assumed to be biologically fixed characteristics, relatively constant among our primate relatives, and persisting throughout human evolution and in human societies. They are products of our hunting past. For more than a million years, man was a hunter and our "innate social responses have been fashioned largely through this lifestyle" (Wilson, 1975:573).

Following in the Hobbesian tradition, Wilson's (1975:573) observations present the non-consoling thought that "some of the 'noblest' traits of mankind, including team play, altruism, patriotism, bravery, and so forth, are the genetic product of warfare."

These earlier renditions of the Man the Hunter hypothesis have been reviewed and discussed in detail elsewhere (e.g., Cartmill, 1993; 1997, Sussman; 1999a; 2004; Marks, 2002; Hart and Sussman, 2005; Fry, 2006).

Chimpanzees and Human Males as Demonic Killers

The most recent claim of the importance of hunting and killing and the biological basis of morality is that of Wrangham and Peterson in their book, *Demonic Males* (1996, see also, Ghiglieri, 1999). The demonic male theory proposes the following. The split between humans and common chimpanzees is much more recent than was once believed, only 6–8 mya. Furthermore, humans may have split from the chimpanzee-bonobo line after gorillas, with bonobos separating from chimps only 2.5 mya. Because the common ancestor of all these forms "was barely distinguishable from chimpanzees" (Wrangham and Peterson, 1996: 49), and because the earliest australopithecine was quite chimpanzee-like, Wrangham (1995:5) speculates that: "The most reasonable view for the moment is that chimpanzees are . . . an amazingly good model for the ancestor of hominids . . . [and if] we know what our ancestor looked like, naturally we get clues about how it behaved . . . that is, like modern-day chimpanzees." Finally, if modern chimpanzees and modern humans share certain behavioral traits, these traits have "long evolutionary roots" and are fixed, biologically-inherited components of our nature and not culturally determined.

Further, the authors of *Demonic Males* claim that only two animal species – chimpanzees and humans – live in patrilineal, male-bonded communities that exhibit intense territorial aggression, including lethal raids that seek vulnerable enemies to kill. Since chimpanzees and humans share these violent urges, the demonic male paradigm emphasizes that chimpanzees and humans also share an inborn morality. Thus, killing and violence are inherited from our ancient relatives. However, Wrangham and Peterson also argue that killing and violence are traits shared by hominids and chimpanzees that are not byproducts of hunting. In fact, it is rather this violent nature and natural blood lust that makes both humans and chimpanzees such good hunters.

The bonobo helps them to this conclusion. They claim that bonobos have lost the desire to kill, as well as losing the desire to hunt; that they have suppressed both personal and predatory aggression; that even though bonobos evolved from a chimpanzee-like ancestor who was both a hunter of monkeys and a hunter of its own kind, during the evolution of bonobos the males lost the desire to kill each other and the desire to kill prey; and, finally, that bonobos and chimps tell us that murder and hunting are very similar. Wrangham and Peterson believe that blood lust ties killing and hunting tightly together, but in this scenario it is the desire to kill that drives the ability to hunt.

However, chimpanzees have been evolving for as long as humans and gorillas, and there is no reason to believe that ancestral chimps were highly similar to present-day chimps. The fossil evidence is extremely sparse for the great apes. It is likely that many forms of apes have become extinct during millions of years – just as many forms of hominids have become extinct. Furthermore, even if chimpanzees were a good model for the ancestor of humans and a conservative representative of this particular branch of the evolutionary bush, it would not follow that humans would necessarily share specific behavioral traits. As the authors of *Demonic Males* emphasize, chimps, gorillas, and bonobos are all very different from one another in their behavior and in their willingness to kill others of their species. It is exactly because of these differences, in fact, that the authors agree that conservative retention of traits alone *cannot* explain the drastic behavioral similarities and differences.

On what data is the *Demonic Males* theory based? By 2004, there had been only 17 suspected and 12 "observed" cases of adult chimpanzee-chimpanzee killings reported from four of nine chimpanzee long-term research sites. This spanned a total of 215 years of combined observer time at these sites and yields a maximum rate of one chimpanzee killing every 7.5 years (see Wilson and Wrangham, 2003; Sussman, 2004; Sanz, 2004). Furthermore, most of the chimpanzee research sites where such data were gathered are highly disturbed by human encroachment and disturbance (Sanz, 2004).

We are not claiming that chimpanzees and humans are not violent under certain circumstances, as we all know they are, but that the claims of inherent "demonism" in both chimpanzees and humans is erroneous. Furthermore, research indicates that the neurophysiology of aggression between species (i.e., predation) is quite different from spontaneous violence linked to intraspecific aggression by humans (i.e., murder) (Worthman and Konner, 1987; Archer, 1988; Hart and Sussman, 2005).

After an examination of ethnographic research on human nomadic foraging societies, Fry (2006) stresses that virtually all the assumptions of pervasive intergroup hostility in these human groups appear to be flawed to some degree. Counter to the assumptions of ongoing hostile intergroup and interindividual relations and recurring warfare over resources, the typical pattern is for humans to get along rather well, relying on resources within their own areas and respecting the resources of their neighbors. *"In sum, an examination of the actual ethnographic information on simple nomadic foragers suggests that the Pervasive Intergroup Hostility Model rests not on fact but a plethora of faulty assumptions and over-zealous speculation"* (Fry, 2006:183; emphasis in original). Again, this is not to say that humans are never aggressive. Fry (2006:xii) goes on:

> We do indeed have propensities to behave assertively and aggressively, but we also have propensities to behave prosocially and cooperatively, with kindness and consideration for others.... the very existence of human societies depends on the preponderance of prosocial tendencies over assertive and aggressive ones.

Early Female Hominids

Were female hominids also killers? Obtaining meat may have been significant in later human evolutionary history, but there is considerable debate concerning the importance of hunting versus scavenging or scavenging versus gathering during various stages of human evolution. Many feminist anthropologists emphasize a "Woman the Gatherer" scenario of human evolution over the "Man the Hunter" scenario.

All Man the Hunter vignettes adhere to male hunters as the leaders, the innovators, the tool makers, and the tool users. Since these aspects of gender specificity may never be revealed in the fossil record, Tanner and Zihlman (1976) examined the way in which tools are used by modern foragers and by chimpanzees. Chimpanzee tools are made mainly by females and used mainly in gathering activities (such as nut-cracking and termite fishing), and it is also the females who teach the next generation how to use these tools (Zihlman, 1997; 2000; Boesch and Boesch-Ackerman, 2000). For both foragers and chimpanzees most food is obtained by gathering plants; hunting is almost always an important but transitional and chance phenomenon and makes up a small proportion of the diet.

Among modern human hunter-gatherers, most tools are used not for hunting large prey, but rather for gathering plants, eggs, honey, small insects, and small burrowing animals. Women's tools include digging sticks, poles for knocking down fruit or nuts, rocks for cracking nuts or tough fruit rinds. Baskets and slings are used for carrying babies and gathered roots, nuts, berries, and grains. And, as with the chimpanzees, most of these tools are made and used by females not males (Zihlman, 1997).

Modern humans, especially in Western cultures, think of themselves as meat-eaters; meat defines that ephemeral status of wealth. Because they, themselves, were rooted in these cultural stereotypes, anthropologists egregiously misnamed the modern forager cultures as hunters and gatherers and initially emphasized only the contributions of male hunters. Nevertheless, more than two-thirds of modern-day foragers' food comes from women gathering plant foods and, in the process, opportunistically capturing small mammals and reptiles. Less than one-third of the diet (the meat portion brought in by dedicated hunting by men) serves to supplement their foraged nutritional intake, except in cold climates or where fishing is prevalent (Marlowe, 2005). Although the Man the Hunter hypothesis may still be conventional wisdom, it does not fit the evidence of living primates or modern hunter-gatherers.

When Japanese macaques were first studied in the wild, it was the young females who started innovative behaviors, such as new ways of processing food (Kawai, 1958). Furthermore, in most primates, females are the repositories of group knowledge concerning home ranges and scarce resources (Zihlman, 1997). Group knowledge and traditions are passed on from mother to offspring, and stability of the group, both in the present and over time, often is accomplished through female associations.

By doing a broad comparative analysis of the literature on primate behavior and modern human gatherers, Zihlman (1997; 2000) envisions early human society as a flexible one in which women carried the young, conducted most of the socialization of the young, were repositories of group knowledge, had cognitive maps of the home range and its resources, were the center of society and the core of group stability, and spread innovations, techniques, and knowledge through the group and onto the next generation. Finally, she poses the theory that female choice of sexual partners was the original pattern, not sex through male coercion or aggression. Successful mating behavior involved being appealing to females, which translates into females choosing less, rather than more, aggressive males with whom to mate.

Reconstructing the Behavior of the Earliest Hominids

> To me, the future of paleoprimatology will prosper the most if researchers would emphasize with new discoveries the anatomical nature of the creatures themselves, their lifestyles and environmental contexts (Simons, 2000:445)

If the evidence for Man the Hunter has not been good, what kinds of evidence should be used in attempts to reconstruct the behavior of our earliest ancestors?

True evidence includes careful examination and understanding of the actual skeletal remains of the creatures. However, it also includes other fossil evidence (such as tools or footprints) left by our earliest relatives and, also, fossil materials that give us clues about the environment in which they lived (such as fauna, flora, or water sources). These fossils provide the most important data for an accurate reconstruction, but Man the Hunter theories fall short (if they are used at all) in the critical examination of the fossil evidence. Craig Stanford, one of the strongest proponents that hunting done by chimps reflects our human origins, noted this lack of fossil evidence in such evolutionary theories: "...models of hominid evolution are, appearances notwithstanding, virtually fossil-free." (Stanford, pers. comm).

We cannot necessarily impute correlation between human ancestors and data based on extant carnivores, modern human foragers, or great apes. For example, even the concept of hunting in chimps and humans is quite different. Present-day human hunters purposely search for animal prey, but chimpanzees do not: "Instead, they forage for plant foods and eat prey animals opportunistically in the course of looking for fruits and leaves" (Stanford, 1999:48). Furthermore, reconstructions must always be compatible with the actual fossil data – the fossils are real but the models we construct are hypothetical and must constantly be tested and reconfirmed. Lastly, when attempting to construct models of our early ancestors' behavior, it is necessary to be precise about timing. If we say our earliest human ancestors (those who lived seven million years ago) behaved in a certain way, we cannot use fossil evidence from two million years ago, nor can we confuse those creatures from two million years ago with those who existed 500,000 years ago. Concerning timing, we pose the question whether hunting could have occurred without tools? The first evidence of stone tools comes from around 2.6 million years ago (Semaw, 2000). The earliest hominid fossils, however, date from almost 7 mya.

When we look at the fossil evidence, it appears that hunting came quite late to our human family. Interpretations of hominid behavior, therefore, should be conservative and cautious, as stated by Klein (1999:306):

> ...the mere presence of animal bones at archaeological sites does not prove that hominids were killing animals or even necessarily exploiting meat. Indeed, as was the case in the earlier South African sites, the hominid remains themselves may have been the meal refuse of large carnivores.

The transition to hunting as a dominant way of life doesn't appear to have started until after the appearance of our own genus, *Homo*, and may not have even begun with the earliest members of our genus. *Homo erectus* has been given credit in the past for existing as a large animal hunter, and dates as far back as 1.75 mya have been hypothesized for such a lifestyle. But the first indications of hunting are amazingly recent. In fact, according to some paleontologists, the first unequivocal evidence of large scale, systematic hunting by humans is available from paleoarchaeological sites possibly only 60,000–80,000 years old (Binford, 1992; Klein, 1999).

No hard archaeological evidence, in other words no fossil evidence of tools designed for hunting, exist earlier in time than a finely-shaped wooden spear excavated at Schöningen, Germany, dated at approximately 400,000 years of age (Dennell, 1997; Theime, 1997). The famous Torralba and Ambrona sites in

Spain, dated at 500,000 years ago, contain huge numbers of large mammal bones. They were thought to represent unquestionable evidence of megafauna killed by Pleistocene hunters. Now these two sites are being reconsidered in light of better archaeological analysis. Elephant bones at these sites could just as likely represent natural deaths or carnivore kills as the remains of human hunting (Klein, 1987; 1999).

Further, no hominids were large-scale hunters before they had the use of fire (because of their dentition and alimentary tract, points we will elucidate below), although insects, small vertebrates, lizards, and birds likely were eaten opportunistically. Pre-digestion by fire had to precede any major meat-eating. The best evidence for the controlled use of fire appears around 800,000 years ago in Israel (Goren-Inbar et al., 2004). Klein (1999:160) states: "The assumption of consistent hunting has been challenged, especially by archaeologists who argue that the evidence does not prove the hunting hypothesis ... it is crucial to remember (although not as exciting) that probably the majority of calories [came] from gathering plant foods."

Dentition and Diet

Obviously, Man the Hunter models of human evolution assume that a significant portion of our earliest ancestors' diets came from killing and eating meat from relatively large mammals. By comparing the characteristics of the dental and jaw morphology of various living primates with those of fossils, inferences can be made about the diets of early hominids. Teaford and Ungar (2000) and Ungar (2004) have carried out such comparisons. Using such features as tooth size, tooth shape, occlusal slope, enamel structure, dental microwear, and jaw biomechanics, they found that the earliest humans had a unique combination of dental characteristics and a diet different from modern apes or modern humans.

Australopithecus afarensis is characterized by thick jawbones, with relatively small incisors and canines in relation to molars. The molars, by comparison with other primates, are huge, flat and blunt, show less slope and relief, and lack long shearing crests necessary to mince flesh. *A. afarensis* also had larger front than back molars. The dental enamel is thick and microwear on the teeth is a mosaic of gorilla-like fine wear striations (indicating leaf-eating) and baboon-like pits and microflakes (indicating fruits, seeds, and tubers in the diet). This evidence all points away from meat-eating.

Given these facts, Teaford and Unger hypothesize that early humans were able to exploit a wide range of dietary resources, including hard, brittle foods (tough fruits, nuts, seeds, and pods), as well as soft, weak foods (ripe fruits, young leaves and herbs, flowers and buds). They may also have been able to eat abrasive objects, including gritty plant parts, such as grass seeds, roots, rhizomes, and underground tubers. As stated by Teaford and Ungar (2000:13508–13509), "this ability to eat both hard and soft foods, plus abrasive and nonabrasive foods, would have left early hominids particularly well suited for life in a variety of habitats, ranging from gallery forest to open savanna."

Dental morphology indicates that the earliest hominids would have had difficulty breaking down tough pliant plant foods, such as soft, fibrous seed coats and mature leaves. Interestingly, Teaford and Ungar stress that another tough pliant food that our early ancestors would have had difficulty processing was meat! They state (2000:13509): "the early hominids were not dentally preadapted to eat meat – they simply did not have the sharp, reciprocally concave shearing blades necessary to retain and cut such foods."

Did the earliest hominids ingest any animal protein? When mammalian omnivore species are compared, body mass is positively correlated with inclusion of plant protein in the diet in lieu of animal protein. If insects are ingested, they are usually social ones that come in large "packages," because it takes a higher amount of energy to capture individual insects (Charles-Dominique, 1977; Kay, 1984). In quantitative studies of dietary intake of mid to large-sized primates, such as macaques, baboons, chimpanzees, and modern human foragers, consumption of animal protein is very low, usually making up less than 5% of time spent feeding (Tanaka, 1980; Garber, 1987; Sussman, 1999b).

Both modern chimpanzees and humans have an alimentary track that is not specialized for eating either leaves or animal protein, but instead is more generalized, similar to the majority of primates who are omnivorous and eat a mixture of food types (Chivers and Hladik, 1980; 1984; Martin et al., 1985; Martin, 1990.) Specialized faunivores typically have large small intestines relative to the digestive system, whereas folivores have relatively larger stomachs or large intestines. Omnivores are intermediate in these characteristics (Chivers and Hladik, 1980; Fleagle, 1999).

We might point out that by the end of the twentieth century there was a full-blown red alert from the medical community warning that meat should be ingested in limited quantities. "Diseases of affluence" caused by high-protein, high-fat diets include raised cholesterol levels, high blood pressure, heart disease, stroke, breast cancer, colon cancer, and diabetes – all correlated to a diet rich in red meat. With colon cancer, in particular, startling data are available: Daily red-meat eaters are two and one-half times more likely to develop this cancer as are people adhering to a mostly vegetarian diet (Willett et al., 1990). T. Colin Campbell, director of one of the longest continuous studies of nutrition and health (Campbell and Campbell, 2005), states: "We're basically a vegetarian species and should be eating a wide variety of plant foods and minimizing our intake of animal foods" (Brody, 1990:C2).

Habitat of our Earliest Ancestor

By far the best known of early australopithecine species is *Australopithecus afarensis,* with many fossil remains dating from between 3.6–2.9 mya and possibly as far back as 5 mya (Hill and Ward, 1988). Collections from Hadar,

Ethiopia alone have yielded 250 specimens, representing at least 35 individuals, and there are a number of other East African sites that contain remains of this species (Conroy, 2005). Specimens include the famous Lucy (dated at 3.2 mya), which is the most complete adult skeleton from this time period and fossil footprints from Laetoli, Tanzania ash deposits (dated at 3.6 mya). Furthermore, most hypotheses concerning human evolution position *A. afarensis* as a pivotal species from which all other later hominids, including *Homo*, evolved (Fleagle, 1999; Conroy, 2005). Given the above, we see *A. afarensis* as a logical species to examine when attempting to reconstruct the ecology and behavior of one of our early human ancestors.

It appears that the combination of skeletal characteristics found in *A. afarensis* enabled these hominids to be very versatile. They were able to use the ground and the trees equally and successfully for a very long time (Susman et al., 1984; Fleagle, 1999; Stern, 2000; Conroy, 2005). We believe these early hominids were well adapted to their environment and not in the least inhibited by switching back and forth from bipedalism on the ground to quadrupedalism in the trees.

Although many early theories on the evolution of our earliest ancestors stress the importance of arid, savanna environments, these do not seem to be the primary habitats, according to the fossil record, until after 2 mya (Conroy, 2005). The African climate was becoming more arid in the time between 12 and 5 mya, and equatorial forests were undoubtedly shrinking. However, the process that led to this climatic phenomenon also greatly enlarged areas of transitional zones between forest and adjacent savanna. Closed woodland forests were still widespread in East Africa 3.5 mya, whereas, the proportion of dry shrub to grassland habitats begins to increase around 1.8 mya (Bonnefille, 1976; Gentry, 1976; Cerling, 1992; Reed and Eck, 1997; Alemseged, 2003; Conroy, 2005).

It is in these transitional zones that the behavioral and anatomical changes were initiated in early hominid evolution. The flora and fauna remains that are found in association with fossil hominids of this time period indicate a mixed, mosaic environment – mosaic in the sense that it was ecologically diverse and subject to seasonal and yearly changes in vegetation (Potts, 1996; Wolpoff, 1998; Conroy, 2005). These environments were wetter than those in which later fossil hominids are found, and most fossil sites of this early time period contained some type of water source, such as rivers and lakes (White et al., 1994; Wolde Gabriel et al., 1994). For example, at Hadar the mammalian fauna remains suggest that a lake existed, surrounded by marshy environments fed by rivers flowing off the Ethiopian escarpment (Conroy, 2005). A mosaic of habitats existed through time that included grassland, bushland, and closed and open woodland.

Thus, the earliest hominids appear to be associated with variegated fringe environments or edges between forest and grassland. These habitats usually contain animal and plant species of both the forest and the grassland, as well as species unique to the borders between the two. The species adapted to these transitional habitats are often referred to as *edge* species. During these earliest times, it appears that hominids began to take advantage of the growing fringe environments, lessening competition with their sibling ape species which were better adapted to exploit the dense forest, and thus partitioning the niche occupied by the parent species of both apes and hominids into two narrower and less overlapping adaptive zones (Klein, 1999; Conroy, 2005). It is this evidence we use to infer that *A. afarensis* was an edge species, able to exploit both the arboreal and the terrestrial habitats, as well as transitional and changing habitats.

From available evidence our early ancestors may have been able to exploit a great variety of food resources but were mainly fruit eaters, probably supplementing this diet with some young leaves and other plant parts, social insects, and a small amount of opportunistically-captured small vertebrate prey – lizards, small snakes, birds, and mammals.

Several other species of primates are intrinsically adapted to edge habitats and also are able to take advantage of changing environments. Ringtailed lemurs in Madagascar, African baboons and vervet monkeys, and some Asian macaques and langurs are non-human primate examples. These, not coincidentally, are some of the most common and numerous of all living primates other than humans. The macaque genus, for example, has the widest geographical distribution of any non-human primate in Asia. Many macaque species in Asia are endangered, but the ones that have the healthiest populations (e.g., long-tailed macaques, *Macaca fascicularis*. and rhesus macaques, *M. mulatta*) are edge-adapted, and often referred to as "weed" species (Richard et al., 1989).

Certain ecological niches may breed certain behavioral repertoires. It could be argued that the closer the DNA comparison, the more similar the behaviors between two related species (but see Marks, 2002). In that case, chimps and bonobos might be the best prototype for early human ancestors. However, if ecology is paramount, then chimps and bonobos may be less suitable prototypes (although some traits between these close relatives may still be important and phylogenetically conservative) and edge species may be the best models for early humans.

Forty years ago, Robin Fox (1967:419) declared:

> But the problem of taking the great apes as models lies in the fact of their forest ecologies. Most modern students of primate evolution agree that we should pay close attention to ecology in order to understand the selection pressures at work on the evolving primate lines. This has been shown to be crucial in understanding . . .evolution.

Even if one were to learn everything about the hominid-ape common ancestor, many of the most crucial questions about distinctively hominid evolution would remain unanswered. Although there is a fairly impressive record of human fossils during the period of 7 to 2 mya there is a lack of chimpanzee fossils at these early sites. It seems likely, therefore, that chimpanzee ancestors did not inhabit these fringe environments and were likely restricted to wetter, closed forest ecosystems – areas where fossils are less likely to be preserved. We would speculate that some populations of chimpanzees moved into more mosaic, open habitats relatively recently, long after humans had moved into more arid environments. Furthermore, modern chimpanzees do not live in habitats in which modern humans lived in the past or are found today. Given the paucity of fossil apes during the period when many of our earliest ancestors have been discovered, we speculate that the historical geographic range of chimpanzees is quite restricted, probably more restricted than even that of early humans before leaving Africa.

The Macaque Model

> Science advances through skepticism and continuous review of available data as it expands and our discipline will be involved with self criticism and change throughout the foreseeable future (Simons, 2000:445).

In our opinion the best primate models to use as a basis for extrapolation about behavioral characteristics of our earliest ancestors are modern primate species living in similar edge habitats. Macaques can be extremely good colonizers of edge habitats. The macaque genus spread throughout Asia before humans reached that continent (Delson, 1980). By the time *Homo erectus* arrived in Asia 1.8 mya, most hominids were exploiting more open habitats and did not displace the macaques. True "weed" species, we propose that the macaques are excellent models for reconstructing the behavioral ecology of our earliest ancestors.

Long-tailed macaques (*Macaca fascicularis*, Fig. 2) are small edge species that spend a good proportion of time both in the trees and on the ground. They are omnivorous and very versatile in their locomotion, although mainly quadrupedal. The most widespread of any Southeast Asian monkey, they occur from Burma through Malaysia and Thailand to Vietnam, while offshore populations are found on Java, Borneo, and numerous smaller islands as far east as the Philippines and Timor. Throughout this area, broadleaf evergreen and other forest types are interspersed with secondary and disturbed habitats, and it is the latter that long-tailed macaques prefer. They are most commonly found in secondary forest habitat, preferably near water (Southwick and Cadigan, 1972; Kurland, 1973; Rijksen, 1978; Rodman, 1981; Wheatley, 1980; Crockett and Wilson, 1980; Sussman and Tattersall, 1986). The success of the long-tailed macaque throughout its extensive Asian distribution is credited to its being an "adaptable opportunist" (Mackinnon and Mackinnon, 1980:187).

The ability of edge species to exploit a wide variety of environments is accompanied by a substantial flexibility of behavior. Long-tailed macaques are primarily arboreal where suitable vegetation exists, but they come to the

Fig. 2 *Macaca fascicularis*, the long-tailed macaque

ground along riverbanks, seashores, and in open areas – and in some portions of their range, such as Mauritius, where they were introduced by Portuguese sailors in the 1500s, they are highly terrestrial (Sussman and Tattersall, 1981; 1986). They are eclectic omnivores with a preference for fruit. But the variety of habitats they exploit is reflected in a wide selection of food items – besides fruit, they feed on leaves, grasses, seeds, flowers, buds, shoots, mushrooms, water plants, gum, sap, bark, insects, snails, shellfish, bird eggs, and small vertebrates (Sussman and Tattersall, 1981; Yaeger, 1996; Shaffer and Sussman, 2005). Human settlements are not avoided; rather, these monkeys tend to live close to humans throughout their range, which results in extensive crop raiding.

Long-tailed macaques live in large multi-male, multi-female groups of up to 80 individuals, although in some areas groups are much smaller. They show distinct flexibility in structure; the large social unit tends to split up into smaller subgroups for daytime foraging activities (Jamieson, 1998; Sussman pers. obs.). Subgroups may be all males, but most often consist of adult males accompanying females and their young. The number and size of subgroups tends to vary with the season and resource availability (Jamieson, 1998). The group reforms and returns to the same sleeping site each night, usually on the edge of a water source. Because of this unique behavior of returning to a home base each evening, these macaques are often referred to as a "refuging" species.

Many of the characteristics described for the long-tailed macaque (such as dietary and habitat flexibility, ability to use the ground and the trees, and many aspects of social organization) are found in other primate edge species.

Fossils and Living Primates

Looking at the fossil evidence, it is apparent that human ancestors, living between 7 and 2.5 mya, were intermediate-sized primates, not smaller than male baboons or larger than female chimpanzees. Given their relative brain size, they were at least as clever as the great apes of today. They had diverse locomotor abilities, exploiting both terrestrial and arboreal habitats. They used climbing and suspensory postures when traveling in the trees and were bipedal when on the ground. We believe that their bipedalism was a preadaptation. Unlike the facultative bipedal behaviors occasionally seen in other primates, human bipedalism is morphologically and physiologically different (Fleagle, 1999). The morphology of these early hominids "forced" them to be habitual bipeds (Tattersall, 2003; Hart and Sussman, 2005). Similar "forced" bipedalism can be seen in sifakas, spider monkeys, and gibbons. However, walking on two feet freed the arms and hands and proved to be advantageous in a number of ways.

Given their relatively small size and small canines, there is no reason to think that our early ancestors were any *less* vulnerable to predation than are modern monkeys – some of which have yearly predation rates comparable to gazelles, antelopes or deer living in similar environments (Hart, 2000; Hart and Sussman, 2005). Indeed, edge species can be *highly* vulnerable to predation and because of this usually live in relatively large social groups with many adult males and adult females; adult males often serve as sentinels and provide protection against predators (Hart, 2000). Because a primate group with only one male and ten females can have the same reproductive potential as a group with ten males and ten females, often the male role in primate groups is to act as first line of defense; if he gets eaten there are other males to take his place. If a mature female gets eaten, she and all her potential offspring (and living dependent infants) are lost.

We propose that, like long-tailed macaques, our human ancestors may have lived in multi-male, multi-female groups of variable size that were able to split up depending on resource availability and to reform each evening at home base refuges. However, certain facts, such as the exact group or subgroup size, which sex migrated, the internal structure of the group (e.g., whether matrilineal or patrilineal), would be impossible to determine accurately. Indications of these social parameters cannot be found in the fossil record and are quite variable even in closely-related living primates.

In sum, the best archetype of early humans may be a multi-male, group-living, mid-sized, omnivorous, quite vulnerable creature living in an edge habitat near a large water source. These primates may well have been a refuging species returning to the same well-protected sleeping site each night. Most modern human foragers are considered central place foragers, focusing their activities around a central place, as are many birds, social carnivores, and primates (Marlowe, 2005). This early hominid was adept at using both the trees and the ground, but when exploiting a terrestrial niche, it had upright posture and was bipedal. It depended mainly on fruit, both soft fruits and some that were quite brittle or hard, but also ate herbs, grasses, and seeds, and gritty foods such as roots, rhizomes, and tubers. A very small proportion of its diet was made up of animal protein, mainly social insects (ants and termites) and, occasionally, small vertebrates captured opportunistically. These early humans did not regularly hunt for meat and could neither process it dentally nor in their digestive tracts. Like all other primates, and especially ground-living and edge species, these early humans were very vulnerable to predators and this trait did not diminish greatly over time (Hart and Sussman, 2005).

Man the Hunted

Given that the earliest hominid ancestors were medium-sized primates who did not have any inherent weapons to fight off the many predators that lived then–and given that they lived in edge environments which incorporate open areas and wooded forests near rivers–then, like other primates, they were vulnerable to predation. We hypothesize that rates of predation were as high in our early ancestors as they are in modern primates and our origins are those of a *hunted* species (Brain, 1981; Hart and Sussman, 2005).

Protection from predation is one of the most important aspects of group-living, and we believe this was true of our earliest ancestors. Based on the long-tailed macaque model, social groups of early hominids may have been organized in a way that allowed efficient exploitation of a highly variable and changing environment and also protected its members of the group from predators. If the human lineage started out as Man the Hunted, we propose a number of strategies for protection from predators based on the behavior and social organization we observe in long-tailed macaques.

- Relatively large groups of 25–75 individuals: Since safety lies in numbers, a major reason all diurnal primates live in groups is predator protection, providing more eyes and ears alert to the presence of predators as a first line of defense. In his research on modern human foragers, Marlowe (2005) found median group size to be 30 individuals.
- Versatile locomotion that exploits both arboreal and terrestrial milieus: The major advantage of agility in the use of diverse habitats is safety in trees and dense underbrush. An added advantage of upright posture is the ability to scan for predators.
- Flexible social organization: Gathering scarce resources in small groups but reuniting as a larger group when predation requires strength in numbers allows small groups to quickly disperse and hide while large groups can mob and intimidate predators. Again, modern human foragers fit this pattern of flexibility (Marlowe, 2005).
- Multi-male social structure: This provides male protection both when traveling through open areas and when the group settles in evening or midday. When large groups break into subgroups, females and young are accompanied by one or more large males.
- Males as sentinels: Males are usually larger in these species. Upright posture adds to the appearance of large size and also allows for better vigilance, as well as waving arms, brandishing sticks, and throwing stones. Males mob or attack predators since they are the more expendable sex.
- Careful selection of sleeping sites: Refuging species bring the whole group together at night in a safe area. During daytime rest periods, staying in very dense vegetation is essential. Males should stay on high alert during these inactive periods and when the group is on the move.
- Out-thinking predators: Intelligence endows primates with the ability to monitor the environment, communicate with other group members, and implement effective anti-predator defenses (Hart and Sussman, 2005).

Our reconstruction of the behavioral ecology of our earliest hominid ancestor reflects the pervasive influence of large ferocious animals throughout the period of human evolution. Many circumstances have been proposed as a catalyst for the evolution of humans, competition for resources, intellectual capacity, male-male conflicts, and hunting. Looking at our primate relatives and the fossil record, however, we believe that predation pressure was a critical component in shaping the evolution of our earliest ancestors.

What do the earliest hominids have to do with living human populations? Modern humans are the result of the next 2.5 million years and are a different genus and several different species later. We believe that a reconstruction of the behavioral ecology of earlier species is important, nevertheless, for two major reasons. First, understanding the events that led to the split between ancestral apes and hominids helps us understand those environmental and behavioral factors that were the catalysts for this separation. These would be seminal, unique events in hominid evolution and the beginning of the evolutionary trajectory that led to future human morphology and behavior; the blocks on which future hominids were built. Secondly, as in earlier reconstructions, such as those of Dart, Washburn, and Wrangham and Peterson, we focus on those characteristics that might separate humans from their nonhuman primate or ape relatives. This approach seems pertinent to us because it is these characteristics that will help us define what is ape and what is human. We, however, leave it up to others to reconstruct the behavioral ecology of such creatures as *Homo erectus, Homo heidelbergensis*, and Cro Magnon. The factors that stimulated the evolution of later hominids were unique to them. We hope that the reconstruction we present here is useful in these future endeavors.

Acknowledgment We would like to thank the organizers of the conference honoring Elwyn Simons for inviting us to participate. We also would like to thank many people for reading earlier versions of this paper and for offering comments on the ideas expressed herein. These include Pam Ashmore, Dennis Bohnenkamp, Glenn Conroy, Charles Hildebolt, Bruce Knauft, Jonathon Losos, Jane Phillips-Conroy, Tom Pilgram, Tab Rasmussen, Ian Tattersall, and Mary Willis. Finally, John Fleagle and two anonymous reviewers provided extremely useful suggestions for improving the paper. The ideas expressed and the errors uncovered, however, are our own.

References

Alemseged, Z. (2003). An integrated approach to taphonomy and faunal change in the Shungura Formation (Ethiopia) and its implication for hominid evolution. *J. Hum. Evol.* 44: 451–478.

Archer, J. (1988). *The Behavioral Biology of Aggression*. Cambridge University Press, New York.

Binford, L. (1992). Subsistence – a key to the past. In: Jones, S., Martin, R., and Pilbeam, D. (eds.), *Cambridge Encyclopedia of Human Evolution*. Cambridge University Press, Cambridge, pp. 365–368.

Boesch, C., and Boesch-Acherman, H. (2000). *The Chimpanzees of the Taï Forest*. Oxford University Press, Oxford.

Bonnefille, R. (1976). Palynological evidence for an important change in the vegetation of the Omo Basin between 2.5 and 2 milion years. In Copens, Y., Howell, F., Isaac, G., and Leakey, R. (eds.), *Earliest man and environments in the Lake Rudolf Basin*. University of Chicago Press, Chicago, pp. 421–431.

Brain, C. K. (1981). *The Hunters or the Hunted?* University of Chicago Press, Chicago.

Brody, J. (1990). Diet study indicts fat and meat. *New York Times 5/8/1990*:C2.

Campbell, T. C., and Campbell II, T. M. (2005) *The China Study*. BenBella Books, Dallas.

Cartmill, M. (1993). *A View to a Death in the Morning: Hunting and Nature through History*. Harvard University Press, Cambridge.

Cartmill, M. (1997). Hunting hypothesis of human origins. In Spencer, F. (ed.), *History of Physical Anthropology: An Encyclopedia*. Garland Publishing, New York, pp. 508–512.

Cerling, T. (1992). Development of grasslands and savannas in East Africa during the Neogene. *Paleogeogr. Paleocl.* 97:241–247.

Charles-Dominique, P. (1977). *Ecology and Behaviour of Nocturnal Primates*. Columbia University Press, New York.

Chivers, D. J., and Hladik, C. M. (1980). Morphology of the gastrointestinal tract in primates: comparisons with other mammals in relation to diet. *J. Morphol.* 166:337–386.

Chivers, D. J., and Hladik, C. M. (1984). Diet and gut morphology in primates. In: Chivers, D. J., Wood, B. A., and Bilsborough, A. (eds.), *Food Acquisition and Processing in Primates*. Plenum Press, New York, pp. 213–230.

Conroy, G. (2005). *Reconstructing Human Origins: A Modern Synthesis*. Second Edition. W. W. Norton, New York.

Crockett, C. M., and Wilson, W. I. (1980). The ecological separation of *Macaca nemestrina* and *Macaca fuscicularis* in Sumatra. In: Lindburg, D. G. (ed.), *The macaques: studies in ecology, behavior, and evolution*. Van Nostrand-Reinhold, New York, pp. 148–181.

Dart, R. (1926). Taungs and its significance. *Nat. Hist.* 3: 315–327.

Dart, R., and Craig, D. (1959). *Adventures with the Missing Link*. Harper, New York.

Delson, E. (1980). Fossil macaques, phyletic relationships and a scenario of deployment. In: Lindburg, D. G. (ed.), *The macaques: studies in ecology, behavior, and evolution*. Van Nostrand-Reinhold, New York, pp. 10–30.

Dennell, R. (1997). The world's oldest spears. *Nature* 385:787–788.

Fleagle, J. (1999). *Primate Adaptation and Evolution*. Second Edition. Academic Press, New York.

Fox, R. (1967). In the beginning: aspects of hominid behavioural evolution. *Man* 2:415–433.

Fry, D. (2006). *The Human Potential for Peace: An Anthropological Challege to Assumptoins about War and Violence*. Cambridge University Press, Cambridge.

Garber, P. A. (1987). Foraging strategies among living primates. *Ann. Rev. Anthropol.* 16:339–364.

Gentry, A. (1976). Bovidae of the Omo Group deposits. In: Coppens, Y., Howell, F., Isaac, G., and Leakey, R. (eds.), *Earliest man and environments in the Lake Rudolf Basin*. University of Chicago Press, Chicago, pp. 293–301.

Ghiglieri, M. (1999). *The Dark Side of Man: Tracing the Origins of Male Violence*. Perseus Books, Reading, Massachusetts.

Goren-Inbar, N., Alperson, N., Kislev, M., Simchoni, O., Melamed, Y., Ben-Nun A., and Werker, E. (2004). Evidence of hominin control of fire at Gesher Benot Ya'aqov, Israel. *Science* 304:725–727

Hart, D. (2000). *Primates as prey: ecological, morphological, and behavioral relationships between primate species and their predators*. Ph.D. Thesis, Washington University, St. Louis, Missouri.

Hart, D., and Sussman, R. W. (2005). *Man the Hunted: Primates, Predators, and Human Evolution*. Westview Press, New York.

Hill, A., and Ward, S. (1988). Origin of the Hominidae: the record of African large hominoid evolution between 14 my and 4 my. *Yrbk. Phys. Anthropol* 31:49–83.

Huxley, T. H. [(1893) 1993]. Evolution and ethics. In: Nitecki M. H., and Nitecki D. V. (eds.), *Evolutionary Ethics*. State University of New York Press, Albany. pp. 29–80.

Jamieson, R. W. (1998). *The effects of seasonal variation in fruit availability on social and foraging behavior in Macaca fascicularis in Mauritius*. Ph.D. Thesis, Washington University, St. Louis, Missouri.

Kawai, M. (1958). On the system of social ranks in a natural group of Japanese monkeys. *Primates* 1:11–48.

Kay, R. F. (1984). The use of anatomical features to infer foraging behavior in extinct primates. In: Rodman, P. S., and Cant, J. G. (eds.), *Adaptations for Foraging in Nonhuman Primates*. Columbia University Press, New York, pp. 21–53.

Kelly, R. C. (2000). *Warless Societies and the Origin of War*. University of Michigan Press, Ann Arbor, Michigan.

Klein, L. F. (2004). *Women and Men in World Cultures*. McGraw Hill, Boston.

Klein, R. G. (1987). Problems and prospects in understanding how early people exploited animals. In: Nitecki, M. H., and Nitecki, D. V. (eds.), *The Evolution of Human Hunting*. Plenum Press, New York, pp. 11–45.

Klein, R. G. (1999). *The Human Career: Human Biological and Cultural Origins*, Second Edition. University of Chicago Press, Chicago.

Kurland, J. A. (1973). A natural history of Kra macaques (*Macaca fascicularis* Raffles, 1821) at the Kutai Reserve, Kalimantan, Timur, Indonesia. *Primates* 14:245–262.

MacKinnon, J. R., and MacKinnon, K. S. (1980). Niche differentiation in a primate community. In: Chivers, D. J. (ed.), *Malayan forest primates*. Plenum Press, New York, pp. 167–190.

Marks, J. (2002). *What It Means to Be 98% Chimpanzee: Apes, People, and Their Genes*. University of California Press, Berkeley.

Marlowe, F. W. (2005). Hunter-gatherers and human evolution. *Evol. Anthropol.*14:54–67.

Martin, R. D. (1990). *Primate Origins and Evolution: A Phylogenetic Reconstruction*. Princeton University Press, Princeton, New Jersey.

Martin, R. D., Chivers, D. J., MacLarnon, A. M., and Hladik, C. M. (1985). Gastrointestinal allometry in primates and other mammals. In: Jungers, W. L. (ed.), *Size and Scaling in Primate Biology*. Plenum Press, New York, pp. 61–90.

Pope, S. J. (2004). Primate sociality and natural law theory. In: Sussman, R. W and Chapman, A. R. (eds.), *The Origins and Nature of Sociality*. Aldine de Gruyter, New York, pp. 313–331.

Potts, R. (1996). Evolution and climate variability. *Science* 273:922–923.

Reed, K., and Eck, G. (1997). Paleoecology of the Plio-Pleistocene Hadar hominid localities. *Amer. J. Phys. Anthropol. Suppl.* 24:194.

Richard, A. R., Goldstein, S. J., and Dewar, R. E. (1989). Weed macaques: the evolutionary implications of macaque feeding ecology. *Int. J. Primatol.* 10:569–594.

Rijksen, H. (1978). *A Field Study on Sumatran Orangutans, (Pongo pygmaeus abelii): Ecology, Behaviour and Conservation*. Veenman and Zonen, Wageningen.

Rodman, P. S. (1981). Structural differentiation of microhabitats of sympatric *Macaca fuscicularis* and *Macaca nemestrina* in East Kalimantan, Indonesia. *Int. J. Primatol.* 12:357–375.

Sanz, C. (2004). *Behavioral ecology of chimpanzees in a central African forest: Pan troglodytes troglodytes in the Goualougo Triangle, Republic of Congo*. Ph.D. Thesis, Washington University, St. Louis, Missouri.

Semaw, S. (2000). The world's oldest stone artifacts from Gona, Ethiopia: Their implication for understanding stone technology and pattern of human evolution between 2.6-1.5 million years ago. *J. Archaeol. Science* 27:1197–1214.

Shaffer, C. A., and Sussman, R. W. (2005). Activity budget and feeding ecology of *Macaca fascicularis* in Mauritius. *Am. J. Phys. Anthropol. Suppl.* 40:188.

Simons, E. L. (2000). A view on the science: Physical anthropology at the millennium. *Am. J. Phys. Anthropol.* 112:441–446.

Southwick, C. H., and Cadigan, F. C. (1972). Population of Malaysian primates. *Primates* 13:1–18.

Stanford, C. (1999). *The Hunting Ape: Meat Eating and the Origins of Human Behavior*. Princeton University Press, Princeton, New Jersey.

Stern, J. T., Jr. (2000). Climbing to the top: A personal memoir of *Australopithecus afarensis*. *Evol. Anthropol.* 9:113–133.

Susman R. L., Stern, J. T. Jr., and Jungers, W. L. (1984). Arboreality and bipedality in the Hadar hominids. *Folia Primatol.* 43:113–156.

Sussman, R. W. (1999a). The myth of man the hunter, man the killer, and the evolution of human morality. *Zygon* 34:453–471.

Sussman, R. W. (1999b). Species-specific dietary patterns in primates and human dietary adaptations. In: Sussman, R. W. (ed.), *The Biological Basis of Human Behavior*. Prentice Hall, Upper Saddle River, New Jersey, pp.143–157.

Sussman, R. W. (2000). Piltdown man: the father of American field primatology. In: Strum, S., and Fedigan, L. (eds.), *Primate Encounters: Models of Science, Gender, and Society*. University of Chicago Press, Chicago, pp.85–103.
Sussman, R. W. (2004). Are humans inherently violent? In: Selig, R., London, M., and Kaupp P. (eds.), *Anthropology Explored: Revised and Expanded.* Smithsonian Books, Washington, D. C., pp.30–45.
Sussman, R., and Tattersall, I. (1981). Behavior and ecology of *Macaca fascicularis* in Mauritius: a preliminary study. *Primates* 22(2):192–205.
Sussman, R.W., and Tattersall, I. (1986). Distribution, abundance, and putative ecological strategy of *Macaca fascicularis* on the island of Mauritius, southwestern Indian Ocean. *Folia Primatol.* 46:28–43.
Tanaka, J. (1980). *The San Hunters-Gatherers of the Kalahari, A Study in Ecological Anthropology*. University of Tokyo Press, Tokyo.
Tanner, N., and Zihlman, A. (1976). Women in evolution: innovation and selection in human origins. *Signs* 1:585–608
Tattersall, I. (2003). Stand and deliver: why did early hominids begin to walk on two feet? *Nat. Hist.* 112:60–64.
Teaford, M., and Ungar, P. (2000). Diet and the evolution of the earliest human ancestors. *P. Natl. Acad. Sci.* 97(25):13506–13511.
Theime, H. (1997). Lower Paleolithic hunting spears from Germany. *Nature* 385:807–810.
Ungar, P. (2004). Dental topography and diets of *Australopithecus afarensis* and early *Homo. J. Hum. Evol.* 46:605–622.
Washburn, S., and. Lancaster, C. (1968). The evolution of hunting. In: Lee, R., and DeVore, I. (eds.), *Man the Hunter*. Aldine, Chicago, pp.293–303.
Wheatley, B. P. (1980). Feeding and ranging of east Bornean *Macaca fascicularis.* In: Lindberg, D. G. (ed.), *The macaques: Studies in ecology, behavior and evolution.* Van Nostrand-Reinhold, New York, pp. 215–246.
White, T. D., Suwa, G., and Asfaw, B. (1994). *Australopithecus ramidus*. a new species of early hominid from Aramis, Ethiopia. *Nature* 375:88.
Willett, W., Stampfer, M., Colditz, G., Rosner, B. A., and Speizer, F. E. (1990). Relation of meat, fat, and fiber intake to the risk of colon cancer in a prospective study among women. *New Eng. J. Med.* 323:1664–1672.
Wilson, E. O. (1975). *Sociobiology: The New Synthesis.* Harvard University Press, Cambridge, Massachusetts.
Wilson, M. L., and Wrangham, R. W. (2003). Intergroup relations in chimpanzees. *Ann. Rev. Anthropol.* 32:363–392.
WoldeGabriel G., White T. D., Suwa, G., Renne, P., de Heinzelin, J., Hart, W. K., and Helken, G. (1994). Ecological and temporal placement of early Pliocene hominids at Aramis, Ethiopia. *Nature* 371:330–333.
Wolpoff, M. (1998). *Paleoanthropology*. Second Edition. McGraw Hill, New York.
Worthman, C. M., and Konner, J. (1987). Testosterone levels change with subsistence hunting effort in !Kung San men. *Psychoneuroendocrinology*. Vol. 12: 449–458.
Wrangham, R. (1995). Ape culture and missing links. *Symbols* (Spring):2–9, 20.
Wrangham, R., and Peterson, D. (1996). *Demonic Males: Apes and the Origins of Human Violence*. Houghton Mifflin, Boston.
Yaeger, C. P. (1996). Feeding ecology of the long-tailed macaques (*Macaca fascicularis*) in Kalimantan, Tengah, Indonesia. *Int. J. Primatol.* 17:51–62.
Zihlman, A. (1997). The Paleolithic glass ceiling: Women in human evolution. In: Hager, L. (ed.), *Women in Human Evolution*. Routledge, London, pp.91–113.
Zihlman, A. (2000). *The Human Evolution Coloring Book*. Second Edition. Harper Collins, New York.

Section 3
Primates from Madagascar and Elsewhere

When Elwyn became Director of the Duke University Primate Center (now the Duke Lemur Center) his boundless energy and creativity extended into the areas of primate behavior and conservation. As described by Patricia Wright in the lead article of this section, "Decades of Lemur Research and Conservation", the efforts took place on many fronts, and involved expanding our understanding of both living and fossil primates on Madagascar. Simons was the Director of the Duke Primate Center from 1977 until 1992 and then held the position of Director of Research for another decade. During that time the center flourished by any criteria one can imagine – the diversity of taxa in residence, the number being bred successfully, grant dollars received, or the amount of research being conducted and published. During that same time, his paleontological expeditions in Madagascar recovered remarkable new fossils in exceptional numbers and completeness. The papers in this section are a reflection of this aspect of his career.

In her recounting of Elwyn's work as Director of the Duke Primate Center, Patricia Wright outlines the broad scope of activities in primate research and conservation that took place at the Primate Center under his leadership, from expanding the colonies at Duke and breeding rare taxa such as the aye-aye, and reintroduction schemes in Madagascar, to the discovery and description of fossil lemurs. Two papers discuss research on living lemurs conducted at the Duke Primate Center under Elwyn's direction. In "Low Fetal Energy Deposition Rates in Lemurs: Another Energy Conservation Strategy", Chris Tilden describes some of his work on maternal investment of lemur mothers in reproduction. He finds that female lemurs do not invest heavily in offspring during gestation compared with other primates. At the other end of life history, in "Old Lemurs: Preliminary Data on Behavior and Reproduction from the Duke University Primate Center" Linda Taylor describes the effects of aging on several species of lemurs. She finds that that age has no effect on either reproduction or dominance relationships and little effect on social interaction among lemurs.

Four papers discuss the evolutionary history of primates on Madagascar. In "Peculiar Tooth Homologies of the Greater Bamboo Lemur (*Prolemur* = *Hapalemur simus*): When is a Paracone not a Paracone?" Jukka Jernvall and colleagues highlight how the peculiar premolars of the greater bamboo lemur came to look very much like molars. In "How Big were the "Giant" Extinct Lemurs of Madagascar?" William Jungers and colleagues provide new size estimates for all of the subfossil lemurs of Madagascar, based in many cases on new fossils collected on Elwyn's expeditions. Absolute size estimates for individual taxa can vary dramatically depending upon the body part being used for the estimates, but Jungers and colleagues find solace in the fact hat the rank order of taxa remains the same. In a delightful effort to reconstruct the diet of the extinct lemurs of Madagascar entitled "Ghosts and Orphans: Madagascar's Vanishing Ecosystems", Godfrey and colleagues attempt to identify residual effects of the extinction of the giant lemurs on the modern flora of Madagascar. Based in part on new data about the morphology of the fossil lemurs obtained by Elwyn Simons' expeditions, they identify living trees that are likely to be orphans whose "normal" seed dispersers were extinct lemurs. In "Vicariance vs. Dispersal in the Origin of the Malagasy Mammal Fauna", Ian Tattersall addresses the perennial question about the origin of the Malagasy fauna. Like most researchers, he agrees that the eclectic nature of the modern fauna suggests some type of a sweepstakes, over-water dispersal. However, the absence of any Cenozoic fossil record prior to the last 10,000 years or so, precludes any knowledge of earlier faunas that may have gone extinct anytime during the last 50 million years.

In the final paper of this section, "On the Brink of Extinction: Research for the Conservation of the Tonkin Snub-nosed Monkey (*Rhinopithecus avunculus*)", Bert Covert and colleagues discuss one of the rarest of all living primates, the Tonkin Snub-nosed Monkey (*Rhinopithecus avunculus*). They review the history of discovery and natural history of this species as well as current conservation efforts.

Decades of Lemur Research and Conservation

The Elwyn Simons Influence

Patricia C. Wright

Introduction: Splendid Isolation, the Lemur Holocaust and Duke Primate Center's Director

Madagascar, the fourth largest island in the world, rifted from Africa more than 150 million years ago, and has been isolated in its present position for over 88 million years (Krause et al., 1997; Tattersall, 2008). The plants and animals became isolated there beginning during the late Cretaceous, and those that arrived in later disersals evolved into species found nowhere else in the world. The first humans did not arrive in Madagascar until approximately 2,000 years ago (Dewar, 1984), and from linguistic and genetic evidence we can pinpoint the origin of the first Malagasy people to the island of Borneo in Southeast Asia (Dewar, 1984). It is hypothesized that these first seafaring Malagasy brought with them fire (Gade, 1996), tools (Perez et al., 2005) and a rice culture. The fauna and flora were naive to the onslaught that ensued from this handful of humans. For example, over sixteen species of giant lemurs, some species as large as chimpanzees or female gorillas, have gone extinct due to anthropogenic influences over the past one thousand years (see Godfrey et al., 2008; Jungers et al., 2008).

As the new Director of the Duke University Primate Center, Dr. Simons became intrigued with this lemur holocaust. Elwyn Simons was trained as a mammalogist, geologist and paleontologist at Rice, Princeton, Oxford, and the position as Director of the Duke University Primate Center allowed him to use all these skills in crafting a broad new vision. The center constituted a superb base for expansion of prosimian research and conservation. In

Patricia C. Wright
Department of Anthropology, Institute for the Conservation of Tropical Environments
Stony Brook University, Stony Brook, NY 11794-4364, 43 Waterview Drive, Sound Beach, NY 11789, USA
PatCWright@aol.com

Fig. 1 Dr. Simons in Director's office, November, 1977

essence, his vision included saving the remaining lemurs from the fate of becoming fossils.

Elwyn Simons' Vision of Lemur Conservation

The vision that Simons conceived was straightforward: 1) Build up populations of the lemur species in captivity by bringing new animals from wild populations; 2) Bring rare species into captivity as a "second line of defense against extinction"; 3) Begin to blur the line between captivity and the wild with natural habitat enclosures and reintroduction programs; 4) Use information on wild living populations to better understand the behavioral ecology of extinct giant lemurs; 5) Discover new species of lemurs - both extant and extinct for a better understanding of the biodiversity and early primate evolution on this island and; 6) Train professors and students to carry on this conservation and research agenda into the future. He knew all these fronts had to be pursued simultaneously (Fig. 1). He also knew that he needed a large team of experts to make the big vision a reality. Elwyn, with his expertise in paleontology, did not have a ready-made team to study lemurs. He realized he would have to build a strong and motivated team from scratch.

The Prosimian Conservation Team

How Elwyn Simons builds a team is mysterious and not very well understood, especially when it is happening. One member of the team was there already, Andrea Katz, a highly motivated, organized and responsible undergraduate who loved lemurs. Soon after his arrival Elwyn hired Michael Stuart, a personable, dedicated, and forward thinking manager, and also hired Kay Izard, a physiologist driven to understand prosimian reproduction. Some of the team arrived as post doctoral researchers, as Michael Pereira and I did, or assistant professors, like Frances White, or Duke professors like South American specialist Ken Glander, or Duke graduate students like Dan Gebo, Claire Hemingway, Joe Macedonia, David Meyers, Deb Overdorff, Joyce Powzyk, Chris Tilden, Tab Rasmussen, and Anne Yoder. Some arrived thanks to Dr. Peter Klopfer who had made an arrangement for student exchange between Tubingen University, Germany and Duke (Renata Foerg, Jorg Ganzhorn, Peter Kappeler) and some came as graduate students with Bob Sussman from Washington University, St Louis (Linda Taylor, Ben Freed, Lisa Gould, Rubin Kaufman). I arrived to accomplish an NSF project awarded to Elwyn to build and study a colony of tarsiers, and with that mission I collected tarsiers in Sabah and the Philippines and brought them back to Duke. David Haring and I took on the task of tarsier husbandry and breeding. All of us became part of Elwyn's team, the Duke Primate Center Team, and all of us signed on to make things happen in Prosimian Research.

Elwyn began a system of numbering Primate Center publications that were based on, or related to, the animals or fossils at the Center. This list now stands at about 1,020 publications. Direct documentation of the remarkable increase in knowledge about the species of lemurs since Elwyn became the Center's director in 1977 is reflected by the change from 14 pages describing them in the 1983 edition of Ankel-Simons, "A Survey of Living Primates" to 21 pages in the second edition, and now encompassing 37 pages in the latest version (Ankel-Simons, 2007).

The Political Situation in Madagascar

The political climate of Madagascar during the 1970s was not open to western views on science and research. The primatologists (I. Tattersall, A. Richard, R. Sussman, A. Jolly, P. Klopfer, and J. Buettner-Janusch) who had begun their careers on lemur biology had been asked to leave in 1972 and were not welcome to return. This was not because of anything they had done. The new president of

Madagascar, D. Ratsirika, had appointed a Revolutionary Council that suspended foreign research while their government reviewed whether they would, in the future, align with researchers from the Communist block or with scholars from the West. The Duke Primate Center, established in 1962 by Peter Klopfer and John Buettner-Janusch to complement research in Madagascar, had lost Buettner-Janusch as director and slipped into a period of decline during the Madagascar socialist isolation. Dr. Elwyn Simons, taking the helm of the Duke Primate Center in 1977, felt that there should be a strong link between the lemurs at Duke and those in Madagascar. In his role as paleontologist Elwyn Simons had become an expert in negotiating with sensitive governments, such as Egypt during the seven-day war and India and Iran during the 1960s.

In September, 1978 Simons traveled to Madagascar in order to attend a colloquium held by the Academie Malgache at Parc Tsimbazaza and then discussed at the Ministry of Education the possibility of initiating protocols of collaboration with Malagasy institutions, but it was not yet the time. It wasn't until the beginning of the 1980s that President Ratsiraka began to reverse his xenophobia and to look outwards for alternative economies and begin to re-open Madagascar to science. Taking advantage of this potential opportunity, DUPC Director Simons journeyed again to Madagascar to begin discussions of possibilities of collaborations between the Duke University Primate Center and the University of Antananarivo. Professor Berthe Rakotosamimanana shared Professor Simons' vision and collaborations began. This was 1981 and the beginning of a decade of successful negotiations and agreements that have resulted in a much better understanding of the geographic distribution, taxonomy, ecology, anatomy, physiology and behavior of lemurs. This research on lemurs rolled easily into a more global problem of forest destruction and the questions of biodiversity loss. Dr. Simons, with his broad vision, ascertained that the Duke Primate Center and his team should take a lead role in the conservation efforts on Madagascar. It was stated at the time that the protocols Simons signed with the University and with Parc Tsimbazaza were the first to be made between their government institutions and any western institution. He had re-opened Madagascar to research. Elwyn's pioneering efforts in this reopening rank as some of his least recognized and most important accomplishments.

Realizing the Simons Vision

Build up Populations of the Lemur Species in Captivity

The first step in the Simons vision was to build up the colonies of lemurs in captivity at DUPC. Since the founders of the lemur colony were brought to Duke in the sixties and early seventies, no new wild lemurs had been brought out of Madagascar because of the sensitive politics. Therefore, the remaining animals were rather inbred, and some, like Nigel, the only member of his species (*Propithecus verreauxi*) in the United States, were leading a solitary life without any possibility of propagating the species. Therefore, Simons reasoned, for genetic health, new lineages should be brought from Madagascar of those species already in captivity. On his arrival Simons found that the center housed a large number of lemurs that were hybrids between various subspecies. These had been produced either to study karyotype evolution or to achieve the inheritance of various transferrins. These animals were not of interest either as zoological exhibits or for species propagation.

Bring Rare Species into Captivity as a "Second Line of Defense" Against Extinction

Almost half of the population at DUPC were varieties or subspecies of brown lemurs used in genetic research by the former Director, John Buettner-Janusch. A small fraction of the Malagasy lemurs (i.e., ringtailed lemurs and ruffed lemurs), were successfully propagated in large numbers in captivity before 1980. As the doors opened in Madagascar, and knowledge of the extent of the destruction of the habitat became known, (Sussman and Richard, 1987; Green and Sussman, 1990), it became obvious that extinction was imminent for the rarest species. Because of the precarious rarity of some species, Simons knew that to establish an effective "second line of defense against extinction", he needed to introduce into captivity the rarest species, and develop a viable colony of them. Captive conservation could also enhance knowledge of vanishing species such as the aye-aye, *Daubentonia madagascariensis,* which had never been brought alive to this country or even to any place in the New World. Tarsiers, the taxonomically most puzzling and controversial of all prosimians were not being propagated in captivity anywhere and it seemed to Simons that an NSF grant to support a tarsier colony would also be important to researchers.

Use Captive Lemurs to Understand Behavior of Living Lemurs and the Behavioral Ecology of Extinct Giant Lemurs

Rebuilding the Colony's Diversity

As mentioned, the diversity and reproductive capacity of the colony had dwindled. Realizing that the center needed both diversity and reproducing animals, Simons was active in securing permission to import new animals. Also, in order to expand research and conservation management, a greater variety of species seemed necessary. When animals are endangered or post-reproductive it is mandatory to acquire stock if captive conservation and study are to succeed. This is an extremely difficult procedure, since permission to capture in the wild and relocate members of endangered species required authorization from both the US and Malagasy governments. All Malagasy lemurs are classified as endangered species. After some years, and with the advice and assistance of founding Director John Buettner-Janusch and various Malagasy colleagues, particularly Mme. Berthe Rakotosamimanina, Simons was able to carry out several importations. Originally the staff of the department of Water and Forests (Eaux et Forêts) captured all lemurs themselves, but by the time of Simons' exploration, the Duke Center was expected to do it themselves. Eventually Elwyn, Kenneth Glander, Andrea Katz, Charlie Welch and I, together with Patrick Daniels and others working from the center, were involved in wild captures of various lemurs in Madagascar. It was extremely fortunate that Duke Anthropology Professor Kenneth Glander was one of the world's most experienced primate darters, since he had been studying howling monkeys in Costa Rica for years. His studies there required frequent captures by darting and subsequent release of these monkeys. His procedures could easily be adjusted for lemur capture. At Duke, *Propithecus verreauxi coquereli* had at one time been considered the "Crown Jewel" of the center's holdings by Director Buettner-Janush. However, before the end of 1977, the year Simons arrived at Duke the colony of *P.v. coquereli* was down to one individual, Nigel, our first captive born *Propithecus* (Fig. 2).

If there was to be more breeding of Coquerel's Sifaka at DUPC there would have to be new animals imported. No other institution had any living sifaka and only at Duke had there been successful rearing of captive born animals such as Nigel. In 1982, Ken Glander and Simons traveled to Madagascar in order to capture more sifaka. Permits had been granted and the mission was carried out together with Voara Randrianasolo of the Malagasy National Zoo: Parc Tsimbazaza (Fig. 3, Fig. 4).

What is not generally recognized is that many endangered lemur species are a NON-RENEWABLE research and conservation resource. Either the Malagasy or the US government can easily raise objections to, or forbid, the exportation or importation of new animals for outbreeding. If a breeding colony ceases to exist because its members are loaned out, sold, contracepted, or allowed to reach reproductive senescence, it may well be impossible to then restore a captive colony. The result is the termination of research, crucial conservation and the end to acquiring husbandry and management skills that would be of use in Madagascar when populations there decline in the wild.

Over the years of his Directorship it was Elwyn Simons' procedure to acquire populations, not only of *Propithecus verreauxi coquereli*, but of *Mirza, Daubentonia, Tarsius, Cheirogaleus, Microcebus* and of *Eulemur rubriventer, Eulemur coronatus, Eulemur collaris, Eulemur macaco flavifrons* and *Varecia variegata rubra*. This was done in order to expand research into the study of lemurs, some of which were nowhere in captivity, and also to attempt captive conservation at the center of the sort that recently had been successfully done for the California Condor and the Black Footed Ferret. It was also part of his vision to use *P. v. coquereli* as a model for later establishing a colony that would maintain captive conservation and study of *Propithecus diadema diadema*. The

Fig. 2 Elwyn Simons with Nigel, the sifaka, photo by Russell Mittermeier

latter subspecies now definitely stands on the verge of extinction. One individual only is now housed at Duke. Many regard this scientifically little known subspecies as the most beautiful of all primates. Continued failure to do anything about the captive conservation of these most extremely endangered lemurs is largely due to a lack of monetary support for such projects. As a result several species that could have been protected as a result of Simons' vision could very well now become extinct. It is improbable that, in the future, any additional members of the species or genera mentioned above will become available through any means in order to add them to captive research and

Fig. 3 Voara Randriansolo, Ken Glander, captured *Propithecus,* and Elwyn Simons in Madagascar

breeding projects. In spite of such problems, recently at DUPC, second generations of both *Propithecus* and *Daubentonia* have been born (see Fig. 5), demonstrating what is possible to achieve.

The DUPC Outdoor Enclosures

In keeping with the vision of understanding behavior from captive animals, and with Ken Glander's astute prompting and knowledge of wild primates, Ken and Elwyn decided to expand the horizons of the caged lemurs. With the strong encouragement and expertise of Bob Sussman, the first group of fifteen brown lemurs was released in Natural Habitat Enclosure #1 on August 4th, 1981 and on October 14th of the same year, eight ringtailed lemurs were also released. This controversial new concept took advantage of the mild climate in North Carolina, and the vast expanse of available forest surrounding the Center, which is part of the Durham Division of Duke Forest. The ringtailed lemurs took to the new habitat immediately, and within days were climbing trees and foraging on leaves and crabapples as if they had always lived in the forest. Other

Fig. 4 1984, Elwyn Simons and Ken Glander hold net in order to catch a darted lemur

forest. Other species followed and within a five-year span over five enclosures ranging from one to twenty acres were added to the DUPC. The trees at the edges were trimmed to prevent escape with a reinforcing electric fence. The lemurs accepted their territorial boundaries and the habitat enclosures were a success.

The success of the natural habitat enclosures revolutionized the ability to do behavioral research at the DUPC and a flood of research projects ensued. Jorg Ganzhorn, a graduate student from Tübingen University, Germany following up on early fieldwork in Madagascar on wild *Lemur mongoz*. began to watch the brown lemurs around the clock, and he documented that they definitely

Fig. 5 Elwyn Simons with “Blue Devil” world’s first captive-born aye aye, born on the day of a big basketball win

Fig. 6 Drs. Patricia Wright and Elwyn Simons (center) at the Inauguration Ceremony for Ranomafana National Park

forage both during the sunshine of the day and when the moon shines in the night of the Duke Forest. Thus, Ganzhorn was the first to note this cathemeral circadian activity pattern. In the outdoor enclosures, Ben Freed, a graduate student from Washington University, observed the brown lemurs cooperatively stalking prey such as cardinals and blue birds (Glander et al., 1985). Joe Macedonia, a Duke graduate student, watched the ringtails react to aerial predators differently from terrestrial predators (a yellow dog disguised as the fossa, a natural predator in Madagascar, was used) (1990; 1993a). Linda Taylor, a graduate student from Washington University first marked one female matriline with red collars and a second matriline with yellow collars and documented the grooming and resting groups segregating by collar color. When the groups grew too large, targeted aggression of one of the subordinate females was dramatic. Slowly, all the yellow collared females were chased over the habitat enclosure fence (Taylor and Sussman, 1985; Taylor, 2008; Vick and Pereira, 1989).

Michael Pereira (Fig. 7), hired as a post doctoral researcher, teamed up with Laura Vick and continued the study of female dominance and found that brown lemurs were much more egalitarian than other lemurs (Pereira et al., 1990; Pereira and MacGlynn, 1997). Michael Pereira and Louise Martin introduced unrelated male ringtailed lemurs to the enclosures and watched mating activities, showing that females did have choice (Pereira and Weiss, 1991). The Coquerel's sifakas were also observed in the outdoor enclosures and shown to be female dominant (Kubzdela et al., 1992). The behavioral repertoire of *Varecia variegata* was described for the first time (Pereira et al., 1988).

Fig. 7 Left: Michael Pereira; center, Kay Izard; right, Andrea Katz in Natural Habitat Enclosure #1

Kindergartens, daycare and infant guarding were first described for *Varecia* from observations in the free-ranging groups in the outdoor enclosures (Pereira et al., 1987). *Varecia* built a nest on the ground, and after triplets were born the male guarded the nest like a dog guards his den, while the mother foraged (Pereira et al., 1987, 1988). These enclosure observations led to later confirmation of these behaviors in wild populations (Vasey, 2003). *Varecia* social organization and male group transfer were also documented for the first time at the DUPC outdoor enclosures (White, 1991; White et al., 1993; Raps and White, 1996). Michael Pereira first described seasonal adjustment of growth rate and adult body weight in ring tails (Pereira, 1993b), a trend found later in the wild (Wright, 1999). In addition, Pereira (1993a) also first described asynchrony within estrous synchrony for ring tailed lemurs, which was later confirmed in the wild for other lemurs too (Wright, 1995, 1999).

Inside the DUPC

Lemurs

Meanwhile Kay Izard and Andrea Katz (Fig. 7) were inside the building organizing several research endeavors. Kay was interested in reproductive physiology and her research corroborated the evidence of seasonality by photoperiodicity, as well as documenting reproduction of prosimians for the first time (Wright et al, 1986.) Tab Rasmussen took up this research theme and examined records of breeding of lemurs kept in zoos at different latitudes. He documented that different levels of light changed the breeding season according to onset of higher light levels (Rasmussen, 1985). Then Chris Tilden began to examine the nutritional content of prosimian milk. This cross species approach showed that lemurs with litters (mouse lemurs, dwarf lemurs and ruffed lemurs) had fat-rich milk, while the lemurs that gave birth to singletons had dilute, not very nutritious milk (Tilden and Oftedal, 1997).

In many species, reproductive behavior was described for the first time at the DUPC. For example, the observed mating of tarsiers was documented with vaginal smears, and descriptions of female sexual swellings and sperm plugs were provided (Wright et al., 1989). For the first time, courtship and sexual behavior was also documented in Coquerel's mouse lemur (Stanger et al., 1995; see Fig. 8), and later confirmed in the wild (Kappeler, 1998). Seasonality in breeding and the estrus cycles of *Varecia variegata* and *Cheirogaleus medius* were described from studies conducted at the DUPC (Foerg, 1982a, 1982b).

Another set of research efforts documented previously unknown prosimian skills in sensory perceptions (Harrington, 1977). Studies of vocalizations of nocturnal primates showed that mouse lemurs and dwarf lemurs (Cherry and Simons, 1989) had ultrasonic vocalizations, as did the tarsier (Wright and Simons, 1989). Carl Erickson showed that aye-ayes could locate prey with percussive foraging (Erickson, 1994; 1995; Erickson et al., 1998). Studies of

Fig. 8 Veterinarian Wayne Kornegay and noted conservationist Lee Durrell, whose Ph. D. is from Duke University, examine a mouse lemur at the center

vocalizations of *Lemur catta* showed they responded differently to aerial and terrestrial predators (Macedonia, 1990; 1993a, b). Stanger and Macedonia (1994) first described the raucous calls of the aye-aye. Pollock (1986) described the vocalizations of the *Indri*.

Morphometrics and infant and juvenile growth patterns of various prosimians were also documented for the first time at the DUPC. Rasmussen and Izard (1988) looked at scaling of growth, life history and metabolic rates in lorises and galagos. Peter Kappeler compared prosimian intrasexual selection and canine dimorphism (1996b), as well as intrasexual selection and testis size (1997a). Ken Glander (1994) described the morphometrics and growth in captive aye-ayes. The Philippine tarsier infant development weights, measurements and behavior were described by David Haring and I (Haring and Wright, 1989), and Miles Roberts described this in *Tarsius bancanus* (Roberts, 1994; Roberts and Kohn, 1993).

Peter Kappeler took advantage of this largest prosimian colony in the world and based most of his earlier theoretical papers on these data. He suggested causes and consequences of life history variation (Kappeler, 1993; 1996a; 1997), examined possible reasons for female dominance in lemurs (Kappeler, 1990), and presented with van Schaik and Heyman thought-provoking hypotheses about the evolution of lemur social behavior (van Schaik and Kappeler, 1996; Kappeler and Heymann, 1996). Around the same time, Anne Yoder began to examine the genetics of the prosimians at the Primate Center to answer some important questions about phylogeny and origins (Yoder, 1992; 1994; Yoder et al., 1996).

Tarsiers

I was originally hired by Elwyn as a post doctoral researcher to establish a breeding colony of tarsiers at Duke. In 1983, the first Philippine tarsier arrived from Skanson Aquarium with Jonas Wahlstrom, the Director. His name was Amos and we did not realize then that he would live to hold the record for longevity of tarsiers (12 years). Elwyn and I forged a collaboration with Dr. Devra Kleiman, Director of the National Zoo and Mike Roberts, National Zoo researcher. As part of that collaboration, in July, 1983 Patrick Andau, Assistant Chief Game Warden of Sepilok Reserve, was invited to visit the Duke Primate Center. We discussed the possibilities of an expedition to Sabah, East Malaysia, to capture 12 *Tarsius bancanus* to establish a breeding center at Duke. This would be the first time that this species had ever been brought to the USA (Wright, 2003).

From September–November 1993 I worked in Sabah, East Malaysia with Patrick Andau and the Sepilok Reserve Game Wardens mist netting 12 Tarsier pairs to bring to the USA (Wright, 2003; Wright et al., 2003). All twelve arrived safely at Duke on November 13, 1983. In May, 1985 I journeyed to the Philippines and captured and carried back 12 pairs of the second species, *Tarsius syrichta*, the Philippine tarsier. At that point at Duke, for the first time we had two species of tarsiers in the same location, an ideal opportunity for a comparative study of their behavior.

With the help of Kay Izard and Andrea Katz, we began to understand tarsier reproductive systems. The gestation length of *T. bancanus* was 176 days, the first time this had been described for *T. bancanus* (Izard et al., 1985). The female reproductive cycle of *T. bancanus* was monitored with observations and vaginal smears and described for the first time (Wright et al., 1986). It was obvious that the two tarsier species had different social systems and this was documented as well (Haring and Wright, 1989; Pochron and Wright, 2003). Brian McNab documented the basal metabolic rate of tarsiers and showed that it was very low for a primate species (McNab and Wright, 1987). We also studied the infant development (Haring and Wright, 1989; see Fig. 9), husbandry (Wright et al., 1989), vocal repetoire (Wright and Simons, 1989), and conservation of tarsiers (Wright et al., 1987; Wright, 2003).

Lorises

Slow and slender lorises were both housed at the Duke Primate Center with pairs of pygmy slow lorises from North Vietnam being sent from Skanson Aquarium in the late 1980s. The reproduction of the slender loris was described by Kay Izard and Tab Rasmussen (Izard and Rasmussen, 1985) and it was determined that they had a six month gestation period. A comparison of the reproductive behavior of two species of *Nycticebus* documented these behaviors for the first time in one of the two species (Weisenseel et al., 1998).

Fig. 9 Juvenile tarsier hand reared at the DUPC by David Haring, photo by Friderun Ankel-Simons

Bushbabies

The research at the DUPC was the first to point out that the reproductive behaviors and breeding season of *Galago senegalensis braccuatus* and *G. s. moholi* were quite different (Izard and Nash, 1988), suggesting species status differences. Kay Izard spearheaded landmark work on the reproduction of galagos, documenting the lactation length in three species (Izard, 1987), infant survival and litter size (Izard and Simons, 1986a), and documenting that mother isolation before parturition saved infants (Izard and Simons, 1986b).

Animal Husbandry

The prosimians housed at the DUPC were rare and many of them never kept in captivity before. The efforts of Andrea Katz and the primate technician staff led to new methods to improve breeding and reduce perinatal mortality in many species. Many excellent papers and husbandry "fact sheets" were produced regarding housing, feeding, veterinary care and breeding of prosimians (Wright et al., 1989; Haring et al., 1985; 1988).

Use Information on Wild Living Populations to Better Understand the Behavioral Ecology of Extinct Giant Lemurs

Behavior of Fossil Lemurs

Elwyn Simons was keenly aware that many of the lemurs had recently become extinct, some as recently as 500 years ago (Simons, 1997). Early in his career he had realized that learning the diet, positional behavior and ecology of living primates in their natural habitat could give insights into the behavior and ecology of fossils (Fleagle and Simons, 1978; Fleagle et al., 1980). In fact, while working at Yale he had been part of a progressive group interpreting fossils by comparison with their closest living relatives. Lemurs serve as important comparative models for ancient extinct primates. Since the extinctions of giant lemurs were so recent, he Simons set about inviting a team that would look at both the living and extinct lemurs to put together a realistic picture of their ecological niche (Godfrey et al., 1997; Simons, 1997; Jungers et al., 2002). Elwyn's holistic way of looking at fossils by comparing them to closest living relatives led to new insights into the behavior and ecology of subfossil lemurs. The lemur subfossils became more than new names, but rather distant members of a living community (Fleagle and Reed, 1996; Godfrey et al., 1997; Simons, 1997; Jungers et al., 2002).

Discover New Species of Lemurs – Both Extant and Extinct – For A Better Understanding of The Biodiversity and Early Primate Evolution on this Island

A perfect example of Elwyn's realization of this part of his vision is the search for the living *Hapalemur simus,* a species once more widespread in Madagascar (*Vuillaume*-Randriamanantena et al., 1985; Wright, 1988 a, b, 1989; Wright et al., 1987a). In that quest for finding the living remnant populations, a new species to science, the golden bamboo lemur was discovered and described (Meier et al., 1987; Wright, 1988a). At the site of this discovery, Ranomafana,

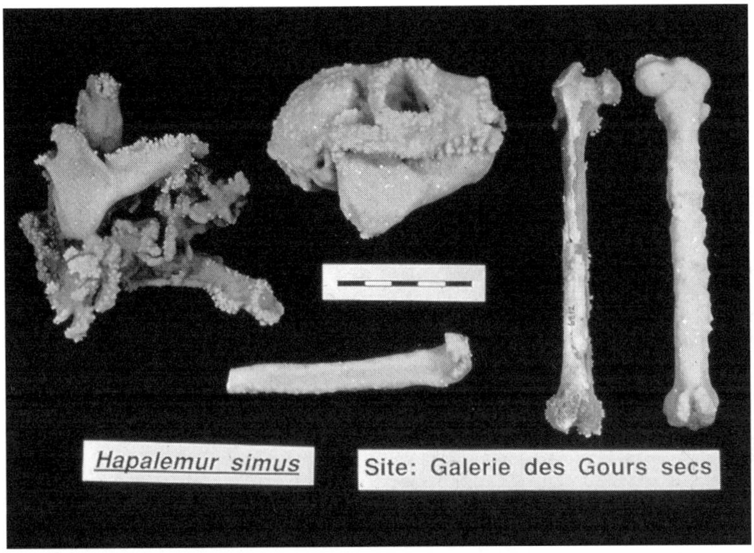

Fig. 10 Specimens of *Hapalemur simus* from the Ankarana massif, Adrafiabe Cave, discovered in the "Galerie des Gours secs", photo by Mario Gagnon

there were three species of bamboo eating lemurs found, literally a discovery of three sympatric "primate pandas", and research on their behavior and ecology began (Wright, 1986; 1988 a, b; Tan, 1999). The fact that two of the three species were tolerating large quantities of cyanide-laden bamboo in their diet was a unique adaptation for primates (Glander et al., 1992; Tan 1999). This lemur species left its traces in the far north where *Hapalemur simus* skulls and jaws litter the crevices of the Ankarana caves (Fig. 10). Applying what we know of the behavior of the living greater bamboo lemur, paleontologists can reconstruct the lifestyle of its subfossil relatives. By measuring the subfossils themselves, with such a large sample size, we can begin to understand morphological variation within those large populations (Jernvall et al., 2008).

Revising the Taxonomy of Lemurs and Naming New Species

Elwyn also has a special love for genealogy, for finding things out, and for naming things. Elwyn's accomplishments include splitting *Lemur catta* from the brown lemur group and naming the remaining brown lemurs *Eulemur* (Simons and Rumpler, 1988). He also recognized *Propithecus tattersalli* as a separate species (Simons, 1988). He was a keen advisor in the description of the new species *Hapalemur aureus*. (Meier et al., 1987).

Simons and his group have discovered many specimens of subfossil primates. While working with Jungers and Godfrey, one of the most dramatic finds was a new genus, *Babakotia,* a large-bodied sloth lemur in the caves of the Ankarana (Simons et al., 1990; 1995a).

Train Professors and Students to Carry on a Conservation and Research Agenda into the Future

The Grandfather of Ranomafana National Park

In 1986, with the advice of Elwyn Simons, I headed a group including Patrick Daniels, David Meyers and Deborah Overdorff to search for the perhaps extinct *Hapalemur simus* in the last sites where it was seen in southeastern Madagascar. We succeeded in not only finding the rare and mysterious *H. simus*, but also a new species to science, later named *Hapalemur aureus* (Wright et al., 1987b; Meier et al., 1987). Questioning how three bamboo lemurs could co-exist in one rainforest, Ken Glander, Dave Seigler (botanist from University of Illinois) and I described the cyanide tolerance of *H. aureus* in eating most of its diet from bamboo shoots containing high amounts of cyanide (Glander et al., 1989). In addition, Ken Glander had convinced me to capture a good sample of the 12 sympatric lemurs in Ranomafana National Park to establish baseline weights and measurements (Glander et al., 1992). Research was progressing well, until the timber exploiters began to hand cut rosewood and pallisandre trees in the area (Wright, 1992; 1997).

In order to save the new species of primate and all the other biodiversity in this splendid rainforest, I became a conservationist. Elwyn encouraged me to negotiate with the Department of Water and Forest, and the Director agreed to assist with gazetting the park boundaries. USAID awarded a large grant to Duke University to assist with setting up the infrastructure for the national park. To make a long story short, 43,500 ha of rainforest was declared a national park in May, 1991 (Wright, 1992, 1995; Wright and Andriamihaja, 2002). Elwyn Simons attended the inauguration (Fig. 6) and was declared the grandfather of the Ranomafana National Park. This park is closely integrated with the local economy, and today stands as an example of conservation that successfully melds research and ecotourism to the advantage of both the local residents and the biodiversity (Lovejoy, 2006).

An Experiment in Reintroduction in Madagascar

Dr. Elwyn Simons was adept in establishing international cooperations in research since the 1960s, and when he began research in Madagascar he had already been expert in the art of negotiations. One of his early goals was to bring more new genetic material back to Duke, in the form of new species, and new individuals of species already in captivity. But the Malagasy government for more than a decade had been closed to exports. Dr. Simons negotiated that if he exported lemurs to Duke, he would return some of them to Madagascar. Transporting live lemurs is not the same as bones, and it took a great deal of effort on Simons' part to achieve each animal importation in order to outbreed captive stocks. First, months had to be spent obtaining an US import permit and then days or weeks living in remote forests on capture missions followed. On returning to Antananarivo, lengthy negotiations with four different government ministries had to be carried out and a veterinarian's health certificate obtained in order to obtain export permits for only a few animals. Dr. Ken Glander's expertise was most important in humane darting of wild individuals. Simons soon became a primary authority on the American and Malagasy regulations concerning acquisition and transport of endangered species, another fact about him that was never well known.

From the start, the government of Madagascar wished all scientists working on the island, even paleontologists, to contribute in some way to conservation and so, beginning in the late 1970s, Simons established a protocol of collaboration with Parc Tsimbazaza, providing them with various supplies and participating with Tsimbazaza staff in capture missions. The Malagasy Department of Water and Forests also wanted a third collaborative project with the Duke Center. They had a few stations where lemurs had been held. One of these was known as Ivoloina located near the city of Tamatave on Madagascar's east coast (Fig. 11).

And thus another contribution from Duke in partnership with the Madagascar Fauna Group consortium was the development of a well-run zoological park and environmental education center at Ivoloina, on the footprint of a collection of cyclone-destroyed cages and pens where the Department of Water and Forests housed confiscated animals, including radiated tortoises and pet lemurs. Andrea Katz and Charlie Welch in 1988 took the helm of establishing the new Parc Ivoloina which is now recognized as a vital conservation center with components of captive breeding, environmental outreach, conservation and ecotourism training for the Malagasy staff and students, agroforestry demonstration, and reforestation research (Katz and Welch, 2003). Without Elwyn Simons' support and encouragement in the early years, as well as the help of his staff at the DUPC, the Parc Ivoloina project could not have succeeded (Fig. 12).

The agreement to return some captive born lemurs from Duke to Madagascar began to build momentum in the mid 1990s, and the DUPC joined with a newly founded consortium of international zoos, Madagascar Fauna Group (MFG), in developing the plan. Black and white ruffed lemurs (*Varecia variegata variegata*) had bred well in captivity, and were well-managed in North America through a Species Survival Plan The first lemurs chosen for reintroduction were a group that had lived for years in one of the DUPC natural habitat enclosures where they had acquired the survival skills needed for life in the wild. After several years of preliminary study and surveys, Betampona

Fig. 11 Lemur enclosure at Parc Ivoloina housing confiscated *Varecia*

Fig. 12 Malagasy students attending a weekly environmental class at Ivoloina

Nature Reserve was chosen as the reintroduction site (Welch and Katz, 1992). This Nature Reserve in eastern Madagascar is located 40 kilometers from Parc Ivoloina in a steep set of mountains, clothed in rainforest, and is one of the few lowland rainforests still existing on the island.

Since November, 1997 the Madagascar Fauna Group (MFG) has released thirteen captive-bred Black and White Ruffed Lemurs *(Varecia variegata variegata)* into the Betampona reserve in eastern Madagascar (Britt, et al., 2000a, b, 2003, 2004, Britt and Iambana, 2004; see Figs. 13–14). The outdoor habitat enclosures at the Duke Center played a most important role in these introductions, since the reintroduced lemurs, even those that did not come from Duke were all first given "boot camp" training in 'living wild" in the center's enclosures. Ten of the released lemurs survived for more than one-year post release, although all required varying levels of supplementary feeding. One Duke-born individual is still known to be surviving nine years post-release and has integrated and reproduced with a wild female. Four offspring were born or sired by the released lemurs, and all continue to thrive in groups with wild individuals. In total four of the captive-bred ruffed lemurs contributed to improving the genetic diversity of Betampona's ruffed lemur population. Five of the released lemurs fell victim to predation by fossa *(Cryptoprocta ferox)*, (Britt et al., 2001), one disappeared from the reserve and one was withdrawn from the release program. All in all, thanks to the hard work and skills of many including Charlie Welch, Andrea Katz, Adam Britt, Grahm Crawford, Ingrid Porton,

Fig. 13 Transport kennels with Duke-born *Varecia* on the way to Betampona Reserve

Fig. 14 *Varecia* in the forest at Betampona Reserve after release

Eva Sargent, David Anderson and Elwyn Simons, the reintroduction was a success, proving that reintroduction of captive lemurs back into the wild is possible and improving the genetic health of a wild lemur population can be accomplished (Britt et al., 2003; Figs. 13–14). Furthermore the lemur reintroduction in Betampona provided the foundation to protect the Reserve, to reach local communities for environmental awareness and also to transform Betapona into an important and valued field research site (Britt et al., 2003).

Summary and Conclusions

Spanning the past 25 years, Dr. Elwyn Simons has been one of the prime movers in primate conservation. He had a vision and implemented it: to make Madagascar an example of a successful and integrated conservation and research endeavor. His broad vision, incredible persistence and motivation, joined with abilities to understand how to negotiate among different cultures has resulted in real progress. His capacity to inspire others, in particular his students and associates, has resulted in an extraordinary legacy.

Known for his patience, combined with determination, he is fond of quoting a remark: “It’s dogged as does it”—that is sometimes used in description of Charles Darwin, but actually comes from a remark in an 1867 Trollope novel.

Fig. 15 Prithijit S. Chatrath with the type and only specimen of *Afrotarsius chatrathi*, described in 1985, and which he discovered at Quarry M in the Fayum

Acknowledgment Special appreciation is expressed to the late Professor Berthe Rakotosamimanana, a close friend and colleague of Elwyn Simons for over twenty-five years. Her death in 2005 is a great loss to Madagascar and her vibrancy, motivation and wisdom will be missed from paleontological, primatological and conservation fronts. Many thanks, as well, to Benjamin Andriamihaja and the MICET/ICTE staff for their logistic help to Elwyn throughout the last twenty years. The manuscript was greatly improved by comments by Andrea Katz, and Friderun Ankel-Simons and photos from Elwyn Simons. My gratitude also goes to Friderun, Cornelia and Verne, who made life possible for Amanda and me in Durham and welcomed us into the Simons family. And special thanks to Prithijit Chatrath, Elwyn's right-hand man in both field and laboratory, who has assisted greatly in making those Madagascar expeditions possible (Fig. 15). His extraordinary eyesight for seeing fossils, his abilities to solve logistic problems – no matter what, his great cooking and his constant good humor make every expedition enjoyable. Together, Prithijit and Elwyn are an amazing field team. This is DLC publication #1028.

References

Ankel-Simons, F. (2007). *Primate Anatomy: An Introduction,* (Third edition). Elsevier, San Diego.

Britt, A., Katz, A., and Welch, C. (2000a). Project Betampona: Conservation and re-stocking of black and white ruffed lemurs (*Varecia variegata variegata*). In: Roth, T. L., Swanson, W. F., and Blattman, L. K. (eds.), *Proceedings of the Seventh World Conference on Breeding Endangered Species, May 22 – 26, 1999*. Cincinnati, Ohio, pp. 87–94.

Britt, A., Welch, C., and Katz, A. (2000b). Ruffed lemur release update. *Lemur News* 5: 36–38.

Britt, A., Welch, C., and Katz, A. (2001). The impact of *Cryptoprocta ferox* on the *Varecia variegata variegata* re-stocking project at Betampona. *Lemur News* 6: 35–37.
Britt, A., Welch, C., and Katz, A. (2003). Can small isolated primate populations be effectively reinforced with the release of individuals from a captive population. *Biol. Conserv.* 115: 319–327.
Britt, A., Welch, C., Katz, A., Iambana, B., Porton, I., Junge, R., Crawford, G., Williams, C., and Haring, D. (2004). The re-stocking of captive bred ruffed lemurs (*Varecia variegata variegata*) into the Betampona Reserve: Methodology and recommendations. *Biodivers. Conserv.* 13: 635–657.
Britt, A., and Iambana, B. (2004). Can captive-bred *Varecia variegata variegata* adapt to a natural diet on release to the wild? *Int. J. Primatol.* 24(5): 987–1005.
Cherry, J. A., Izard, M. K., and Simons, E. L. (1987). Description of ultrasonic vocalizations of the mouse lemur (*Microcebus murinus*) and the fat-tailed dwarf lemur (*Cheirogaleus medius*). *Am. J. Primatol.* 13: 181–185.
Dewar, R. E. (1984). Recent extinctions in Madagascar: The loss of the subfossil fauna. In: Martin, P. S., and Klein, R. G. (eds.), *Quaternary Extinctions: A Prehistoric Revolution*. University of Arizona Press, Phoenix, pp. 574–593.
Erickson, C. J. (1994). Tap-scanning and extractive foraging in aye-ayes, *Daubentonia madagascariensis Folia Primatol.*. 62: 125–135.
Erickson, C. J. (1995). Feeding sites for extractive foraging by the aye-aye, *Daubentonia madagascariensis*. *Am. J. Primatol.* 35: 235–240.
Erickson, C. J., Nowicki, C., Dollar, L., and Goehring, N. (1998). Percussive foraging: Stimuli for prey location by aye ayes (*Daubentonia madagascariensis*). *Int. J. Primatol.* 19: 111–122.
Fleagle, J. G., and Reed, K. E. (1996). Comparing primate communities: A multivariate approach. *J. Hum. Evol.* 30: 489–510.
Fleagle, J. G., and Simons, E. L. (1978). Humeral morphology of earliest apes. *Nature* 276: 705–707.
Fleagle, J. G., Kay, R. F., and Simons, E. L. (1980). Sexual dimorphism in early anthropoids. *Science* 287: 328–330.
Foerg, R. (1982a). Reproduction in *Cheirogaleus medius*. *Folia Primatol.* 39: 49–55.
Foerg, R. (1982b). Reproductive behavior in *Varecia variegata*. *Folia Primatol.* 38: 108–121.
Gade, D. W. (1996). Deforestation and its effects in highland Madagascar. *Mountain Research* 16:101–116.
Glander, K. E. (1994). Morphometrics and growth in captive aye-ayes. *Folia Primatol.* 62: 108–114 .
Glander, K. E., Freed, B. Z., and Ganzhorn, J. (1985). Meat-eating and predation in captive-born *Lemur fulvus* and caged *Lemur macaco*. *Zoo Biol.* 4: 361–365.
Glander, K. E., Wright, P. C., Seigler, D. S., Randrianosolo, V., and Randrianosolo, B. (1989). Consumption of cyanogenic bamboo by a newly discovered species of bamboo lemur. *Am. J. Primatol.* 19: 119–124.
Glander, K. E., Wright, P. C., Merenlander, A., and Daniels, P. S. (1992). Morphometrics and testicle size of rainforest lemur species from southeastern Madagascar. *J. Hum. Evol.* 22: 1–17.
Godfrey, L. R., Jungers, W. L., Reed, K. E., Simons, E. L., and Chatrath, P. S. (1997). Subfossil lemurs: Inferences about past and present primate communities in Madagascar In: Goodman, S. M., and Patterson, B. D. (eds.), *Natural Change and Human Impact in Madagascar*. Smithsonian Press, Washington, DC, pp. 218–256.
Godfrey, L. R., Jungers, W. L., Schwartz, G. T., and Irwin, M. T. (2008). Ghosts and orphans: Madagascar's vanishing ecosystems. In: Fleagle, J. G. and Gilbert, C. C. (eds.), *Elwyn Simons: A Search for Origins*. Springer, New York, pp. 361–395
Green, G. M., and Sussman, R. W. (1990). Deforestion history of the eastern rain forests of Madagascar from satellite images. *Science* 248: 212–215.

Haring, D. M., Wright, P. C., and Simons, E. L. (1985). Social behaviors of *Tarsius syrichta* and *Tarsius bancanus*. *Am. J. Phys. Anthropol.* 66(2): 179.

Haring, D. M., Wright, P. C., and Simons, E. L. (1988). Conservation of Madagascar's sifakas *(Propithecus)* in captivity and in the wild. In: Dresser, B. (ed.), *Proceedings of Fifth World Conference on Breeding Endangered Species in Captivity*. Cincinnati Zoo, Cincinnati, pp. 67–81.

Haring, D. M. and Wright, P. C. (1989). Hand raising a Philippine tarsier (*Tarsius syrichta*). *Zoo Biol.* 8: 265–274.

Harrington, J. E. (1977). Discrimination between males and females by scent in *Lemur fulvus*. *Anim. Behav.* 25: 147–151.

Izard, M. K. (1987). Lactation length in 3 species of *Galago*. *Am. J. .Primatol.*13(1): 73–76.

Izard, M. K., and Rasmussen, D. T. (1985). Reproduction in the slender loris (*Loris tardigradus malabaricus*). *Am. J. Primatol.* 8: 153–155.

Izard, M. K., Wright, P. C., and Simons, E. L. (1985). Gestation length in *Tarsius bancanus*. *Am. J. Primatol.* 9: 327–331.

Izard, M. K., and Simons, E. L. (1986a). Isolation of females prior to parturition reduces neonatal mortality in *Galago*. *Am. J. Primatol.* 10: 249–255.

Izard, M. K., and Simons, E. L. (1986b). Infant survival and litter size in Primigravid and Multigravid *Galagos*. *J. Med. Primatol.* 15: 27–35.

Izard, M. K., and Nash, L. T. (1988). Contrasting reproductive parameters in *Galago senegalensis braccatus* and *G. s. moholi*. *Int. J. Primatol.* 9(6): 519–527.

Jernvall, J., Gilbert, C. C., and Wright, P. C. (2008). Peculiar tooth homologies of the greater bamboo lemur (*Prolemur = Hapalemur simus*): When is a paracone not a paracone? In: Fleagle, J. G. and Gilbert, C. C. (eds.), *Elwyn Simons: A Search for Origins*. Springer, New York, pp: 335–342.

Jungers, W. L, Godfrey, L. R., Simons, E. L., Wunderlich, R. E., Richmond, B. G., and Chatrath, P. S. (2002). Ecomorphology of behavior of giant extinct lemurs from Madagascar. In: Plavcan, J. M., Kay, R. F., Jungers, W. L., and van Schaik, C. P. (eds.), *Reconstructing Behavior in the Primate Fossil Record*. Kluwer Academic/Plenum Publishers, New York, pp. 371–411.

Jungers, W. L., Demes, B., and Godfrey, L. R. (2008). How big were the "giant" extinct lemurs of Madagascar? In: Fleagle, J. G. and Gilbert, C. C. (eds.), *Elwyn Simons: A Search for Origins*. Springer, New York pp.343–360.

Kappeler, P. M. (1990). Female dominance in *Lemur catta*: more than just female feeding priority? *Folia Primatol.* 55: 92–95.

Kappeler, P. M. (1993). Sexual selection and lemur social systems. In: Kappeler, P. M. and Ganhorn, J. U. (eds.), *Lemur Social Systems and Their Ecological Basis*. Plemum Press, New York, pp. 223–240.

Kappeler, P. M. (1996a). Causes and consequences of life-history variation among strepsirhine primates. *Am. Nat.* 148: 868–891.

Kappeler, P. M. (1996b). Intrasexual selection and phylogenetic constraints in the evolution of sexual canine demorphism in strepsirhine primates. *J. Evolution. Biol.* 9: 43–65.

Kappeler, P. M. (1997a). Determinants of primate social organization: Comparative evidence and new insights form Malagasy lemurs. *Biol. Rev. Cambridge Phil. Soc.* 72(1): 111–151.

Kappeler, P. M. (1997b). Intrasexual selection and testes size in strepsirhine primates. *Behav. Ecol.* 8: 10–19.

Kappeler, P. M. (1998). To whom it may concern: the transmission and function of chemical signals in *Lemur catta*. *Behav. Ecol. Sociobiol.* 42: 411–421.

Kappeler, P. M., and Heyman, E. W. (1996). Nonconvergence in the evolution of primate life history and socio-ecology. *Biol. J. Linn. Soc.* 59: 297–326.

Katz, A. S., and Welch, C. R. (2003). Parc Ivoloina. In: Goodman, S., and Benstead, J. (eds.), *The Natural History of Madagascar*. University of Chicago Press, Chicago, pp. 1555–1559

Krause, D. W., Hartman, J. H., and Wells, N. A. (1997). Late Cretaceous vertebrates from Madagascar: Implications for biotic change in deep time. In: Goodman, S. M., and Patterson, B. D. (eds.), *Natural Change and Human Impact in Madagascar*. Smithsonian Press, Washington, DC, pp. 3–43.

Kubzdela, K. S., Richard, A. D., and Pereira, M. E. (1992). Social relations in semi-free-ranging sifakas (*Propithecus verreauxi coquereli*) and the question of female dominance. *Am. J. Primatol.* 28: 139–145.

Lovejoy, T. (2006). Protected areas: A prism for a changing world. *TRENDS Ecol. Evol.* 21(6): 329–333.

Macedonia, J. M. (1990). What is communicated in the antipredator calls of lemurs: evidence from playback experiments with ringtailed and ruffed lemurs. *Ethology* 86:177–190.

Macedonia, J. M. (1993a). Adaptation and phylogenetic constraints in the antipredator behavior of ringtailed and ruffed lemurs. In: Kappeler, P. M. and Ganzhorn, J. U. (eds.), *Lemur Social Systems and Their Ecological Basis*. Plenum Press, New York, pp. 67–84.

Macedonia, J. M. (1993b). The vocal repertoire of the ringtailed lemurs *(Lemur catta)*. *Folia Primatol.* 61(4): 186–217.

McNab, B., and Wright, P. C. (1987). Temperature regulation and oxygen consumption in the Philippine tarsier, *Tarsius syrichta*. *Physiol. Zool.* 60(5): 596–600.

Meier, B., Albignac, R., Peyrieras, A., Rumpler, Y., and Wright, P. C. (1987). A new species of *Hapalemur* (Primates) from southeast Madagascar. *Folia Primatol.* 56: 57–63.

Pereira, M. E. (1993a). Agonistic interaction, dominance relation and ontogenetic trajectories in ringtailed lemurs. In: Pereira, M. E., and Fairbanks, L. A. (eds.), *Juvenile Primates Life History, Development, and Behavior*. Oxford University Press, New York, pp. 285–305.

Pereira, M. E. (1993b). Seasonal adjustment of growth rate and adult body weight in ringtailed lemurs. In: Kappeler, P. M., and Ganzhorn, J. U. (eds.), *Lemur Social Systems and Their Ecological Basis*. Plenum, New York, pp. 205–221.

Pereira, M. E., Klepper, A., and Simons, E. L. (1987). Tactics of care for young infants by forest living ruffed lemurs (*Varecia variegata variegata*): Ground nests, parking, and biparental guarding. *Am. J. Primatol.* 13(1): 129–144.

Pereira, M. E., Seeligson, M. L., and Macedonia, J. M. (1988). The behavioral repertoire of the black-and-white ruffed lemur, *Varecia variegata variegata* (Primates: Lemuridae). *Folia Primatol.* 51(1): 1–32.

Pereira, M. E., Kaufman, R., Kappeler, P. M., and Overdorff, D. (1990). Female dominance does not characterize all the Lemuridae. *Folia Primatol.* 55: 96–103.

Pereira, M. E., and Weiss, M. L. (1991). Female mate choice, male migration, and the threat of infanticide in ringtailed lemurs. *Behav. Ecol. Sociobiol.* 28: 141–152.

Pereira, M. E. and McGlynn, C. A. (1997). Special relationships instead of female dominance for redfronted lemurs, *Eulemur fulvus rufus*. *Am. J. Primatol.* 43: 239–258.

Perez, V., Godfrey, L., Nowak-Kemp, M., Burney, D., Ratsimbazafy, J., and Vasey, N. (2005). Evidence of early butchery of giant lemurs in Madagascar. *J. Hum. Evol. 49(6):* 722–742.

Pochron, S. T., and Wright, P. C. (2003). Flexibility in adult group compositions of a prosimian primate. *Behav. Ecol. Sociobiol.* 54: 285–293.

Pollock, J. I. (1986). The Song of the *Indri* (*Indri indri*; Primates: Lemuroidea) natural history form and function. *Int. J. Primatol.* 7:225–265.

Raps, S., and White, F. (1996). Female social dominance in semi-free-ranging ruffed lemurs *(Varecia variegata)*. Folia Primatol. 65: 163–168.

Rasmussen, D. T. (1985). A comparative study of breeding seasonality and litter size in eleven taxa of captive lemurs (*Lemur* and *Varecia*). *Int. J. Primatol.* 6: 501–511.

Rasmussen, D. T, and Izard, M. K. (1988). Scaling of growth and life history traits relative to body size, brain size, and metabolic rate in lorises and galagos (Lorisidae, primates). *Am. J. Phys. Anthropol.* 75(3): 357–367.

Roberts, M. (1994). Growth, development, and parental care in the western tarsier (*Tarsius bancanus*) in captivity: Evidence for a slow life-history and nonmonogamous mating system. *Int. J. Primatol.* 15(1): 1–28.
Roberts, M., and Kohn, F. (1993). Habitat at use, foraging behavior, and activity patterns in reproducing western tarsiers, *Tarsius bancanus*, in captivity: A management synthesis. *Zoo Biol.* 12(2): 217–232.
Simons, E. L. (1988). A new species of *Propithecus* (Primates) from northeast Madagascar. *Folia Primatol.* 50: 14–21.
Simons, E. L. (1997). Lemurs: Old and new. In: Goodman, S. M., and Patterson, B. D. (eds.), *Natural Change and Human Impact in Madagascar*. Smithsonian Press, Washington, DC, pp. 142–166.
Simons, E. L., and Rumpler, Y. (1988). *Eulemur:* New generic name for species of *Lemur* other than *Lemur catta*. *C. R. Acad. Sci. Paris* 307: 547–551.
Simons, E. L., Godfrey, L. R., Vuillaume-Randriamanantena, M., Chatrath, P. S., and Gagnon, M. (1990). Discovery of new giant subfossil lemurs Ankarana Mountains of northern Madagascar. *J. Hum. Evol.* 19: 311–319.
Simons, E. L., Burney, D. A., Chatrath, P. S., Godfrey, L. R., Jungers, W. L., and Rakotosamimanana, B. (1995a). AMS C-14 dates for extinct lemurs from caves in the Ankarana Massif, northern Madagascar. *Quaternary Res.* 43: 249–254.
Simons, E. L., Godfrey, L. R., Jungers, W. L., Chatrath, P. S., and Ravaoarisoa, J. (1995b). A new species of *Mesopropithecus* (Primates, Palaeopropithecidae) from northern Madagascar. *Int. J. Primatol.* 16: 653–682.
Stanger, K. F., and Macedonia, J. M. (1994). Vocalizations of aye-ayes (*Daubentonia madagascariensis*) in captivity. *Folia Primatol.* 62(1–3): 160–169.
Stanger, K. F., Coffman, B. S., and Izard, M. K. (1995). Reproduction in Coquerels dwarf lemur (*Mirza coquereli*). *Am. J. Primatol.* 36(3): 223–237.
Sussman, R. and Richard, A. (1987). Coservation priorities in Madagascar. In: Marsh, C. and Mittermeier, R. A. (eds.), Primate Conservation. Cambridge University Press, Cambridge, pp. 121–134.
Tan, C. L. (1999). Group composition, home range size and diet of three sympatric bamboo lemur species (genus *Hapalemur*) in Ranomafana National Park, Madagascar. *Int. J. Primatol.* 20: 547–566.
Tattersall, I. (2008). Vicariance vs. Dispersal in the Origin of the Malagasy Mammal Fauna. In: Fleagle, J. G. and Gilbert, C. C. (eds.), *Elwyn Simons: A Search for Origins*. Springer, New York, pp. 397–408.
Taylor, L. (2008). Old lemurs: preliminary data on behavior and reproduction from the Duke University Primate Center. In: Fleagle, J. G. and Gilbert, C. C. (eds.), *Elwyn Simons: A Search for Origins*. Springer, New York, pp. 319–334.
Taylor, L., and Sussman, R. W. (1985). A preliminary study of kinship and social organization in a semi free-ranging group of *Lemur catta*. *Int. J. .Primatol.* 6: 601–614.
Tilden, C. D., and Oftedal, O. T. (1997). Milk composition reflects pattern of maternal care in prosimian primates. *Am. J. Primatol.* 41: 195–211.
Vasey, N. (2003). *Varecia*, Ruffed Lemurs. In: Goodman, S., and Benstead, J. (eds.), *Natural History of Madagascar*. University of Chicago Press, Chicago, pp. 1132–1136.
van Schaik, C. P., and Kappeler, P. M. (1996). The social systems of gregarious lemurs: Lack of convergence with anthropoids due to evolutionary disequilibrium? *Ethology* 102(11): 915–941.
Vick, L. G., and Pereira, M. E. (1989). Episodic targeted aggression and the histories of lemur social groups. *Behav. Ecol. Sociobiol.* 25: 3–12.
Vuillaume-Randriamanantena, M., Godfrey, L. R., and Sutherland, M. R. (1985). Revision of *Hapalemur* (*Prohapalemur*) *gallieni* (Standing, 1905). *Folia Primatol.* 45: 89–116.
Weisenseel, K. A., Izard, M. K., Nash, L. T., Ange, R. L., and Poorman-Allen, P. (1998). A comparison of reproduction in two species of *Nycticebus*. *Folia Primatol.* 69(1): 321–324.

Welch, C. R. and Katz, A. S. (1992). Survey and census work on lemurs in the natural reserve of Betampona in eastern Madagascar with a view to reintroductions. *DODO* 28: 45–58.

White, F. (1991). Social organization, feeding ecology, and reproductive strategy of ruffed lemurs, *Varecia variegata*. In: Ehara, A., Kimura, T., Takenaka, O., and Iwamoto, M. (eds.), *Primatology Today, Proceedings of the XIIIth Congress of the International Primatological Society Nagoya and Kyoto, 18–24 July 1990*. Elsevier Science Publishers, Amsterdam, pp.81–84.

White, F. J., Balko, E. A., and Fox, E. A. (1993). Male transfer in captive ruffed lemurs, *Varecia variegata variegata*. In: Kappeler, P. M. and Ganzhorn, J. G. (eds.), *Lemur Social Systems and their Ecological Basis*. Plenum Press, New York, pp. 41–50.

Wright, P. C. (1986). Diet, ranging behaviour and activity pattern on the gentle lemur (*Hapalemur griseus*) in Madagascar. *Am. J. Phys. Anthropol.* 69: 283.

Wright, P. C. (1988a). Lemurs lost and found. *Nat. Hist.* 97(7): 56–66.

Wright, P. C. (1988b). Lemurs' last stand. *Animal Kingdom* 91(1): 12–25.

Wright, P. C. (1989). Comparative ecology of three sympatric bamboo lemurs in Madagascar. *Am. J. Phys. Anthropol.* 78: 327.

Wright, P. C. (1992). Primate ecology, rainforest conservation and economic development: Building a national park in Madagascar. *Evol. Anthropol.* 1: 25–33.

Wright, P. C. (1995). Demography and life history of free-ranging *Propithecus diadema edwardsi* in Ranomafana National Park, Madagascar. *Int. J. Primatol.* 16: 835–854.

Wright, P. C. (1997). The future of biodiversity in Madagascar: A view form Ranomafana National Park. In: *Patterson, B. D. and Goodman, S. M. (eds.), Natural Change and Human Impact in Madagascar*. Smithsonian Institution Press, Washington, DC, pp. 381–405.

Wright, P. C. (1999). Lemur traits and Madagascar ecology: Coping with an island environment. *Yearb. Phys. Anthropol.* 42: 31–72.

Wright, P. C. (2003). Are tarsiers leaping into extinction? In: Wright, P., Simons, E., and Gurksy, S. (eds.), *Tarsiers: Past, Present and Future*. Rutgers University Press, New Jersey, pp. 296–308.

Wright, P. C. and Simons, E. L. (1989). Calls of the Mindanao tarsier (Tarsius syrichta). American Journal of Physical Anthropology 66: 236.

Wright, P. C., Toyama, L., and Simons, E. L. (1986). Courtship and copulation in *Tarsius bancanus*. *Folia Primatol.* 46: 142–148.

Wright, P. C., Haring, D., Simons, E. L., and Andau, P. (1987). Tarsiers: A conservation perspective. *Primate Conservation* 8: 51–54.

Wright, P. C., Haring, D., Izard, M. K., and Simons, E. L. (1989). Psychological well-being of nocturnal primates in captivity. In: E. Segal (ed.), *Housing Care and Psychological Well-being of Captive and Laboratory Primates*. Noyes Publications, New York, pp. 61–74.

Wright, P. C., and Andriamihaja, B. A. (2002). Making a rain forest national park work in Madagascar: Ranomafana National Park and its long-term research commitment. In: Terborgh, J., van Schaik, C., Rao, M., and Davenport, L. (eds.), *Making Parks Work: Strategies for Preserving Tropical Nature*. Island Press, Covelo, California, pp. 112–136.

Wright P. C., Pochron, S. T., Haring, D. H., and Simons, E. L. (2003). Can we predict seasonal behavior and social organization from sexual dimorphism and testes measurements? In: Wright, P., Simons, E. and Gurksy, S. (eds.), *Tarsiers: Past, Present and Future*. Rutgers University Press, New Jersey, pp. 260–273.

Yoder, A. D. (1992). The applications and limitations of ontogenetic comparisons for phylogeny reconstruction: The case of the strepsirhine internal carotid artery. *J. Hum. Evol.* 23: 183–195.

Yoder, A. D. (1994). Relative position of the Cheirogaleidae in strepsirhine phylogeny: A comparison of morphological and molecular methods and results. *Am. J. Phys. Anthropol.* 94: 25–46.

Yoder, A. D., Cartmill, M., Ruvolo, M., Smith, K., Vagalys, R. (1996). Ancient single origin for Malagasy primates. *P. Natl Acad. Sci. U.S.A* 93: 5122–5126.

Low Fetal Energy Deposition Rates in Lemurs

Another Energy Conservation Strategy

Chris Tilden

Introduction

Group-living Malagasy lemurs, whose social systems historically were viewed as convergent with those of anthropoid primates (Jolly, 1966; Sussman, 1992), exhibit a suite of unique social, demographic and morphological traits that contrast with those found among anthropoids (Van Schaik and Kappeler, 1993, 1996; Pereira et al., 1999; Wright, 1999). Strict seasonal breeding, high infant mortality, sperm competition in conjunction with male-male aggression, targeted female aggression, cathemerality and female social dominance are exhibited by many of the diurnal, gregarious lemurs of Madagascar. A number of hypotheses have been offered to explain one of more of these unique traits found in lemuriform primates.

Among these hypotheses is a long-standing premise that female dominance in lemurs evolved in response to unusually high reproductive stress faced by females (Jolly, 1984; Richard and Nicoll, 1987; Young et al., 1990). Young et al. (1990) provided empirical data on prenatal growth rates among primates, and suggested that the relatively high rates in lemurs are indicative of a high degree of prenatal investment. They concluded that lemurs' "energy intensive strategy" is facilitated behaviorally through female feeding priority and social dominance.

Pereira et al. (1990) questioned the role of prenatal investment in the evolution of female social dominance, noting that not all species of lemurs with higher rates of prenatal growth rates exhibit female dominance. Kappeler (Kappeler, 1996; Van Schaik and Kappeler, 1996) conceded that prenatal investment rates were high in lemurs, but also pointed out that postnatal growth rates are not comparably high in lemurs. Kappeler suggested that female dominance would more likely arise as an adaptation to high lactation costs, since the energetic demands of lactation are far greater than those of

Chris Tilden
1121 Williamsburg Court, Lawrence, KS, 66049, (785) 296-7439
christilden@hotmail.com

J. G. Fleagle, C. C. Gilbert (eds.), *Elwyn Simons: A Search for Origins.*

gestation. Because lactation costs do not appear to be relatively high in lemurs, he questioned Young's hypothesis that female feeding priority was a response to an energy-intensive reproductive strategy. Pereira et al. (1999) and Wright (1999) both argued that female dominance does not function to relieve unusually high costs of reproduction, but rather that female dominance is one strategy to avoid energetic stress and maximize the use of scarce resources.

This paper presents data that suggest that lemur females do not have unusually high energy investment in their fetuses during gestation. The data support the view that lemur females minimize energy expenditure and avoid energetic stress during gestation.

Methods

Chemical analyses were conducted on cadavers of captive-born strepsirhines at the Duke University Primate Center that died on the day of birth. Only intact cadavers of normal birth weight with no readily discernable physical defects were analyzed, with the exception of several animals that had the eyes removed for other research purposes. Some of these neonates apparently died as a result of injuries suffered from an accident such as a fall, or by a physical attack by an adult animal. Others died of unknown causes. Cadavers were homogenized using a whole body grinder. The homogenized tissue was analyzed for water concentration by oven drying, fat concentration by ether extraction and protein concentration by Kjeldahl methods. These standard procedures have been previously described (Association of Official Analytical Chemists, 1990; Tilden, 1993). Analyses were performed on specimens of three lemur species: the brown lemur *Eulemur fulvus*, the black lemur *Eulemur macaco*, and the ring-tailed lemur *Lemur catta*, and on two lorisiform strepsirhines: the thick-tailed bushbaby *Otolemur crassicaudatus* and Garnett's bushbaby *Otolemur garnettii*. Energy concentration was calculated as the sum of protein and fat energy concentration using standard energy values of 5.65 and 9.3 kilocalories/gram (kcal/g) respectively (Brouwer, 1965). Species means for energy concentration were then multiplied by mean litter weight at birth (average neonatal weight multiplied by average litter size) to estimate total litter energy content. Clearly, litter size is a primary variable influencing maternal investment. It was possible to use actual litter sizes of the analyzed specimens rather than species means. However, I opted to use species means as a conservative measure. All lemurs and *Otolemur garnettii* analyzed in the study were singletons, whereas all *Otolemur crassicaudatus* were twins; comparisons based on actual litter size would have only made interspecific comparisons more pronounced.

Total estimated litter energy content was divided by known gestation lengths to derive average daily litter energy deposition rates (kcal/day). All life history values (female mass, neonatal weight, gestation length, and litter size) were derived from analyses of records at the Duke Primate Center.

Results

Despite small sample sizes, body composition values were remarkably homogenous within species and even genera (Table 1). Statistical analyses were not conducted, but body water concentration appears to be greater in the three lemurs (80.4–82.6%) than in the bushbabies (74.6–74.7%). The concentration of fat in lemurs (1.5–1.9%) is one-third that of bushbabies (3.4–4.3%), and protein concentration is also lower in lemurs (10.9–13.9% in lemurs and 16.4–16.5% in bushbabies). Based on unplanned pairwise comparisons of means among genera (Sokal and Rohlf, 1981), lemurs neonates were lower in energy concentration than bushbaby neonates, most likely due in large part to the high concentration of energy-rich fat in bushbaby neonates. *Otolemur* data were remarkable in their homogeneity. There was very little difference between *O. crassicaudatus* and *O. garnettii* even though the *O. crassicaudatus* specimens were all twins and the *O. garnettii* were all singletons.

When allometrically corrected by expressing values relative to maternal metabolic mass (female mass$^{0.75}$), both total litter energy content and the average rates of litter energy deposition appear greater in bushbabies than in lemurs (Table 2). Energy deposition rates are highest in *O. crassicaudatus*, the species with the highest average litter size and greatest tendency for multiple-litter births. However, litter size is not the only story. Both litter energy and energy deposition relative to maternal metabolic mass appear higher in *O. garnettii* than in lemur species, despite the fact the average litter size is lower in *O. garnettii* than in any of the lemur species. Energy deposition rates appear nearly identical in *Lemur* and *Eulemur* species.

Table 1 Neonatal mass and mean body compsition values

	Species				
	(1)	(2)	(3)	(4)	(5)
Sample Size	2	1	2	2	1
Weight	45.1	33.3	63.4	67.0	56.8
Body water (%)	74.6	74.7	80.4	80.4	82.6
	(76.3, 72.8)		(80.6, 80.2)	(79.1, 81.6)	
Fat (%)	3.4	4.3	1.8	1.9	1.5
	(3.2, 3.5)		(1.5, 2.1)	(1.7, 2.0)	
Protein (%)	16.4	16.5	13.9	10.9	11.1
	(16.2, 16.6)		(14.1, 13.8)	(10.8, 10.9)	
Lean body mass (%)	96.7	95.7	98.2	98.2	98.5
	(96.8, 96.5)		(98.5, 97.9)	(98.3, 98.0)	
Gross energy (kcal/g)	1.24	1.33	0.98	0.78	0.76

Species: *O. crassicaudatus* (1); *O. garnettii* (2); *L. catta* (3); *E. fulvus* (4); *E. macaco* (5).

Table 2 Life history data and prenatal energy deposition rates

	Species				
	(1)	(2)	(3)	(4)	(5)
Female mass (kg)	1.07	1.09	2.68	2.3	2.83
Litter size (mean)	1.6	1.1	1.3	1.2	1.4
Gestation length (days)	136	130	135	120	127
Mean litter mass (g)	72	36	82	80	80
Litter energy (kcal)	96	48	79	62	62
Litter energy per MS (kcal/kg$^{0.75}$)	91	45	39	33	28
Energy deposition (kcal/day)	0.71	0.37	0.59	0.52	0.49
Energy deposition per MS (kcal/day/kg$^{0.75}$)	0.67	0.35	0.27	0.28	0.22

Species: *O. crassicaudatus* (1); *O. garnettii* (2); *L. catta* (3); *E. fulvus* (4); *E. macaco* (5).
MS = female mass (in kg)$^{0.75}$

Discussion

While lemurs gain weight more rapidly in utero than do bushbabies and other lorisiform strepsirhines (Young et al., 1990; Kappeler, 1996), these analyses show that much of the lemur weight gain is due to water gain. On average, bushbaby fetuses deposit fat and protein (and thus energy) more rapidly during fetal development. One early reviewer of this manuscript suggested that interspecific differences in fat concentration at birth could be affected by dissection of the eyes in some cadavers, since these organs are high in lipids. However, the two *O. crassicaudatus* cadavers were among those whose eyes had been removed prior to chemical analyses. Despite the fact the eyes were removed, their fat concentration was similar to that of *O. garnettii*, not low like lemurs. This suggests that removal of the eyes for other analyses did not have a confounding affect on comparative analyses. These data show that during pregnancy the average energy transfer rates from mother to the prenatal litter are not unusually great for lemurs. The fat and protein concentration of lemur neonates is similar to that found in other mammals (but unfortunately data are not available for other non-human primates), while bushbabies appear to have relatively high concentrations of protein and fat compared to most mammals, with the notable exception of *Homo sapiens* (Table 3). Humans are born with a remarkably high fat concentration compared to other mammals. The data presented here add to a growing body of knowledge suggesting that maternal energetic investment is not greater in lemurs than other primate species.

Lemur females appear to minimize energy transfer to their young during gestation, producing infants of low energy investment after relatively short gestations (Table 2). A low rate of prenatal energy transfer is one of many energy-conserving physiological and behavioral characteristics of gregarious, diurnal lemur species. In fact, diurnal and nocturnal lemurs both exhibit

Table 3 Neonatal body composition in selected mammals

	Water (%)	Protein (%)	Fat (%)
O. crassicaudatus	74.6	16.4	3.4
O. garnettii	74.7	16.5	4.3
L. catta	80.4	13.9	1.8
E. fulvus	80.4	10.9	1.9
E. macaco	82.6	11.1	1.5
Pig	84.1	11.3	1.1
Rabbit	84.6	11.1	2.0
Rat	86	10.8	1.1
Human	69.1	11.9	16.1

Data for other species from Widdowson (1950).

suites of energy-conserving traits. Features that appear to be adaptations for the conservation of energy in lemurs include hypometabolism, seasonal torpor, small group living, sun bathing and group huddling (Jolly, 1966; Charles-Dominique et al., 1980; Daniels, 1984; Müller, 1985; Richard and Nicoll, 1987; Pereira et al., 1999; Wright and Martin, 1995; Ortmann et al., 1997). In addition to these energy-conserving traits, lemurs also exhibit features that allow for maximal use of scarce resources: cathemerality (Engqvist and Richard, 1991; Colquhoun, 1993; Overdorff and Rasmussen, 1995; Curtis and Zaramody, 1999; Rasmussen, 1999), seasonal breeding (Pereira, 1991; Richard, 1974; Sauther, 1991; Wright, 1995; Wright et al., 2005), circannual rhythms of somatic growth and body fat deposition (Pereira, 1993; Pereira et al., 1999), and female feeding priority and social dominance. In total, this suite of features exhibited by Malagasy lemurs appear to represent adaptations to the harsh energetic stress that results from the marked seasonality of Madagascar, low quality diets, and direct and indirect food competition as a result of living in large, cohesive social groups (Pereira, 1993; Petter-Rousseaux, 1968; Van Schaik, 1989). The results presented here add to the list of energy conservation strategies found among lemurs and support the "energy frugality hypothesis."

The data presented here also should sound a note of caution. The study demonstrates that the weight gain of offspring *in utero*, particularly in the absence of data on the chemical composition of the fetus(es), is not an appropriate comparative measure of maternal reproductive investment. Mammals whose young gain weight rapidly in utero may not have high fetal energy deposition rates. Like lemurs, mammals that are considered altricial typically have low neonatal energy concentration (Adolph and Heggeness, 1971; Moulton, 1923; Widdowson, 1950) and low fetal energy deposition rates, and thus may not be characterized by high maternal investment as has been claimed (Martin and MacLarnon, 1985, 1988). Bushabies are often "parked" by their mothers after birth, whereas lemur infants constantly "ride" with their mothers in the first few months postpartum. Riding provides lemur infants with the opportunity for frequent suckling, and the body warmth offered by the mother may minimize the cost of thermoregulation for the lemur infant. Mammals with frequent suckling patterns generally produce dilute milks, and this is certainly true of lemurs who produce some of the most dilute milks of all mammals studied to date (Tilden and oftedal, 1998). It is interesting to note that the most energy-rich milk found among diurnal lemurs is that of *Varecia*, which, unlike most diurnal lemurs, parks it infants. The milk of bushbabies is also richer than that of *Eulemur* species. Neonatal fat stores and energy-rich milks may help bushbaby infants meet the demands of thermoregulation when they mother leaves them parked during foraging bouts.

Further research is needed to enhance our understanding of the relationship between lemurs' reproductive energy requirements and their unique social, demographic and ecological features. In particular, little is known of energy and nutrient requirements in wild, free-ranging primates (and the effect of reproduction on these requirements) or of resource availability in their native habitats (Oftedal, 1991), although new information is emerging from intensive studies of food intake in free-ranging lemurs (Powzyk and Mowry, 2003; Wright et al., 2005; Irwin, 2005; Arrigo-Nelson, 2006). It is important that these new data be supplemented with more detailed quantification of specific resource availability in native habitats, as well as direct analyses of female energy requirements in varying reproductive states Research on lemuriform "parkers," such as *Varecia*, *Microcebus*, and *Cheirogaleus* would be of greatest interest for ongoing comparative studies.

Acknowledgment I would like to thank Dr. Olav Oftedal of the Division of Conservation Biology at the National Zoological Park (Smithsonian Institution) for providing access to laboratory facilities at the National Zoo and for being an ongoing source of inspiration and support over the years. Michael Jakubasz at the Nutrition Laboratory of the National Zoo provided technical assistance with many of the chemical analyses. I've discussed the ideas presented here with many thoughtful individuals that I feel fortunate to call dear friends including (but not limited to) Pat Wright, Ken Glander, Deborah Overdorff, Friderun Ankel-Simons, Mike Power, Carel Van Schaik, and Tab Rasmussen. I now work in health care policy and have not spoken directly to many of these individuals in several years, so hello to them all! Michael Pereira and Alison Richard have provided me with much food for thought, including valuable comments on earlier versions of this manuscript. I would like to thank my family. My wife Kathy and children Aaron and Caitlyn have been ever-so-patient with my self-indulgence over the years in trying to maintain a modest research agenda in primatology while having another full-time career. Finally, I owe much gratitude to Dr. Elwyn Simons, who was willing to take a chance on a naïve young student who could barely spell prosimian and helped me gain entry to graduate school at Duke. From day one Elwyn served as my advisor, even after I decided to limit my time in the fossil lab so I could begin dabbling in the study of reproductive energetics, which eventually became the focus of my research energies. Elwyn, Friderun and their children will always be part of my family.

This work was conducted with financial support of the Chicago Zoological Society and by a Grant-in-Aid from Sigma Xi, the Scientific Research Society. This is DLC publication #1030.

References

Adolph, E. F., and Heggeness, F. W. (1971). Age changes in body water and fat in fetal and infant mammals. *Growth* 35: 55–63.

Arrigo-Nelson, S. J. (2006). The impact of habitat disturbance on the feeding ecology of the Milne-Edwards' sifaka (*Propithecus edwardsi*) in Ranomafana National Park, Madagascar. PhD. Thesis, Stony Brook University.

Association of Official Analytical Chemists. (1990). *Official Methods of Analysis. 15th edition*, Association of Official Analytical Chemists, Arlington, VA

Brouwer, E. Report of subcommittee on constants and factors. (1965). In: Blaxter, K. L. (ed.), *Energy Metabolism: Proceedings of 3rd Symposium*. Academic Press, London, pp. 441–443.

Charles-Dominique, P., Cooper, H. M., Hladik, A., Hladik, C. M., Pages, E., Pariente, G. F., Petter-Rousseaux, A., Petter, J. J., and Schilling, A. (1980). *Nocturnal Malagasy Prosimians*, Academic Press, New York.

Colquhoun, I. C. (1993). The socioecology of *Eulemur macaco*: A preliminary report. In: Kappeler, P. M., and Ganzhorn, J. U. (eds.), *Lemur Social Systems and their Ecological Basis*. Plenum Press, New York, pp. 11–23.

Curtis, D. J., and Zaramody, A. (1999). Social structure and seasonal variation in behavior of *Eulemur mongoz*. *Folia Primatol.* 70: 79–96.

Daniels, H. L. (1984). Oxygen consumption in *Lemur fulvus*: deviation form the ideal model. *J. Mammal.* 65(4): 584–592.

Engqvist, A., and Richard, A. F. (1991). Diet as a possible determinant of cathemeral activity patterns. *Folia Primatol.* 57: 169–176.

Irwin, M. T. (2005). Ecological Impacts of Forest Fragmentation on Diademed Sifakas (*Propithecus diadema*) at Tsinjoarivo, Eastern Madagascar: Implications for Conservation in Fragmented Landscapes. PhD. thesis, Stony Brook University.

Jolly, A. (1966). *Lemur Behavior*, University of Chicago Press, Chicago.

Jolly, A. (1984). The puzzle of female feeding priority. In: Small, M. E. (ed.), *Female Primates: Studies by Women Primatologists*. Alan Liss, New York, pp. 197–215.

Kappeler, P. M. (1996). Causes and consequences of life-history variation among strepsirhine primates, *Am. Nat.* 148: 868–891.

Martin, R.D. and MacLarnon, A.M. (1985). Gestation period, neonatal size and maternal investment in placental mammals, *Nature* 313: 220–223.

Martin, R. D. and MacLarnon, A. M. (1988). Comparative quantitative studies of growth and reproduction, *Symp. Zool. Soc. Lond.* 60: 39–80.

Moulton, C. R. (1923). Age and chemical development in mammals, *J. Biol. Chem.* 57: 79–97.

Müller, E. F. (1985). Basal metabolic rates in primates: The possible role of phylogenetic and ecological factors, *Comp. Biochem. Physiol.* A91: 707–711.

Oftedal, O. T. (1991). The nutritional consequences of foraging in primates: The relationship of nutrient intakes to nutrient requirements, *Phil. Trans. R. Soc. Lond.* B334: 161–170.

Ortmann, S., Heldmaier, G., Schmid, J., and Ganzhorn, J. U. (1997). Spontaneous daily torpor in Malagasy mouse lemurs. *Naturwissenshaften*. 84: 28–32.

Overdorff, D. J., and Ramussen, M. A. (1995). Determinants of nighttime activity in "diurnal" lemurid species. In: Alterman, L. A., Doyle, G. A., and Izard, M. K. (eds.), *Creatures of the Dark*. Plenum, New York, pp. 61–74.

Pereira, M. E. (1991). Asynchrony within estrous synchrony among ringtailed lemurs (Primates: Lemuridae). *Physiol. Behav.* 49: 47–52.

Pereira, M. E. (1993). Seasonal adjustment of growth rate and adult body weight in ringtailed lemurs. In: Kappeler, P. M., and Ganzhorn, J. U. (eds.), *Lemur Social Systems and their Ecological Basis*. Plenum Press, New York, pp. 205–211.

Pereira, M. E., Kaufman, R., Kappeler, P. M., and Overdorff, D. J. (1990). Female dominance does not characterize all the Lemuridae. *Folia Primatol.* 55: 96–103.

Pereira, M. E., Strohecker, R. A., Cavigelli, S., Hughes, C. L. and Pearson, D. D. (1999). Metabolic strategy and social behavior in Lemuridae. In: Rakotosamimanana, B., Rasamimanana, H., Ganzhörn, J., and Goodman, S. M. (eds.), *New Directions in Lemur Studies*. Kluwer Academic/Plenum Publishers, New York, pp. 93–118.

Petter-Rousseaux, A. (1968). Cycles genitaux saisonniers des lemuriens malagaches. In: Canivenc, R. (ed.), *Cycles Genitaux Saisonniers du Mammiferes Sauvages*. Masson, Paris, pp. 11–22.

Powzyk, J. A. and Mowry, C. B. (2003). Dietary and feeding differences between sympatric *Propithecus diadema diadema* and *Indri indri*. *Int. J. Primatol.* 24: 1143–1162.

Rasmussen, M. A. (1999). Ecological influences on activity in two cathemeral primates, the mongoose lemur (*Eulemur mongoz*) and the common brown lemur (*Eulemur fulvus fulvus*). PhD. Thesis, Duke University.

Richard, A. F. (1974). Patterns of mating in *Propithecus verreauxi* Grandidier 1867. In: Martin, R. D., Doyle, G. A., and Walker, A. C. (eds.), *Prosimian Biology*. Duckworth, London, pp. 49–74.

Richard, A. F. and Nicoll, M. E. (1987). Female social dominance and basal metabolism in a Malagasy primate, *Propithecus verreauxi, Am. J. Phys. Anthropol.* 12: 309–314.

Sauther, M. L. (1991). Reproductive behavior of free-ranging *Lemur catta* in Beza Mahafaly Special Reserve, Madagascar. *Int. J. Primatol.* 10: 595–606.

Sokal, R. R. and Rohlf, F. J. (1981). *Biometry*. W.H. Freeman and Co., New York.

Sussman, R. W. (1992). Male life histories and inter-group mobility among ringtailed lemurs (*L. catta*). *Am. J. Primatol.* 13: 395–413.

Tilden, C.D. (1993). *Reproductive energetics of prosimian primates*. PhD. Thesis, Duke University.

Tilden, C. D. and Oftedal, O. T. (1998). Milk composition reflects patterns of maternal care in prosimian primates. *Am. J. Primatol.* 41: 195-211.

Van Schaik, C. P. (1989). The ecology of social relationships amongst female primates. In: Standen V. and Foley, R. A. (eds.), *Comparative Socioecology: The Behavioural Ecology of Humans and Other Mammals*. Blackwell, Boston, pp. 195–218.

Van Schaik, C. P., and Kappeler, P. M. (1993). Life history, activity period, and lemur social systems. In: Kappeler, P. M., and Ganzhorn, J. U. (eds.), *Lemur Social Systems and their Ecological Basis*. Plenum Press, New York, pp. 241–160.

Van Schaik, C. P., and Kappeler, P. M. (1996). The social systems of gregarious lemurs: lack of convergence with anthropoids due to evolutionary disequilibrum? *Ethology*. 102: 915–941.

Widdowson, E. M. (1950). Chemical composition of newly born mammals, *Nature* 166: 625–628.

Wright, P. C. (1995). Demography and life history of free-ranging *Propithecus diadema edwardsi* in Ranomafana National Park, Madagascar. *Int. J. Primatol.* 16: 835–854.

Wright, P.C. (1999). Lemur traits and Madagascar ecology: Coping with an island environment, *Yrbk. Phys. Anthropol.* 42: 31–72.

Wright, P. C., and Martin, L. B. (1995). Predation, pollination and torpor in two nocturnal primates: *Cheirogaleus major* and *Microcebus rufus* in the rain forest of Madagascar. In: Alterman, L. A., Doyle, G. A., and Izard, M. K. (eds.), *Creatures of the Dark*. Plenum, New York, pp. 45–60.

Wright, P. C., Razafindratsita, T., Pochron, S. T., and Jernvall, J. (2005). The key to Madagascar Frugivores. In: Dew, J. and Boubli, H. (eds.), *Tropical Fruits and Frugivores: the Search for Strong Interactors*. New York, Springer, pp. 121–138.

Young, A. L., Richard, A. F., and Aiello, L. C. (1990). Female dominance and maternal investment in strepsirhine primates, *Am. Nat.* 135: 473–488.

Old Lemurs

Preliminary Data on Behavior and Reproduction from the Duke University Primate Center

Linda Taylor

Introduction

Nonhuman primate gerontology is a relatively new area of inquiry in primatology. As DeRousseau (1994) points out, studies of nonhuman primate aging are "...particularly important for their relevance to human aging and for their evolutionary content." (p. 153). To date, such studies are limited (see summary Table 1 in Corr et al., 2002:197; iPAD, 2005) and are usually focused on the popular biomedical models, like chimpanzees (*P. troglodytes*) and macaques (*Macaca* sp.). Likewise, anatomical studies of aged primates are based primarily on captive anthropoid primates, like rhesus monkeys (*M. mulatta*) and the great apes, especially those regularly used as models for aspects of human aging (e.g., Coleman and Binkley, 2002; Erwin et al., 2002; Roberts, 2002). Some highlight the differences in locomotion and other physically demanding behaviors in older subjects, but overall results are conflicting about how "active" aged anthropoids may be in comparison to their younger conspecifics.

Investigators have used these nonhuman primate models to test theories about changes in social behavior and "social engagement" (Achenbaum and Bengston, 1994) drawn from the literature on humans (see discussion in Corr et al., 2002; Tarou et al., 2002). Roles for older animals have also been examined in nonhuman primates, modeled on roles in human groups (e.g., Borries, 1988). Pavelka and co-workers (2002) tested the correlation between living post-reproductive female Japanese macaques (*M. fuscata*) and the reproductive success of their daughters, i.e., the grandmother role. Their study (Pavelka et al., 2002) found that "the few postreproductive females that have unweaned grandchildren available appear to have a positive influence on their survival" (p. 407). Others focused on changes in reproduction that accompany age in macaques, with special emphasis on reproductive quiescence (Pavelka and Fedigan, 1991; Walker, 1995; Johnson and Kapsalis, 1998).

Linda Taylor
Department of Anthropology, University of Miami, P. O. Box 248106, Coral Gables, FL 33124-2005
LTaylor@Miami.edu

J. G. Fleagle, C. C. Gilbert (eds.), *Elwyn Simons: A Search for Origins.*

Table 1 Sample of species used in the study of aging

Taxon	Females Birth year	Males Birth year
Eulenur mongoz	74, 77, 78	76,77,78
Eulemur fulvus rufus	78	80
E. f. collaris		74
E. coronatus	77	80
Propithecus verreauxi		72, 80
Varecia variegata variegata	70, 76	71, 78

A small number focused on the intersection of anatomy and behavior, examining, for example, cognitive function and social behavior in apes (Corr et al., 2002).

The paucity of data is due, in part, to the lack of suitable study animals of known age. Often, animals are classed by general age groups, e.g., juvenile, subadult, adult, aged adult, etc., because their exact birth dates were not known. Few investigations have encompassed the entire lifespan for individuals of known birth date and/ or pedigree. Documenting an entire life is particularly difficult in long-lived taxa, like the great apes, because researchers and subjects may have similar life spans. Few data have been published on how advanced age may change the body and behavior of nonhuman primates in their native habitats (e.g., Huffman, 1990 for behavior; Morbeck et al., 2002 for skeletal anatomy).

Free-ranging, or island-bound, populations offer significant benefits and drawbacks for investigations of age-related changes. Sites like Cayo Santiago (e.g., DeRousseau, 1988; Corr, 2000) or the Japanese islands (e.g., Nakamichi, 1991; Kato, 1999) reliably determine the exact birth dates of all study animals and offer deep life-history data for several groups. Life history data can continue even when the monkeys have been translocated great distances from their native habitats (e.g., Pavelka, 1991, 1993).

Captive colonies may have accurate documentation of birth and pedigree data, although wild-caught individuals were sometimes imported with few data other than an age classification (e.g., "wild caught as adult"). In these cases, the age may be approximated by combining the number of years in captivity with the known age of maturity for those taxa. Because captive animals have the benefit of recent advances in veterinary care, have daily supplies of appropriate feed, and experience reduced predation, it is not unusual for them to live long, healthy lives, and perhaps longer lives than their wild conspecifics.

The results of previous investigations have provided data that are difficult to interpret, at times, because of variation in the definition of "aged", variation in methodology, and variation in conclusions within species, but at different sites. Almost all have a very small sample size of subjects. Tarou et al. (2002) report longitudinal data for a single individual. Others were relatively brief in comparison to the life spans of the study animals (Corr et al., 2002, Table 1), making species-specific conclusions difficult to draw.

Macaques, especially the rhesus macaque (*M. mulatta*), are often the biomedical model of choice, and are well represented in a large body of research on all aspects of their biology, behavior and development. DeRousseau (1988) and Coleman and Binkley (2002) detail how age affects the body of rhesus monkeys. There may even be some old females who live a few years after they cease to exhibit signs of reproductive cycling (Walker, 1995; Johnson and Kapsalis, 1998). Old rhesus, those 20 years of age and older, exhibit many of the same age-related changes in their bones as have been documented in old humans, i.e., osteoporosis and osteoarthritis. Corr et al. (2002) found significant sex-related differences in the behavior of aged rhesus, namely increased social interaction among old males whereas old females "spent significantly less time grooming and being engaged in general social contact than younger females." (p. 199).

In other studies, older monkeys are reported to be less engaged in social behaviors than younger conspecifics (e.g., Hauser and Tyrrell, 1984), scored less often in social proximity (e.g., Nakamichi, 1984), or in active locomotion (Veenema et al., 1997). Older monkeys were scored as resting more often than younger conspecifics (Hauser and Tyrrell, 1984; Nakamichi, 1984).

In other macaque species, the results are also mixed. Nakamichi (1984, 1991) and Kato (1999) describe a decreased frequency of social interactions for old Japanese macaque females in the wild (*M. fuscata*). Hauser and Tyrell (1984) published similar findings for Japanese and stumptail macaques (*M. arctoides*), but with considerable individual variation linked to rank, kinship affinities, and season.

Pavelka (1991, 1999), on the other hand, could not find similar trends in the translocated population of Japanese macaques she observed in Texas. She stated that "old" Japanese macaques, those age 20+ years (the last trimester of life), cannot be distinguished behaviorally as a unique subset with the group (see comparative calculation of age in Pavelka and Fedigan, 1999:462). Maxim (1979) found no decrease in sociality in his study of aged pigtail macaques (*M. nemestrina*). Behavioral plasticity, within species and across sites and taxa, plays a significant role throughout the lifespan in many species.

Aged apes are increasingly studied as models for human biology and behavior (Erwin et al., 2002). Their bodies appear to age as ours do (DeRousseau, 1988; Morbeck et al., 2002). Tarou et al. (2002) state that "old" apes (*P. troglodytes*, *G. gorilla*) are not behaviorally different from younger conspecifics.

Senescence in Lemurs

To date, very little is known about growing "old" and any accompanying changes that occur in locomotion or social behavior in strepsirrhines, including the lemurs. For this paper, I define aged animals as being in the last trimester of life, in accordance with much of the literature on aging in other taxa (e.g., Pavelka and Fedigan, 1999). For the diurnal taxa I observed, the last trimester of life begins at age 18 years. This age threshold conforms well with maximum life span of 27 for wild diurnal lemurs, as discussed in King et al. (2005). Very few diurnal lemurs live beyond this age, even in captivity (DUPC records). For example, the DUPC longevity record for the sifaka species *Propithecus verreauxi* is 31 years (see Fig. 1).

Studies of aging and age-related changes in behavior of strepsirrhine primates are few in number, perhaps due in part to the perceived lack of their suitability as models for human primates, especially the nocturnal taxa. iPAD, (http://ipad.primate.wisc.edu), the internet primate aging database produced in conjunction with the National Institute on Aging (NIA) and the Wisconsin National Primate Research Center, focuses on studies of biomarkers of aging in nonhuman primates. It features more than 15,000 data points for the single genus of squirrel monkeys (*Saimiri* sp.) but none for any prosimian or strepsirrhine taxa. Aging is thus a rich field of potential inquiry in this branch of our family tree.

There is a body of literature, growing from Jolly's (1966) pioneering study, that has documented the first weeks of life, including the development of independence, kin-based bonds, acquisition of rank, vocal communication, and the social lives of lemurs in the wild and in captivity (e.g., Taylor and Sussman, 1985; Macedonia, 1986, 1993; Taylor, 1986; Gould, 1990;

Fig. 1 Nigel, the Verreaux's sifaka (*Propithecus verreauxi*) with the record for longevity at the DUPC. On the left, Nigel at age 11. On the right, Nigel at age 30. Note the effects of aging

Zimmerman,1991; Sussman, 1992; Kappeler and Ganzhorn, 1993; Overdorff and Strait, 1995; Pereira, 1995; Oda, 2002). None have examined such features in the lives of aged lemurs, i.e., animals age 18 years or older. Life history data for wild lemurs is only recently available through longitudinal field studies (e.g., Wright, 1995, 1999; Richard et al., 2002; Gould et al., 2003; Pochron and Wright, 2003; King et al., 2005).

Of the few studies to date, several have focused on the small, nocturnal grey mouse lemur (*Microcebus murinus*), e.g., Aujard and Perret, 1998. The maximum lifespan in captivity is approximately 13 years of age for this taxon (DUPC records). Therefore, the threshold for being aged, in the last trimester of life, is approximately 8–9 years. Gilissen et al. (2001) documented the development of amyloid deposits in old grey mouse lemurs. Picq (1992) also focused on this small, nocturnal lemur and reported that social contact, like allogrooming, was less frequent in his 12 aged female subjects in comparison to younger conspecifics. Decline in olfactory communication abilities may help explain the changes in mouse lemur social behavior (Aujard and Némoz-Bertholet, 2004). Taylor (1988a, 1988c, 2000) and her co-workers (e.g., Pasin et al., 1998; Taylor and Pasin, 1999) are among the very few to examine age related changes in diurnal lemurs.

Many lemur taxa exhibit female dominance over males, especially in feeding situations. (e.g., Jolly, 1984; Taylor, 1986; Richard, 1987; Sauther, 1993; Gould, 1999; Pochron et al., 2003). Because the lemur taxa and their adaptive suites are so varied, it follows that there is also significant variation in the presence and prevalence of female dominance among species lemurs (e.g., Pereira et al., 1990; Pereira and McGlynn, 1997). However, this aspect of lemur social life has yet to be studied in animals of advanced age.

Study Site and General Methods

The Duke University Primate Center (DUPC) was founded in 1966 and is located on 85 acres in Durham, NC (Fig. 2). The site has been described in detail by various authors (e.g., Taylor, 1986; Taylor and Pasin, 1999, Wright, 2008). During its history, the collection has included many lemur taxa, tarsiers, and lorises and galagos (DUPC husbandry records, personal observation). The site has been renamed the Duke University Lemur Center, but I will continue to refer to it as DUPC herein as that was its name when this research was done. Current information on the site and collection may be found at: http://lemur.duke.edu.

Because of Duke University Primate Center's significant advances in lemur husbandry, breeding management, and veterinary care, they housed a unique collection of aged individuals in several taxa. No comparable population exists. In this chapter I present data on locomotion, social behavior and dominance for aged lemurs of six diurnal taxa.

Fig. 2 An aerial view of the Duke University Primate Center (now the Duke Lemur Center) in Durham, NC

Questions yet to be answered include whether or not these lemurs are behaviorally distinct from younger individuals of the same species in similar housing. Are they still fully engaged in social life? Do behaviors unique to lemurs, like female dominance, diminish with age? Does aging have an impact on basic locomotion and positional behaviors? Two initial hypotheses were tested:

1. Aged lemurs in captivity would be behaviorally distinct from younger conspecifics in similar housing.
2. Female dominance would not diminish with age.

To test these hypotheses, I used continuous focal animal sampling to gather data on 17 aged individuals in six diurnal taxa (Table 1), approximately 6% of the total DUPC population. Aged animals were age 18 years and above. Data on 17 younger lemurs (adults age 2 to 17 years) were analyzed for comparison between "older" and "younger" as specific age classes. All were socially housed at the Duke University Primate Center in a variety of settings ranging from large, naturalistic forest enclosures to smaller, conventional enclosures. Study animals lived in species-typical social groups or were housed with at least one cage mate. Each old animal was observed for 30 minutes, followed immediately by a 30 minute session on a younger animal in the same enclosure. Because of the unequal sample size, frequency data were taken from the focal animal sampling and converted to acts per hour (APH) for comparison of rates of behavior between older lemurs as one analytical group and a group of their younger counterparts in the same enclosures. Gould (1999) used a similar method in her analysis of female dominance and male behavior for a wild population of lemurs. For this study, 540 hours of focal animal data were analyzed. Results from different individuals, different species and both sexes were pooled to provide an overall comparison between older lemurs and their younger conspecifics.

Results

Age and Locomotion

Table 2 shows the rates of locomotion for 5 major categories of locomotor behavior: leap, climb, run, walk, and movement in place. Moving in place is a locomotor category that encompasses movement, like adjusting a resting posture, but does not include traveling more than a body length in distance. The younger lemurs engaged in more physically demanding behavior, like leaping and running, much more often than did the old lemurs. The old lemurs climbed more often, perhaps as an alternative means of moving in a three-dimensional world without leaping. Climbing also may have been the best way to navigate across the chain link walls of conventional enclosures (Pasin et al., 1998; Taylor, 1998b; Pasin and Taylor, 2000). Oldsters were also more apt to be walking than were the younger lemurs. These composite results are complicated by the use of one very unique locomotion style by just one species in the study – the vertical clinging and leaping (VCL) of sifakas. Both young and old sifakas move almost exclusively by leaping, which certainly increases the frequency of scores for that category overall. Similarly, sifakas were never recorded as walking, probably due to the difficulty translating their VCL locomotion style for a terrestrial setting. Nevertheless, these results give a general view of the differences between young and old lemurs.

Table 2 Summed locomotor differences between aged and younger lemurs

Locomotion Scored	Aged	Younger
Leap	2.1	8.9
Climb	6.7	5
Run	1.1	7.2
Walk	4.2	3
Move in place	3.5	4.9

Age and Social Behavior

As has been suggested with older primates of other species, there may be a tendency for old lemurs to be less engaged in the daily life of their social group. The question arises as how to measure social engagement. Nakamichi (1984) initially documented declining rates of social interaction associated with aging in Japanese macaques in Japan. Pavelka (1991, 1993) did not report similar findings for the same species, in an entire group translocated to south Texas. Some diurnal lemurs share social traits with macaques – they live in female-bonded groups with male dispersal. Does increasing age correlate with increasing social isolation in lemurs? To answer this question, I used one very simple measure of sociality – whether or not the focal animal was scored in proximity to another lemur (i.e., within one body length).

Overall, aged animals of both sexes were scored as being alone more often, a mean of 38.8% of the 687 scores, in comparison to the mean of 30.8% of the 810 scores for their younger cage mates. Aged females were scored as alone (35.6% of scores) less often than aged males, who were scored as alone more often (42% of scores) (Fig. 3). Likewise, younger females were scored alone in only 26.7% of samples, in contrast to males of the same age (35% of samples). Aged animals are apparently less apt to be in close proximity to another animal in comparison to the younger animals of the same sex in the groups or pairs.

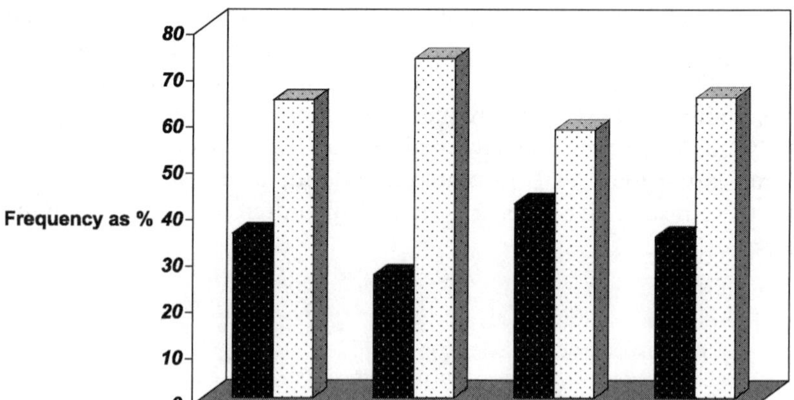

	Aged Female	Younger Female	Aged Male	Younger Male
■ Alone	35.6	26.7	42	35
□ Proximity	64.4	73.3	58	65

Fig. 3 Oldest animals of were scored as being alone more often than younger animals of the same sex. Females were all scored as being in proximity more often than all males

Initiating and participating in affiliative behaviors is a measure of degree of involvement in a social group. In female-bonded groups like lemurs, one would predict that females would continue to be active in the social life of the group. Older females initiated friendly social interactions with others at the rate of 3.2 APH, whereas old males initiated only 1.8 APH. Young females initiated 9.8 APH and young males initiated 5.5 APH of observation. Older females are still active in soliciting social interaction, although at a reduced rate in comparison to younger females. Older males seem to be the least apt to initiate social interaction.

Simultaneous mutual grooming characterizes many of the diurnal taxa and occurs much more often than allogrooming. Because grooming has such an important social function, the frequency of participation in grooming bouts between adults is a good measure of sociability in lemurs. Old animals were scored in mutual grooming at a rate of 2.6 APH, in comparison to the rate of 3.4 APH for younger animals. Clearly, oldsters are still very engaged in this aspect of social life. They are less apt to initiate the grooming bouts, however. Younger females initiated 5.1 APH in contrast to the oldest females who initiated mutual grooming at the rate of at 2.1 APH.

Age and Dominance

No data on dominance were gathered during the birth or breeding season, when same-sex and female-male adult aggression rates rise (e.g., Taylor, 1986). It may be that rates would change dramatically if infants were present, or competition for access to estrous females were involved. Likewise, no groups were sampled that contained both old animals and multigenerational matrilines. The presence or absence of these two factors could change the rates and targets of aggressive interactions and affiliative interactions, although they remain untested to date.

Old females were always able to displace males of any age (100% of 1520 observations). They won all agonistic interactions, including those with males relating to access to choice food items (100% of 865 interactions) – just as their younger female peers did. Thus, the unique female dominance over males persists undiminished as females age. I saw no observations of agonism between young females and old females over access to food because of the way food was presented (multiple feeding sites per enclosure) and the composition of the study groups. Priority of access to food may be a source of agonism between older and younger adults of the same in species-typical groups, although additional data are needed to address this question.

Targeted aggression is a fact of life in some lemur taxa. For example, in a ringtailed lemur (*Lemur catta*) group at DUPC, one matriline selectively targeted unrelated females for intense aggression until they eliminated them from the group (Taylor, 1995). Old animals could be targeted as a means for younger animals to rise in rank, especially among incoming and resident males, or among females in competing matrilines in a social group (Taylor, 1986). Were old lemurs initiators of aggression or were they targets?

Old females initiated a mean of 2 APH of agonistic behavior and yet were targeted at a rate lower than 1 APH. Older males initiated 1.2 APH of observation and were the recipients of only 1 act per hour – most of these scores were between an aged father and his son housed in the same enclosure. Younger females were targeted at the rate 1.3 APH (always from other females) and they initiated 2 APH. Younger adult males were apt to initiate aggression at a rate of 3 APH but were the targeted at 7 APH, by adults of both sexes.

Based on this sample, it appears that the older lemurs retain dominance status, perhaps by means other than aggressive interactions. These data also suggest that younger lemurs do not target the older lemurs as a means to rise in rank, although this idea has yet to be tested in species-typical groups across the study taxa. Pochron and Wright (2003) suggest that older animals have an increased chance of winning aggressive interactions with same-sex individuals, but that age does not increase the chances of winning aggressive interactions between the sexes. If younger lemurs were targeting older animals of the same sex as a way to rise in rank, then these findings from the field suggest that targeting older animals might not be an effective strategy since older animals would most likely be clear winners of the interaction.

All females, regardless of age, had social dominance as measured in feeding, agonism, and by displacing. Dominance over males appears to be constant across the life span. This aspect of female dominance may be related to continued reproduction into advanced age.

Age and Reproduction

Pavelka and Fedigan (1991), Caro et al. (1995), Walker (1995), and Johnson and Kapsalis (1998) raise the suggestion that there may be an age-related decline or cessation in reproduction. However, the evidence for true menopause in all nonhuman primates is scant and inconsistent within or across taxa, especially among prosimians, including strepsirrhines. I used the single criterion for human-like menopause that would require that all females, of the same taxon, cease reproductive cycling before or during the last third to the last half of their adult lives.

Data on reproduction senescence are complicated by the fact that DUPC is a captive population with a long-term plan to promote genetic diversity while controlling the size of the collection, and by the fact that the available data is drawn from on only a few taxa with a very small sample size for each. Older animals that have had many living offspring in the captive population are prevented from breeding. Others may be prevented from breeding because they are isolated from suitable partners, either intentionally or as a by-product of captive demography. For example, some of the study animals were pair-housed with same-sex partners, rather than in species-typical groups. Rare specimens, like the lone male diademed sifaka (*P. diadema*), may not have any members of the opposite sex of any age in the collection with which to breed.

Clear answers about reproductive changes with aging were difficult to find because of the above. Many of the older lemurs at the DUPC were not in a situation that would permit breeding. In man y instances they were in same-sexing living situations or were on contraception medicines. A notable exception was Coquerel's sifaka. Both males and females that were in breeding situations during this study continued to produce offspring up to age 24.

For other taxa, many lemurs of both sexes continued to produce infants up to and including the year they died (DUPC colony records). There doesn't appear to be an upper limit at which otherwise healthy animals are "too old" to breed. Mouse lemurs were still producing litters at age 13 – an advanced age for them, roughly equivalent to a human female giving birth at age 98 (using the methods of Pavelka and Fedigan, 1999). One female black and white ruffed lemur, wild caught as an adult in 1970, successfully produced a litter of babies every year until she was removed from the breeding population at the age of 20. One aged sifaka, born in captivity in 1972, became a father again and grandfather in the same birth season when he was 24 years old. He sired additional offspring outside of the study when he was age 30, just a few months before his death (B. Grossi, personal communication).

Parga and Lessnau (2005) report that there appears to be no upper age limit on reproduction in the free ranging lemurs on St. Catherine's Island, GA. Interestingly, they report a significant decline in survivorship in infants born to oldest mothers – perhaps based on the fact that worn teeth are having a negative impact on the oldest females' milk. An analysis of the decades of breeding and birth records for the DUPC colony could provide a more complete picture of the intersection of reproductive rates, infant survival, and age across taxa.

In sum, it appears that there is little evidence for reproductive senescence in the lemurs in the DUPC collection, although data on captive animals may be confounded by factors above. I found no evidence of "normal" animals (i.e., healthy, not contracepted or isolated) having any significant end-of-life years without reproduction (Taylor, 1998a).

Discussion and Conclusions

By some measures, older lemurs appear to be less actively involved in social groups, if one considers only rates per hour of interactions. However, since social dominance did not wane, it may be that declining rates of interactions alone do not accurately reflect the position of old lemurs in their social group.

It is unlikely that old lemurs are less socially engaged for many of the same reasons Pavelka (1999) presents in her essay on anthropoid taxa. They may be less physically close to their social companions, but proximity may not be equal to social engagement. In addition, to withdraw from social life is problematic for diurnal nonhuman primates because they live in social groups that are always on the move. Withdrawing from a group, or being left by it, is a potentially dangerous situation. There are apparently no adult lemurs of either sex who become solitary late in life.

Gould (personal communication) describes three old female ringtailed lemurs in her focal groups in the population at Beza-Mahafaly Reserve (Madagascar) at different time periods between 1992–2003. All three of these females could be described as not as engaged in the social life of the group when compared to younger females. Why? Because they repeatedly lagged behind the main body of the group as they traveled and foraged, often losing their group altogether despite continuous contact calling. Does this behavior set them apart as a distinct class from all younger lemurs in the group? Can a species-specific conclusion be drawn based on only three animals? It may be that a single criterion may not be sufficient to judge whether or not aged lemurs constitute a behaviorally-distinct subgroup.

Clearly, the interpretation of age-related differences in various nonhuman primates is hampered by the lack of uniformity across studies and the fact that there are few old animals to study for whom exact birth year is known. What constitutes an "aged" or "very old" nonhuman primate? I suggest that the first year of the last trimester of life be the threshold for classifying an animal as "aged". The timing will likely vary across taxa, based on the average maximum life span of known individuals. The sifakas observed by Wright and her co-workers live to approximately age 27 years (Wright, 1999; King et al., 2005). These sifakas would enter old age in their native habitat at 18 years – the same age threshold I used in this study of captive lemurs. The data in this chapter may be confounded by other variables – animals not always being housed in species-typical social groups or by the fact that they are free from the daily demands of foraging travels and the dangers of predation.

There appears to be some suggestion that older lemurs in captivity can be distinguished behaviorally from younger conspecifics. Lack of reproduction in later years appears to be a function of management or poor health. Lemurs will reproduce at species-appropriate intervals, even including the year of their death. Female success in rearing an infant to weaning or puberty appears to be more closely correlated with dental attrition than any endogenous hormonal influence (e.g., Parga and Lessnau, 2005; King et al., 2005). Future studies, of known individuals with well-documented life spans, will shed new light on the questions of life-long sociality and reproduction in lemurs.

Acknowledgment I thank the conference organizers for all their labors in celebrating the life and work of Elwyn Simons in fine style. A special debt is owed to Dr. Simons and the Mysterious Dr. Axle for adopting me into their Duke Troop. My thanks are also due to

Bob Sussman, Pat Wright, Jorg Ganzhorn, Ken Glander, Andrea Katz, Mike Stuart, Kay Izard, Dean Gibson, Carol Holman, Brian Grossi, and all of the staff at DUPC who care of a most special and unique lemur population. My thanks to the Editor for his patience, as well as two reviewers for their improvements to the manuscript. This research was supported in part by a James W. McLamore Summer Award in Business and Social Sciences and a General Research Support Award, both from the University of Miami. This is DLC publication #1029.

References

Achenbaum, W. A., and Bengston, V. L. (1994). Re-engaging the disengagement theory of aging: On the history and assessment of theory development in gerontology. *Gerontologist* 34:756–763.

Aujard, F., and Perret, M. (1998). Age-related effects of reproductive function and sexual competition in the male prosimian primate, *Microcebus murinus*. *Physiol. Behav.* 64(4):513–519.

Aujard, F., and Némoz-Bertholet, F. (2004). Response to urinary volatiles and chemosensory function decline with age in a prosimian primate. *Physiol. Behav.* 81(4):639–644.

Borries, C. (1988). Patterns of grandmaternal behavior in free-ranging Hanuman langurs (*Presbytis entellus*). *Human Evolution* 3:239–260.

Caro, T. M., Sellen, D. W., Parish, A., Frank, R., Brown, D. M., Voland, E., and Borgerhoff, M. M. (1995). Termination of reproduction in nonhuman and human female primates. *Int. J. Primatol.* 16(2):205–220.

Coleman, R. J., and Binkley, N. (2002). Skeletal aging in macaque monkeys. In: Erwin, J. M., and Hof, P. R. (eds.), *Aging in Nonhuman Primates. Interdisciplinary Topics in Gerontology, Volume 31*. Karger, Basel, pp. 32–47.

Corr, J. (2000). *The Effects of Aging on Social Behavior in Male and Female Rhesus Macaques of Cayo Santiago*. PhD Dissertation, Ohio State University, Columbus.

Corr, J. A., Martin, L. J., and Boysen, S. T. (2002). Comparative Models of Cognitive Decline in aging great apes. In: Erwin, J. M., and Hof, P. R. (eds.), *Aging in Nonhuman Primates. Interdisciplinary Topics in Gerontology, Volume 31*. Karger, Basel, pp. 196–208.

DeRousseau, C. J. (ed.). (1988). *Osteoarthritis in Rhesus Monkeys and Gibbons: A Model of Joint Degeneration. Contributions to Primatology, Volume 25*. Karger, Basel.

DeRousseau, C. J. (1994). Primate gerontology: An emerging discipline. In: Crews, D. E., and Garruto, R. M. (eds.), *Biological Anthropology and Aging: Perspectives on Human Variation Over the Life Span*. Oxford University Press, New York, pp. 127–153.

Erwin, J. M., Hof, P. R., Ely, R. J., and Perl, D. P. (2002). One gerontology: Advancing understanding of aging through studies of great apes and other primates. In: Erwin, J. M., and Hof, P. R. (eds.), *Aging in Nonhuman Primates. Interdisciplinary Topics in Gerontology, Volume 31*. Karger, Basel, pp. 1–21.

Gilissen, E. P., Dhenian, M., and Allman, J. M. (2001). Brain aging in strepsirhine primates. In: Hof, P. R., and Mobbs, C. V. (eds.), *Functional Neurobiology of Aging*. Academic Press, San Diego, pp. 421–434.

Gould, L. (1990). The social development of free-ranging infant *Lemur catta* at Berenty Reserve, Madagascar. *Int. J. Primatol.* 11:297–317.

Gould, L. (1999). How female dominance and reproductive seasonality affect the social lives of adult male ringtailed lemurs. In: Dolinhow, P., and Fuentes, A. (eds.), *The Nonhuman Primates*. Mayfield Press, Mountain View, pp. 133–139.

Gould, L., Sussman, R. W., and Sauther, M. L. (2003). Demographic and life-history patterns in a population of ring-tailed lemurs (*Lemur catta*) at Beza Mahafaly Reserve, Madagascar: A 15-year perspective. *Am. J. Phys. Anthropol.* 120(2):182–194.

Hauser, M. D., and Tyrrell, G. (1984). Old age and its manifestations: A study on two species of macaque. *Folia Primatol.* 43:24–35.
Huffman, M. A. (1990). Some socio-behavioral manifestations of old age. In: Nishida, T. (ed.), *The Chimpanzees of the Mahale Mountains: Sexual and Life History Strategies*. University of Tokyo Press, Tokyo, pp. 237–255.
iPAD (2005). The Internet Primate Aging Database. http://ipad.primate.wise.edu. The National Institute on Aging (NIA) and the Wisconsin National Primate Research Center.
Johnson, R. L., and Kapsalis, E. (1998). Menopause in free-ranging macaques: Estimated incidence, relation to body condition, and adaptive significance. *Int. J.Primatol.* 4:751–765.
Jolly, A. (1966). *Lemur Behavior*. University of Chicago Press, Chicago.
Jolly, A. (1984). The puzzle of female feeding priority. In: Small, M. (ed.), *Female Primates: Studies by Women Primatologists*. Alan R. Liss, New York, pp. 197–215.
Kappeler, P. M., and Ganzhorn, J. U. (eds.) (1993). *Lemur Social Systems and their Ecological Basis*. Plenum Press, New York.
Kato, E. (1999). Effects of age, dominance, and seasonal changes on proximity relationships in female Japanese macaques in a free-ranging group at Katsuyaya. *Primates* 40:291–300.
King, S. J., Arrigo-Nelson, S. J., Pochron, S. T., Semprebon, G. M., Godfrey, L. R., Wright, P. C., and Jernvall, J. (2005). Dental senescence in a long-lived primate links infant survival to rainfall. *Proc. Natl. Acad. Sci.* 102(46):16579–16583.
Macedonia, J. M. (1986). Individuality in a contact call of the ringtailed lemur (*Lemur catta*). *Am. J.Primatol.* 11:163–179.
Macedonia, J. M. (1993). The vocal repertoire of the ringtailed lemur (*Lemur catta*). *Folia Primatol.* 61:186–217.
Maxim, P. E. (1979). Social behavior. In: Bowden, D. M. (ed.), *Aging in Nonhuman Primates*. Van Nostrand Reinhold, New York, pp. 56–70.
Morbeck, M. E., Galloway, A., and Sumner, D. R. (2002). Getting old at Gombe: Skeletal aging in wild-ranging chimpanzees. In: Erwin, J. M., and Hof, P. R. (eds.), *Aging in Nonhuman Primates. Interdisciplinary Topics in Gerontology, Volume 31*. Karger, Basel, pp. 48–62.
Nakamichi, N. (1984). Behavioral characteristics of old female Japanese monkeys in a free-ranging group. *Primates* 25:192–201.
Nakamichi, N. (1991). Behavior of old females: Comparison of Japanese monkeys in the Arashiyama East and West groups. In: Fedigan, L., and Asquith, P. (eds.), *The Monkeys of Arashiyama: Thirty-Five Years of Research in Japan and the West*. State University of New York, Albany, pp. 175–193.
Oda, R. (2002). Individual distinctiveness of the contact calls of ring-tailed lemurs. *Folia Primatol.* 73:132–136.
Overdorff, D. J., and Strait, S. G. (1995). Life history and predation in *Eulemur rubriventer* in Madagascar. *Am. J.Primatol.* 100:487–506.
Parga, J. A., and Lessnau, R. G. (2005). Female age-specific reproductive rates, birth seasonality, and infant mortality of ring-tailed lemurs on St. Catherine's Island: 17-year reproductive history of a free-ranging colony. *Zoo Biol.* 24(4):295–309.
Pasin, M. S., Taylor, L. L., and Leon, B. S. (1998). Age and locomotion in captive Coquerel's sifakas (*Propithecus verreauxi coquereli*). *Am. J.Primatol.* 45(2):199.
Pasin, M. S., and Taylor, L. L. (2000). Intraspecific variation in locomotor and postural behavior among three captive species of lemur (*Propithecus verreauxi coquereli, Varecia variegata variegata*, and *Eulemur coronatus*). *Am. J. Phys.Anthropol.* Suppl. 30:246.
Pavelka, M. S. M. (1991). Sociability in old female Japanese monkeys: Human versus nonhuman primate aging. *Am. Anthropol.* 93:588–598.
Pavelka, M. S. M. (1993). *Monkeys of the Mesquite: The Social Life of the South Texas Snow Monkey*. Kendall-Hunt Publishing, Dubuque.

Pavelka, M. S. M. (1999). Primate gerontology. In: Dolhinow, P., and Fuentes, A. (eds.), *The Nonhuman Primates*. Mayfield Publishing, Mountain View, pp. 220–224.
Pavelka, M. S. M., and Fedigan, L. M. (1991). Menopause: A comparative life history perspective. *Yearb. Phys. Anthropol.* 34:13–38.
Pavelka, M. S. M., and Fedigan, L. M. (1999). Reproductive termination in Japanese monkeys: A comparative life history perspective. *Am. J. Phys. Anthropol.* 109:455–464.
Pavelka, M. S. M., Fedigan, L. M., and Zohar, S. (2002). Availability and adaptive value of reproductive and postreproductive Japanese macaque mothers and grandmothers. *Anim. Behav.* 64:407–414.
Pereira, M. E. (1995). Development of social dominance among group-living primates. *Am. J. Primatol.* 37:143–175.
Pereira, M. E., Kaufman, R., Kappeler, P. M., and Overdorff, D. J. (1990). Female dominance does characterise all of the Lemuridae. *Folia Primatol.* 55:96–103.
Pereira, M. E., and McGlynn, C. (1997). Special relationships instead of female dominance for redfronted lemurs, *Eulemur fulvus rufus*. *Am. J. Primatol.* 43:239–258.
Picq, J. (1992). Aging social behavior in captivity in *Microcebus murinus*. *Folia Primatol.* 59:217–220.
Pochron, S. T., and Wright, P. C. (2003). The effect of age on intra- and intersexual aggression in a wild Malagasy primate: The Milne-Edward's sifaka (*Propithecus diadema edwardsi*). *Am. J. Primatol.* Suppl. 1 60:142.
Pochron, S. T., Fitzgerald, J., Gilbert, C. C., Lawrence, D., Grgas, M., Rakotonirina,G., Ratsimbazafy, R., Rakotosoa, R., and Wright, P. C. (2003). Patterns of female dominance in *Propithecus diadema edwardsi* in Ranomafana National Park. *Am. J. Primatol.* 61(4):173–185.
Richard, A. F. (1987). Malagasy prosimians: Female dominance. In: Smuts, B. B., Cheney, D. L., Seyfarth, R. M., Wrangham, R. W., and Struhsaker, T. T. (eds.), *Primate Societies*. University of Chicago Press, Chicago, pp. 25–33.
Richard, A. F., Dewar, R. E., Schwartz, M., and Ratsirarsoh, J. (2002). Life in the slow lane? Demography and life histories of male and female sifaka (*Propithecus verreauxi verreauxi*). *J. Zool.* 356:421–436.
Roberts, J. A. (2002). The aged rhesus macaque in neuroscience research: Importance of the nonhuman primate model. In: Erwin, J. M., and Hof, P. R. (eds.), *Aging in Nonhuman Primates. Interdisciplinary Topics in Gerontology, Volume 31*. Karger, Basel, pp. 155–177.
Sauther, M. L. (1993). Resource competition in wild populations of ringtailed lemurs (*Lemur catta*): Implications for female dominance. In: Kappeler, P. M., and Ganzhorn, J. U. (eds.), *Lemur Social Systems and their Ecological Basis*. Plenum Press, New York, pp. 135–152.
Schaaf, C. D., and Stuart, M. D. (1983). Reproduction of mongoose lemurs (*Lemur mongoz*) in captivity. *Zoo Biol.* 2:23–28.
Sussman, R. W. (1992). Male life histories and inter-group mobility among ringtailed lemurs (*Lemur catta*). *Int. J. Primatol.* 13:395–413.
Tarou, L. R., Bloomsmith, M. A., Hoff, M. P., Erwin, J. M., and Maple, T. L. (2002). The behavior of aged great apes. In: Erwin, J. M., and Hof, P. R. (eds.), *Aging in Nonhuman Primates. Interdisciplinary Topics in Gerontology, Volume 31*. Karger, Basel, pp. 209–321.
Taylor, L. L. (1986). *Kinship, Dominance and Social Organization in a Semi-Free Ranging Group of Ringtailed Lemurs (Lemur catta)*. PhD. Dissertation, Washington University, St. Louis.
Taylor, L. L. (1995). The Importance of Kinship in the Management of Ringtailed Lemurs (*Lemur catta*). *Regional Conference Proceedings of the American Association of Zoological Parks and Aquariums*, pp. 467–472.
Taylor, L. L. (1998a). Behavior and reproduction in aged lemurs. *Am. J. Phys. Anthropol.* Suppl. 26:217.

Taylor, L. L. (1998b). Promoting species typical behavior in Coquerel's sifakas (*Propithecus verreauxi coquereli*). *Regional Conference Proceedings of the Association of Zoos and Aquariums*, pp. 599–603.
Taylor, L. L. (1998c). Social isolation and aged lemurs. *Am. J. Primatol.* 45(2):210–211.
Taylor, L. L. (2000). Social behavior of aged lemurs. *Am. J. Phys. Anthropol.* Suppl. 30:300.
Taylor, L. L., and Pasin, M. S. (1999). Age, locomotion and positional behavior in Coquerel's sifakas (*Propithecus verreauxi coquereli*). *Am. J. Primatol.* Suppl. 28:263.
Taylor, L. L., and Sussman, R. W. (1985). Grooming matrices and social structure in *Lemur catta*. *Int. J. Primatol.* 6:601–614
Veenema, H. C., Spruijt, B. M., Gispen, W. H., van Hooff, J. A. R. A. M. (1997). Aging, dominance history, and social behavior in Java-monkeys (*Macaca fuscata*). *Neurobiol. Aging* 18:509–515.
Walker, M. (1995). Menopause in female rhesus monkeys. *Am. J. Primatol.* 35:59–72.
Wright, P. C. (1995). Demography and life history of free-ranging *Propithecus diadema edwardsi* in Ranomafana National Park. *Int. J. Primatol.* 16(5):835–854.
Wright, P. C. (2008). Decades of lemur research and conservation: The Elwyn L. Simons Influence. In: Fleagle, J. G. and Glibert, C. C. (eds.), *Elwyn Simons: A Search for Origins*. Springer, New York, pp. 283–310.
Wright, P. C. (1999). Lemur traits and Madagascan ecology: Coping with an island environment. *Yearb. Phys. Anthropol.* 42:31–72.
Zimmerman, E. (1991). Ontogeny of acoustic communication in prosimian primates. In: Kimura, T., Takenada, O., and Iwamoto, M. (eds.), *Primatology Today: Proceedings of the XIIIth Congress of the International Primatological Society, Nagoya and Kyoto*. Elsevier, Amsterdam, pp. 337–340.

Peculiar Tooth Homologies of the Greater Bamboo Lemur (*Prolemur* = *Hapalemur simus*)

When is a Paracone Not a Paracone?

Jukka Jernvall, Christopher C. Gilbert and Patricia C. Wright

Introduction

Among living primates, bamboo lemurs provide a rare case of adaptation to feeding on fibrous vegetation. Bamboo makes up a significant portion of the diet of all three living species, *Hapalemur griseus*, *H. aureus*, and *H. simus* (now also known as *Prolemur simus*, Groves, 2001). The latter, the greater bamboo lemur, has perhaps the most monotonous plant diet of all primates. Long term studies on wild greater bamboo lemurs (Tan, 1999) have shown that 95% of its diet is composed of one species of bamboo (*Cathyiostachyum madagascariensis*). *H. simus* is also the only species that routinely breaks open and eats the culms (the bamboo's trunk) of mature bamboo. Even the remaining portion of the diet is made mostly of other bamboo species, with fruit and other foods (mostly soil and mushrooms) making 2% of the diet.

Bamboos, which are grasses belonging to the family Poaceae, are highly fibrous. Bamboo culms contain close to 50% cellulose with the remaining material made mostly of lignin and polyose (hemicellulose) (Fengel and Shao, 1984). Additionally, bamboo contains silica bodies (or phytoliths). Taken together, these factors make bamboo a challenging food to consume and digest. It is then perhaps to be expected that the challenge of bamboo feeding should be reflected in the dentition of bamboo lemurs. Indeed, H. O. Forbes in the Allen's Naturalist's Library (1894) described the *Hapalemur* dentition to be "peculiar and characteristic". While several "peculiarities" have been documented in the bamboo lemur dentition (Tattersall, 1982), in this paper we will focus on one, namely the molarization of premolars.

Jukka Jernvall
Developmental Biology Program, Institute of Biotechnology, University of Helsinki
PO Box 56, FIN-00014, Helsinki, Finland
jernvall@fastmail.fm

J. G. Fleagle, C. C. Gilbert (eds.), *Elwyn Simons: A Search for Origins.*

Molarization of Premolars but how?

Several mammalian groups have species with molariform premolars. Of the living species, elaborate molariform premolars are found in species with a fibrous, plant-based diet. For example, horses and hyraxes, which rely heavily on grass, have molars and premolars which form a single cheek-tooth battery. Based on their namesake diet, bamboo lemurs provide further evidence of the relationship between a dietary reliance on fibrous plant material and molarized premolars. Furthermore, as might be expected based on the extreme specialization to bamboo feeding, *H. simus* shows the greatest degree of premolar molarization among the bamboo lemurs (Tattersall, 1982).

In contrast to other Lemuriformes (Ankel-Simons, 2007), all species of bamboo lemur have molariform features on their posterior, upper fourth, premolars (P^4) (see Fig. 1). Molariform features, in this context, refer to the addition of buccal and lingual cusps on the premolars as well as the position of these cusps. In general, lemur taxa have tribosphenic molars with three cusps: the paracone, protocone, and metacone (e.g., see Swindler, 2002). Some lemur taxa, such as *Hapalemur*, add a fourth cusp, the hypocone, as well. The more closely a premolar mimics the number and position of these cusps as typically observed on a molar, the higher the degree of molarization.

In most lemur taxa, the anterior upper premolars (P^2–P^3) are marked by only one prominent cusp, the buccally-positioned paracone (Swindler, 2002). The upper third premolar (P^3) of *H. simus*, however, has a well-developed lingual cusp, the protocone, making it more molariform than the P^3 of any other lemur taxon. Additionally, the prominent buccal cusp (i.e., the paracone) of the P^3 is flanked by shorter mesial and distal cusps (Fig. 1).

While lemur species are variable in P^4 morphology, this tooth typically displays two cusps, a prominent paracone and a smaller protocone (Swindler, 2002). In contrast, the P^4 of *H. simus* has a squared-off occlusal shape and appears almost completely molariform. In addition to a paracone and protocone, the P^4 of *H. simus* has also developed a second buccal cusp, the metacone. In many individuals one can even detect incipient development of a second lingual cusp, the hypocone, from the postprotocingulum (Fig. 1).

Whereas the ecological context of premolar molarization appears to be diet, the way premolars acquire molar shapes is not necessarily clear. One of the best studied cases is the molarization of premolars in a lineage of Eocene perissodactyls. Beginning in *Hyracotherium*, the Eocene "dawn horse", the premolars began to expand lingually and attain molariform features in this lineage (Granger, 1908; Butler, 1952; Van Valen, 1982). On the P^4, the protocone has been reconstructed to have evolved first, followed by the hypocone distal to the protocone. This mode of cusp evolution is to be expected because it follows the ancestral mode of lingual expansion of molars themselves. Futhermore, the identity of cusps and their names are based on their relative location to other cusps and crown features. Regardless of the shape, the single or mesial lingual

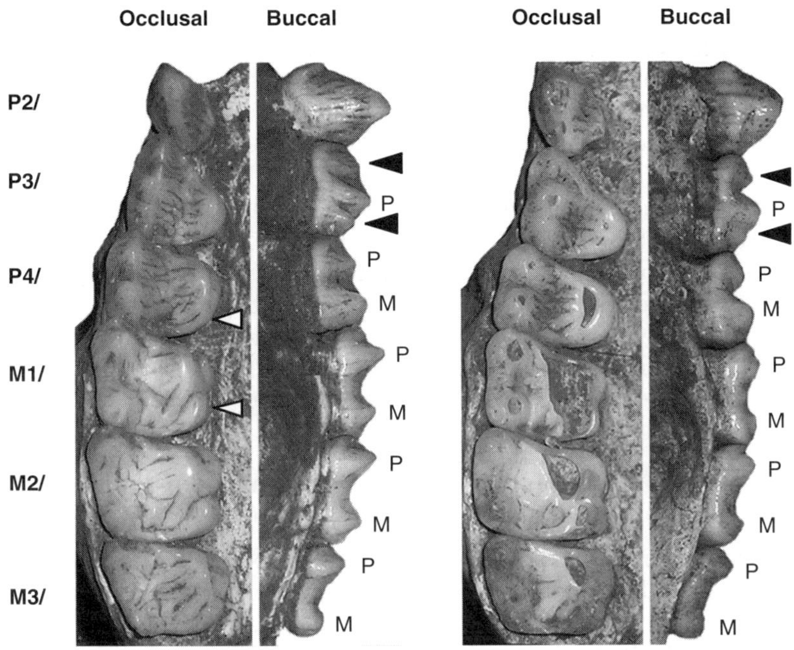

Fig. 1 Two examples of the greater bamboo lemur postcanine tooth rows. The buccal views have the paracone (P) and the metacone (M) cusps marked. The open arrow heads mark the postprotocingulum on P4/ and the hypocone on M1/. The closed arrow heads mark the two accessory cusps mesial and distal to the paracone on P3/ (see text for details). Note the presence of the lingual cusp (the protocone) on the premolars of both specimens. Also note the presence of the homologous trough between the paracone and the mesial accessory cusp on P3/ and the metacone and paracone on P4/ in the specimen on the right (buccal view). DPC 7939 on the left, DPC 7854 on the right

cusp is called the protocone, but this is also where the *Hyracotherium* P^3 has turned out to be interesting. When the P^3 of *Hyracotherium* evolved a new lingual cusp, this cusp appeared not distal, but mesial to the protocone. Nevertheless, both P^4 and P^3 eventually evolved similar molariform morphologies in perissodactyls. In this instance, the molarization of P^3 happened by the protocone shifting distally and changing its "identity" to the hypocone whereas the new, mesial lingual cusp, by definition, became the protocone.

This shift in cusp position during molarization between P^3 and P^4 was first noted by Granger (1908) and discussed in connection to cusp homology by Van Valen (1982; 1994). The problem is that whereas we could label the single lingual cusp of the *Hyracotherium* P^3 as the "hypocone", and thus solve the shift in cusp identity, this is only possible by the benefit of hindsight. That is, if the P^3 molarization had stopped at the single lingual cusp, we would be content in calling this cusp the protocone.

Why knowing about shifts in cusp homology may matter is in the connection to phylogenetic studies in which one relies on comparison of homologous characters across species. As the molarization of *Hyracotherium* premolars suggests, in certain situations the historical continuity of cusp homologies may break, a view that gains further support from developmental biology studies. Evidence of mammalian tooth development indicates that molar cusps may appear during ontogeny as an iterative replay of the same set of developmental regulatory genes (Salazar-Ciudad and Jernvall, 2002). Experimental work on mice has shown that decreasing or increasing the production of a single signaling molecule required for cusp formation causes many of the existing cusps to fuse or new cusps to appear, respectively (Kangas et al., 2004). Thus there might not be a unique genetic code for the protocone, for example, but cusps evolve as a result of changes in the overall interplay between molecular signaling and tissue growth (Salazar-Ciudad and Jernvall, 2002; Kangas et al., 2004).

Molarization of Premolars and the Discontinuity of Homology

In contrast to the molarization of premolars in the extinct evolutionary lineage of Eocene perissodactyls, the premolars of the greater bamboo lemur provide a living example of what may happen to cusp homologies during molarization. Theoretically, studying an extant taxon allows for a better appraisal of morphological variation as well as the opportunity to study developmental genetic correlates in a laboratory setting. Unfortunately, extant *H. simus* is too rare and endangered to study in the laboratory and only one social group is habituated enough to obtain dental measurements for the study of subtle morphological variation. However, the field work of Elwyn Simons has provided us with an extensive sample of recently extinct *H. simus* populations. His work demonstrates that *H. simus* was widespread and probably present throughout Madagascar, except in the south (Godfrey et al., 2004). One specimen has been dated to 4560±70 BP (Simons et al., 1995) and, at least from a paleontological perspective, it is likely that all the studied specimens are drawn from a relatively short time interval. We therefore focused on a sample of *H. simus* specimens from Ankarana Massif, where a large number of individuals have been recovered. By examining correlated morphological variation within a large sample, we can record the pattern of cusp change during the premolar molarization process.

Because *H. simus* shows a slightly lesser degree of molarization than the Eocene perissodactyls, we focused on the variation in buccal cusps. From a sample of 15 relatively unworn specimens, we measured the distance between buccal cusps of P^3, P^4, and M^1. As a proxy for molarization of the P^4, we measured the distance between the mesial and distal cusps, the paracone and the metacone. The distances between these cusps in our sample of P^4s was then compared to the distances between these cusps in our sample of corresponding M^1s. The closer a distance measured on P^4 approximates the same distance measured on a corresponding M^1, the higher the degree of premolar molarization. The values seen within P^4 range from 65 to 80% of the corresponding distance in M^1, showing that in some individuals the P^4 is indeed very close to the M^1 in cusp spacing.

Next we tested whether the increased molarization of P^4 is associated with a corresponding increase in the molarization of P^3. The third premolar has one buccal cusp, the paracone, often with one mesial and one distal accessory cusp. We measured whether the relative P^4 distance between paracone and metacone correlates with the relative distance between the P^3 paracone and its distal cusp or its mesial cusp (Fig. 2). The former correlation would indicate that increased molarization of P^4 results in increased molarization of P^3 by distal extension, and the formation of the metacone in the normal distal position. The latter option would show that the increasing P^3 molarization is obtained by further development of the mesial part of the tooth and by formation of a new mesial cusp. Hence, in this latter case a new paracone is formed and the "original" paracone becomes the metacone (Fig. 2). These hypotheses assume that the molarization affects all the premolars but at a decreasing strength in the anterior teeth. Developmentally, this kind of shift in cusp homology may stem from development affecting the overall high-level complexity of teeth rather than individual cusp positions (Evans et al., 2007).

The results demonstrate that the P^3 paracone to distal cusp distance shows no correlation with the P^4 molarization ($r_s = -0.31, p = 0.252$). In contrast, the P^3 paracone to mesial cusp distance increases with increasing P^4 molarization ($r_s = 0.70$, $p = 0.009$, Fig. 3). This indicates that, developmentally, the paracone-metacone pair of the P^4 are homologous to the mesial cusp-paracone pair of the P^3. Thus, in *H. simus*, the P^3 paracone is in the process of becoming the metacone by shifting distally. This shift represents a break in homology.

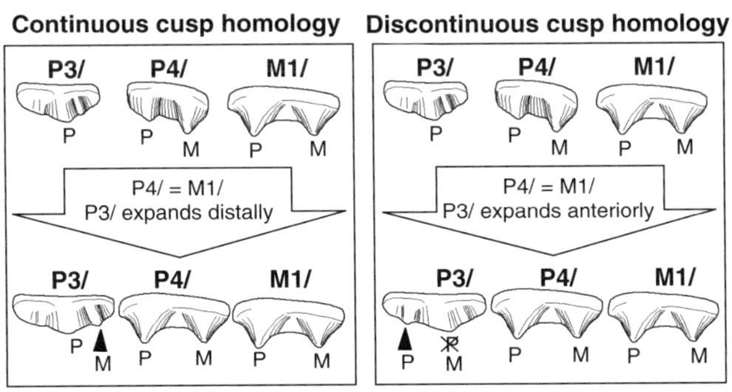

Fig. 2 Two possibilities for the molarization of P3/ in *H. simus*. Illustrations show buccal views with the paracone (P) and the metacone (M) cusps marked. On the left the P4/ becomes fully molarized and the posterior part of P3/ expands giving rise to the metacone. On the right the P4/ becomes fully molarized but the mesial portion of P3/ expands giving rise to a new paracone whereas the old paracone becomes the metacone, braking the continuity of historical homology

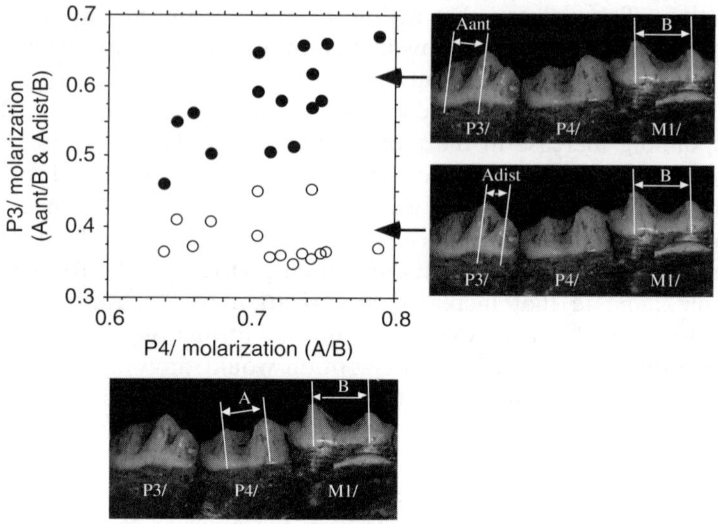

Fig. 3 The relative molarization of P4/ and P3/. As the paracone-metacone distance (A) of P4/ approaches that of M1/ (B), the relative P3/ paracone-anterior accessory cusps distance (Aant) increases (see dark circles, $r_s = 0.70$, $p = 0.009$). In contrast, the relative P3/ paracone-distal accessory cusps distance (Adist) shows no molarization effect (see open circles, $r_s = -0.31$, $p = 0.252$). These correlation analyses indicate that the anterior accessory cusp on P3/ is homologous to the paracone on P4/. Measurements were taken from digital buccal views (DPC 6755, 6811, 7854, 7938, 7939, 10942, 10956a, 10956b, 10978a, 10978b, 10978c, 10980, 10983, 11714, 20705)

The interpretation presented here is further supported by one *H. simus* specimen in which a well developed trough between the P^4 paracone and the metacone is paired with a corresponding trough between the P^3 mesial cusp and the paracone (Fig. 1, specimen on the right). It is also quite possible that the shift in homology of cusps is not limited to P^3 because in every studied case on P^4, the P^4 paracone was shorter than the P^4 metacone. This is noteworthy because in mammals relative cusp height correlates with the initiation of cusp development (Berkovitz, 1967, Kangas et al., 2004) and thus the P^4 metacone may well have been the paracone in the ancestor of *H. simus*. This prediction could be testable with a Tertiary fossil record, which is, unfortunately, currently unavailable in Madagascar (see Tattersall, 2008).

Conclusions

Our study indicates that the molarization of upper premolars in *H. simus* takes place through a process in which the paracone is shifted distally to become the metacone and a new mesial buccal cusp (paracone) is developed. This process represents a discontinuity in cusp homology. Although we often assume the development of cusps follows a standard sequence, it is clear from Eocene perissodactyls and *H. simus* that cusp homologies are not always as they appear. The paracone of one tooth may become the metacone of another tooth. Similarly, the protocone of one tooth may become the hypocone of another tooth. These changes can occur within species, such as *H. simus*, and between species as documented by the Eocene perissodactyl lineage. Researchers must be cautious in assuming cusp homologies without studies of cusp development, whether through experimental lab work or through morphological observations such as those presented here.

The available material from *H. simus* also shows that there is subtle but detectable variation in the molarization of its premolars. Without further study of the temporal range of the specimens, it is not possible to test whether this variation may indicate differences in selective regimes through time and microhabitat. However, the effect of habitat might be testable in *H. griseus*, the smallest and widest ranging of the extant bamboo lemur species. Our observation regarding the discontinuity in homology of buccal cusps is a further indication of the dynamic nature of tooth development and evolution. This, together with the high degree of premolar molarization in *H. simus*, underscores the evolutionary uniqueness of this taxon and adds yet another reason to protect the species from extinction (Fig. 4). Finally, this study was largely made possible by the work of Elwyn Simons in Madagascar and his continued emphasis on studying both fossil and living primates.

Fig. 4 The extant greater bamboo lemur, *Hapalemur simus*. Photo by J. Jernvall

Acknowledgment We thank Prithijit Chatrath and Elwyn Simons for access to the subfossil *H. simus* material at the Division of Fossil Primates. We thank Elwyn Simons for his tireless fieldwork and inspiration. We also thank John Fleagle for his patience and helpful comments on an earlier version of this manuscript. This is DLC publication #1091.

References

Ankel-Simons, F. (2007). *Primate Anatomy*. Elsevier, Academic Press, Burlington, MA.

Berkovitz, B. K. B. (1967). The dentition of a 25-day pouch-young specimen of *Didelphis virginiana* (Didelphidae: Marsupialia). *Arch. Oral Biol.* 12: 1211–1212.

Butler, P. M. (1952). Molarization of the premolars in the Perissodactyla. *Proc. Zool. Soc. London* 121: 819–843.

Evans, A. R., Wilson, G. P., Fortelius, M., and Jernvall, J. (2007). High-level similarity of dentitions in carnivorans and rodents. *Nature* 445: 78–81.

Fengel, D., and Shao, X. (1984). A chemical and ultrastructural study of the Bamboo species *Phyllostachys makinoi*. *Wood Sci. Technol.* 18: 103–112.

Forbes, H. O. (1894). *A Handbook to the Primates: Allen's Naturalist's Library*. W. H. Allen & Co, London.

Godfrey, L. R., Simons, E. L., Jungers, W. L., De Blieux, D. D., and Chatrath, P. S. (2004). New discovery of subfossil *Hapalemur simus*, the greater bamboo lemur in western Madagascar. *Lemur News* 9: 9–11.

Granger, W. (1908). A revision of the American Eocene horses. *B. Am. Mus. Nat. Hist.* 24: 221–264.

Groves, C. P. (2001). *Primate Taxonomy*. Smithsonian Institution Press, Washington, DC.

Kangas, A. T., Evans, A. R., Thesleff, I., and Jernvall, J. (2004). Nonindependence of mammalian dental characters. *Nature* 432: 211–214.

Salazar-Ciudad, I., and Jernvall, J. (2002). A gene network model accounting for development and evolution of mammalian teeth. *P. Natl Acad. Sci. USA* 99: 8116–8120.

Simons, E. L., Burney, D. A., Chatrath, P. S., Godfrey, L. R., Jungers, W. L., and Rakotosamimanana, B. (1995). AMS C-14 dates for extinct lemurs from caves in the Ankarana Massif, northern Madagascar. *Quaternary Res.* 43: 249–254.

Swindler, D. R. (2002). *Primate Dentition: An Introduction to the Teeth of Non-Human Primates*. Cambridge University Press, Cambridge.

Tan, C. L. (1999). Group composition, home range size, and diet of three sympatric bamboo lemur species (genus *Hapalemur*) in Ranomafana National Park, Madagascar. *Int. J. Primatol.* 20: 547–566.

Tattersall, I. (1982). *The Primates of Madagascar*. Columbia University Press, New York.

Tattersall, I. (2008). Vicariance vs. dispersal in the origin of the Malagasy mammal fauna. In: Fleagle, J. G. and Gilbert, C. C. (eds.), *Elwyn Simons:* A Search for Origins. Springer, New York, pp. 397-408.

Van Valen, L. (1982). Homology and causes. *J. Morphol.* 173: 305–312.

Van Valen, L. (1994). Serial homology: The crests and cusps of mammalian teeth. *Acta Palaeontol. Pol.* 38: 145–158.

How Big were the "Giant" Extinct Lemurs of Madagascar?

William L. Jungers, Brigitte Demes and Laurie R. Godfrey

Introduction

The primate community in Madagascar has changed dramatically since people arrived just over two millennia ago (Burney et al., 2004). At least nine genera and up to sixteen species were extinguished throughout the Holocene (Godfrey et al., 1997a; Godfrey and Jungers, 2002), and these "subfossil" lemurs all shared one distinguishing feature: they were much larger than their living counterparts, and some can be fairly characterized as "giants". Just how big they were depends on the choice of proxy for body mass and/or on the choice of predictor variables when regression estimates of mass are made.

The Megaladapidae (sometimes called "koala lemurs") consists of three extinct species, *Megaladapis edwardsi*, *M. grandidieri*, and *M. madagascariensis*. All three species were very large, even by anthropoid standards, but their crania appear to be "over-sized" for the rest of their bodies. The relationship of the megaladapids to living lemurs remains controversial. The Palaeopropithecidae (also known colloquially as "sloth lemurs") is the most speciose family of subfossil lemurs, with four genera and seven recognized species: *Archaeoindris fontoynontii*, *Palaleopropithecus maximus*, *P. ingens*, *Babakotia radofilai*, *Mesopropithecus dolichobrachion*, *M. pithecoides*, and *M. globiceps*. This group is ripe for taxonomic revision, but we defer that exercise for another place and time. Regardless, the palaeopropithecids include some very large-bodied taxa, and *Archaeoindris* is one of the largest primates to ever evolve anywhere in the world. Based on ancient DNA, morphology and development, it is now quite clear that the sloth lemurs are the sister group to living indriids (Godfrey and Jungers, 2003; Karanth et al., 2005). The Archaeolemuridae (dubbed the "monkey lemurs") includes two genera, *Archaeolemur* and *Hadropithecus*, with three currently recognized species, *A. edwardsi*, *A. majori*, and *H. stenognathus*. There is also a large unnamed taxon of *Archaeolemur* from the northwest and extreme northern tip of the island, and we believe it has close

---

William L. Jungers,
Department of Anatomical Sciences, School of Medicine, Stony Brook University, Stony Brook, NY 11794-8081

affinities to *A. edwardsi*. Precisely where the archaeolemurids fit in a phylogenetic sense with living lemurs has yet to be resolved unequivocally. *Daubentonia robusta*, as its name implies, is a much larger, more robust version of the living aye-aye, *D. madagascariensis*. *D. robusta* is known primarily from the southwest, where the living aye-aye was, and remains, conspicuously absent. Finally, *Pachylemur* (two species, *P. jullyi* and *P. insignis*) is a "giant" lemurid with especially close affinities to the living *Varecia*.

Qualitative, "ballpark" estimates of body size for subfossil lemurs are colorful and wide-ranging. *Megaladapis* has been compared to a calf (Lavauden, 1931; Kurten, 1971), a bear cub (Simons, 1972) and a Saint Bernard dog (Simons, 1972). Impressed by its very long skull (>30 cm for some *M. edwardsi*), Radinsky (1970) incorrectly opined that *Megaladapis* was the largest of the extinct lemurs. The poorly known *Archaeoindris fontoynontii* is said to have been larger than a human (Lavauden, 1931), perhaps even gorilla-sized (Lamberton, 1934; Vuillaume-Randriamanantena, 1988), and possibly the largest primate ever (Villette, 1911 – reporting on Lavauden's verbal comments at the Academie Malgache). The highly suspensory *Palaeopropithecus* has been compared in size to the common chimpanzee (Vuillaume-Randriamanantena, 1988), and images of baboons have been invoked repeatedly for the Archaeolemuridae (e.g., Jolly, 1970).

More specific estimates of body mass also exist in the literature. Burney et al., (1987) suggested that *Palaeopropithecus* was perhaps 60 kg, and Walker (2002) has recently offered a mass estimate of 79.5 kg for *M. edwardsi*. Jungers (1978) had earlier bracketed the masses of *M. edwardsi* and *M. grandidieri*, respectively, at 50–100 kg and 40–75 kg based on approximate skeletal trunk lengths. Jungers (1990) subsequently estimated body mass for five different species using both dental and postcranial joint dimensions; the differences between craniodental and postcranial predictions were marked in some instances. For example, *M. edwardsi* was reconstructed to weigh in as large as 156.5 kg based on maxillary first molar size (cf. Gingerich et al., 1982), but only tipped the scales at 67.5 kg based on femoral head diameter. Similarly, *M. madagascariensis* was estimated to weigh either 66.2 kg (molar area) or only 36.0 kg (femoral head diameter). *P. maximus* was estimated dentally at 110.4 kg (molar area) but only 47.4 kg postcranially (femoral head diameter). *A. fontoynontii* appeared huge in Jungers' study no matter which predictor variable was used: 230.5 kg (molar area) or 244.1 kg (femoral head diameter)! Relying on cranial dimensions, Martin (1990) estimated the body masses of *Archaeolemur edwardsi*, *M. edwardsi*, and *P. maximus*. Martin's average estimates ranged from a reasonable 26.0 kg for *A. edwardsi* to an incredible 390 kg for *M. edwardsi*; the *P. maximus* average was also quite high at 90 kg. Differences between Martin's estimates based on skull length and those he calculated from the size of the foramen magnum were uniformly enormous. Also using a regression-based methodology, but enlisting midshaft circumferences of the humerus and femur as the predictor variables, Godfrey et al., 1995 (also cf. Jungers et al., 2002) offered new body mass estimates for all recognized subfossil lemurs, and many of these published values differ quite dramatically from textbook estimates based on tooth size (Fleagle, 1988, after Gingerich et al., 1982). The three species of *Mesopropithecus* (9.4–10.6 kg), the smallest of the sloth lemurs, the two species of *Pachylemur* (10–12.8 kg), *Daubentonia robusta* (13.5 kg), and *A. majori* (13. 9 kg) were the smallest of the extinct lemurs in the Godfrey et al. analysis, but all of them still weighed much more than any of the living species on Madagascar today (Smith and Jungers, 1997). *Babakotia radofilai* was reconstructed in the Godfrey et al. study as slightly larger at 16.2 kg, but the remaining subfossil taxa were suggested to be ***much*** bigger: 24.0 kg for *A. edwardsi*, 27.1 kg for*Hadropithecus stenognathus*, 25.8–52.3 kg for the species of *Palaeopropithecus*, 38–75.4 kg for the three megaladapids, and a massive 197.5 kg for *A. fontoynontii*.

The diversity of values and opinions apparent in this brief review suggests that a systematic re-evaluation of body size in subfossil lemurs is warranted, especially if we want to evaluate other data in the context of comparative body size (e.g., life histories, brain size, body shape, community structure, etc.). Following the biomechanical rationale of Ruff (1990, 2003), we opt here to use postcranial predictor variables linked fundamentally to the support of an animal's body mass. Although the accuracy of predictive equations is ultimately an empirical determination, we are attracted to the logic that those structures that actually support body mass should be especially reliable and useful in our endeavor (Hylander, 1985; Jungers, 1990; Ruff, 2003). Our cross-sectional geometrical variables of choice are cortical areas (CA) of the humerus and femur at midshaft. The distribution of cortical bone in the femora of some subfossil lemurs is quite unusual; for example, the cortical shell of the femur in *Megaladapis* is very thin both proximally and distally, but there is a pronounced thickening of the cortex around midshaft. We assume that even in this extreme case, one can take the cortical bone in cross-section at midshaft to be "homologous" among species of mammals. We also note that all efforts to reconstruct body mass (or virtually anything else) in the fossil record require certain uniformitarian assumptions, and those based on the biomechanics of weight support by the locomotor sekeleton seem more reasonable to us than most others.

Raw body mass and geometrical data for our extant reference sample derive from a diverse sample of primates and other mammals published by Polk et al. (2000) and supplemented by us here (see below). We are happy to share this data base with anyone who requests it. Because long bones are relatively abundant for some of the extinct lemur taxa, we can also evaluate variation in body size estimates for the first time and thereby establish approximate "confidence intervals" to mean estimates (see below). Assessing this ***individual variation*** is one of the primary goals of this study, and our statistical approach is novel if perhaps less than orthodox. In order to better assess the impact of choosing among competing regression equations and their body mass estimates for subfossils, we also compare our results to predictions based on orbit height (Kappelman, 1996), area of the first mandibular molar (Gingerich et al., 1982) and diaphyseal circumferences (Godfrey et al., 1995; Jungers et al., 2002). We do not wish to imply that our estimates are perfect, nor should they be regarded necessarily as "gold standards." Rather, we believe that we have taken biological variation into account more fully than any prior analysis, and that we have done so consistently within a well validated and explicitly biomechanical framework.

Materials and Methods

The raw data for the subfossil lemurs derive from measurements taken on biplanar radiographs of 123 long bones distributed across nine genera and fifteen species (Table 1). Most taxa of interest are represented by both humeri and femora, with the exceptions of *Archaeoindris fontoynontii* (a single femur) and *Hadropithecus stenognathus* (several humeri only; Godfrey et al. (1997b) clarified prior hind-limb misattributions and identified other femora appertaining to *Hadropithecus*, but radiographs of this material in Paris and Antananarivo were not available for this study). Midshaft humeral cortical area (HCA) and midshaft femoral cortical area (FCA) are calculated using the asymmetrical or eccentric model (e.g., Ohman, 1993).

The expanded Polk et al. (2000) data base includes 93 nonprimate species, dominated by carnivores and rodents, to which we have added sloths and koalas. Sixty nonhuman primate species are also included, and some of the very sexually dimorphic taxa are separated by sex. Although subfossil lemurs are indisputably primates, aspects of their skeletal design are unusual for primates (Jungers et al., 2002), and it seems prudent therefore to consider other mammals as well. We utilize two reference data bases, one that consists

Table 1 Sample of biplanar radiographs for subfossil lemurs

(species, humeral n, femoral n, total n for that species)
Megaladapis edwardsi (6, 6, 12)
Megaladapis grandidieri (2, 5, 7)
Megaladapis madagascariensis (4, 6, 10)
Daubentonia robusta (3, 2, 5)
Pachylemur insignis (9, 9, 18)
Pachylemur jullyi (4, 4, 8)
Archaeolemur edwardsi (7, 8, 15)
Archaeolemur majori (2, 3, 5)
Hadropithecus stenognathus (3, 0, 3)
Archaeoindris fontoynontii (0, 1, 1)
Palaeopropithecus maximus (4, 3, 7)
Palaeopropithecus ingens (4, 3, 7)
Babakotia radofilai (4, 4, 8)
Mesopropithecus globiceps (5, 6, 11)
Mesopropithecus dolichobrachion (1, 1, 2)

of primates only and a second that includes a wider range of mammals (but which also includes the primate sample).

Using all mammals and primates only, a total of eight different least squares regressions of mass on cortical area were developed and applied to the subfossil sample in Table 1. Both raw and logged data were examined (Table 2), and all logged regresssion estimates (as well as the standard errors of the estimate, SEE) were adjusted for detransformation bias via the quasi-maximum likelihood estimator (Smith, 1993). The resulting eight correlations are similar and highly statistically significant. The standard errors of the estimates are also comparable, although femoral regressions do have consistently smaller standard errors than humeral regressions. All eight equations exhibit similar levels of significant residual autocorrelation. Accordingly, we believe that there is no rigorous, objective statistical criterion to prefer logged regressions over raw ones in our analyses, and we have elected to combine the results (see below). Figures 1 and 2 are the bivariate plots of body mass on femoral cortical area in log-log and raw data spaces for all mammals, respectively. Ninety-five percent confidence bands around the regressions are indicated. The log-log regression probably looks more familiar to most readers and does possess a certain aesthetic appeal, but as noted, the correlations, standard errors and overall good-of-fit are remarkably similar. We have elected not to provide "prediction intervals" for every single individual fossil (cf. Giles and Klepinger, 1988), although we do so for the one representative of *A. fontoynontii* in our sample (one femur) in order to illustrate the predictable result (also see Smith, 2005, for a sobering treatment of confidence intervals in paleoanthropology). In addition to our "mean of the means", we do provide the smallest and largest estimate obtained for any individual in a species, using all regression/sample possibilities. This alone

Table 2 Least squares regressions of body mass (in grams) on cortical area (in mm^2) used to predict body mass in subfossil lemurs

1. All mammal: mass in grams on humeral cortical area in mm^2
   - raw ($r = 0.952$; SEE = 5.0 kg) : $y = 263.8\,x - 2919$
   - log-log ($r = 0.973$; SEE = 4.4 kg): $\log y = 1.234 \log x + 1.749$
2. All mammal: mass in grams on femoral cortical area in mm^2
   - raw ($r = 0.976$; SEE = 3.4 kg): $y = 217.4\,x - 2891$
   - log-log ($r = 0.979$; SEE = 2.9 kg) : $\log y = 1.301 \log x + 1.542$
3. Primates only: mass in grams on humeral cortical area in mm^2
   - raw ($r = 0.954$; SEE = 6.6 kg): $y = 272.1\,x - 4233$
   - log-log ($r = 0.985$; SEE = 6.3 kg): $\log y = 1.255 \log x + 1.689$
4. Primates only: mass in grams on femoral cortical area in mm^2
   - raw ($r = 0.983$; SEE = 3.9 kg): $y = 221.2\,x - 4358$
   - log-log ($r = 0.987$; SEE = 3.1 kg): $\log y = 1.288 \log x + 1.499$

Note: All 8 regressions exhibit comparable degrees of positive serial autocorrelation of residuals (Durban-Watson statistic <1.6). Standard errors of the estimates (SEEs) are reported in kg; the log-based SEEs are based on detransformed data.

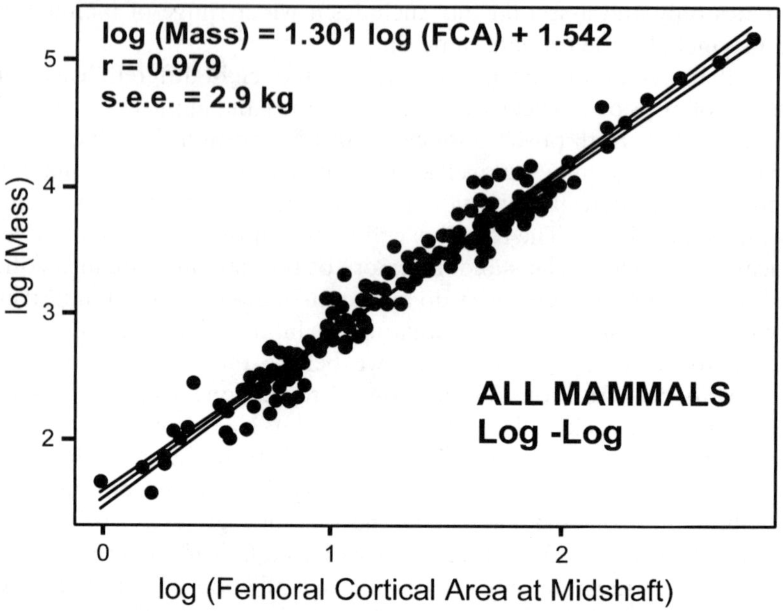

Fig. 1 Log-log regression of body mass (grams) on femoral cortical area (mm^2) in a diverse sample of mammals (including primates). The 95% confidence bands on the least-squares regression line are indicated

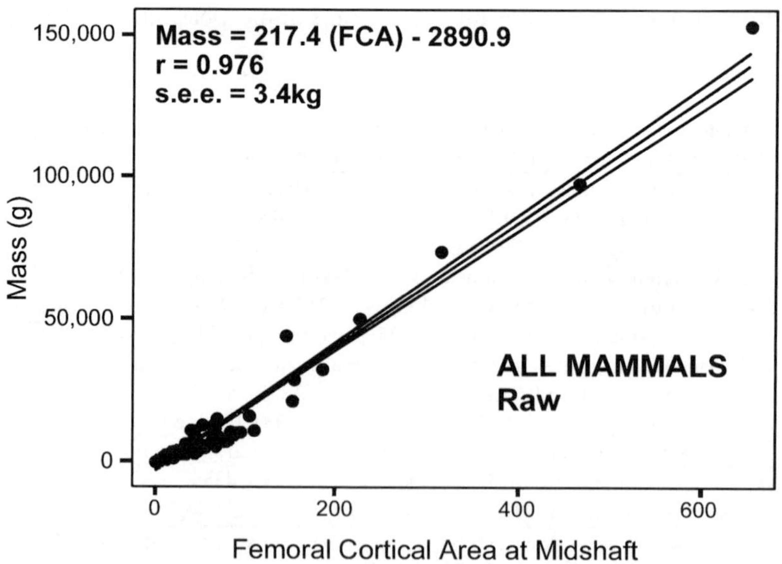

Fig. 2 Regression of body mass on femoral cortical area in raw data space for the same mammalian sample as in Fig. 1

results in surprisingly large ranges, and prediction intervals would obviously serve to expand these even further. Most workers are probably interested in the "best" estimates rather than the less likely extremes, no matter how constructed; we think our mean of means is currently our best prediction, but we also recommend that the use of our preferred values in other analyses and comparisons be qualified by the very substantial ranges that we provide for all species (also see Delson et al., 2000). As Gingerich (1996) has noted, broad intervals around an estimate do not mean it is either wrong or a poor prediction; they do, however, make it difficult to reject alternative estimates.

We are well aware of the development of comparative methods in regression analysis that incorporate information about phylogeny to (hopefully) better estimate regression and other statistical parameters (Felsenstein, 1985; Pagel, 1993; Rohlf, 2001). The goals and results of these methods are frequently misconstrued (Rohlf, 2006), and we believe they are inappropriate for our purposes here. If our objectives were the analysis and interpretation of trait covariation and scaling, and if giant lemurs were in the comparative data base of interest, then we might well elect to examine phylogenetically independent contrasts (PICs) or employ phylogenetic generalized least squares (PGLS) to estimate allometric regression coefficients and correlation coefficients. However, the giant lemurs are obviously not in the reference samples, so their phylogenetic relationships and branch lengths cannot inform the regression estimates made here. We are explicitly **not** trying to control for phylogenetic effects in our regression estimates. To the contrary, we have cast our taxonomic and phylogenetic net very wide and far outside Strepsirrhini and Primates for the reasons stated above.

Results

We present our results in Table 3 and graphically. Table 3 provides the estimates for *Megaladapis edwardsi* based on all eight regressions. Each bony element generates four estimates (e.g., a given femur will have estimates based on raw and logged mammal data and raw and logged primate data); the mean of the four is taken as the predicted value for that individual element. Individual element means are then combined to calculate the species mean (i.e., the aforementioned "mean of the means"). Box-and-whiskers plots of mean predictions for each regression/element exhibit a good deal of variation as well as considerable overlap, but estimates made from femoral cortical areas tend to be slightly higher in this case. With respect to *M. edwardsi*, if all twelve individual mean estimates are pooled across the eight regressions, the species-specific average body mass based on the six femora and six humeri is 85.1 kg, with a standard deviation of 11.8 kg. Similarly, in Table 3 and all the graphical summaries to follow, the element-specific means are the "raw data" used in the calculations of central tendencies and variation for each species. Obviously, for *M. edwardsi*

Table 3 Body mass estimates of the subfossil lemurs

Species	Mean of means* (kg)	Range of individuals (kg)
Megaladapis edwardsi	85.1	58–110
Megaladapis grandidieri	74.3	60–98
Megaladapis madagascariensis	46.5	31–61
Archaeoindris fontoynontii	161.2	151–188
Palaeopropithecus maximus	45.8	27–56
Palaeopropithecus ingens	41.5	22–56
Babakotia radofilai	20.7	12–30
Mesopropithecus dolichobrachion	13.7	9–18
Mesopropithecus globiceps	11.3	6–18
Archaeolemur edwardsi	26.5	15–35
Archaeolemur majori	18.2	13–26
Hadropithecus stenognathus	35.4	29–41
Daubentonia robusta	14.2	11–18
Pachylemur jullyi	13.4	8–18
Pachylemur insignis	11.5	6–21

* We regard this value as our "best" estimate. See the text for its calculation.

and all other species examined in this study, the range of individual predictions (58–110 kg) is necessarily greater than the range of means (72–101 kg).

"Average" body masses for the megaladapids (koala lemurs) range from the aforementioned 85.1 kg for *M. edwardsi* to 74.3 kg for *M. grandidieri* to 46.5 kg for *M. madagascariensis* (Fig. 4). Approximate coefficients of variation are similar for each species (~14–15). *M. edwardsi* from the south and southwest of Madagascar overlaps *M. grandidieri* from the high plateau, but *M. madagascariensis* from the south and southwest does not overlap with either of the other two larger species. An approximately 2 to 1 ratio of maximum to minimum **individual** estimates underscores the reality of substantial intraspecifc variation in size (Table 3, last column). Walker's (2002) estimate of 79.5 kg for *M. edwardsi* is quite close to our own "best" prediction.

The palaeopropithecids (sloth lemurs) were a diverse clade closely allied with the living indriids. Average body masses range from a low of 11.3 kg in *M. globiceps* to 20.7 kg in *Babakotia*, whereas *Palaeopropithecus ingens* and *P. maximus* both weighed in excess of 40 kg across the island (41.5 kg and 45.8 kg, respectively). Burney et al.'s (1987) estimate of 60 kg for *Palaeopropithecus* seems to be an over-estimate, as does that of Martin (1990). Jungers' (1990) estimate based on femoral head diameter is much closer to the values predicted by our new regressions. *Archaeoindris* was indeed huge at an estimated mass of 161.2 kg (Fig. 5), a value seen today only among the largest male gorillas (Smith and Jungers, 1997). Even with only one element (the femur), there is variation in mean estimates for *Archaeoindris*; the range of four predicted values is 150–187.8 kg. Percentage-wise, relative to their respective mean of means, there is even greater variation in individual/element mean estimates

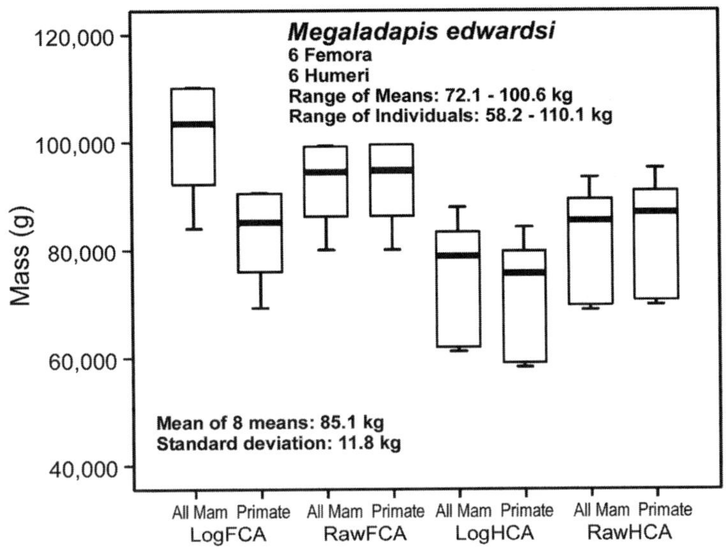

Fig. 3 Box-and-whiskers plots of body mass estimates for *Megaladapis edwardsi* using both humerus and femur, the different reference samples, and both raw and log-log regressions. The horizontal line in each box represents the median value; each box encloses 50% of the observed data. The whiskers encompass the remainder of the observed values except for extreme outliers

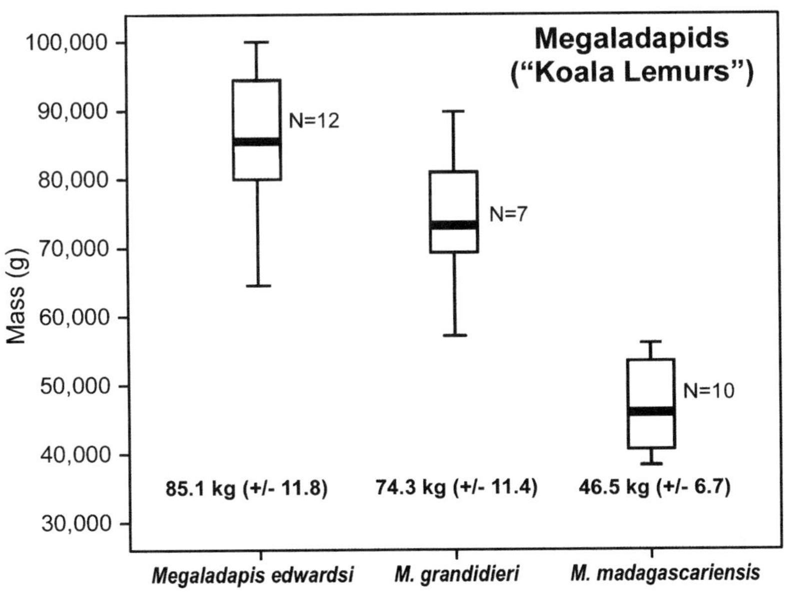

Fig. 4 New body mass estimates for the megaladapids (koala lemurs)

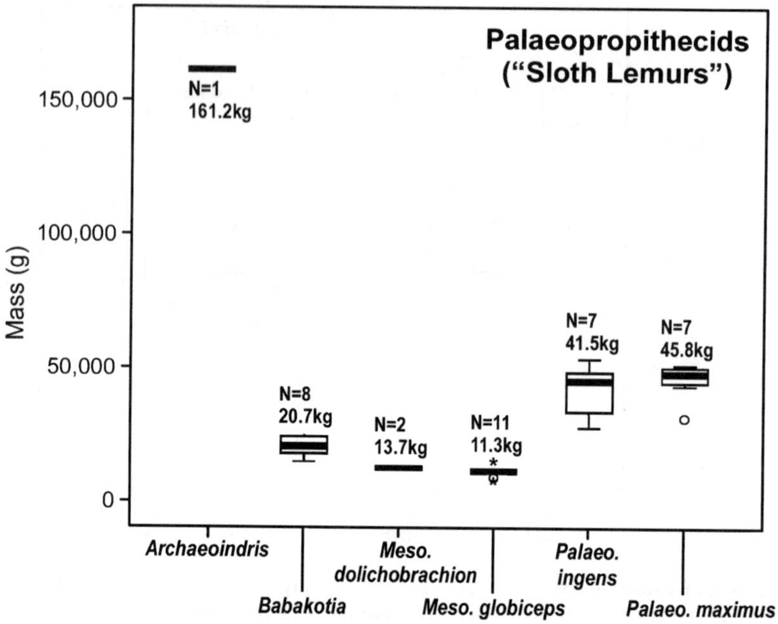

Fig. 5 New body mass estimates for the palaeopropithecids (sloth lemurs)

for the smaller sloth lemurs (e.g., 9–18 kg for *M. dolichobrachion*, 12–30 kg for *Babakotia*). The predictable impact of considering prediction intervals for a single individual can be illustrated by examining the range of values estimated for *A. fontoynontii* using the all-mammal, raw femur regression: this equation predicts a value of 151 kg and carries with it a prediction interval of 146 to 156 kg. Although far from trivial, this additional source of statistical variation is small relative to inter-individual differences (max/min values exceed 200% for most sloth lemurs).

The archaeolemurids (monkey lemurs) range from 18.2 kg in *A. majori* to 35.4 kg in *Hadropithecus*. *A. edwardsi* from the high plateau and the northern morph of *Archaeolemur* sp. cf. *edwardsi* are similar in their estimated masses at 26.5 kg and 27.3 kg, respectively (Fig. 6). There appear to be several small outliers in the *A. edwardsi* sample, and this could represent natural variation within the species, possibly secular trends (Godfrey et al., 1999), or it could signal a few *A. majori* interlopers in the *A. edwardsi* sample (i.e., the exclusively geographic criterion used to sort the species may be inadequate). Regardless, there is substantial variation in mean predictions for all the monkey lemurs (e.g., 19–33 kg for *A. edwardsi*), and the range of individual predictions are wider still (e.g., 15–35 kg for *A. edwardsi*). On the basis of the size and robusticity of the newly attributed femora of *Hadropithecus* (Godfrey et al., 1997b), we anticipate that our best estimates of body mass for this genus would not be any smaller if these specimens had been incorporated into

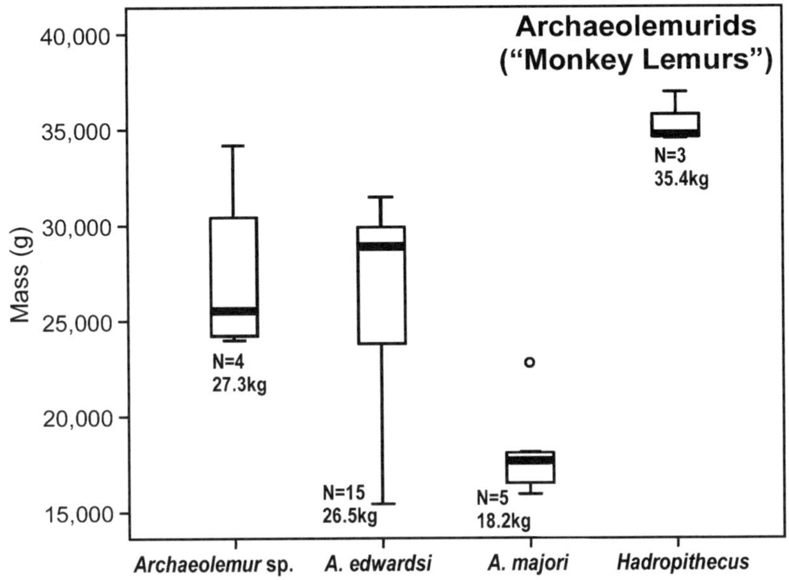

Fig. 6 New body mass estimates for the archaeolemurids (monkey lemurs)

our analysis, although the range of individual predictions would probably expand somewhat.

Pachylemur jullyi from the high plateau was slightly bigger at 13.4 kg than its southern congener, *P. insignis* (11.5 kg), but both dwarf their living sister taxon, *Varecia*. Absolute variation in mean estimates is 5 kg for *P. insignis* and 7 kg for *P. jullyi;* individual ranges are considerably greater at 15 and 10 kg, respectively (Table 3). The extinct aye-aye, *D. robusta*, deserves its specific nomen, weighing in at 14.2 kg (Fig. 7), with individual/element mean estimates ranging from 11–18 kg; our best estimate is more than five times the mass of the living aye-aye (Smith and Jungers, 1997).

Discussion

It isn't too difficult to rank order the subfossil lemurs by body size using most skeletal surrogates for body mass. However, if accurate estimates of body mass are required for other analyses (e.g., considerations of brain size/body size relationships and life histories), we might wish to investigate just how different the estimates are for body mass in extinct lemurs derived from competing predictors (also see Smith, 1996). Here we compare our new estimates to those derived from orbital height (after Kappelman, 1996), first mandibular molar area (Gingerich et al., 1982), and external diaphyseal circumferences (Godfrey et al., 1995; Jungers et al., 2002).

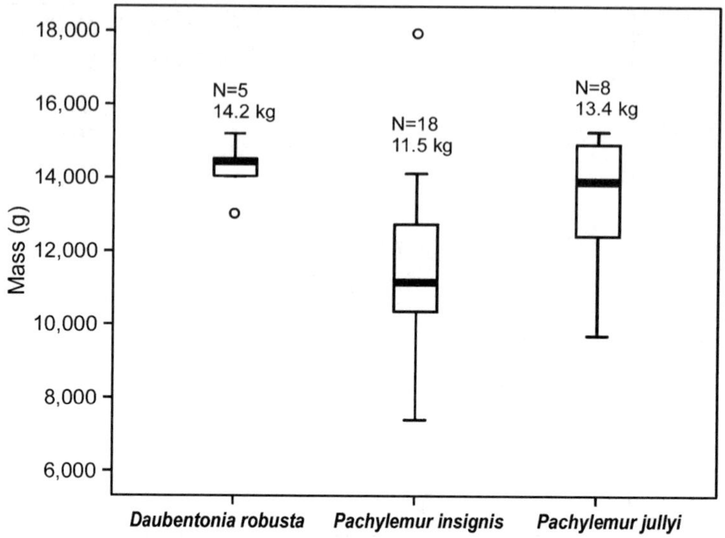

Fig. 7 New body mass estimates for *Pachylemur* and *Daubentonia robusta*

Figure 8 is a bivariate plot of mass estimates based on orbit height versus our own average predictions. The solid line is a slope of 1.0, which would indicate identical estimates if all points fell on that line; we also provide the 95% confidence intervals on the mean (dashed lines) to see if other surrogates produce predictions within reasonable limits of our own We acknowledge that Kappelman (1996) did not develop his equations to estimate body size in subfossil lemurs, but the differences between the estimates are still instructive. Despite a highly significant rank order correlation between predicted values (rho = 0.92), orbit height appears to underpredict body mass in most taxa; except for *Archaeolemur*, most points fall below the line of equality (slope = 1.0). The three megaladapids and *A. fontoynonti* also fall below the lower 95% boundary. This result suggests to us that most of the giant lemurs have relatively small orbits for their size, a finding not incompatible with a reconstructed diurnal activity cycle for most, if not all (Godfrey et al., 1997a; Godfrey et al., 2006; Jungers et al., 2002). Unpublished analyses (WLJ) reveal that orbit height often does a relatively poor job of predicting mass in large-bodied anthropoids (cercopithecines and some apes) of known mass at death. Other cranial measures such as skull length also seem to do relatively well for *Archaeolemur* (i.e., they are similar to our results), but most "fail" miserably for other species like *Megaladapis* and *Palaeopropithecus* (Martin, 1990). As a case in point, it seems safe to conclude that *M. edwardsi* did not weigh in excess of 300 kg!

Given the abundance of dental remains in the fossil record, the desire to employ tooth size to estimate the body mass of extinct species is understandable

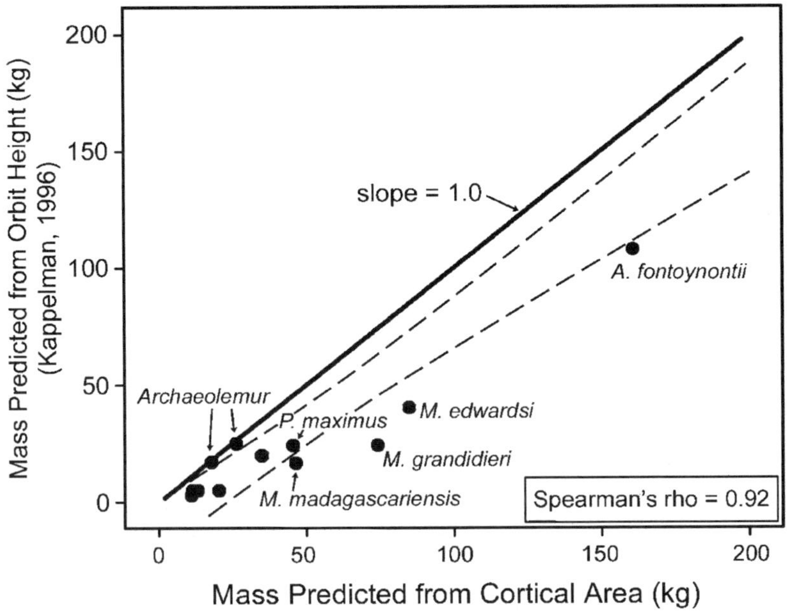

Fig. 8 Bivariate scatterplot of body mass estimates derived from orbit height versus the average estimates in this study. A solid line of equality is indicated (slope = 1.0). Note that most points fall below the isometric line. The rank order correlation (rho) is provided. 95% confidence bands for the prediction of mean Y on mean X are also provided (dashed lines)

(Gingerich et al., 1982; Conroy, 1987). Figure 9 compares mass estimates based on molar size (first mandibular molar area) to our new ones derived from long bone cortical areas. The rank order correlation between the two is a significant and respectable 0.96, but there are several noticeable differences in absolute values. *Palaeopropithecus* falls far above the line of equality and above the 95% upper band, and is probably best regarded as a megadont lemur. Compared to other subfossil lemurs, including the other megaladapids, *M. grandidieri* appears to be a microdont folivore that falls well below the isometric line and below the 95% band. First mandibular molar area predicts a value for *Archaeoindris* within 15 kg of our new estimate, but first maxillary molar area greatly overpredicts the mass of this truly giant lemur (230.5 kg; Jungers, 1990). Unpublished analyses (WLJ) indicate that dental dimensions often predict body masses substantially different from known masses.

Our final bivariate scatterplot compares values based on external diaphyseal circumferences at the midshafts of humeri and femora to ours based on cortical area at the same locations (Fig. 10). The correspondence between the two sets of predictions is very high indeed, with rho = 0.99. Perhaps this result should not be too surprising in view of the fact that cortical area in cross-section is partially a function of external circumference. Nevertheless, if

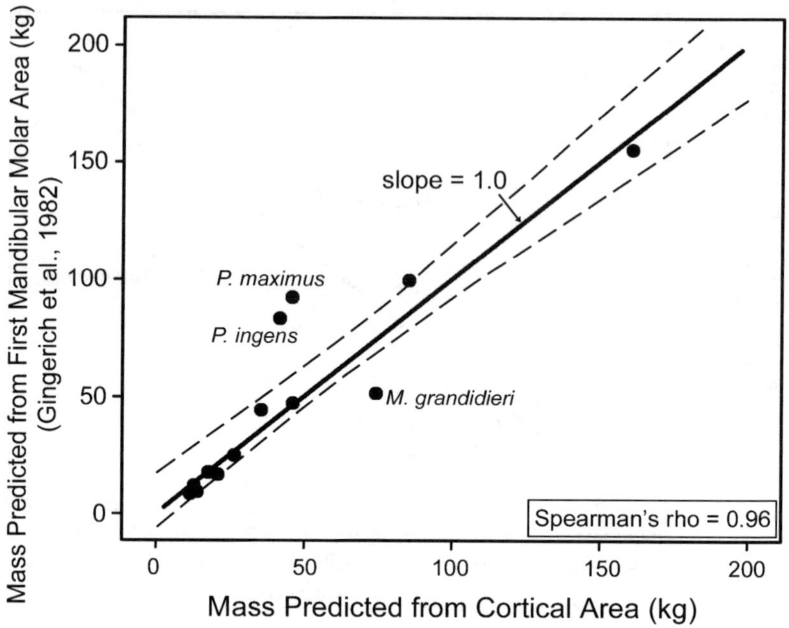

Fig. 9 Bivariate scatterplot of body mass estimates derived from first mandibular molar area versus the average estimates in this study. *Palaeopropithecus* and *M. grandidieri* are clear outliers

our best estimates are valid, then megaladapids are consistently underpredicted by external shaft measures, whereas *Archaeoindris* is greatly overpredicted. Of the sixteen taxa in common between these two studies, if the ranges of **variation in individuals** offered here are taken into account, the mean values offered by Godfrey et al. (1995) and Jungers et al. (2002) can be accommodated with only one exception. At 197.5 kg, *Archaeoindris* falls outside (above) our range of 151–188 kg.

It would be desirable to validate our regressions on an independent sample of skeletons of known body mass at death, but we do not have access to such data at this time. For this reason, we cannot proclaim that our results necessarily approach the "truth" any closer than other regression-based estimates, but theoretical reasons suggest that they are reasonable and probably preferable to those based on cranial, orbital, and dental dimensions (which do in fact fail attempts at validation). Our results reinforce prior observations that estimates of body mass depend greatly on the choice of predictor variables, and that convenience of data collection or relative abundance of certain elements in the fossil record may be inadequate rationales for choosing among the alternatives. Variation in body mass estimates within a species, even when using the same predictors, is real and all too often ignored. "Best" estimates have their place, but we also need to move beyond point estimates and the illusion of precision in

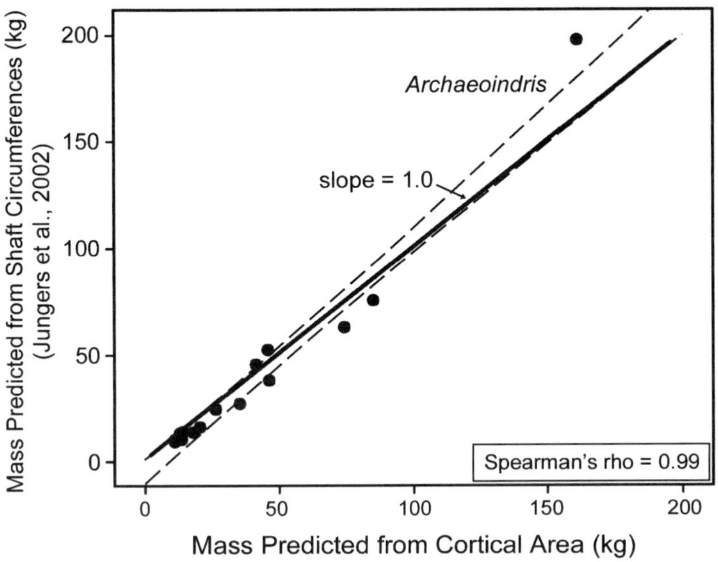

Fig. 10 Bivariate scatterplot of body mass estimates derived from external diaphyseal circumferences versus the average estimates in this study. The correspondence is quite high between the two data sets (rho = 0.99), but *Archaeoindris* is far above the line of equality

mass estimates for fossils. We should acknowledge and explore **intraspecific variation** in our future comparative analyses, whether we are interested in relative brain size (e.g., Godfrey et al., 2008), relative tooth size (Vinyard and Hanna, 2005), relative limb lengths (Jungers et al., 2002), life histories (Godfrey et al., 2002) or almost anything else presumably linked to overall body size. We want to stress that our treatment of body mass variation is driven here by skeletal variation within our fossil samples and is probably slightly too conservative. As noted above for *A. fontoynontii*, if one were to also take into account statistical prediction intervals for each individual with respect to each regression estimate, the approximate ranges of variation offered here would be expanded even more. Given the relatively good rank order correspondence among predictor variables examined here, however, it might also be wise to consider using ranked data or nonparametric alternatives whenever possible. At the very least, we need to bracket our results, guided by more realistic considerations of individual variation in fossil body size.

Acknowledgment It is a genuine pleasure to offer this contribution to celebrate the scientific career of Elwyn Simons, a dear friend and stimulating colleague. We also thank the many curators of fossil and extant skeletal collections for access to material in their care and for their help in arranging to acquire radiographs. Luci Betti-Nash prepared the illustrations and K. S. Lamm assisted in part of the data collection. We wish to thank Richard Smith for his constructive criticism and thoughtful suggestions. This is DLC publication #1027.

References

Burney, D. A., Burney, L. P., Godfrey, L. R., Jungers, W. L., Goodman, S. M., Wright, H. T., and Jull, A. J. T. (2004). A chronology for late prehistoric Madagascar. *J. Hum. Evol.* 47: 25–63.

Burney, D. A., MacPhee, R. D. E., Dewar, R. E., Wells, N. A., Andriamanantena, E. H., and Vuillaume-Randriamanantena, M. (1987). L'environment au cours de l'Holocene et la disparition de la megafaune a Madagascar: Quel rapport avec la conservation de la nature? In: Mittermeier, R. (ed.), *Proceedings of the IUCN/SCC Symposium on Species Conservation Priorities in Madagascar, Antananrivo: Occasional Papers, IUCN Species Survival Commission*, pp. 137–143.

Conroy, G. C. (1987). Problems of body-weight estimation in fossil primates. *Int. J. Primatol.* 8: 115–137.

Delson, E., Terranova, C. J., Jungers, W. L., Sargis, E. J., Jablonski, N. G., and Dechow, P. C. (2000). Body mass in Cercopithecidae (Primates, Mammalia): Estimation and scaling in extinct and extant taxa. *Anthropol. Pap. Am. Mus.* 83: 1–159.

Felsenstein, J. (1985). Phylogeneies and the comparative method. *Am. Nat.* 125: 1–15.

Fleagle, J. G. (1988). *Primate Adaptation and Evolution.* First Edition, Academic Press, New York.

Giles, E. and Klepinger, L. L. (1988). Confidence intervals doe estimates based on linear regression in forensic anthropology. *J. Forensic. Sci.* 33: 1218–1222.

Gingerich, P. D. (1996). Comment on R. J. Smith's "Biology and body size in human evolution." *Curr. Anthropol.* 37: 464–465.

Gingerich, P. D., Smith, B. H., and Rosenberg, K. (1982). Allometric scaling in the dentition of primates and predictions of body weight from tooth size in fossils. *Am. J. Phys. Anthropol.* 58: 81–100.

Godfrey, L. R. and Jungers, W. L. (2002). Quaternary fossil lemurs. In: Hartwig, W. C. (ed.), *The Primate Fossil Record.* Cambridge University Press, Cambridge, pp. 97–121.

Godfrey, L. R. and Jungers, W. L. (2003). The extinct sloth lemurs of Madagascar. *Evol. Anthropol.* 12: 252–263.

Godfrey, L. R., Jungers, W. L., Reed, K. E., Simons, E. L., and Chatrath, P. S. (1997a). Subfossil lemurs: Inferences about past and present primate communities. In: Goodman, S. M. and Patterson, B. D. (eds.), *Natural Change and Human Impact in Madagascar.* Smithsonian Institution Press, Washington, DC, pp. 218–256.

Godfrey, L. R., Jungers, W. L., and Schwartz, G. T. (2006). Ecology and extinction of Madagascar's subfossil lemurs. In: Gould, L. and Sauther, M. L. (eds.), *Lemurs: Ecology and Adaptation.* Springer, New York, pp. 41–65.

Godfrey, L. R., Jungers, W. L., Wunderlich, R. E. and Richmond, B. G. (1997b). Reappraisal of the postcranium of *Hadropithecus* (Primates, Indroidea). *Am. J. Phys. Anthropol.* 103: 529–556.

Godfrey, L. R., Jungers, W. L., Simons, E. L., and Chatrath, P. S. (1999). Past and present distributions of lemurs in Madagascar. In: Rakotosamimanana, B., Rasamimanana, H., Ganzhorn, J. U., and Goodman, S. M. (eds.), *New Directions in Lemur Studies.* Kluwer Academic / Plenum, New York, pp. 19–53.

Godfrey, L. R., Jungers, W. L., Schwartz, G. T., and Irwin, M. T. (2008). Ghosts and orphans: Madagascar's vanishing ecosystems. In: Fleagle, J. G. and Gilbert, C. C. (eds.), *Elwyn Simons: A Search for Origins.* Springer, New York, pp. 361–396.

Godfrey, L. R., Petto, A. J., and Sutherland, M. R. (2002). Dental ontogeny and life-history strategies: The case of the giant extinct indroids of Madagascar. In: Plavcan, J. M., Kay, R. F., Jungers, W. L., and van Schaik, C. P. (eds.), *Reconstructing Behavior in the Primate Fossil Record.* Kluwer Academic / Plenum Publishers, New York, pp. 113–157.

Godfrey, L. R., Sutherland, M. R., Paine, R. R., Williams, F. L., Boy, D. S., and Vuillaume-Randriamanantena, M. (1995). Limb joint surface areas and their ratios in Malagasy lemurs and other mammals. *Am. J. Phys. Anthropol.* 97: 11–36.
Hylander, W. L. (1985). Mandibular function and biomechanical stress and scaling. *Am. Zool.* 25: 315–330.
Jolly, C. J. (1970). *Hadropithecus*: A lemuroid small-object feeder. *Man* 5: 619–626.
Jungers, W. L. (1978). The functional significance of skeletal allometry in *Megaladapis* in comparison to living prosimians. *Am. J. Phys. Anthropol.* 49: 303–314.
Jungers, W. L. (1990). Problems and methods in reconstructing body size in fossil primates. In: Damuth, J. and MacFadden, B. J. (eds.), *Body Size in Mammalian Paleobiology*. Cambridge University Press, Cambridge, pp. 103–118.
Jungers, W. L., Godfrey, L. R., Simons, E. L., Wunderlich, R. E., Richmond, B. G., and Chatrath, P. S. (2002). Ecomorphology and behavior of giant extinct lemurs from Madagascar. In: Plavcan, J. M., Kay, R. F., Jungers, W. L., and van Schaik, C. P. (eds.), *Reconstructing Behavior in the Primate Fossil Record*. Kluwer Academic/ Plenum Publishers, New York, pp. 371–411.
Kappelman, J. (1996). The evolution of body mass and relative brain size in fossil hominids. *J. Hum. Evol.* 30: 243–276.
Karanth, K. P., Delefosse, T., Rakotosamimanana, B., Parsons, T. J., and Yoder, A. D. (2005). Ancient DNA from giant extinct lemurs confirms single origin of Malagasy primates. *P. Natl. Acad. Sci USA* 102: 5090–5095.
Kurten, B. (1971). *The Age of Mammals*. Weidenfeld and Nicolson, London.
Lamberton, C. (1934). Contribution a la connaissance de la faune subfossile de Madagascar. L'*Archaeoindris fontoynontii* Stand. *Mem. Acad. Malgache* 17: 9–39.
Lavauden, L. (1931). Animaux disparus et legendaires de Madagascar. *Rev. Scientifique* 10: 297–308.
Martin, R. D. (1990). *Primate Origins and Evolution. A Phylogenetic Reconstruction*. Princeton University Press, Princeton.
Ohman, J. C. (1993) Computer software for estimating cross-sectional geometric properties of long bones with concentric and eccentric elliptical models. *J. Hum. Evol.* 25: 217–227.
Pagel, M. (1993). Seeking the evolutionary regression coefficient: A analysis of what comparative methods measure. *J. Theor. Biol.* 164: 191–205.
Polk, J. D., Demes, B., Jungers, W. L., Biknevicius, A. R., Heinrich, R. E., and Runested, J. A. (2000). A comparison of primate, carnivoran and rodent limb bone cross- sectional properties: are primates really unique? *J. Hum. Evol.* 39: 293–325.
Radinsky, L. B. (1970). The fossil evidence of prosimian brain evolution. In: Noback, C. R. and Montagna, W. (eds.), *The Primate Brain. Advances in Primatology – Volume 1*. Appleton-Century-Crofts, New York, pp. 209–224.
Rohlf, F. J. (2001). Comparative methods for the analysis of continuous variables: Geometric interpretation. *Evolution* 55: 2143–2160.
Rohlf, F. J. (2006). A comment on phylogenetic correction. *Evolution* 60: 1509–1515.
Ruff, C. B. (1990). Body mass and hindlimb cross-sectional and articular dimensions in anthropid primates. In: Damuth, J. and MacFadden, B. J. (eds.), *Body Size in Mammalian Paleobiology*, Cambridge University Press, Cambridge, pp. 119–149.
Ruff, C. B. (2003). Long bone articular and diaphyseal structure in Old World monkeys and apes. II. Estimation of body mass. *Am. J. Phys. Anthropol.* 120: 16–37.
Simons, E. L. (1972). *Primate Evolution: An Introduction to Man's Place in Nature*. MacMillan, New York.
Smith, R. J. (1993). Logarithmic transformation bias in allometry. *Am. J. Phys. Anthropol.* 90: 215–228.
Smith, R. J. (1996). Biology and body size in human evolution. *Curr. Anthropol.* 37:451–481.
Smith, R. J. (2005). Species recognition in paleoanthropology: implications of small sample sizes. In: Lieberman, D. E., Smith, R. J. and Kelley, J. (eds.), *Interpreting the Past: Essays*

on Human, Primate and Mammal Evolution. Brill Academic Publishers, Boston, pp. 207–219.

Smith, R. J. and Jungers, W. L. (1997). Body mass in comparative primatology. *J. Hum. Evol.* 32: 523–559.

Villette, T. (1911). Proces-verbal de la séance en 22 septembre 1910. *Bull. Acad. Malgache* [anc. ser.] 8: 18–19.

Vinyard, C. J. and Hanna, J. (2005). Molar scaling in strepsirrhine primates. *J. Hum. Evol.* 49: 241–269.

Vuillaume-Randriamanantena, M. (1988). The taxonomic attributions of giant subfossil lemur bones from Ampasambazimba: *Archaeoindris* and *Lemuridotherium*. *J. Hum. Evol.* 17: 379–391.

Walker, A. (2002). Looking for lemurs on the great red island. In: Bowen, T. (ed.), *Back Country Pilot: Flying Adventures with Ike Russell*, The Univeristy of Arizona Press, Tucson, pp. 99–104.

Ghosts and Orphans

Madagascar's Vanishing Ecosystems

Laurie R. Godfrey, William L. Jungers, Gary T. Schwartz and Mitchell T. Irwin

Introduction: Ghosts and Orphans

Nineteenth century naturalist-explorer Alfred Grandidier was perhaps the first European to gaze upon the ghost bones of Madagascar. The Malagasy knew them well: some belonged to fierce but stupid birds with eggs large enough to hatch a human infant; others belonged to ogres that could be rendered helpless and unable to move when marooned on smooth rocky surfaces. Still others belonged to ferocious, cow-like creatures that might charge unsuspecting humans from a distance, spraying urine and making minced meat out of those who failed to escape. These killer cows might take refuge in crater lakes or water holes; they might steal corn at dawn but retreat into their aqueous abodes when pursued by vigilant farmers (Raybaud, 1902). They had many names, including "omby rano" (water cows), "tsy-aomby-aomby", or "song-aomby" (literally, the "not-cow-cow" or the "cow that isn't a cow"; see Ferrand, 1893; Godfrey, 1986; Burney and Ramilisonina, 1998).

Grandidier was fascinated by accounts of these creatures, and he made a point of inquiring about them wherever he traveled. A village headman whom he met in 1868 at Ambolisatra (southwest Madagascar) was more than obliging (Mantaux, 1971). He led Grandidier to a marsh where he knew that bones of the "song-aomby" could be had. Barefoot and sporting pants cut at the knees, Grandidier ventured into the muck, delicately feeling for treasures beneath his toes. His first prize was an enormous femur of an elephant bird – previously known only through its 8-liter eggs. There were indeed bones of the song-aomby – Madagascar's pygmy hippopotamus. And there was more, much more, including bones of some gigantic lemurs.

The word "lemures" means ghosts, specters, or creatures of the night. Linnaeus (1758) gave it to the ring-tailed lemur – ironically one of the relatively few diurnal species of living lemur. The bones of the giant lemurs are ghosts of

Laurie R. Godfrey
Department of Anthropology, 240 Hicks Way, University of Massachusetts, Amherst, MA 01003-9278 Phone: 413-545-2064 Fax: 413-545-9494
lgodfrey@anthro.umass.edu

J. G. Fleagle, C. C. Gilbert (eds.), *Elwyn Simons: A Search for Origins.*

a different ilk – evolution's true phantoms. Giant lemurs and associated "megafauna" disappeared during the last wave of Quaternary extinctions. They suffered a slow but inexorable decline beginning soon after the first human colonists arrived (around 2,300 years ago) and continuing into the European (or historic) period (Simons et al., 1995a; Simons, 1997; Burney, 1999, 2005; Perez et al., 2003; Burney et al., 2004). Indeed, there is ethnohistoric evidence of megafaunal survival well into the 19th or 20th centuries AD (Godfrey, 1986; Burney and Ramilisonina, 1998; Burney et al., 2004).

In 1982, Daniel Janzen and Paul Martin published a landmark article in Science magazine. Following the lead provided by Stan Temple's research on plant-animal mutualism in Mauritius (1977), Janzen and Martin argued that human-induced megafaunal extinctions in the late Quaternary had left numerous plants effectively orphaned – i.e., lacking important agents of seed dispersal. Their focus was the megafaunal extinctions in Central America 10,000 years ago, hence their title, "Neotropical anachronisms: the fruits the gomphotheres ate." Redford's (1992) "The empty forest" expands on the same theme, as does Barlow's (2000) semipopular book, "The Ghosts of Evolution: Nonsensical Fruit, Missing Partners, and Other Ecological Anachronisms." Richard and Dewar (1991: 167) were among the first to raise the question for Madagascar, proclaiming that "Conservation will not work if forests are protected but lack the fauna needed to ensure their persistence." More recently, aspects of this problem have been addressed by Dransfield and Beentje (1995a, b), Baum (1995, 2003), Ganzhorn et al. (1999) and Birkinshaw (2001), among others. If orphaned plants can survive 10,000 years after the extinction of their partners in Central America, they can certainly survive the few hundred years (or less) since the last of the megafaunal extinctions in Madagascar.

Our task in this chapter is to identify Madagascar's ghosts and orphans. Which of Madagascar's surviving plants are anachronistic, and who are their missing partners? Were some of those partners giant lemurs, and can we identify likely candidates? Can our current knowledge of life history profiles and trophic adaptations of giant lemurs help us in this endeavor?

The Paleontologists' Toolkit

Reconstructing Life Ways of Extinct Species

Paleontologists have the tools to resurrect the ghosts, at least in virtual reality (Jungers et al., 2002; Godfrey et al., 2006a, b). Knowledge of the extinct lemurs of Madagascar has grown rapidly over the past several decades, due in large part to the success of Elwyn Simons in forging a new era of paleontological discovery there (Simons et al., 1990; Simons, 1997; Fig. 1). Simons's expeditions resulted in the discovery of previously unknown skeletal elements belonging to recognized species, and several entirely new species (Godfrey et al., 1990;

Fig. 1 Elwyn Simons, 1988, searching for subfossil lemur bones at Antsiroandoha Cave, Ankarana Massif (northern Madagascar). This was the year of our discovery of the type specimens of the sloth lemur, *Babakotia radofilai*

Simons et al., 1995b). They also resulted in the discovery of virtually whole hands and feet of giant lemurs (Hamrick et al., 2000; Jungers et al., 2005). Long-standing errors of taxonomic attribution have now been corrected (e.g., Wunderlich et al., 1996; Godfrey et al., 1997a; Godfrey and Jungers, 2002; Godfrey et al., 2006c). And, with the discovery of new specimens of immature individuals, we have been able to embrace a new challenge: reconstructing the life history profiles of the giant lemurs.

Life history can be thought of as a set of adaptations that governs the allocation of an organism's energy towards growth, maintenance, reproduction, etc., played out within an ecological context. It is described by such milestones as gestation length, dental and somatic growth rates, weaning age, the ontogeny of ecological independence, age at first reproduction, and so on. Life history features do not fossilize and therefore cannot be examined directly in extinct species. However, microstructural analysis of teeth provides a lens through which life history parameters of extinct species can be observed. Molar crown formation times (CFTs) can be measured as can dental precocity at birth, and life history parameters such as age at M1 emergence and even weaning can be estimated (assuming they relate in a simple manner to age at crown completion). The specific methods of analysis are reviewed elsewhere (see, for example, Bromage, 1987; Beynon et al., 1991; Dean et al., 1993; Schwartz et al., 2002). Chronologies for the development of the entire molar row are now available for *Archaeolemur* and *Palaeopropithecus*; partial molar chronologies are available for *Megaladapis* (first two molars) and *Hadropithecus* (second molar only) (Schwartz et al., 2002, 2005, 2007; Godfrey et al., 2005, 2006c); additional work on *Mesopropithecus* and *Babakotia* is in progress.

Reconstructing Diets

Tools for reconstructing diet in extinct primates are also diverse, and include comparative analysis of dental morphology, craniofacial architecture, dental microwear, stable isotopes, and trace elements (Tattersall, 1973; Kay, 1975, 1984; Ungar, 1998; Jungers et al., 2002; Rafferty et al., 2002; Burney et al., 2004; Godfrey et al., 2004). Various levels of magnification, from 35x (low magnification stereomicroscopy) to 500x (scanning electron microscopy), have been used in the interpretation of dental microwear of subfossil lemurs (Rafferty et al., 2002; Godfrey et al., 2004; Semprebon et al., 2004). A study using confocal microscopy is in progress (Scott et al., 2007; for the method, see Ungar et al., 2003). Only preliminary stable carbon isotope data ($\delta^{13}C$) are available (see especially Burney et al., 2004); additional research in this realm is also in progress (Crowley et al., 2007). Trace elements have yet to be studied. Table 1 summarizes these data for extinct lemurs from southern Madagascar, thus controlling for variation in abiotic factors (such as climate).

Brains and Body Size

We have measured the cranial capacities of subfossil lemurs, including representatives of all known genera (except for *Daubentonia robusta*, which is unknown cranially). Elsewhere in this volume, we estimate body masses for these same extinct taxa (Jungers et al., 2008), so we can now briefly consider the relationship between body size and brain size in subfosssil lemurs (also see Jungers, 1999), within a comparative context of living strepsirrhines (37 species) and nonhuman anthropoid primates (43 species, sexes plotted separately for dimorphic species) (Fig. 2). Nonparametric (LOESS) lines are plotted for each infraorder in ln–ln space, and the extinct lemurs are enclosed in a convex polygon. The well-known separation ("grade effect" = clade effect) between living strepsirrhine and anthropoid primates (Stephan et al., 1977, 1981; Martin, 1990) is readily observed, and this separation is increased at the larger body masses of subfossil lemurs (all of whom exceed 10 kg). The apparent curvature or kink created by the addition of the giant lemurs indicates that allometric scaling of brain size in strepsirrhines is more complex than linear; regardless, most of the subfossils are much less encephalized than size-matched anthropoids.

Table 1 Dietary inferences for extinct lemurs from southern Madagascar*

Taxon	$\delta^{13}C$ values	Primate species with most similar microwear profiles	Additional observations	Inferred diet
Palaeopropithecus ingens	−19.0, −21.0, −20.6	*Propithecus verreauxi, Indri indri, Mesopropithecus globiceps, Macaca fascicularis*	High shearing quotient suggests that shearing leaves is important, as in sifakas.	Mixed fruit and foliage (with seed processing). Stable carbon isotopes suggest mainly C3 plant foods.
Mesopropithecus globiceps	−15.8, −19.3, −20.3,	*Indri indri, Palaeopropithecus ingens, Propithecus verreauxi, Macaca fascicularis*	High shearing quotient suggests that shearing leaves is important, as in sifakas.	Mixed fruit and foliage (with seed processing). Stable carbon isotopes suggest mainly C3 plant foods.
Daubentonia robusta	−16.3	[No known molars]	Incisors and postcranial skeleton reveal trophic adaptations similar to those of living aye-ayes.	Omnivore, hard-object processor. Stable carbon isotopes suggest mainly C3 plant foods.
Pachylemur insignis	−20.2, −20.6	*Eulemur collaris, Lemur catta, Pachylemur jullyi, Eulemur rubriventer, Varecia variegata*	Pattern of enamel erosion comparable to that observed in other lemurids, such as *Lemur catta*	Mixed fruit and foliage (without seed processing). Stable carbon isotopes suggest mainly C3 plant foods.
Megaladapis madagascariensis	−18.9, −22.4	*Megaladapis grandidieri, Lepilemur mustelinus, Avahi laniger, Megaladapis edwardsi, Alouatta Palliate*	Both SEM and low magnification microwear suggest consumption of leaves. This is consistent with the signal from dental morphology (the high shearing quotient).	Leaves. Stable carbon isotopes suggest mainly C3 plant foods.
Megaladapis edwardsi	−20.1, −20.2, −20.4			

(continued)

Table 1 (continued)

Taxon	$\delta^{13}C$ values	Primate species with most similar microwear profiles	Additional observations	Inferred diet
		Alouatta palliata, *Avahi laniger*, *Lepilemur leucopus*, *Megaladapis grandidieri*, *Megaladapis madagascariensis*, *Lepilemur mustelinus*	Both SEM and low magnification microwear suggest consumption of leaves. This is consistent with the signal from dental morphology (the high shearing quotient).	Leaves. Stable carbon isotopes suggest mainly C3 plant foods.
Archaeolemur majori	−18.3, −19.9, −20.5	*Archaeolemur edwardsi*, *Hadropithecus stenognathus*, *Cebus apella*, *Chiropotes satanas*, *Cacajao melanocephalus*, *Daubentonia madagascariensis*	Coprolites of a related species, *Archaeolemur* sp. cf. *edwardsi* from Anjohikely in the northwest, reveal the presence of bones of small animals (such as frogs) and shells of gastropods in addition to plant material.	Seed predator, hard object processor, omnivore. Stable carbon isotopes suggest mainly C3 plant foods.
Hadropithecus stenognathus	−8.4, −9.1, −13.22	*Archaeolemur majori*, *Archaeolemur edwardsi*, *Cebus apella*	Coarser microwear than in *Archaeolemur*. Both SEM and low magnification microwear studies suggest that *Hadropithecus* was not graminivorous, as had been previously thought.	Seed predator, hard object processor. Stable carbon isotopes suggest mixed C4 and CAM plants and/or animals consuming CAM and C4 plants. Likely omnivorous.

*Sources: Burney et al. (2004) for carbon isotopes; Godfrey et al. (2004) and Rafferty et al. (2002) for microwear. For additional information, see text.

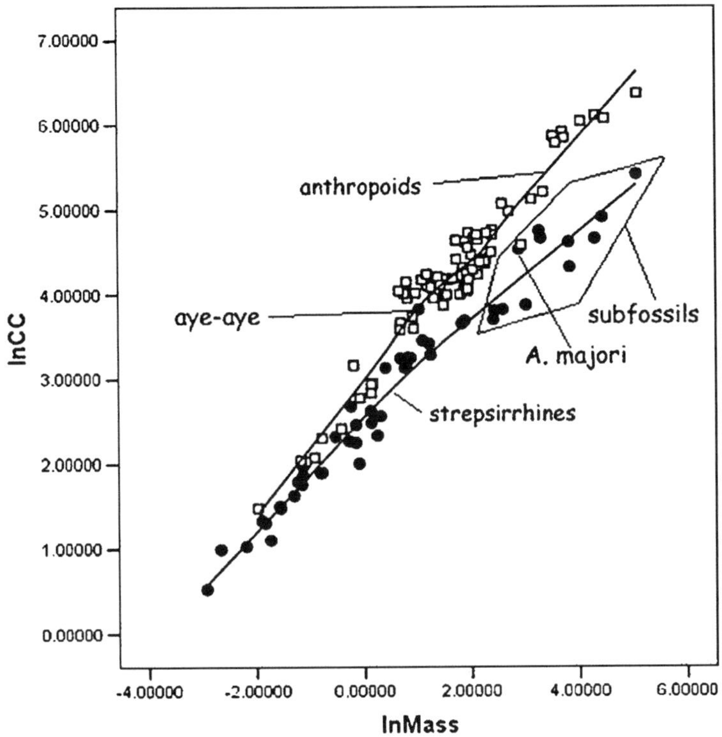

Fig. 2 Bivariate plot of logged cranial capacity (lnCC in ml) and logged body mass (lnMass in kg) in strepsirrhine primates (closed circles) and living anthropoids (open squares). The extinct giant lemurs are enclosed by a convex polygon, and nonparametric (LOESS) lines are fit to each group (tension = 0.7). Strepsirrhines fall consistently below size-matched anthropoids in brain size, although there are exceptions (e.g., the living aye–aye, *Daubentonia madagascariensis*, falls on the anthropoid line; *Archaeolemur majori* is virtually indistinguishable in its brain:body ratio from male proboscis monkeys (*Nasalis larvatus*)

Portraits of the Ghosts

We now present portraits of the extinct lemurs, focusing mainly on their trophic adaptations, as inferred from data on dental microwear, stable isotopes, and general morphology, as well as the pattern of growth and development, with special attention to age at M1 and M2 emergence, if known.

Palaeopropithecidae, or Sloth Lemurs

There are seven recognized species of sloth lemurs (Godfrey and Jungers, 2002), some apparently more folivorous than others, but all showing microwear evidence of a mixed diet of foliage and fruit, and likely deliberate processing of seeds (Godfrey et al., 2004; Table 1). The seed-processing signal is strongest in *Babakotia radofilai* and *Mesopropithecus dolichobrachion*; *Mesopropithecus pithecoides* appears to have been more folivorous (Godfrey et al., 2004). In general, the sloth lemurs have microwear signals most similar to those of extant indriids and colobines. Their teeth resemble those of indriids, and their high shearing quotients suggest a similar ability to process leaves (Jungers et al., 2002).

Collagen-derived stable carbon isotopes ($\delta^{13}C$) range from −15.8 to −23.5, with less negative values in the arid south (*Mesopropithecus globiceps*) and more strongly negative values in the central highlands (*Archaeoindris fontoynontii*); see Burney et al. (2004). This range of $\delta^{13}C$ values suggests diets dominated mainly by C3 plant foods except in the more arid regions of Madagascar, where CAM plants were likely more important.

The sloth lemurs were poorly encephalized in comparison to size-matched anthropoids. The largest member of this clade, *Archaeoindris*, was similar in body size to male gorillas, but had a brain less than 40% of a gorilla brain in volume. The cranial capacity of *Palaeopropithecus ingens* barely exceeds 100 cc despite a body mass (~45 kg) in the bonobo-chimpanzee range; this is less than one-third the brain size of the African apes. The smaller sloth lemurs (*Babakotia* and *Mesopropithecus*) were similarly "small-brained." In terms of dental development, the palaeopropithecids were perhaps the most remarkable of the subfossils. Their rates of molar crown growth (measured here as the square root of the molar crown occlusal area divided by the crown formation time, or CFT, in days) are about six times as rapid as those of roughly like-sized

Table 2 Reconstructed body masses and average molar crown growth rates for extinct lemurs and like-sized anthropoid primates*

Taxon	Adult Female Body Mass (kg)	M1 Crown Growth Rate (mm^2/day)	M2 Crown Growth Rate (mm^2/day)
Palaeopropithecus ingens	41.5	0.071	0.063
Pongo pygmaeus	35.8	0.012	0.010
Megaladapis edwardsi	85.1	0.038	0.035
Gorilla gorilla	71.5	0.012	0.013
Archaeolemur majori	18.2	0.015	0.017
Papio hamadryas	11.4	0.017	0.019
Hadropithecus stenognathus	35.4	–	0.012
Pan troglodytes	33.7	0.010	0.008

*Crown growth rate = square root of molar occlusal area (md × bl in mm^2)/CFT in days. Dental measurements taken from unpublished data and from Swindler (2002).

anthropoids (such as adult female orang-utans; Table 2). Despite weighing over 40 kg as an adult (Jungers et al., 2008), *Palaeopropithecus ingens* would have had its first molar erupt only a few months after birth, just as in much smaller-bodied indriids, and earlier than in much smaller-bodied lemurids (Schwartz et al., 2002, Godfrey et al., 2006b); Table 3. Accelerated dental development is associated here with small adult brain size, early processing of solid food, and early acquisition of ecological independence.

Archaeolemuridae, or Monkey Lemurs

A very different developmental pattern characterizes the monkey lemurs (*Archaeolemur* spp. and *Hadropithecus*). These are the only giant lemurs whose molar CFTs approach those of like-sized anthropoids. In addition, encephalization is relatively high in the Archaeolemuridae. Indeed, *Archaeolemur majori* lies adjacent to the single anthropoid captured in the subfossil convex polygon seen in a plot of ln cranial capacity vs. ln mass (Fig. 2), *Nasalis larvatus* (males). *A. edwardsi* and *Hadropithecus stenognathus* are similar to *A. majori* in being the most encephalized subfossil lemurs, lying just to the right of *A. majori* and *Nasalis* on Fig. 2. In other words, the whole clade is relatively "big-brained" relative to other extinct lemurs. Among the living lemurs, only the aye–aye is similarly highly encephalized (Fig. 2; see also Bush and Allman, 2004; Kaufman et al., 2005).

Dental microstructural research on *Archaeolemur majori* and *Hadropithecus stenognathus* underscores developmental differences between the archaeolemurids and other giant lemurs (Godfrey et al., 2005, 2006b, c). The archaeolemurids are the only giant lemurs studied to date with molar crown growth rates roughly similar to those of like-sized anthropoids (Table 2). Furthermore, there is variation within the Archaeolemuridae: A hominoid-like dental developmental schedule characterizes only *Hadropithecus* (Table 3). Estimated crown formation time of an isolated M^2 of *Hadropithecus stenognathus* is 945 days (2.59 years) – twice as long as the reconstructed CFT for the second molar of *Archaeolemur majori*. Our samples of *Archaeolemur* and *Hadropithecus* studied to date suggest that the second molars did not begin to form until after birth, and that molar emergence occurred at a relatively advanced age – considerably later than in larger bodied subfossil species). We estimate M1 emergence in *A. majori* at around one and a half years, and M2 emergence by age 2 years (Table 3). In *Hadropithecus*, molar emergence was later yet. We only have a single molar (an M2) on which to base our dental chronology, but if we can assume a similar crown growth rate for both M1 and M2 (a reasonable assumption given the pattern observed in other subfossil lemurs; see Table 2), as well as a prenatal age for initiation of M1 formation similar to that of *Archaeolemur*, then we can estimate M1 CFT (987.5 days) and the age of M1 crown completion (902.5 days) in *Hadropithecus* on the basis of the size of the

Table 3 Life history inferences for extinct lemurs derived from dental microstructure, with comparative data for like-sized anthropoids*

Taxon	M1 CFT (days)	Age at M1 Crown Completion (days, years)	Age at M1 Emergence (years)	M2 CFT (days)	Age at M2 Crown Completion (days, years)	Age at M2 Emergence (years)
Palaeopropithecus ingens	221	34 days (0.09 yrs.)	~0.2–0.5	247	135 days (0.37 yrs.)	> 0.6
Pongo pygmaeus	993	969 days (2.66 yrs.)	~4.6	1267	(~5.0–5.5)	~5.0
Megaladapis edwardsi	380	248 days (0.68 yrs.)	~0.75–1.1	517	442 days (1.21 yrs.)	>1.4
Gorilla gorilla	1237	1212 days (3.32 yrs.)	3.3–3.5	~1250	(~5.0–6.0 yrs)	~6.6–6.8
Archaeolemur majori	522	437 days (1.20 yrs.)	~1.25–1.6	460	552 days (1.51 yrs.)	>1.7
Papio hamadryas	536	496 days (1.36 yrs.)	1.71	595	1095 days (3.00 yrs)	3.78–4.94
Hadropithecus stenognathus	–	–	–	945	~1037 days (2.84 yrs.)†	>3.0
Pan troglodytes	1004	968 days (2.65 yrs.)	2.7–4.1	1333	1820 days (4.99 yrs.)	5.3–8.3

*Sources: Schwartz et al. (2002, 2005), Godfrey et al. (2005, 2006b, c).

†Assuming a postnatal delay in the timing of initiation of M2 crown mineralization similar to that exhibited by *Archaeolemur*.

first molar. M1 was not crown-complete until almost 2.5 years after birth. This suggests an age for emergence (allowing for some root development) of the first molar not before 2.5 years and indeed probably closer to 3 years (approaching that of *Pan troglodytes*). Data for the second molar suggest an age of crown completion at around 2.8 years, and a likely age at M2 emergence between 3 and 3.5 years (Table 3).

There is strong evidence that all species of the genus *Archaeolemur* were generalists (i.e., mixed feeders consuming fruit with seeds, some foliage, and animal matter) as well as hard-object processors (Jungers et al., 2002; Godfrey et al., 2004, 2005; see Table 1). That evidence derives from the morphology of the teeth (Tattersall, 1973), the thickness and internal structure of the enamel (Godfrey et al., 2005), coprolites (Burney et al., 1997), and low magnification microwear (Godfrey et al., 2004, but see Rafferty et al., 2002) which resembles most closely that of *Cebus apella*, pitheciins, and, among lemurs, *Daubentonia* (Table 1). The molars of *Archaeolemur* are bilophodont, very like those of *Cebus apella* (King et al., 2005). The diet of *Cebus apella* includes, in addition to fruits and seeds (including palm nuts), insects, small vertebrates including titi monkeys (Sampaio and Ferrari, 2005), oysters (Fernandes, 1991), and crabs (Port-Carvalho et al., 2004; see also Rosenberger, 1992; Simmen and Sabatier, 1996). Nut-cracking is a skill acquired by infants through observational learning (Ottoni and Mannu, 2001). *Daubentonia madagascariensis* is also an extractive forager and omnivore with a diverse diet (Iwano and Iwakawa, 1988; Iwano et al., 1991; Sterling, 1994); delayed weaning allows youngsters to acquire a complex set of foraging skills, again through observation (Krakauer and van Schaik, 2005). Coprolites of *Archaeolemur* sp. cf. *edwardsi* reveal omnivory, with consumption of small vertebrates (frogs) and invertebrates (gastropods) as well as plants (Burney et al., 1997; Natalia Vasey and David Burney, Pers. comm.). Stable carbon isotope values ranging from −18.3 to −20.5 for *A. majori* in the south and west suggest mainly C3 plant consumption; more strongly negative values for *A. edwardsi* (−27.5) suggest C3 browsing perhaps in more densely wooded habitats, or on different plant species (Burney et al., 2004). Both species of *Archaeolemur* have stable carbon isotope values in the range of C3 plant consumers.

Like *Archaeolemur*, *Hadropithecus* shows evidence of hard-object processing and seed predation. Indeed the microwear signal (very heavy pitting and lots of scratches) bears testimony to an even coarser diet than in *Archaeolemur* (Rafferty et al., 2002; Godfrey et al., 2004, 2005), and it refutes the long-favored hypothesis that *Hadropithecus* was a specialized grazer and consumer of grass seeds (although it does not refute the possibility that grass was a component of its diet). Further evidence that the particular foods preferred by *Hadropithecus* were not the same as those preferred by *Archaeolemur* is provided by their very different stable carbon isotope values and dental morphology. Collagen-derived stable carbon isotopes of *Hadropithecus* range from −8.4 to −13.2 (Burney et al., 2004; Table 1), signaling primary consumption of C4 or CAM plants and/or animals consuming C4 or CAM plants. The morphology of the cheek teeth of *Hadropithecus* is distinctive, and "designed" to produce a macrowear pattern that is unusual for primates. The posterior premolar and two anterior molars of *Hadropithecus* are disproportionately enlarged. When unworn, the molars exhibit a superficial bilophodonty, but the transverse crests remain sharp for a relatively short period. As they wear, dentine is exposed first at the tips of the cusps and then along the entirety of the crests. Each "crest" becomes a flat band or loop surrounding a dentine basin that soon coalesces with others to form a complex dentine lattice or network. These teeth are ideally shaped to trap objects (such as seeds) in the dentine basins, holding them in place while they are subjected to high crushing and grinding forces. Overall, the macrowear pattern is reminiscent of those of certain ungulates. The cranial architecture (especially the mandibular corpus) is extremely robust, as might be expected of a seed predator or hard-object processor (Godfrey et al., 2005; Norconk et al., 2006). Burney et al. (2004) and Godfrey et al. (2005) suggested that consumption of terrestrial snails (that in turn consumed C4 or CAM plants) may have contributed to the weakly negative stable carbon isotope signal exhibited by *Hadropithecus*. *Paranthropus* was suggested as a trophic analogue (Godfrey et al., 2005); it is noteworthy that *Paranthropus* was a likely omnivore and hard-object processor, athough its stable carbon isotope values signal a higher percentage of C3 foods in its diet (see, for example, Lee-Thorp et al., 1994, 2000; Peters and Vogel, 2005). Other possible analogues may be peccaries (Tayassuidae) – omnivores that feed selectively on CAM plants (including the hard seeds of Euphorbiaceae; Ilse and Hellgren, 1995; Benirschke et al., 1990), have a similar macrowear pattern, and similar microwear pattern (Semprebon, unpubl. data). Peccary stable carbon isotope signatures (MacFadden and Cerling, 1996) are somewhat different from those of *Hadropithecus*, which may simply reflect different combinations of foods with CAM or C4 metabolism. Peccaries and suids are also among the many omnivorous mammals that have been reported to consume land snails (Benirschke et al., 1990; Allen, 2004). Peccaries are frugivore/herbivores, eating mainly fruits and other fleshy plant parts. They are both seed predators and dispersers.

Among extant primate species, extractive foragers and hard-object processors tend to have the most protracted life histories and the largest brains. Gibson (1986) related exceptionally high encephalization (within respective clades) to omnivory and extractive foraging in *Homo*, *Pan*, *Cebus* and *Daubentonia* (see also Parker and Gibson, 1977). The archaeolemurids contrast strikingly with the palaeopropithecids in exhibiting much higher encephalization and much slower dental development, as well as a much coarser dietary signal. Slow dental development may signal delayed weaning and a prolonged period of infant and juvenile dependency, during which time the foraging skills associated with a diverse diet and hard object processing must have been acquired (much as in *Cebus apella* or *Daubentonia madagascariensis*) (Fragaszy et al., 1991; Fragaszy and Adams-Curtis, 1997; Krakauer and van Schaik, 2005).

Megaladapidae, or Koala Lemurs

Although the three recognized species of *Megaladapis* vary considerably in estimated body size (from ~45 to 85 kg), they share a similar and unique Bauplan. They were specialized arboreal folivores with some morphological convergences on *Phascolarctos*, the Australian koala (Tattersall, 1972; Walker 1974; Jungers, 1978). In terms of dental development and life history strategies, they appear to have been more like the large-bodied sloth lemurs than the monkey lemurs (Schwartz et al., 2005; 2007; Godfrey et al., 2006b) (Tables 2 and 3). Body proportions and their huge grasping hands and feet (Wunderlich et al., 1996; Jungers et al., 2002) suggest that they were slow-moving climbers at home in the trees.

They share with *Lepilemur*, convergently it would appear (Karanth et al., 2005), a suite of craniodental adaptations linked to a predominately leaf-eating diet (e.g., loss of upper central incisors as adults and "cresty" molars with marked ectolophs). Both SEM and reflected light microscopy microwear analyses of the molars reveal a predominance of scratches and very few pits (Rafferty et al., 2002; Godfrey et al., 2004), a pattern characteristic of dedicated browsers (e.g., howler monkeys, *Avahi,* sportive lemurs). Data on stable isotopes also corroborate the reconstruction of forest browsing in megaladapids; C3 plant consumption is consistent with $\delta^{13}C$ values ranging from −18.9 to −22.4 (Table 1). As primary folivores, it seems highly unlikely that the koala lemurs were major players in seed dispersal.

Compared to like-sized anthropoids (e.g., orang-utans, chimpanzees and female gorillas), all three species had relatively very small brains. Whereas the endocranial volumes of female gorillas average around 460 cc, the cranial capacity of the similarly sized *M. edwardsi* is only ~30% of this value (also see Radinsky, 1970). Permanent teeth erupted quite early and weaning probably also occurred relatively early in koala lemurs (although not as precociously as in sloth lemurs), and there is little to suggest a prolonged period of infant learning prior to their becoming "ecological adults" (Schwartz et al., 2005; Godfrey et al., 2006b).

Other Ghosts

Daubentonia robusta appears to have been a much larger, more robust version of its congener, the living aye–aye (Grandidier, 1929). It shares the highly derived complex of skeletal features linked in *D. madagascariensis* to extractive foraging on structurally-defended resources (Sterling, 1994), including chisel-like, continually-growing incisors and an elongate, filiform third digit of the hand (Lamberton, 1934). Living aye-ayes are seed predators par excellence, supplementing a diet of nuts and fruits with insects and larvae. In view of its much larger body mass (ca. 15 kg), it is likely that *Daubentonia robusta* relied even more on seeds than its congener. Limited stable carbon isotope data (one $\delta^{13}C$ value of −16.3; Burney et al., 2004) suggest predominantly C3 foraging. Although the postcranial skeleton is well known (Simons, 1994) – and points to a deliberate, arboreal quadruped with relatively short and robust limbs (Jungers et al., 2002) – there are no known crania or cheek teeth. On the basis of its remarkable similarity to the highly encephalized living aye–aye and a comparable adaptation to extractive foraging, we suspect that its cranial capacity was also relatively large. Food-hoarding or caching behavior may or may not have characterized *D. robusta*; it has been observed in *D. madagascariensis* in captivity (Iwano, 1991) but not in the wild (E. Sterling, pers. commun.). Nuts are sometimes retrieved from epiphytes or the ground, but generally where they could have fallen.

Although roughly three times its body size, *Pachylemur* is clearly the sister taxon of *Varecia* (Szalay and Delson, 1979; Crovella et al., 1994). The two recognized species of *Pachylemur* (*insignis* and *jullyi*) are similar in morphology and body size, but can be distinguished reliably by both geography and details of the dentition (Vasey et al., 2005). Dental anatomy and microwear are consistent with a predominantly frugivorous diet, similar to those of frugivorous lemurs (Table 1; Seligsohn and Szalay, 1974; Godfrey et al., 2004). The macrowear pattern and presence of caries further corroborate this reconstruction (Vasey et al., 2005). Relatively low pit counts suggest that *Pachylemur*, like other lemurids, tended to spit seeds or swallow them whole; it is therefore an excellent candidate for being an agent of seed-dispersal in the recent past. Available stable isotope data are compatible with C3 foraging ($\delta^{13}C$ values of −20.0 to −20.2). As is characteristic of most extinct lemurs, the limb bones of *Pachylemur* are relatively short and robust, and the overall form of the postcranium implies arboreal quadrupedalism (with little evidence for leaping but some suspension as in *Varecia*). With a cranial capacity of ~45 cc, both species had brains roughly half (or less) of body-size matched anthropoids (e.g., some macaques and odd-nosed colobines). Although much remains to be done on the dental and somatic development of this genus, preliminary data suggest a developmental pattern similar to those manifested in *Varecia* and other lemurids.

Madagascar: Past and Present

Our task in this section is to place the ghosts into their paleoenvironmental contexts, and to review what is known about changes in those environments from the past to the present. We ask: What is unique about Madagascar's ecosystems? What has changed since the arrival of humans? Do the plant communities of Madagascar today have anachronistic species? Our discussion is intended not to be exhaustive but rather suggestive. Clearly much more research is required to elucidate details, and to identify additional plant groups of interest.

What's Special about Madagascar's Ecosystems?

It is common knowledge that the lemurs of Madagascar display a unique suite of behavioral and physiological characteristics (see reviews by Wright, 1999; Gould and Sauther, 2006). These have been understood not merely as accidents of history but as adaptations to the unusual structure and low productivity of Madagascar's forests, in turn affected by Madagascar's unpredictable climate and nutrient-poor soils (Ganzhorn, 1995; Reed and Fleagle, 1995; Fleagle and Reed, 1996; Goodman and Ganzhorn, 1997; Wright, 1999; Grubb, 2003; Reed and Bidner, 2004; Gould and Sauther, 2006). The peak fruit production season in Malagasy rain forests is approximately three months shorter than in rain forests in the Amazon or continental Africa (Wright et al., 2005). Certain plant groups, such as figs (Moraceae), are poorly represented on Madagascar; elsewhere, they are important keystone species for many animal groups, including primates (Terborgh, 1986a, b; Shanahan et al., 2001). Goodman and Ganzhorn (1997) have posited that the rarity of figs contributes to the relatively low number of primate frugivores on the island.

Yet, in some ways, the habitats of Madagascar are quite suitable for nonhuman primates – at least they were so until the arrival of humans. An extraordinary number of primate species, ranging in body size from 30 g to 160 kg, evolved on an island less than 600,000 km^2 in area, and primates comprise one of the dominant elements of Madagascar's mammalian fauna. There are 16 recognized extinct lemur species (and an additional one in the process of being described) and more than 50 extant ones. More than 20 primate species may have coexisted in single forests prior to the megafaunal extinctions of the Holocene (Godfrey et al., 1997b). Despite the dearth of figs species on the great red island, other plants that are favorite resources for primates are well-represented. Madagascar has many tree species with dull-colored, fiber-rich fruits that appeal to lemurs or other mammals with poor color vision. This contrasts sharply with other regions (such as South Africa) sporting many more sugar-rich, "bird-colored" fruits appealing to frugivorous birds with excellent color vision (Dominy et al., 2003; Voigt et al., 2004). A number of Madagascar's plant families have representatives elsewhere in the world that are well-exploited by primates, not merely for their leafy vegetation, but for their edible fruits and/or seeds. Even regions receiving very low annual rainfall (southern and southwestern Madagascar) are able to support many species of primates, due to unusually high humidity, which promotes the establishment of thicket rather than open savanna (Grubb, 2003). The thickets of the south are the so-called spiny forests, dominated by succulents. Gallery forests with a wide range of deciduous plant species line the rivers of the southwest, some of which are dry for all but a few months out of the year. The south is today inhabited by primates belonging to the families Indriidae, Lepilemuridae, Lemuridae and Cheirogaleidae (often at higher densities than in the eastern forests; Irwin, 2006b), and in the past also by members of the Megaladapidae, Daubentoniidae, Archaeolemuridae, and Palaeopropithecidae.

As might be expected given its long geographic isolation, Madagascar harbors a unique biota (i.e., with some taxa absent or poorly represented and others unusually speciose) and virtually unparalleled levels of endemism, both regionally and island-wide. The biotic diversity of Madagascar was recently reviewed by Goodman and Benstead (2005) on the basis of contributions to their edited volume, *The Natural History of Madagascar* (Goodman and Benstead, 2003). Levels of endemism are high for all groups of Malagasy vertebrates (for example, 100% of the total of 101 species of non-volant mammals, 99% of 199 species of frogs, over 90% of about 340 reptiles, and over 50% of more than 200 species of birds; Goodman and Benstead, 2005). Some invertebrate groups have equally high endemism; for example, terrestrial gastropods, with 671 species, have 100% endemicity (Goodman and Benstead, 2005). Mollusc and crustacean species are poorly known (Goodman and Benstead, 2005), but they number in the thousands, and do exhibit regional endemism.

Levels of endemism are also exceptional for major groups of plants. Euphorbiaceae, with 700 species, is the most speciose terrestrial plant family on the island; most are endemic (Hoffmann and McPherson, 2003; Goodman and Benstead, 2005). Along with the Didiereaceae, euphorbs are the dominant flora of the arid south. Other plants common in the spiny forests include the families Crassulaceae (*Kalanchoe* spp.) and Cucurbitaceae. All four use CAM photosynthetic pathways (Winter, 1979).

The Rubiaceae, or coffee family, is the second most speciose family of terrestrial plants on Madagascar, with more than 650 species and 98% endemicity (Davis and Bridson, 2003). Also extremely speciose on Madagascar is the Fabaceae (legumes), with 573 species and 80% endemicity (Du Puy et al., 2002; Labat and Moat, 2003; Goodman and Benstead, 2005), including the subfamilies Papilionoideae and Mimosoideae.

Palms (Arecaceae, including 170 species with 98% endemicity) are the 4th most speciose family of Malagasy terrestrial (non-marine) plants, excluding ferns and fresh-water aquatic plants (Goodman and Benstead, 2005). Palms are common in arid regions of Madagascar (including the palm savannas of the north and northwest), but also in the lowland rain forests of the east. Palms are also found in the highland rain forests.

The Pandanaceae (pandans, or screw pines) are another dominant floral component of the lowland rain forests, and are particularly common along rivers and streams. There are 99 pandan species in Madagascar, and 100% are endemic (Goodman and Benstead, 2005). Also speciose in Madagascar is the Clusiaceae (a large pantropical family including *Calophyllum* and *Garcinia*, two favorites of many lemur species; Birkinshaw and Colquhoun, 2003).

Other plant groups are less speciose, but are nevertheless important elements of the flora of Madagascar. They include the bamboos (Poaceae: Bambusoideae), the baobabs (Malvaceae), and the Strelitziaceae (Madagascar's "Traveller's tree" and its relatives). There are 34 species of bamboos on Madagascar; all are endemic (Goodman and Benstead, 2005). Only seven species of baobabs exist on Madagascar, but these too are mostly endemic (Goodman and Benstead, 2005).

Most of the speciose terrestrial plant groups are important resources for living lemurs (Birkinshaw and Colquhoun, 2003) and most were probably eaten by the extinct lemurs as well. The plants most exploited (and dispersed) by living lemurs belong to the families Euphorbiaceae, Fabaceae, Rubiaceae, Clusiaceae and Moraceae (Birkinshaw, 2001; Birkinshaw and Colquhoun, 2003). Of these, only the Moraceae (figs) is not speciose in Madagascar; however, figs are selectively favored by many lemurids and may, for them, serve as keystone species, providing food during times of otherwise scarce resources. Bamboos (Poaceae) comprise the primary, year-round, foods for one clade of living lemurs (*Hapalemur* spp.), but are more rarely consumed by others, if at all. Other groups that are not speciose in Madagascar, but are sought out by some lemur species, include the Sapindaceae and the Combretaceae.

Some parasitic plants (mistletoe, genus *Bakerella*, family Loranthaceae) are important keystone resources for both small- and larger-bodied extant lemurs (see Atsalis, 1999, on *Microcebus rufus*; Irwin, 2006a, on *Propithecus diadema*). Arecoid palms with relatively small seeds (e.g., *Dypsis decaryi*) are dispersed by lemurids (Ratsirarson and Silander, 1997; Overdorff and Strait, 1998; Birkinshaw, 2001; Ratsirarson, 2003) and other vertebrates; palms of various sizes are keystone species for primates outside Madagascar (Terborgh, 1986a, b; Dominy et al., 2003).

Along with figs (Moraceae), the Euphorbiaceae, Papilionoideae, Mimosoideae, and Sapindaceae are among the plant groups that are universally exploited by primate seed predators across the Neotropics, continental Africa, and Asia (Norconk et al., 2006), and it is likely that they were eaten by the extinct seed predators of Madagascar as well. Living-lemur seed predators do in fact exploit these families; for example, seeds of palms (Arecaeae) are consumed by *Daubentonia madagascariensis*, and seeds of euphorbs as well as many other plant families are eaten by sifakas. Other families favored by aye-ayes include the Burseraceae (ramy) and the Combretaceae; these families are also exploited by other living lemurs.

Baobabs have a wide range of dispersal mechanisms. On continental Africa, some are dispersed by baboons and other mammals (Peters, 1993). Some pandan fruits are eaten and the seeds dispersed by living lemurids (Birkinshaw, 2001; Callmander and Laivao, 2003). Lemurids also exploit the nectar of the Traveller's tree (Strelitziaceae); in doing so, they become important pollinators (Kress et al., 1994; Birkinshaw and Colquhoun, 2003). Seeds of the Traveller's tree have been reported only in the feces of the largest-bodied living lemurid, *Varecia* (Dew and Wright, 1998).

What's Changed since the Arrival of Humans?

In the past 2000 years, Madagascar lost its "megafauna" – its giant lemurs, pygmy hippopotamuses, giant tortoises and crocodiles, and flightless elephant birds. Many other native or endemic species disappeared, generally the largest-bodied members of their respective orders or families on Madagascar, including rodents, carnivores, bats, fishes, and flighted birds. Humans introduced animals from southeast Asia and Africa, including horses, cattle, bush pigs (*Potamochoerus larvatus*), domestic pigs (*Sus scrofa*), goats, sheep, dogs, cats, poultry, *Viverricula indica* (small Indian civet), and, of course, rats and mice.

Figure 3 provides a summary of the trophic changes in Malagasy primates that have occurred over the past two thousand years (Godfrey and Irwin, in press). Of a total of 64 species analyzed, 17 are extinct and 47 extant.[1] Each species is classified as belonging to one of five guilds according to the observed or reconstructed major components of their diets: hard-object processors/seed predators; mixed feeders/seed predators; specialized folivores; medium to large-bodied frugivores (dispersers of seeds of various sizes, including some large ones); and small-bodied insect, fruit, and/or gum consumers (potential dispersers of small seeds).

(1) Almost 30% of the primate species that became extinct were hard-object processors; they comprised almost 10% of the primates living on Madagascar in the past. Included in this guild are all of the Archaeolemuridae and the Daubentoniidae, along with one palaeopropithecid – *Babakotia radofilai* (identified by Rafferty et al., 2002, as a hard-object processor, and by Godfrey et al., 2004, as a *Pithecia*-like seed predator, with the coarsest diet of all palaeopropithecids). The sole living member of this guild is *Daubentonia madagascariensis*.

(2) Approximately 35% of the primate species that became extinct were mixed feeders on foliage and fruit, and likely seed predators. This guild, including most of the Indriidae, is moderately well represented in Madagascar today (8.5% of Madagascar's extant primates), but was much better represented in the past (with over 15% of the primate species living in the recent past). Living members of this guild are physiological folivores and efficient predators on seeds. Most of the Palaeopropithecidae (the most speciose of the families of extinct lemurs) are included in this guild.

(3) The percentage of Madagascar's primate species that can be considered specialized folivores has remained roughly constant from the past to the present, at almost 30%. This guild includes all of the Lepilemuridae and Megaladapidae, and the most folivorous indriids (*Avahi* spp.),

[1] The taxonomy of extant lemurs is under revision as new species, particularly of cheirogaleids, are being discovered and named. For a complete list of species analyzed, see Godfrey and Irwin (in press).

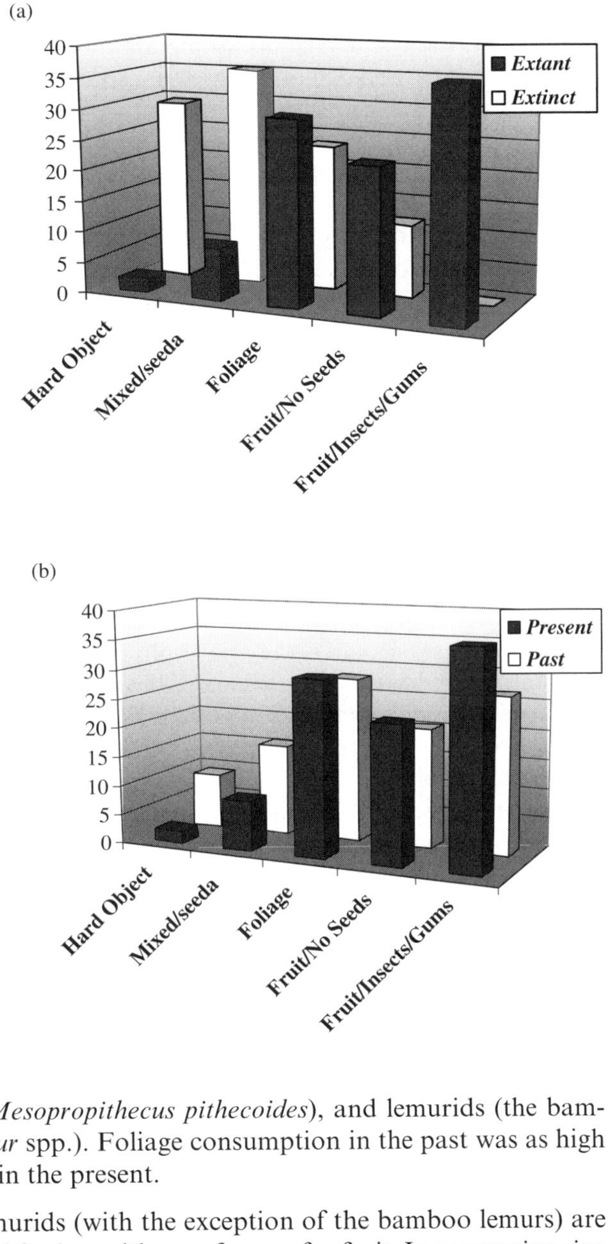

Fig. 3 Relative percentages of lemur species in each of five trophic guilds. (a) Extinct vs. extant lemurs. (b) Species living in the recent past (i.e., extinct plus extant) vs. those still extant. Data from Godfrey and Irwin (in press)

palaeopropithecids (*Mesopropithecus pithecoides*), and lemurids (the bamboo lemurs, *Hapalemur* spp.). Foliage consumption in the past was as high among lemurs as it is in the present.

(4) Living and extinct lemurids (with the exception of the bamboo lemurs) are classified here as mixed feeders with a preference for fruit. In consuming ripe fruit without chewing the seeds, living members of this guild disperse small, medium, and large seeds, thus the label "Fruit/No seeds." Approximately 20% of the primate species in the past and the present belonged to this guild. The percentage has changed trivially with the extinction of *Pachylemur* spp. This guild is relatively small today, as it was in the past.

(5) All of the Cheirogaleidae are classified here as members of a guild consuming fruit, insects, and gums. Some are more frugivorous, others are more insectivorous, and still others are gum specialists. Leaves are a rather trivial component of their diets. The relative percentage of cheirogaleids has increased from the past to the present by virtue of the loss of species belonging to other trophic guilds (there are no known extinct cheirogaleids). They now comprise more than 35% of living lemurs. The more frugivorous cheirogaleids are important dispersers of plants with small fruits and seeds, but they are not effective large-seed dispersers. Food selection by potential seed dispersers is correlated with fruit size and seed size (Bollen et al., 2004a, b); today on Madagascar, larger-bodied frugivores (such as *Eulemur* spp.) disperse larger-seeded plant species, while bats, frugivorous birds, and cheirogaleids disperse smaller-seeded plants (Bollen et al., 2004a).

Ecological Anachronisms, or Plants without Partners?

The orphans of the plant world have a number of identifying characteristics (Barlow, 2000; Chapman and Chapman, 2002). First, they must be native to the affected forests. Secondly, they have physical adaptations for, or geographic distributions indicative of, zoochory, but no obvious extant seed dispersers. Their fruits may be underexploited (e.g., rotting in large numbers under their parent trees). If sufficiently impacted, orphaned species may have limited (non-viable) geographic distributions. Their primary dispersal agents may be recently-introduced (non-endemic) animals. Many plants that were dispersed by megafauna have unusually large seeds.

It is usually easy to tell whether fruits are adapted for mechanical, wind, water, or animal dispersal. Those that depend on endozoochory (dispersal after passage through an animal's gut) have mechanisms to attract a particular set of dispersers (e.g., special colors or odors), even if they do not rely on a single, co-evolved partnership. Generally, there will be a nutritious, edible pulp (it may be rich in lipids or sugar, again, depending on the intended consumer); the seeds themselves may be hard or protected by a thick testa to survive transport through a gut. The pericarp may be brittle or easy to crack open when the seeds are mature and ready for ingestion. Often, the fruits will be "indehiscent" (they will not open to release their seeds). The seeds will show no adaptations for wind (e.g., wings or parachutes) or water (e.g., buoyancy) dispersal; the plants themselves may occur far from waterways.

We know of no comprehensive review of possible Malagasy plant anomalies, but the question has certainly been raised in the literature with regard to particular plant groups. Dransfield and Bentje (1995a, b), for example, discuss apparently orphaned palms – species that (1) are found at a distance from water, (2) appear to require zoochory but have no known seed disperser other than introduced species, such as bush pigs, (3) survive far from rivers or streams only by virtue of their resistance to fire, or (4) have very restricted geographic distributions (e.g., *Ravenea*, the single palm of the spiny forest). The dull color of palm fruits suggests predominant mammal dispersal. The large size of the seeds of some species suggests prior dispersal by primate or other megafauna.

Living lemurids are known to disperse seeds as large as 10–20 mm (or even more – up to 40 mm) in at least one dimension (Ganzhorn et al., 1999; Birkinshaw, 2001). These include *Strychnos decussata* (Loganiaceae), *Cordyla madagascariensis* (Fabaceae), *Capurodendron rubrocostatus* (Sapotaceae), *Sorindeia madagascariensis* (Anacardiaceae), *Canarium madagascariense* (Burseraceae), *Terminalia* spp. (Combretaceae), *Plagioscyphus* sp. (Sapindaceae), *Macphersonia madagascariensis* (Sapindaceae), and *Chrysophyllum perrieri* (Sapindaceae).

There are, however, living on Madagascar today, some plants with fruits and seeds too large to be dispersed by any extant Malagasy animal. One such tree with no known seed disperser is *Dilobeia* (Proteaceae) (Turk, 1997). This eastern rain forest tree produces single-seeded fruits with seeds 3–4 cm by 2–2.5 cm. Other examples include the borassoid palm genera *Borassus*, *Hyphaene*, *Bismarckia* and *Satranala*, and the arecoid palms *Orania* and *Lemurophoenix* (Dransfield and Bentje, 1995b). Elsewhere in the Paleotropics, relatives of *Borassus* are dispersed by orang-utans, bats, elephants, and baboons (Zona and Henderson, 1989). In the Neotropics today, large-seeded palms are dispersed by capuchin monkeys, peccaries (Tayassuidae), and tapirs (Tapiridae) (Terborgh 1986a, b; Zona and Henderson, 1989; Fragoso, 1997; Fragoso and Huffman, 2000; Quiroga-Castro and Roldán, 2001).

Some Malagasy baobabs (genus *Adansonia*, Malvaceae) are likely orphans (Baum, 1995, 2003). *Adansonia grandidieri* and *A. suarezensis* in particular have fruit with fragile pericarps and tasty, nutritious pulp, and their seeds are protected by tough, thick testa. These species have restricted geographic distributions today. Whereas they are clearly adapted for animal dispersal, no extant animal disperser is known, and Baum (1995, 2003) suggests that the seeds of *Adansonia grandidieri* and *A. suarezensis* may have been dispersed by extinct lemurs (possibly *Archaeolemur*). *Adansonia* seeds on continental Africa are mammal-dispersed (elephants, baboons; Peters, 1993; Baum, 1995).

Ramy nuts (*Canarium madagascariense*, Burseraceae) are a favorite resource for aye-ayes (Iwano and Iwakawa, 1988). The whole fruits are about 60–70 mm long and 40–50 mm wide, and have substantial flesh. Each has a single seed about 40 mm long and 20 mm wide. These seeds are eaten by aye-ayes, who remove just enough flesh to expose the top of the seed and carve into it using their incisors. It is likely that a fruit with this much flesh evolved to attract a large-bodied seed disperser, and that the predatory aye–aye is in essence an intruder on the system. The only frugivorous lemur large enough to swallow these seeds intact today is *Varecia variegata* (Dew and Wright, 1998). Relatives of the Malagasy *Canarium* species in Thailand are dispersed by very large birds (hornbills and one large parrot; see Kitamura et al., 2006); large birds or mammals (most likely, including *Pachylemur*) certainly did the same in Madagascar.

Other Malagasy plants may be underexploited by animals today, but may well have served Madagascar's megafauna. Little is known about the reproductive biology of Malagasy pandans. Lemurs do sometimes disperse their seeds (Callmander and Laivao, 2003), although species common near waterways are also dispersed by water. The seeds themselves are nutritious but well adapted for animal dispersal, as they are well protected by a hard endocarp, and difficult to cut even with a knife (Callmander, pers. commun.). Most Malagasy pandan seeds do not exceed 20 mm in width, but the largest are 30–40 mm wide (Callmander, pers. comm.). However, even the largest Malagasy pandan seeds easily could have been swallowed whole and dispersed by frugivores only slightly larger than the largest living lemurids. Given their hardness, pandan seeds would have been vulnerable to predation only by skilled extractive foragers.

Madagascar's flagship tree is the "Traveller's tree," *Ravenala madagascariensis*, (Strelitziaceae); it thrives in monocultures in degraded forests of eastern Madagascar largely by virtue of its vegetative reproduction. The seeds are moderately large (generally over 10 mm) and are not adapted for wind or water dispersal (Calley et al., 1993). *Ravenala* produces a "fan" with brown bracts and hard seeds surrounded by odoriferous light blue integuments or arils. Similarly-adapted plants provide nourishment in the aril; the seeds are adapted to pass through the gut unharmed (Calley et al., 1993). However, Dew and Wright (1998) have found viable seeds of *Ravenala* only in the dung of the largest-bodied living lemurid, *Varecia variegata*. The primary disperser for the Traveller's tree may have been the larger-bodied *Pachylemur*. How such a plant might fare following extinction (or local extirpation) of its endemic seed dispersers is difficult to predict, given its obvious success at colonizing areas that have been denuded by humans. Aggressive vegetative reproduction can create huge stands, albeit lacking in genetic diversity. Ultimately, however, without dispersal partners, the fate of the Traveller's tree may parallel that of the "double coconut" or "coco-de-mer" (*Lodoicea maldivica*) – a borassoid palm today endemic to two small islands in the Seychelles, where it is highly endangered but well represented in a few lingering stands. With fruit weighing in excess of 20 kg and bearing the largest known seed (Edwards et al., 2002), this species has apparently survived long after the demise of its natural dispersers (Corner, 1966). Mabberley (1983, 1988) considers it the earth's most spectacular dispersal anachronism. Unlike the coconut, viable fruit of the double coconut do not float and the seeds are killed by sea water.

Primate Portraits in Ecological Perspective

We now come full circle to the question posed at the start: Can we use our knowledge of life history and trophic characteristics of giant lemurs to identify likely partners for Madagascar's orphaned plants? Which primate niches, if any, have been vacated during the recent extinctions?

Birkinshaw (2001: 484, following Richard and Dewar, 1991) suggests that the extinct lemurs *Pachylemur* spp. and *Archaeolemur* spp. probably belonged to the large-seed dispersal guild. To this we add *Hadropithecus stenognathus*, with the caveat that *Archaeolemur* and *Hadropithecus* were probably primarily seed predators and occasional seed dispersers, while *Pachylemur* was undoubtedly primarily a seed disperser and occasional seed predator. Ecologically, these primates served critical roles in the wooded ecosystems of Madagascar, much like the capuchins, peccaries, atelines, tapirs (and formerly, the gomphotheres, ground sloths, and equids) of the Neotropics, as well as the cercopithecines, apes, and elephants of the Paleotropics. We believe that the palaeopropithecids and megaladapids would have played a lesser role. The former were probably more like colobines and indriids (i.e., mixed feeders and sometimes deliberate seed predators, but very destructive to seeds); the latter were more strictly folivorous. *Daubentonia robusta* would have been, like its congener, a seed predator par excellence, and probably a seed disperser only if it was also a hoarder.

We recognize that seed dispersal is commonly achieved through diverse, essentially redundant, systems (Chapman and Chapman, 2002; Bollen et al., 2004a, b); plant species rarely depend on single co-evolved dispersal partners (Bodmer, 1991; Garber and Lambert, 1998). Certainly on Madagascar, non-primates, including large-bodied reptiles and elephant birds, as well as the omnivorous carnivores endemic to the island, would have contributed to large-seed dispersal. Lord et al. (2002) make a convincing case that moas were important seed dispersers in New Zealand. Collagen-derived stable carbon isotopes of elephant bird bone range from −13.25 to −24.9 (Burney et al., 2004), suggesting strong dietary diversity, and, at the less negative end of the range, a mixed diet including C4 or CAM plants. Dransfield and Bentje (1995a) suggest that *Satranala* may have been dispersed by elephant birds. But primates process seeds differently than birds, and some large-seeded plants may depend on large-bodied frugivores to swallow and defecate seeds; they may not do as well when seeds are spit or regurgitated (Dominy and Duncan, 2005). Endozoochory contributes to long-distance seed dispersal as seed spitting cannot; according to the Janzen-Connell hypothesis (see Ganzhorn et al., 1999), seeds dropped near parent plants are unlikely to be successful. Ingestion can enhance seed germination by several means (Rick and Bowman, 1961; Traveset and Verdú, 2002; Robertson et al., 2006): 1) scarification (slight damage to seed coats increases permeability to water and gases); 2) deinhibition (some seeds will not germinate without pulp removal, because pulp can function as an inhibitor of germination); and 3) fertilization (faecal material nourishes seeds, and can therefore trigger germination).

Pachylemur was undoubtedly the most effective large-seed disperser in the Order Primates on Madagascar. It was the largest-bodied member of a family of effective dispersers of medium or large seeds; lemurids tend to swallow seeds whole or with minimal damage, and seeds passing through their guts have demonstrated high germination success (Dew and Wright, 1998; Birkinshaw, 2001). In the eastern rain forest, fruits with large seeds (>10 mm in length) are dull-colored (green, brown, tan, purplish, or black) and primate-dispersed, while fruits with seeds smaller than 10 mm are generally red, yellow, orange, pink, blue, or white and not primate-dispersed (Dew and Wright, 1998, on *Eulemur rubriventer*, *Eulemur fulvus*, and *Varecia variegata*). The same pattern holds for *Eulemur macaco* in the northwest (Birkinshaw, 2001), and *Eulemur collaris* in the southeast (Bollen et al, 2004a, b). In the dry forest in western Madagascar today, only the introduced bush pig *Potamochoerus larvatus* and endemic *Eulemur fulvus rufus* (the largest-bodied lemurid remaining in the west) ingest seeds larger than 11 mm in length (Ganzhorn et al., 1999). Species belonging to the genus *Pachylemur* were widespread in Madagascar, from the southern to the northern tips of Madagascar and into the central highlands (Godfrey and Jungers, 2002). They occupied sites with extant lemur communities much like those of modern eastern rainforests (Godfrey et al., 1999); this suggests that *Pachylemur* also occupied at least some eastern rain forests (where there are no subfossil sites). Thus, most, if not all, of Madagascar's forests would have had large and smaller-bodied lemurids acting as their primary seed dispersers. We know that *Varecia* can swallow seeds intact with diameters of over 30 mm (Dew and Wright, 1998) and that *Eulemur* can do the same for seeds with diameters of 20–30 mm (Birkinshaw, 2001). If the ratio of seed width to the cube root of body mass were similar for *Pachylemur* and *Eulemur*, then *Pachylemur* would have been able to swallow seeds with diameters up to $\sim$50 mm – perhaps more.

It is likely that Madagascar's primate seed predators also played an essential role in seed dispersal. We hypothesize that this was the case for the Archaeolemuridae, who were considerably larger in body size than the largest species of *Pachylemur* and had much more robust jaws. Seed predators may contribute to the process of seed dispersal in a number of ways (Norconk et al., 1998). First, and importantly, they may regularly open fruits that are difficult for species with less robust jaws to process, making available to other species (including benign frugivores) discarded fruit pulp with untouched seeds. Elsewhere in the world, palm nut exocarp and mesocarp are eaten by many species (including very small-bodied species), but the fruits must be opened by animals with jaws powerful enough to do so, or the ability to use tools to crack them open (Boesch and Boesch, 1983; Visalberghi, 1990; Anderson, 1990; Daegling, 1992; Anapol and Lee, 1994; Inoue-Nakamura and Matsuzawa, 1997; Ottoni and Mannu, 2001; Fragaszy et al., 2004; Wright, 2005). Secondly, seed predators disperse seeds by passing through their guts the occasional seed that is minimally injured (or not at all). Germination success of intact seeds varies by species, and may depend on internal digestive processing, including gut transit time. Among living species, the ability to function as an effective seed disperser seems to be inversely correlated with transit time; thus, for example, the most effective seed dispersers among primates (e.g., the more frugivorous atelines and lemurids) have short transit times, whereas species with longer gut transit time (such as indriids) tend to be less effective. Thirdly, if seed predators hoard fruits or seeds (in the manner of rodents), then seeds in forgotten caches will have been inadvertently dispersed.

Faunivores tend to have fast transit times, followed by omnivores and frugivores; folivores have slow transit times (Lambert, 1998). Diet is more important than body size in determining transit time (Lambert, 1998). This is because, to ensure effective nutrient processing, foods (such as fruit) that are high in soluble carbohydrates must pass through the gut more quickly than foods (such as leaves) that require long fermentation for effective nutrient uptake (Milton, 1984; Lambert, 1998). Indeed, Milton (1986) described an efficiency-velocity tradeoff in primates and other animals: species that maximize nutrient extraction and thus gut efficiency (folivores, with long gut retention) necessarily compromise their capacity to exploit fruit regularly, whereas frugivores, with short gut transit time can maximize the total mass of food processed per unit time at the expense of being able to process leaves effectively. The combination of digestive anatomy (the length and complexity of components of the digestive tract) and transit time constrains diet; frugivores actually require fast transit, while folivores require prolonged fermentation and slower transit (Milton, 1993; Lambert, 1998). Among extant lemurs, the relevant contrast is between lemurids (with the exception of the bamboo specialists) with very short transit time, and the indriids with long transit time (Campbell et al., 2004).

This suggests that if one can reconstruct extinct species as primarily frugivorous or primarily folivorous, one can estimate their potential for seed dispersal. Many characteristics of the Archaeolemuridae point to a key ecological role as predominantly frugivorous seed predators and hard-object processors. Especially when combined with omnivory (as in *Cebus apella* or *Daubentonia madagascariensis*), hard-object processing allows individuals to gain access to high-quality foods that are otherwise unavailable. As reviewed above, *Archaeolemur* and *Hadropithecus* had robust jaws and dental architectural features characteristic of hard-object feeders. Coprolites of *Archaeolemur* demonstrate omnivory. Evidence for high encephalization and relatively slow dental development supports the inference that the archaeolemurids depended on complex extractive foraging; high encephalization is characteristic of extractive foragers and correlated with prolonged maternal investment. If the archaeolemurids did indeed have "high-quality" diets, then it is also likely that they had relatively short gut retention, which may have increased their effectiveness as seed dispersers, despite their propensity for seed predation.

The other extinct lemurs were probably either highly destructive to seeds or rarely consumed them at all. The Megaladapidae, as specialized folivores, would have consumed them only incidentally. We hypothesize that the Palaeopropithecidae (as predominantly folivorous seed predators) were also destructive to seeds, as are their closest living relatives, the Indriidae, or their ecological vicars on continental Africa and Asia, the colobines. Indriids today are *physiological* folivores. Some do consume a lot of fruit, but they tend to consume unripe fruits whole. They also deliberately extract and masticate seeds to the point that the seeds are virtually never distinguishable in feces (Hemingway, 1996; Ralisoamalala, 1996; Dew and Wright, 1998; Overdorff and Strait, 1998; Ganzhorn et al., 1999). Occasional intact seeds (generally small ones; see Dew and Wright, 1998) suffer prolonged fermentation due to slow gut transit time, which may compromise the probability of successful germination. Colobines are similarly destructive to seeds; cercopithecines and particularly apes are far better seed dispersers (Davies, 1991; Peters, 1993; Wrangham et al., 1994; Lambert, 1999; Dominy and Duncan, 2005).

Summary and Conclusions

In summary, a simple picture is beginning to emerge from the combined analysis of life history profiles and trophic adaptations of extinct lemurs. It is one that has parallels in other regions of the world. We agree with Bollen et al. (2004b) that the nearly exclusive dependence of some larger-seeded trees on the remaining large-bodied lemurids today is likely a byproduct of the extinction of larger, frugivorous birds and lemurs, and that megafaunal extinctions have indeed put the largest-seeded trees in jeopardy. To this, we add the following observations:

First, within the Order Primates at least, the benign-frugivore guild on Madagascar was not much larger in the past than it is today, including only two additional members – the two species of the genus *Pachylemur*. However, there is excellent reason to believe that these species played a pivotal role in large-seed dispersal, and that their disappearance is having an adverse effect on Madagascar's forest regeneration.

Secondly, Madagascar had a large number of specialized primate folivores in the past, just as it is does today. The high percentage of primate folivores on Madagascar today is not an artifact of recent extinctions.

Thirdly, Madagascar did have many mixed feeders (large-bodied folivore/frugivores), including most of the palaeopropithecids – close relatives of indriids with similar developmental profiles and microwear signaling similar diets. High folivory in sloth lemurs is also suggested by their high shearing capacity. The largest-bodied extinct lemurs belonged to this guild, which suffered a decline in the recent past. The extent to which the extinction of these species adversely affected seed dispersal is unknown. If, in fact, palaeopropithecids were physiological folivores, it is likely that they had slow gut transit times, and were similarly destructive to seeds.

However, there existed a guild of large-bodied frugivore/omnivores that suffered an even-more-major decline, and that, we believe, did play a critical role in the long-distance dispersal of large seeds. These were Madagascar's hard-object processors, including the archaeolemurids and daubentoniids. These species had jaws powerful enough to crush the hardest foods, or they had other means (chisel-like incisors) of gaining access to well-protected resources. The combination of omnivory and exploitation of hard or protected plant parts likely allowed year-round access to high-quality foods. These species were apparently able to maximize quality-resource use even in forests with low productivity and during the least productive times of the year. That access came with a price, however: prolonged maternal investment.

Within this guild, the archaeolemurids were almost certainly less destructive to seeds than the specialized daubentoniids. There is also strong evidence that members of this guild targeted different plant and/or animal species; they exhibit very different stable carbon isotope signals even where sympatric. *Hadropithecus* was the single lemur (extinct or extant) with a distinct (though likely not exclusive) preference for CAM or C4 plants, and/or for animals consuming such plants. The ecological roles assumed by the Archaeolemuridae would have been comparable to those served by large-bodied cercopithecines, apes and elephants in the Paleotropics, or by peccaries and capuchins in the Neotropics.

Finally, there are indeed tree species in Madagascar that appear to have been orphaned by the recent disappearance of Madagascar's megafauna. A more thorough review of possible ecological anachronisms (especially large-seeded trees with adaptations for endozoochory) is warranted. In addition, ecological investigations of the dispersal and recruitment patterns of Madagascar's large-seeded flora will elucidate their reproductive strategies and more clearly reveal which species may be suffering most from recent faunal extinctions. This information will contribute not merely to our understanding of past changes in Malagasy ecosystems, but to preventing further extinctions and ecological disruptions in the future.

Acknowledgment We are delighted to contribute this chapter to honor our friend and colleague, Elwyn Simons. LRG and WLJ in particular owe a great debt to Elwyn, his lab and field partner, Prithijit Chatrath, and their families, whose love and generosity we will always cherish. Thank you, Elwyn and Prithijit, for inviting us to join you in the field in Madagascar. This research would not have been possible without the help of many other people (too numerous to mention), including our colleagues in Madagascar (especially the late Mme. Berthe Rakotosamimanana and Mme. Gisèle Randria), additional partners in the field (particularly David Burney) and partners in the laboratory (particularly Marina Blanco, Kierstin Catlett, Frank Cuozzo, Stephen King, Patrick Mahoney, Karen Samonds, and Gina Semprebon). We are exceedingly grateful, also, for comments on an earlier draft by Joerg Ganzhorn, Bob Sussman, Matt Sponheimer, and Ian Tattersall, and for additional information provided by colleagues, including David Baum, Kurt Benirschke, Martin Callmander, Neal Hockley, and Eleanor Sterling, in response to our queries. Grant support from the National Science Foundation for various aspects of this research is gratefully acknowledged. Finally, we thank John Fleagle and Chris Gilbert for their skilled editing, and for putting together this wonderful tribute to Elwyn's career.

References

Allen, J. A. (2004). Avian and mammalian predators of terrestrial gastropods. In: Barker, G. M. (ed.), *Natural Enemies of Terrestrial Molluscs*. CABI Publishing, Cambridge, MA, pp. 1–36.

Anapol, F., and Lee, S. (1994). Morphological adaptation to diet in platyrrhine primates. *Am. J. Phys. Anthropol.* 94(2):239–261.

Anderson, J. R. (1990). Use of objects as hammers to open nuts by capuchin monkeys (*Cebus apella*). *Folia Primatol.* 54:138–145.

Atsalis, S. (1999). Diet of the brown mouse lemur (*Microcebus rufus*) in Ranomafana National Park, Madagascar. *Int. J. Primatol.* 20:193–229.

Barlow, C. (2000). *The Ghosts of Evolution: Nonsensical Fruit, Missing Partners, and Other Ecological Anachronisms.* Basic Books, New York.

Baum, D. A. (1995). A systematic revision of *Adansonia* (Bombacaceae). *Ann. Mo. Bot. Gard.* 82:440–471.

Baum, D. A. (2003). Bombacaceae, *Adansonia*, Baobab, *Bozy*, *Fony*, *Renala*, *Ringy*, *Za*. In: Goodman, S. M., and Benstead, J. P. (eds.), *The Natural History of Madagascar.* University of Chicago Press, Chicago, pp. 339–342.

Benirschke, K., Byrd, M. L., and Meritt, D. (1990). New observations on the Chacoan peccary, *Catagonus wagneri*. *Internationalen Symposiums über die Erkrankungen der Zoo- und Wildtiere* 32:341–347.

Beynon, A. D., Dean, M. C., and Reid, D. J. (1991). Histological study on the chronology of the developing dentition in gorilla and orangutan. *Am. J. Phys. Anthropol.* 86:189–203.

Birkinshaw, C. R. (2001). Fruit characteristics of species dispersed by the black lemur (*Eulemur macaco*) in the Lokobe Forest, Madagascar. *Biotropica* 33(3):478–486.

Birkinshaw, C. R., and Colquhoun, I. C. (2003). Lemur food plants. In: Goodman, S. M., and Benstead, J. P. (eds.), *The Natural History of Madagascar.* University of Chicago Press, Chicago, pp. 1207–1220.

Bodmer, R. E. (1991). Strategies of seed dispersal and seed predation in Amazonian ungulates. *Biotropica* 23(3):255–261.

Boesch, C., and Boesch, H. (1983). Optimization of nut cracking with natural hammers by wild chimpanzees. *Behaviour* 83:265–286.

Bollen, A., Van Elsacker, L., and Ganzhorn, J. U. (2004a). Relations between fruits and disperser assemblages in a Malagasy littoral forest: A community-level approach. *J. Trop. Ecol.* 20(Part 6):599–612.

Bollen, A., Van Elsacker, L., and Ganzhorn, J. U. (2004b). Tree dispersal strategies in the littoral forest of Sainte Luce (SE Madagascar). *Oecologia* 139(4):604–616.

Bromage, T. G. (1987). The biological and chronological maturation of early hominids. *J. Hum. Evol.* 16:257–272.

Burney, D. A. (1999). Rates, patterns, and processes of landscape transformation and extinction in Madagascar. In: MacPhee, R. D. E. (ed.), *Extinction in Near Time.* Kluwer/Plenum, New York, pp. 145–164.

Burney, D. A.(2005). Finding the connections between paleoecology, ethnobotany, and conservation in Madagascar. *Ethnobotany Research & Applications* 3:385–389.

Burney, D. A., and Ramilisonina. (1998). The kilopilopitsofy, kidoky, and bokyboky: Accounts of strange animals from Belo-sur-Mer, Madagascar, and the megafaunal "extinction window." *Am. Anthropol.* 100:957–966.

Burney, D. A., James, H. F., Grady, F. V., Rafamantanantsoa, J.-G., Ramilisonina, Wright, H. T., and Jull, A. J. T. (1997). Environmental change, extinction, and human activity: Evidence from caves in NW Madagascar. *J. Biogeogr.* 24:755–767.

Burney D. A., Burney, L. P., Godfrey, L. R., Jungers, W. L., Goodman, S. M., Wright, H. T., and Jull, A. J. T. (2004). A chronology for late prehistoric Madagascar. *J. Hum. Evol.* 47:25–63.

Bush, E. C., and Allman, J. M. (2004). The scaling of frontal cortex in primates and carnivores. *P. Natl. Acad. Sci. USA* 101:3962–3966.

Calley, M., Braithwaite, R. W., and Ladd, P. G. (1993). Reproductive biology of *Ravenala madagascariensis* Gmel. as an alien species. *Biotropica* 25(1):61–72.

Callmander, M. W., and Laivao, M. O. (2003). Pandanaceae, *Pandanus*. In: Goodman, S. M., and Benstead, J. P. (eds.), *The Natural History of Madagascar.* University of Chicago Press, Chicago, pp. 460–467.

Campbell, J. L., Williams, C. V., and Eisemann, J. H. (2004). Use of total dietary fiber across four lemur species (*Propithecus verreauxi coquereli*, *Hapalemur griseus griseus*, *Varecia*

variegata, and *Eulemur fulvus*): Does fiber type affect digestive efficiency? *Am. J. Primatol.* 64(3):323–335.
Chapman, C. A., and Chapman, L. J. (2002). Plant-animal coevolution: Is it thwarted by spatial and temporal variation in animal foraging? In: Levey, D. J., Silva, W. R., and Galetti, M. (eds.), *Seed Dispersal and Frugivory: Ecology, Evolution and Conservation*. CABI Publishing, New York, pp. 275–290.
Corner, E. J. H. (1966). *The Natural History of Palms*. Weidenfeld and Nicolson, London.
Crovella, S., Montagnon, D., Rakotosamimanana, B., and Rumpler, Y. (1994). Molecular biology and systematics of an extinct lemur, *Pachylemur insignis*. *Primates* 35: 519–522.
Crovella, S., Montagnon, D., and Rumpler, Y. (1993). Highly repeated DNA analysis and systematics of the Lemuridae, a family of Malagasy prosimians. *Primates* 34:61–69.
Davies, G. (1991). Seed-eating by red leaf monkeys (*Presbytis rubicunda*) in dipterocarp forest of northern Borneo. *Int. J. Primatol.* 12(2):119–144.
Davis, A., and Bridson, D. (2003). Introduction to the Rubiaceae. In: Goodman, S. M., and Benstead, J. P. (eds.), *The Natural History of Madagascar*. University of Chicago Press, Chicago, pp. 431–434.
Daegling, D. J. (1992). Mandibular morphology and diet in the genus *Cebus*. *Int. J. Primatol.* 13(5):545–570.
Dean, M. C., Beynon, A. D., Thackery, J. R., and Macho, G. A. (1993). Histological reconstruction of dental development and age at death of a juvenile *Paranthropus robustus* specimen, SK 63, from Swartkrans, South Africa. *Am. J. Phys. Anthropol.* 91:401–419.
Dew, J. L., and Wright, P. C. (1998). Frugivory and seed dispersal by primates in Madagascar's eastern rainforest. *Biotropica* 30:425–437.
Dominy, N. J., Svenning, J.-C., and Li, W.-H. (2003). Historical contingency in the evolution of primate color vision. *J. Hum. Evol.* 44:25–45.
Dominy, N. J., and Duncan, B. W. (2005). Seed-spitting primates and the conservation and dispersal of large-seeded trees. *Int. J. Primatol.* 26(3):631–649.
Dransfield, J., and Beentje, H. (1995a). *Satranala* (Coryphoideae: Borasseae: Hyphaeninae), a new palm genus from Madagascar. *Kew Bull.* 50:85–92.
Dransfield, J., and Beentje, H. (1995b). *The Palms of Madagascar*. Royal Botanic Gardens and the International Palm Society, Kew, Great Britain.
Du Puy, D. J., Labat, J.-N., Rabevohitra, R., Villiers, J.-F., Bosser, J., and Moat, J. (2002). *The Leguminosae of Madagascar*. Royal Botanic Gardens, Kew, Great Britain.
Edwards, P. J., Kollmann, J., and Fleischmann, K. (2002). Life history evolution in *Lodoicea maldivica* (Arecaceae). *Nordic J. Bot.* 22(2):227–237.
Ferrand, G. (1893). *Contes Populaires Malgaches*. Ernest Leroux, Paris.
Fernandes, M. E. B. (1991). Tool use and predation of oysters (*Crassostrea* Rhizophorae) by the tufted capuchin, *Cebus apella apella*, in brackish water mangrove swamp. *Primates* 32(4):529–531.
Fleagle, J. G., and Reed, K. D. (1996). Comparing primate communities: A multivariate approach. *J. Hum. Evol.* 30:489–510.
Fragaszy, D. M., and Adams-Curtis, L. E. (1997). Developmental changes in manipulation in tufted capuchins (*Cebus apella*) from birth through 2 years and their relation to foraging and weaning. *J. Comp. Psych.* 111:201–211.
Fragaszy, D. M., Baer, J., and Adams-Curtis, L. (1991). Behavioral development and maternal care in tufted capuchins (*Cebus apella*) and squirrel monkeys (*Saimiri sciureus*) from birth through 7 months. *Dev. Psychobiol.* 24:375–393.
Fragaszy, D., Izar, P., Visalberghi, E., Ottoni, E. B., and De Oliveira, M. G. (2004). Wild capuchin monkeys (*Cebus libidinosus*) use anvils and stone pounding tools. *Am. J. Primatol.* 64(4):359–366.

Fragoso, J. M. V. (1997). Tapir-generated seed shadows: Scale-dependent patchiness in the Amazon rain forest. *J. Ecol.* 85(4):519–529.

Fragoso, J. M. V., and Huffman, J. M. (2000). Seed dispersal and seedling recruitment patterns by the last Neotropical megafaunal element in Amazonia, the tapir. *J. Trop. Ecol.* 16(Part 3):369–385.

Ganzhorn, J. U. (1995). Cyclones over Madagascar: Fate or fortune? *Ambio* 24:124–125.

Ganzhorn, J. U., Fietz, J., Rakotovao, E., Schwab, D., and Zinner, D. (1999). Lemurs and the regeneration of dry deciduous forest in Madagascar. *Conserv. Biol.* 13(4):794–804.

Garber, P. A., and Lambert, J. E. (1998). Primates as seed dispersers: Ecological processes and directions for future research. *Am. J. Primatol.* 45(1):3–8.

Gibson, K. R. (1986). Cognition, brain size and the extraction of embedded food resources. In: Else, J. G. (ed.), *Primate Evolution*. Cambridge University Press, New York, pp. 93–103.

Godfrey, L. R. (1986). The tale of the tsy-aomby-aomby. *The Sciences* 1986, 49–51.

Godfrey, L. R., and Irwin, M. T. (in press). The evolution of extinction risk: Past and present anthropogenic impacts on the primate communities of Madagascar. In: Soligo, C. (guest ed., special issue), *Folia Primatol.*

Godfrey, L. R., and Jungers, W. L. (2002). Quaternary fossil lemurs. In: Hartwig, W. C. (ed.), *The Primate Fossil Record*. Cambridge University Press, Cambridge, pp. 97–121.

Godfrey, L. R., Simons, E. L., Chatrath, P. S., and Rakotosamimanana, B. (1990). A new fossil lemur (*Babakotia*, Primates) from northern Madagascar. *C. R. Acad. Sci.*, Paris 310:81–87.

Godfrey, L. R., Jungers, W. L., Wunderlich, R. E., and Richmond, B. G. (1997a). Reappraisal of the postcranium of *Hadropithecus* (Primates, Indroidea). *Am. J. Phys. Anthropol.* 103:529–556.

Godfrey, L. R., Jungers, W. J., Reed, K. E., Simons, E. L., and Chatrath, P. S. (1997b). Inferences about past and present primate communities in Madagascar. In: Goodman, S. M., and Patterson, B. D. (eds.), *Natural Change and Human Impact in Madagascar*. Smithsonian Institution Press, Washington DC, pp. 218–256.

Godfrey, L. R., Jungers, W. L., Simons, E. L., and Chatrath, P. S. (1999). Past and present distributions of lemurs in Madagascar. In: Rakotosamimanana, B., Rasamimanana, H., Ganzhorn, J. U., and Goodman, S. M. (eds.), *New Directions in Lemur Studies*. Kluwer Academic / Plenum, New York, pp. 19–53.

Godfrey, L. R., Semprebon, G. M., Jungers, W. L., Sutherland, M. R., Simons, E. L., and Solounias, N. (2004). Dental use wear in extinct lemurs: Evidence of diet and niche differentiation. *J. Hum. Evol.* 47:145–169.

Godfrey, L. R., Semprebon, G. M., Schwartz, G. T., Burney, D. A., Jungers, W. L., Flanagan, E. K., Cuozzo, F. P., and King, S. J. (2005). New insights into old lemurs: The trophic adaptations of the Archaeolemuridae. *Int. J. Primatol.* 26(4):825–854.

Godfrey, L. R., Jungers, W. L., and Schwartz, G. T. (2006a). Ecology and extinction of Madagascar's subfossil lemurs. In: Gould, L., and Sauther, M. L. (eds.), *Lemurs: Ecology and Adaptation*. Springer, New York, pp. 41–64.

Godfrey, L. R., Schwartz, G. T., Samonds, K. E., Jungers, W. L., and Catlett, K. K. (2006b). The secrets of lemur teeth. *Evol. Anthropol* 15:142–154.

Godfrey, L. R., Jungers, W. L., Burney, D. A., Vasey, N., Ramilisonina, Wheeler, W., Lemelin, P., Shapiro, L. J., Schwartz, G. T., King, S. J., Ramarolahy, M. F., Raharivony, L. L., and Randria, G. F. N. (2006c). New discoveries of skeletal elements of *Hadropithecus stenognathus* from Andrahomana Cave, southeastern Madagascar. *J. Hum. Evol* 51:395–410.

Goodman, S. and Benstead, J. P. (eds.) (2003). The Natural History of Madagascar. University of Chicago Press, Chicago.

Goodman, S. M., and Benstead, J. P. (2005). Short communication: Updated estimates of biotic diversity and endemism for Madagascar. *Oryx* 39:73–77.

Goodman, S. M., and Ganzhorn, J. U. (1997). Rarity of figures (*Ficus*) on Madagascar and its relationship to a depauperate frugivore community. *La Terre et La Vie* 52(4):321–329.

Gould, L., and Sauther, M. L. (eds.), (2006). *Lemurs: Ecology and Adaptation*. Springer, New York.

Grandidier, G. (1929). Une varieté du *Cheiromys madagasacariensis* actuel et un nouveau *Cheiromys* subfossile. *Bull. Acad. Malgache* 11:101–107.

Grubb, P.J.(2003). Interpreting some outstanding features of the flora and vegetation of Madagascar. *Perspectives in Plant Ecology, Evolution and Systematics* 6(1–2):125–146.

Hamrick, M. W., Simons, E. L., and Jungers, W. L. (2000). New wrist bones of the Malagasy giant subfossil lemurs. *J. Hum. Evol.* 38(5):635–650.

Hemingway, C. A. (1996). Morphology and phenology of seeds and whole fruit eaten by Milne-Edwards' sifaka, *Propithecus diadema edwardsi*, in Ranomafana National Park, Madagascar. *Int. J. Primatol.* 17(5):637–659.

Hoffman, P., and McPherson, G. (2003). Euphorbiaceae – Overview. In: Goodman, S. M., and Benstead, J. P. (eds.), *The Natural History of Madagascar*. University of Chicago Press, Chicago, pp. 379–383.

Ilse, L. M., and Hellgren, E. C. (1995). Resource partitioning in sympatric populations of collared peccaries and feral hogs in southern Texas. *J. Mammal.* 76(3):784–799.

Inoue-Nakamura, N., and Matsuzawa, T. (1997). Development of stone tool use by wild chimpanzees (*Pan troglodytes*). *J. Comp. Psychol.* 111(2):159–173.

Irwin, M. T. (2006a). Leaner diet, less energy: Diademed sifakas (*Propithecus diadema*) in forest fragments show reduced ranging and altered activity patterns. *Am. J. Phys. Anthropol.* Suppl. 42:108.

Irwin, M. T. (2006b). Ecologically enigmatic lemurs: The sifakas of the eastern forests (*Propithecus candidus, P. diadema, P. edwardsi, P. perrieri* and *P. tattersalli*). In: Gould, L., and Sauther, M. L. (eds.), *Lemurs: Ecology and Adaptation*. Springer, New York, pp. 305–326.

Iwano, T. (1991). The usage of the digits of a captive aye–aye (*Daubentonia madagascariensis*). *African Study Monographs* 12(2):87–98.

Iwano, T., and Iwakawa, C. (1988). Feeding behavior of the aye–aye (*Daubentonia madagascariensis*) on nuts of ramy (*Canarium madagascariensis*). *Folia Primatol.* 50:136–142.

Iwano, R., Randalana, R., and Rakotoarisoa, G. (1991). Ecology of the aye–aye (*Daubentonia madagascariensis*) I. Distribution. In: Ehara, A., Kimura, T., Takenaka, O., and Iwamoto, M. (eds.), *Primatology Today: Proceedings of the XIIIth Congress of the International Primatological Society, Nagoya and Nyoto, 18–24 July (1990)*. Elsevier Science Publishers, Amsterdam, pp. 41–42.

Janzen, D. H., and Martin, P. S. (1982). Neotropical anachronisms: The fruits the gomphotheres ate. *Science* 215:19–27.

Jungers, W. L. (1978). The functional significance of skeletal allometry in *Megaladapis* in comparison to living prosimians. *Am. J. Phys. Anthropol.* 49:303–314.

Jungers, W. L. (1999). Brain size and body size in subfossil Malagasy lemurs. *Am. J. Phys. Anthropol.* Suppl. 28:163.

Jungers, W. L., Lemelin, P., Godfrey, L. R., Wunderlich, R. E., Burney, D. A., Simons, E. L., Chatrath, P. S., James, H. F., and Randria, G. F. N. (2005). The hands and feet of *Archaeolemur*: Metrical affinities and their functional significance. *J. Hum. Evol.* 49(1):36–55.

Jungers, W. L., Godfrey, L. R., Simons, E. L., Wunderlich, R. E., Richmond, B. G., and Chatrath, P. S. (2002). Ecomorphology and behavior of giant extinct lemurs from Madagascar. In: Plavcan, J. M., Kay, R. F., Jungers, W. L., and van Schaik, C. P. (eds), *Reconstructing Behavior in the Primate Fossil Record*. Kluwer Academic / Plenum Publishers, New York, pp. 371–411.

Jungers, W. L., Demes, B., and Godfrey, L. R. (2008). How big were the "giant" extinct lemurs of Madagascar? In: Fleagle, J. G. and Gilbert, C. C. (eds.), *Elwyn Simons: A Search for Origins*. Springer, New York, pp. 343–360.

Karanth, K. P., Delefosse, T., Rakotosamimanana, B., Parsons, T. J., and Yoder, A. D. (2005). Ancient DNA from giant extinct lemurs confirms single origin of Malagasy primates. *P. Natl. Acad. Sci. USA*. 102(14):5090–5095.

Kay, R. F. (1975). The functional adaptations of primate molar teeth. *Am. J. Phys. Anthropol.* 43:195–216.

Kay, R. F. (1984). On the use of anatomical features to infer foraging behavior in extinct primates. In: Rodman, P. S., and Cant, J. G. H. (eds.), *Adaptations for Foraging in Nonhuman Primates*. Columbia University Press, New York, pp. 21–53.

Kaufman, J. A., Ahrens, E. T., Laidlaw, D. H., Zhang, S., and Allman, J. M. (2005). Anatomical analysis of an aye–aye brain (*Daubentonia madagascariensis*, Primates: Prosimii) combining histology, structural magnetic resonance imaging, and diffusion-tensor imaging. *Anat. Rec.* 287A (1):1026–1037.

King, S. J., Blanco, M. B., and Godfrey, L. R. (2005). Dietary reconstruction of *Archaeolemur* using dental topographic analysis. *Am. J. Phys. Anthropol. Suppl.* 40:129.

Kitamura, S., Suzuki, S., Yumoto, T., Poonswad, P., Chuailua, P., Plongmai, K., Maruhashi, T., Noma, N., and Suckasam, C. (2006). Dispersal of *Canarium euphyllum* (Burseraceae), a large-seeded tree species, in a moist evergreen forest in Thailand. *J. Trop. Ecol.* 22:137–146.

Krakauer, E., and van Schaik, C. P. (2005). Independent and social learning in the development of aye–aye tap-foraging skills. *Am. J. Phys. Anthropol. Suppl.* 40:132–133.

Kress, W. J., Schatz, G. E., Andrianifahanana, and Morland, H. S. (1994). Pollination of *Ravenala madagascariensis* (Strelitziaceae) by lemurs in Madagascar: Evidence for an archaic coevolutionary system. *Am. J. Bot.* 81(5):542–551.

Labat, J.-N., and Moat, J. (2003). Leguminosae (Fabaceae). In: Goodman, S. M., and Benstead, J. P. (eds.), *The Natural History of Madagascar*. University of Chicago Press, Chicago, pp. 346–373.

Lambert, J. E. (1998). Primate digestion: Interactions among anatomy, physiology, and feeding ecology. *Evol. Anthropol.* 7:8–20.

Lambert, J. E. (1999). Seed handling in chimpanzees (*Pan troglodytes*) and redtail monkeys (*Cercopithecus ascanius*): Implications for understanding hominoid and cercopithecine fruit-processing strategies and seed dispersal. *Am. J. Phys. Anthropol.* 109(3):365–386.

Lamberton, C. (1934). Contribution a la connaissance de la faune subfossole de Madagascar. Lemuriens et Ratites. Chiromys robustus sp. nov. Lamb. Memories de I'Academic Malgache 17:40–46.

Lee-Thorp, J., van der Merwe, N., and Brain, C. K. (1994). Diet of *Australopithecus robustus* at Swartkrans from stable carbon isotopic analysis. *J. Hum. Evol.* 27:361–372.

Lee-Thorp, J., Thackeray, J. F., and van der Merwe, N. (2000). The hunters and the hunted revisited. *J. Hum. Evol.* 39:565–576.

Linnaeus, C. (1758). *Systema naturae*. 10th edition. Holmiae.

Lord, J. M., Markey, A. S., and Marshall, J. (2002). Have frugivores influenced the evolution of fruit traits in New Zealand? In: Levey, D. J., Silva, W. R., and Galetti, M. (eds.), *Seed Dispersal and Frugivory: Ecology, Evolution and Conservation*. CABI Publishing, Oxford, pp. 55–68.

Mabberley, D. J. (1983). *Tropical Rain Forest Ecology*. Blackie, Glasgow.

Mabberley, D. J. (1988). The living past: Time state of the tropical rain forest. In: Reynolds, E. R. C., and Thompson, F. B. (eds.), *Forests, Climate, and Hydrology*. United Nations University Press, printed by Kefford Press, Singapore [http://www.unu.edu/unupress/unupbooks/80635e/80635E03.htm]

MacFadden, B. J., and Cerling, T. E. (1996). Mammalian herbivore communities, ancient feeding ecology, and carbon isotopes: A 10 million year sequence from the Neogene of Florida. *J. Vertebr. Paleontol.* 16(1):103–115.

Mantaux, C. G. (1971). Il y a cent ans à Madagascar: Un grand savant Alfred Grandidier. *Revue de Madagascar* Sér. 4(53–54):33–62.

Martin, R. D. (1990). *Primate Origins and Evolution: A Phylogenetic Reconstruction.* Princeton University Press, Princeton, NJ.

Milton, K. (1984). The role of food processing factors in primate food choice. In: Rodman, P. S., and Cant, J. G. H. (eds.), *Adaptations for Foraging in Nonhuman Primates: Contributions to an Organismal Biology of Prosimians, Monkeys and Apes*, Columbia University Press, New York, pp. 249–279.

Milton, K. (1986). Digestive physiology in primates. *News Physiol. Sci.* 1:76–79.

Milton, K. (1993). Diet and primate evolution. *Sci. Am.* 269:86–93.

Norconk, M. A., Grafton B. W., and Conklin-Brittain, N. L. (1998). Seed dispersal by neotropical seed predators. *Am. J. Primatol.* 45(1):103–126.

Norconk, M. A., Grafton, B. W., and McGraw, W. S. (2006). The international context: Pitheciins and patterns of seed predation in the Neotropics, Africa, and Asia. *Int. J. Primatol.* 27(Suppl. 1): Abstract # 512.

Ottoni, E. B., and Mannu, M. (2001). Semifree-ranging tufted capuchins (*Cebus apella*) spontaneously use tools to crack open nuts. *Int. J. Primatol.* 22:347–358.

Overdorff, D. J., and Strait, S. G. (1998). Seed handling by three prosimian primates in southeastern Madagascar: Implications for seed dispersal. *Am. J. Primatol.* 45(1):69–82.

Parker, S. T., and Gibson, K. R. (1977). Object manipulation, tool use and sensorimotor intelligence as feeding adaptations in cebus monkeys and great apes. *J. Hum. Evol.* 6:623–641.

Perez, V. R., Godfrey, L. R., and Nowak-Kemp, M. (2003). Box 4. Butchered sloth lemurs. *Evol. Anthropol.* 12(6):260.

Peters, C. R. (1993). Shell strength and primate seed predation of nontoxic species in eastern and southern Africa. *Int. J. Primatol.* 14(2):315–344.

Peters, C. R., and Vogel, J. C. (2005). Africa's wild C-4 plant foods and possible early hominid diets. *J. Hum. Evol.* 48:219–236.

Port-Carvalho, M., Ferrari, S. F., and Magalhaes, C. (2004). Predation of crabs by tufted capuchins (*Cebus apella*) in eastern Amazonia. *Folia Primatol.* 75(3):154–156.

Quiroga-Castro, V. D., and Roldán, A. I. (2001). The fate of *Attalea phalerata* (Palmae) seeds dispersed to a tapir latrine. *Biotropica* 33(3):472–477.

Radinsky, L. B. (1970). The fossil evidence of prosimian brain evolution. In: Noback, C. R., and Montagna, W. (eds), *The Primate Brain.* Appleton-Centrury-Crofts, New York, pp. 209–222.

Rafferty, K. L., Teaford, M. F., and Jungers, W. L. (2002). Molar microwear of subfossil lemurs: Improving the resolution of dietary inferences. *J. Hum. Evol.* 43:645–657.

Ralisoamalala, R. C. (1996). Role de *Eulemur fulvus rufus* (Audeberg 1799) et de *Propithecus verreauxi verreauxi* (A Grandier 1867) dans des dissemination des graines. In: Ganzhorn, J. U., and Sorg, J.-P (eds.), *Primate Report 46-1 Special Issue: Ecology and Economy of a Tropical Dry Forest in Madagascar*. German Primate Center, Gottingen, Germany, pp. 285–293.

Ratsirarson, J. (2003). *Dysis decaryi*, triangle palm, lafa. In: Goodman, S. M., and Benstead, J. P. (eds.), *The Natural History of Madagascar*. University of Chicago Press, Chicago, pp. 457–459.

Ratsirarson, J., and Silander, Jr., J. A. (1997). Factors affecting the distribution of a threatened Madagascar palm species *Dypsis decaryi* Jumelle. *Principes* 41:100–111.

Raybaud (1902). Rapport de M. l'administrateur Raybaud sur les gisements fossilifères d'Ampasambazimba. *Bull. Acad. Malgache* 1(2):64–67.

Redford, K. H. (1992). The empty forest. *Bioscience* 42(6):412–422.

Reed, K. E., and Fleagle, J. G. (1995). Geographic and climatic control of primate diversity. *P. Natl. Acad. Sci. USA* 92:7874–7876.

Reed, K. E., and Bidner, L. R. (2004). Primate communities: Past, present, and possible future. *Am. J. Phys. Anthropol. Suppl.* 39:2–39.

Richard, A. F., and Dewar, R. E. (1991). Lemur ecology. *Annu. Rev. Ecol. Syst.* 22:145–175.

Rick, C. M., and Bowman, R. I. (1961). Galápagos tomatoes and tortoises. *Evolution* 15:407–417.

Robertson, A. W., Trass, A., Ladley, J. J., and Kelly, D. (2006). Assessing the benefits of frugivory for seed germination: The importance of the deinhibition effect. *Funct. Ecol.* 20:58–66.

Rosenberger, A. (1992). Evolution of feeding niches in New World monkeys. *Am. J. Phys. Anthropol.* 88(4):525–562.

Sampaio, D. T., and Ferrari, S. F. 2005. Predation of an infant titi monkey (*Callicebus moloch*) by a tufted capuchin (*Cebus apella*). *Folia Primatol.* 76(2):113–115.

Schwartz, G. T., Samonds, K. E., Godfrey, L. R., Jungers, W. L., and Simons, E. L. (2002). Dental microstructure and life history in subfossil Malagasy lemurs. *P. Natl. Acad. Sci. USA* 99:6124–6129.

Schwartz, G. T., Mahoney, P., Godfrey, L. R., Cuozzo, F. P., Jungers, W. L., and Randria, G. F. N. (2005). Dental development in *Megaladapis edwardsi* (Primates, Lemuriformes): Implications for understanding life history variation in subfossil lemurs. *J. Hum. Evol.* 49:702–721.

Schwartz, G. T., Godfrey, L. R., and Mahoney, P. (2007). Inferring primate growth, development, and life history from dental microstructure: The case of the extinct giant Malagasy lemur, *Megaladapis*. In: Bailey, S. E., and Hublin, J.-J. (eds.), *Dental Perspectives on Human Evolution: State of the Art Research in Dental Paleoanthropology*. Springer, Dordrecht, The Netherlands, pp. 147–162.

Seligsohn, D., and Szalay, F. S. (1974). Dental occlusion and the masticatory apparatus in *Lemur* and *Varecia*: Their bearing on the systematics of living and fossil primates. In: Martin, R. D., Doyle, G. A., and Walker, A. C. (eds.), *Prosimian Biology*. Duckworth, London, pp. 543–561.

Semprebon, G. M., Godfrey, L.R., Solounias, N., Sutherland, M. R., and Jungers, W. L. (2004). Can low-magnification stereomicroscopy reveal diet? *J. Hum. Evol.* 47:115–144.

Shanahan, M., So, S., Compton, S. G., and Corlett, R. (2001). Fig-eating by vertebrate frugivores: A global review. *Biol. Rev.* 76:529–572.

Simmen, B., and Sabatier, D. (1996). Diets of some French Guianan primates: Food composition and food choices. *Int. J. Primatol.* 17:661–693.

Simons, E. L. (1994). The giant aye–aye *Daubentonia robusta. Folia primatol.* 62:14–21.

Simons, E. L. (1997). Lemurs: Old and new. In: Goodman, S. M., and Patterson, B. D. (eds.), *Natural Change and Human Impacts in Madagascar*. Smithsonian Press, Washington, D. C., pp. 142–168.

Simons, E. L., Godfrey, L. R., Vuillaume-Randriamanantena, M., Chatrath, P. S., and Rakototsamimanana, B. (1990). Discovery of new giant subfossil lemurs in the Ankarana Mountains of Northern Madagascar. *J. Hum. Evol.* 19(3):311–319.

Simons, E. L., Burney, D. A., Chatrath, P. S., Godfrey, L. R., Jungers, W. L., and Rakotosamimanana, B. (1995a). AMS ^{14}C dates for extinct lemurs from caves in the Ankarana Massif, northern Madagascar. *Quat. Res.* 43(2):249–254.

Simons, E. L., Godfrey, L. R., Jungers, W. L., Chatrath, P. S., and Ravaoarisoa, J. (1995b). A new species of *Mesopropithecus* (Primates, Palaeopropithecidae) from Northern Madagascar. *Int. J. Primatol.* 16:653–682.

Stephan, H., Frahm H., and Bauchot, R. (1977). Ergleichende Untersuchungen an den Gehirnen madagassischer Halbaffen. Encephalization und Macromorphologie. *J. Hirnforsch* 18:115–147.

Stephan, H., Frahm, H., and Baron, G. (1981). New and revised data on volumes of brain structures in insectivores and primates. *Folia Primatol.* 35:1–29.

Sterling, E. J. (1994). Aye-ayes: Specialists on structurally defended resources. *Folia Primatol.* 62:142–154.
Szalay, F. S, and Delson, E. (1979). *Evolutionary History of the Primates.* Academic Press, New York.
Swindler, D. R. (2002). *Primate Dentition: An Introduction to the Teeth of Non-human Primates.* Cambridge University Press, Cambridge, U. K.
Tattersall, I. (1972). The functional significance of airorynchy in *Megaladapis*. *Folia Primatol.* 18:20–26.
Tattersall, I. (1973). Cranial anatomy of the Archaeolemurinae (Lemuroidea, Primates). *Anthropol. Pap. Am. Mus. Nat. Hist.* 52:1–110.
Temple, S. A. (1977). Plant-animal mutualism: Coevolution with Dodo leads to near extinction of plant. *Science* 197:885–886.
Terborgh, J. (1986a). Keystone plant resources in the tropical forest. In: Soulé, M. E. (ed.), *Conservation Biology: The Science of Scarcity and Diversity.* Sinauer Associates, Sunderland, MA, pp. 330–344.
Terborgh, J. (1986b). Community aspects of frugivory in tropical forests. In: Estrada, A., and Fleming, T. H. (eds.), *Frugivores and Seed Dispersal.* Dr. W. Junk, Dordrecht, pp. 371–384.
Traveset, A., and Verdú, M. (2002). A meta-analysis of the effect of gut treatment on seed germinaton. In: Levey, D. J., Silva, W. R., and Galetti, M. (eds.), *Seed Dispersal and Frugivory: Ecology, Evolution and Conservation.* CABI Publishing, Wallingford, U. K., pp. 339–350.
Turk, D. (1997). *A Guide to Trees of Eastern Madagascar and Ranomafana National Park.* Missouri Botanical Garden St. Louis, MO.
Ungar, P. S. (1998). Dental allometry, morphology, and wear as evidence for diet in fossil primates. *Evol. Anthropol.* 6:205–217.
Ungar, P. S., Brown, C. A., Bergstrom, T. S., and Walker, A. (2003). Quantification of dental microwear by tandem scanning confocal microscopy and scale-sensitive fractal analyses. *Scanning* 25:185–193.
Vasey, N., Godfrey, L. R., and Perez, V. R. (2005). The paleobiology of *Pachylemur*. *Am. J. Phys. Anthropol.* Suppl. 40:212.
Visalberghi, E. (1990). Tool use in *Cebus*. *Folia Primatol.* 54:146–154.
Voigt, F. A., Bleher, B., Fietz, J., Ganzhorn, J. U., Schwab, D., and Bohning-Gaese, K. (2004). A comparison of morphological and chemical fruit traits between two sites with different frugivore assemblages. *Oecologia* 141:94–104.
Winter, K. (1979). δ^{13}C values of some succulent plants from Madagascar. *Oecologia* 40:103–112.
Walker, A. C. (1974). Locomotor adaptations in past and present prosimian primates. In: Jenkins, F. A., Jr. (ed.), *Primate Locomotion.* Academic Press, New York, pp. 349–381.
Wrangham, R. W., Chapman, C. A., and Chapman, L. J. (1994). Seed dispersal by forest chimpanzees in Uganda. *J. Trop. Ecol.* 10(3):355–368.
Wright, B. W. (2005). Craniodental biomechanics and dietary toughness in the genus *Cebus*. *J. Hum. Evol.* 48(5):473–492.
Wright, P. C. (1999). Lemur traits and Madagascar ecology: Coping with an island environment. *Yrbk. Phys. Anthropol.* 42:31–72.
Wright, P. C., Razafindratsita, V. R., Pochron, S. T., and Jernvall, J. (2005). The key to Madagascar frugivores. In: Dew, J. L., and Boubli, J. P. (eds.), *Tropical Fruits and Frugivores: The Search for Strong Interactors.* Springer, Dordrecht, The Netherlands, pp. 121–138.
Wunderlich, R. E., Simons, E. L., and Jungers, W. L. (1996). New pedal remains of *Megaladapis* and their functional significance. *Am. J. Phys. Anthropol.* 100(1):115–138.
Zona, S., and Henderson, A. (1989). A review of animal-mediated seed dispersal of palms. *Selbyana* 11:6–21.

Vicariance vs. Dispersal in the Origin of the Malagasy Mammal Fauna

Ian Tattersall

Introduction

Both Africa and Madagascar have loomed large in the long career of Elwyn Simons, whose extraordinary diversity of achievements this volume celebrates. So it seems appropriate here to try to combine these two geographical foci into a consideration of one of the most intriguing questions in all of biogeography: the origin of Madagascar's unusual mammalian fauna. Madagascar, the world's largest oceanic island, has long been renowned as the home of one of the planet's most bizarre and endemic biotas. In the two and a half centuries since Joseph-Philibert Commerson first wrote to his patron in 1771 that "For naturalists Madagascar is the true Promised Land. Nature seems to have withdrawn there into a private sanctuary, to work on models different from any she has used elsewhere. There you meet the most unusual and marvelous forms at every step... What an admirable country, this Madagascar!" (author's trans.), his words have become something of a cliché. Yet, like most clichés, they express a profound truth. There really is no place like Madagascar, and the few countries that are comparable in terms of overall floral and faunal diversity are all vastly larger in area.

As Sanmartín and Ronquist (2004: 216) recently remarked, "the Southern Hemisphere has traditionally been considered as having a fundamentally vicariant [biogeographic] history," linked to the breakup of the ancient super continent of Gondwana. Except, that is, for the curious landmass of Madagascar, where, among the terrestrial vertebrates at least, there is a remarkably low count of surviving Gondwanan elements in today's fauna. Even the chameleons, more abundant and diverse in Madagascar than anywhere else, and long viewed as the prime example of biogeographic congruence with the Gondwanan break-up, have now been implicated in a Malagasy origin with multiple dispersal events to the adjacent islands and the African continent (Raxworthy et al., 2002). Overwater dispersal has thus been a major ingredient

Ian Tattersall
Division of Anthropology, American Museum of Natural History, New York, NY 10024, USA

J. G. Fleagle, C. C. Gilbert (eds.), *Elwyn Simons: A Search for Origins.*

in virtually all explanations of the origin of Madagascar's current biotic diversity. Further, the rather anomalous biogeographical situation that this huge island occupies may stem in large part from Madagascar's rather complicated journey to its present position (both latitudinal and relative to Africa) subsequent to the initial mid-Mesozoic breakup of Gondwana.

Geographical Origins

During much of the Mesozoic Madagascar lay deep within Gondwana, hemmed in by modern India to its east and Africa to its west (Fig. 1). As east and west Gondwana began to separate in the mid-Jurassic a split began to form between Africa and Madagascar, and by about 160 Ma the island's western edge was already underwater (Fig. 1; see review by Wells, 2003). Initial graben faulting along this edge was eventually replaced by a south-southwesterly movement of Indiagascar relative to Africa along a slip-strike fault whose modern remnants are identified today in the Mozambique Channel as the Davie Ridge. By about 130 Ma the entire landmass of Madagascar lay south of 40° S latitude (Rabinowitz et al., 1983), taking the island well out of the tropical belt and plausibly accounting for substantial attrition within the tropical Gondwanan fauna, of which there are rather few reminders in the Malagasy fauna of today. By around 120 Ma, the island had assumed essentially its modern position vis-à-vis a very southerly-positioned Africa (Rabinowitz et al., 1983), although at this time it was probably separated from India by only a very narrow channel (Fig. 1). The whole assemblage subsequently drifted northward in unison, India itself possibly maintaining a connection with Africa via Antarctica until the later Mesozoic (Krause et al., 1997). Finally, India parted company with Madagascar in the late Cretaceous, ca. 88 Ma (Storey, 1995; Storey et al., 1997), following which it departed northward with extreme rapidity, leaving the Seychelles in its wake (Fig. 1). Scars from much of this geological history can be seem in the current geography of the western part of the Indian Ocean (Fig. 2).

Flynn and Weiss (2003) have reported increasing faunal endemism and provinciality through the later part of the Mesozoic in Madagascar, although to Krause et al. (1997) the discovery of Malagasy gondwanatheres in the late Cretaceous suggested some lingering cosmopolitanism putatively due to a land connection maintained via Antarctica. The Mesozoic mammal record currently known from Madagascar is thin but surprisingly diverse, including a handful of late Cretaceous teeth that include marsupials, a multituberculate, and two different gondwanatheres (Krause, 2000, 2001). What is most notable, though, is that the collection lacks a plausible ancestor for any of today's terrestrial Malagasy mammals. Further, since Madagascar has been widely separate from both Africa and India since well before the beginning of the Age of Mammals, it seems certain that dispersal must have been involved in the origin of the modern Malagasy terrestrial mammal fauna.

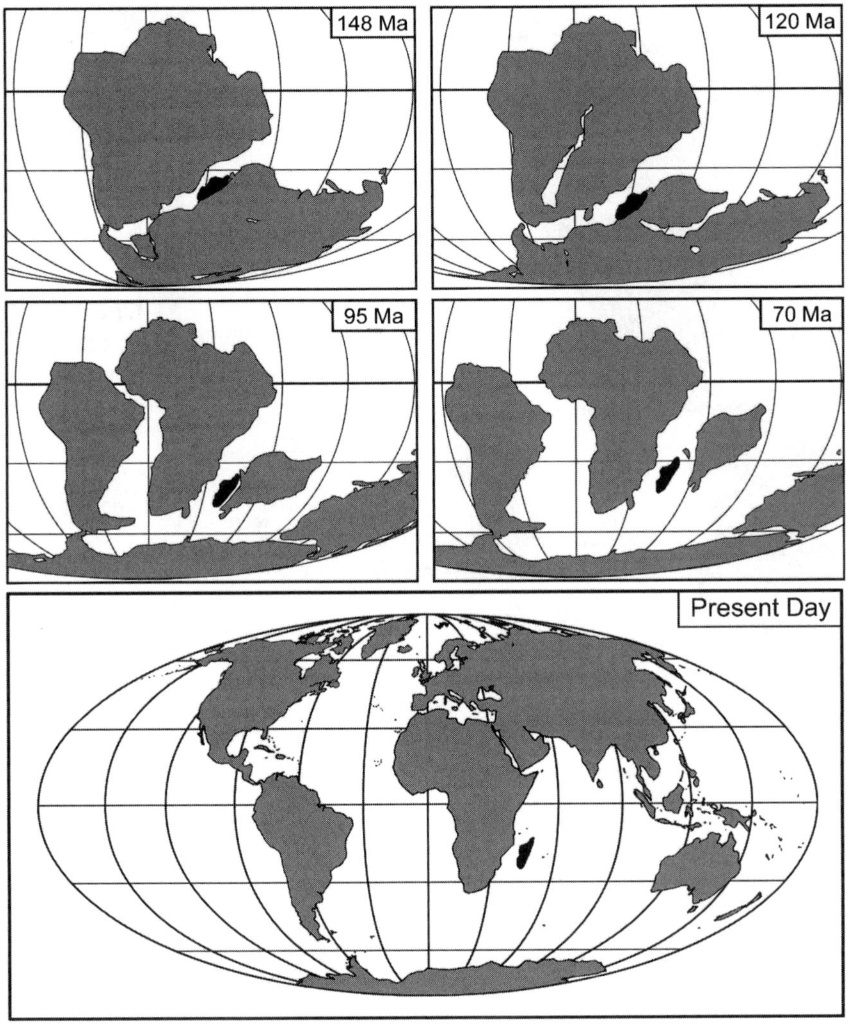

Fig. 1 Map illustrating the position of Madagascar relative to the other continents over the past 150 million years. Madagascar is shaded in black

Madagascar's Terrestrial Mammals

The question thus arises of what kind of dispersal. Madagascar has evidently been separated from its neighboring Gondwanan landmasses by very substantial water barriers throughout the Tertiary. And terrestrial mammals are notoriously poor overwater dispersers (Lawlor, 1986). Yet, given their affinities, and in the absence of any putative fossil ancestry in Madagascar itself,

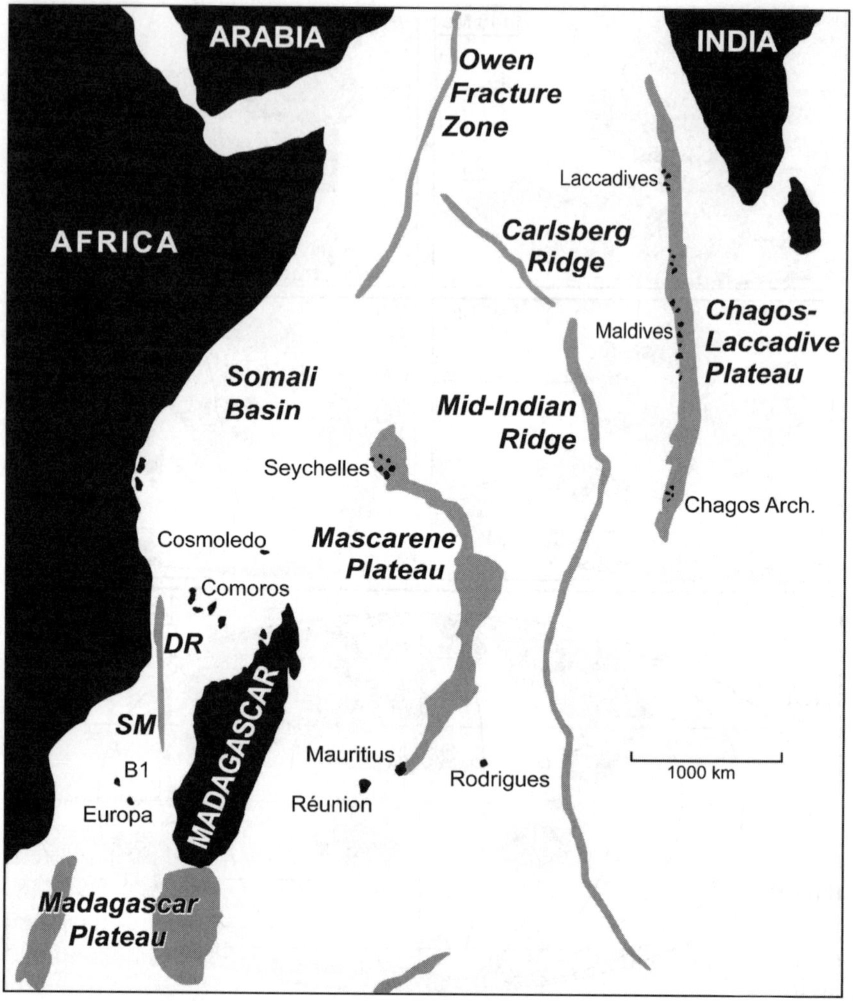

Fig. 2 Map of the western Indian Ocean showing the current position of Madagascar relative to Africa and India and the location of major topographic and submarine features mentioned in the text. DR: Davie Ridge; SM: Sakalava Seamounts. Drawn by Luci Betti-Nash

it is virtually certain that the ancestors of all of Madagascar's surviving terrestrial mammals, at the ordinal level a highly depauperate assemblage that includes only primates, rodents, carnivorans and lipotyphlans, must have arrived from elsewhere. The timing of such arrivals could potentially have a powerful bearing on the means of dispersal, but evidence intrinsic to the living Malagasy terrestrial mammal fauna itself is suggestive rather than conclusive, as the brief review below illustrates.

Although the morphological indications are less than clear on this subject (see Schwartz and Tattersall, 1985), most authorities today consider on the basis of molecular comparisons that Madagascar's amazingly diverse primates, the lemurs, form a monophyletic group that lies within the Afro-Asian-Malagasy suborder Strepsirhini and is descended from a single common ancestor that somehow contrived to cross the Mozambique Channel (Karanth et al., 2005). Yoder and Yang's (2004) latest molecular estimate for the age of that ancestor is in the region of 62–65 Ma, though other estimates have been somewhat more recent, i.e. 47 Ma (Porter et al., 1997), >54 Ma (Yoder et al., 1996), and 60–50 Ma (Poux et al., 2005). Like the lemurs, Madagascar's tenrecid lipotyphlans are highly diverse, and their monophyly, like that of the Afro-Malagasy family Tenrecidae as a whole, is also supported on molecular grounds (Olson, 1999; Douady et al., 2002; Olson and Goodman, 2003; Poux et al., 2005). Douady et al. (2002) provide a molecular estimate of about 53 Ma for the divergence of the Malagasy tenrec clade from the otter shrews, and suggest a window for the tenrecid colonization of Madagascar of between <53 and 37 Ma (the latter being the age of the basal split [between Orizoryctinae and Tenrecinae] among the Malagasy tenrecs). In contrast, Poux et al. (2005) postulate a rather later window, at between 42 and 26 Ma. Madagascar's endemic carnivorans are also both ecologically and morphologically diverse, and are additionally of much-debated affinities. On morphological grounds, various Malagasy carnivorans have at one time or another been regarded as cats or civets or mongooses (see reviews by Yoder and Flynn, 2003; Tattersall, 2005); but molecular analyses by Yoder et al. (2003) and Yoder and Flynn (2003) have revealed strong evidence that the Malagasy carnivorans form a single monophyletic clade that is sister to the continental herpestids. The estimated age of the ancestral Malagasy carnivoran is on the order of 26–18 Ma (Yoder et al., 2003; Poux et al., 2005). Madagascar's astonishingly varied native rodents have traditionally all been assigned to the single endemic subfamily Nesomyinae, though on the basis of morphological characters the affinities of this group as a whole have been much debated. And while molecular studies (Du bois et al., 1996; Jansa et al., 1999; Jansa and Wecksler, 2004) are far apart on matters of detail, they agree in their broad implication of a natural grouping that falls within the highly cosmopolitan family Muridae All permit the conclusion that Madagascar's modern rodent radiation is the result of a single ancestral immigration to the island, which Poux et al. (2005) estimate at 24–20 Ma.

On balance, then, the evidence from the internal cohesion of the four endemic groups of Malagasy terrestrial mammals argues for a minimum of introductions, which may certainly be taken as strengthening the case for sweepstakes-style arrivals across a water barrier. But which water barrier? This is where recent fossil discoveries in adjacent continents, not least by Elwyn Simons and his co-workers, can help shed light on the situation.

Fossil Evidence from Neighboring Continents

Madagascar's lemurs belong to the euprimate suborder Strepsirhini, with living representatives also in subsaharan Africa and southern Asia. These primates have no known fossil record in Madagascar prior to the latest Pleistocene. Perhaps the most striking synapomorphy of this group is the possession of a procumbent toothcomb at the front of the lower jaw. Until recently there was no pre-Miocene record anywhere of primates with toothcombs; and it has long been known that by the early Miocene both modern Afro-Asian strepsirhine families were clearly in existence (Simpson, 1967). Now, however, that picture is changing.

The earliest primate that is directly documented to have possessed a toothcomb is *Karanisia clarki*, recently described from the Fayum of Egypt by Seiffert et al. (2003). The jaw fragments and teeth on which this new genus and species was based come from Eocene sediments that are dated to around 37 Ma, and they are interpreted by their describers as not simply lorisiform but lorisid, and even as possibly representing the sister genus to the living west African *Arctocebus*. From the same stratigraphic level, but represented only by a couple of molars, is *Saharagalago misrensis*, interpreted as galagid. If this reading is correct, and there is no evident reason to dispute it at present, then the two living non-Malagasy strepsirhine families had already diverged by the middle Eocene; and if the Malagasy strepsirhines are indeed monophyletic, then the basal strepsirhine divergence must have preceded that event, perhaps by a substantial length of time. These conclusions are also bolstered by the recent reinterpretation (Seiffert et al., 2005) of the Fayum *Wadilemur*, 35 myr old, as a stem galagid rather than as a non-strepsirhine anchomomyin, a finding that further supports the notion that Strepsirhini originated in Africa.

This new African evidence for early strepsirhine primates is paralleled by somewhat later Asian finds, in the Bugti Hills of Pakistan, that were reported by Marivaux et al. (2001). The isolated teeth concerned are dated to the early Oligocene, around 30 Ma, and are assigned to the new strepsirhine genus *Bugtilemur*. They include a lower canine which, though shortish and not extremely procumbent, appears in its crown morphology to bear the hallmarks of a toothcomb. On the basis of molar morphology Marivaux and colleagues assigned the *Bugtilemur* fossils to the extant Malagasy family Cheirogaleidae, and they invoked a putative sweepstakes/filter route via a Chagos/Laccadive (Fig. 2) paleoridge system to explain this apparent affinity (but see Marivaux et al., 2006; Seiffert, 2007).

Among the other major terrestrial mammal taxa represented in Madagascar, the only possible claimant to pre-Miocene representation on an adjacent mainland is Tenrecidae. This claim is founded on the discovery, in the late Eocene of the Egyptian Fayum, of jaw fragments and teeth that were attributed to species of the new genus *Widanelfarasia* by Seiffert and Simons (2000). In their publication Seiffert and Simons were quite tentative as to the affinities of *Widanelfarasia*, simply noting similarities to tenrecs as well as to chrysochlorids and solenodontids. Subsequently, however, Seiffert et al. (in press) argue that the form of the mandible and anterior dentition of *Widanelfarasia*, in particular, indicate that it is indeed a stem tenrecid. The stumbling-block to this view, as Seiffert points out, is that the molars of *Widanelfarasia* are not fully zalambdodont. Zalambdodonty is a derived state present in both tenrecids and their sister group the golden moles, and it would thus be expected in the ancestral tenrec – unless this character were actually convergent in the two taxa. That this might well be the case is suggested by the fact that African early Miocene undoubted tenrecs such as *Protenrec* are not fully zalambdodont either. And if *Widanelfarasia* is indeed a stem tenrecid, rather than a stem afrosoricid predating the divergence of tenrecs from golden moles, this would permit an early (pre-late Eocene) origin, from Africa, of the Malagasy tenrecs. And it would also reinforce the conclusion of Olson and Goodman (2003: 1239) that these insular forms are indeed monophyletic, with "a single colonization event followed by an extensive adaptive radiation."

As concerns close relatives of the remaining Malagasy native terrestrial mammal groups, the earliest African murid rodents appear in the early Miocene (Lavocat, 1973; McKenna and Bell, 1997), with only phiomyids occurring in the Fayum. This would suggest a limiting lower date of only about 24 Ma for the rodent colonization of Madagascar: about the same as the date for invasion by the ancestral carnivore suggested by Yoder et al. (2003; see above). This latter date is consonant with Van Couvering's notion (Madden and Van Couvering, 1976; Van Couvering and Harris, 1991) that aeluroid carnivorans first entered Africa during the Grande Coupure, an episode of greatly reduced sea-levels that occurred at the beginning of the Oligocene, some 34 years ago. This timing would, however, push the later Fayum faunas, usually considered Oligocene, back into the Eocene (Van Couvering and Harris, 1991). And the earliest record in Africa of any herpestid dates only to the early Miocene (McKenna and Bell, 1997).

Thus the African evidence, such as it is, seems to be pointing currently toward two specific periods for the colonization of Madagascar by its endemic modern terrestrial mammals (ignoring the extinct, enigmatic and poorly known *Bibymalagasia* [MacPhee, 1994] and the recently arrived semiaquatic hippopotamuses). The first date range is very early in the Tertiary for the lemurs and tenrecs, and the later one is the early Miocene for the carnivorans and also plausibly the rodents. Both of these ranges are at least broadly compatible with molecular estimates; and if the arrivals were indeed concentrated in this way, the question must be asked what factors might have facilitated invasion at these points in time.

A Static or a Dynamic Paleogeography?

The physical evidence is compelling that Madagascar has lain far from Africa and other Gondwanan continents throughout the Tertiary. Land mammals, it is well established, are not effective overwater dispersers (Lawlor, 1986), and on current evidence no land-bound mammal has managed to make the crossing to Madagascar in the last 15–20 million years. Nonetheless, given an effectively static geography of Africa and Madagascar over the Cenozoic these considerations make a compelling case both for some form of oceanic dispersal in the origin of the modern Malagasy biota, and for the existence in the remote past of epemeral factors that facilitated such dispersal. It is thus more or less mandatory to inquire whether geography has indeed remained constant subsequent to the establishment of the Mozambique Channel, for the bathymetry of the Channel is such that no normal sea level excursion would have appreciably lessened its 400-mile width.

McCall (1997) has suggested that compressional forces, ultimately resulting from the collision of India with the Asian plate, provoked mid-Tertary uplift along the line of the Davie Ridge (DR) in the Mozambique Channel (Fig. 2). This uplifting ultimately resulted in the exposure of a large swath of the Channel's floor during the period between about 45 and 26 Ma. McCall's evidence for this claim lies in apparently terrigenous sediments found in core samples taken along the Ridge and studied by Leclaire et al. (1989) and Bassias (1992). These authors suggested that a chain of topographic highs broke the surface of the Channel before tensional conditions resumed with the activation of the East African rifting system, and the resulting subsidence returned the Ridge to the seafloor. The extent of subaerial seafloor exposure at the time of topographic highs is uncertain, but it was potentially extensive. And although a continuous land bridge was clearly not formed between Africa and Madagascar (since otherwise at least a substantial sampling of the African fauna would presumably have found its way to Madagascar, whereas the Mozambique Channel quite evidently continued to exert a very strong filter effect), the scene might have been set for an island-hopping scenario. Krause (2003) dismissed such facilitation of passage by pointing to the "extreme dissimilarity" of the African and Malagasy faunas, and Yoder et al. (2003) raised the issue that the 45–26 Ma window of time quoted by McCall falls between the molecular date ranges for the arrivals of primates and carnivorans in Madagascar. And Poux et al. (2005), while noting that penetration of Madagascar by terrestrial mammals appeared to have been concentrated in two distinct sectors of the Tertiary, nevertheless favored transoceanic dispersal as the mechanism. Still, these authors did not entirely rule out a scenario involving a land connection of some kind, and in 2004 de Wit and Masters raised afresh the possibility of episodic connections between Madagascar and its neighboring continent along the Davie Ridge, or even along the series of similar seafloor structures that lie between Antarctica and Madagascar and Africa (Fig. 2). And they also suggested that in certain respects it might be profitable to view Madagascar's faunal composition as somewhat more cosmopolitan than generally admitted.

Similarly, as long ago as 1975 Gingerich proposed that the ancestral lemurs might have rafted to Madagascar from India, a suggestion more recently echoed by Marivaux et al. (2001) to explain the apparent affinities between *Bugtilemur* and the cheirogaleids. These latter authors suggested an invasion of Madagascar via exposed portions of the Eocene Mascarene and Chagos-Laccadive paleoridge systems (Fig. 2). de Wit and Masters (2004) have also raised the possibility of immigration into Madagascar via a sweepstakes/filter route from India, and Masters et al. (2005) allude to the potential existence, between about 65 and 43 Ma, of a subaerial "Deccan Plateau" connection between India and Afro-Arabia that might have played a role in faunal transfer to the former from the latter (and ultimately to Madagascar).

None of the possibilities raised above is proven, but the fact is that a great deal can happen to the Earth's crust in a time span equivalent to the Tertiary, and the geology of the ocean floor surrounding Madagascar is so far rather poorly known. Yoder et al. (2003) argue that the small number of colonization events needed to populate Madagascar with its modern mammal groups was low, and thus that it is superfluous to propose Tertiary landbridges. On the other hand, there is a fair *a priori* possibility that a Mozambique Channel of today's configuration, let alone the southern Indian Ocean, is an insuperable obstacle to purely terrestrial mammals (certainly, none has contrived a successful crossing in at least the last 15 myr or so). And, if so, there is an active possibility that at the very minimum a partial landbridge or some form of island-hopping was necessary to achieve any transfer at all of strictly terrestrial mammals to Madagascar at any point during the Tertiary. Clearly, given all these uncertainties it would be premature to abandon the notion of some periodic ephemeral land exposure in the Mozambique Channel or even in the southwestern Indian Ocean during this period. In which case, some of the most compelling future evidence bearing on the mode of Madagascar's colonization by terrestrial mammals may well emerge not from biological studies, but from growing knowledge of the history of the island's surrounding sea floor.

Conclusion

Clearly, then, the last word has yet to be written on the manner of colonization of Madagascar by terrestrial mammals, although the strongly filtered nature of the island's mammal fauna clearly implies some degree of overwater crossing by the founding stocks. One of the major impediments to clarifying the picture is, of course, the virtually complete lack of a terrestrial Tertiary fossil record in Madagascar. In the absence of such a record we have, for example, no idea whether or not failed colonizations of the island by mammal groups other than those represented there today might have occurred at some point during the Tertiary: Something that would bear strongly on the matter of the *a priori* probabilities of an oceanic crossing, as well as on the timing of any such potential crossings. And without a local fossil record we have no direct calibration of the diversification of today's astonishingly varied Malagasy mammal groups, or anything but inferential knowledge of the island's Tertiary faunal history. To date, effectively no fossiliferous Tertiary terrestrial sediments have been located in Madagascar; but in the spirit of inspired exploration that Elwyn Simons so amply exemplifies, it is not unreasonable to hope that at some point they may be.

Acknowledgment It is a pleasure to have been given this opportunity to acknowledge my debt to Elwyn Simons, my mentor ever since I entered graduate school in 1967, and to express my admiration for all he has achieved in a long, illustrious and energetically continuing career. I thank Freddie Oakley for organizing the celebration of Elwyn's seventy-fifth birthday, John Fleagle for inviting me to contribute to this volume, two reviewers for insightful comments, and Luci Betti-Nash for drawing the maps.

References

Bassias, Y. (1992). Petrological and geochemical investigations of rocks from the Davie Fracture Zone (Mozambique Channel) and some tectonic implications. *J. Afr. Earth Sci.* 15: 321–339.

de Wit, M. and Masters, J. C. (2004). The geological history of Africa, India and Madagascar, dispersal scenarios for vertebrates. *Folia Primatol.* 75: 117.

Douady, C. J., Catzeflis, F., Kao, D. J., Springer, M. S., and Stanhope, M. J. (2002). Molecular evidence for the monophyly of Tenrecidae (Mammalia) and the timing of the colonization of Madagascar by Malagasy tenrecs. *Mol. Phylogenet. Evol.* 22: 357–363.

Du Bois, J-Y., Rakotondravony, D., Hanni, C., Sourouille, P., and Catzeflis, F. M. (1996). Molecular evolutionary relationships of three genera of Nesomyinae, endemic rodent taxa from Madagascar. *J. Mammal. Evol.* 3: 239–259.

Flynn, J. J. and Wyss, A. R. (2003). Mesozoic terrestrial vertebrate faunas: The early history of Madagascar's vertebrate diversity. In: Goodman, S. M. and Benstead, J. P. (eds.), *The Natural History of Madagascar.* University of Chicago Press, Chicago, pp. 34–40.

Gingerich, P. D. (1975). Dentition of *Adapis parisiensis* and the evolution of lemuriform primates. In: Tattersall, I. and Sussman, R., (eds.), *Lemur Biology*. Plenum Press, New York, pp. 65–80.

Jansa, S. A., Goodman, S. M., and Tucker, P. K. (1999). Molecular phylogeny and biogeography of the native rodents of Madagascar (Muridae: Nesomyinae): A test of the single-origin hypothesis. *Cladistics* 15: 253–270.

Jansa, S. A. and Weksler, M. (2004). Phylogeny of muroid rodents: Relationships within and among major lineages as determined by IRBP gene sequences. *Mol. Phylogenet. Evol.* 31: 256–276.

Karanth, K. P., Delefosse, T., Rakotosamimanana, B., Parsons, T. J., and Yoder, A. D. (2005). Ancient DNA from giant extinct lemurs confirms single origin of Malagasy primates. *P. Natl. Acad. Sci. USA* 102: 5090–5095.

Krause, D. W. (2000). New mammalian specimens from the late Cretaceous of Madagascar. *J. Vertebr. Paleontol.* 20: 52A–53A.

Krause, D. W. (2001). Fossil molar from a Madagascan marsupial. *Nature* 412: 497–498.

Krause, D. W. (2003). Late Cretaceous vertebrates of Madagascar: A window into Gondwanan biogeography at the end of the Age of Dinosaurs. In: Goodman, S. M. and Benstead, J. P. (eds.), *The Natural History of Madagascar*. University of Chicago Press, Chicago, pp. 40–47.

Krause, D. W., Hartman, J. H., and Wells, N. A. (1997). Late Cretaceous vertebrates from Madagascar: Implications for biotic change in deep time. In: Goodman, S. M. and Patterson, B. D., (eds.), *Natural Change and Human Impact in Madagascar*. Smithsonian Institution Press, Washington, DC, pp. 3–43.

Lavocat, R. (1973). Les rongeurs du Miocène d'Afrique Centrale I – Miocène inférieur. *Mém. Et Trav. Inst. Montpellier Haut-Etudes*, 1, 1–284.
Lawlor, T. E. (1986). Comparative biogeography of mammals on islands. *Biol. J. Linn. Soc.* 28: 99–125.
Leclaire, L., Bassias, Y., Clocchiatti, M., and Ségoufin, J. (1989). Le Ride de Davie dans le Canal de Mozambique: Approche stratigraphique et géodynamique. *C.R. Acad. Sci.* II 308: 1077–1082.
MacPhee, R. (1994). Morphology, adaptations, and relationships of *Plesiorycteropus*, and a diagnosis of a new order of eutherian mammals. *B. Am. Mus. Nat. Hist.* 220: 1–214.
Madden, C. T. and Van Couvering, J. A. (1976). The proboscidean datum event: Early Miocene migration from Africa. *Geol. Soc. Am., Abstr. Progr.*, 8, 992–993.
Marivaux, L., Chaimanee, Y., Tafforeau, P., and Jaeger, J-J. (2006). New Strepsirrhine primate from the late Eocene of peninsular Thailand (Krabi Basin). *Am. J. Phys. Anthropol.* 130: 425–434.
Marivaux, L., Welcomme, J. L., Antoine, P. O., Metais, G., Baloch, I. M., Bennami, M., Chaimanee, Y., Ducrocq, S., and Jaeger, J. J. (2001). A fossil lemur from the Oligocene of Pakistan. *Science* 294: 587–591.
Masters, J. C., Anthony, N. M., de Wit, M. J., and Mitchell, A. (2005). Reconstructing the evolutionary history of the Lorisidae using morphological, molecular, and geological data. *Am. J. Phys. Anthropol.* 127: 465–480.
McCall, R. A. (1997). Implications of recent geological investigations of the Mozambique Channel for the mammalian colonization of Madagascar. *P. Roy. Soc. Lond. B. Bio.* 264: 663–665.
McKenna, M. C. and Bell, S. K. (1997). *Classification of Mammals above the Species Level.* Columbia University Press, New York.
Olson, L. E. (1999). *Systematics, evolution and biogeography of Madagascar's tenrecs (Mammalia: Tenrecidae).* Unpublished PhD thesis, University of Chicago.
Olson, L. E. and Goodman, S. M. (2003). Phylogeny and biogeography of tenrecs. In: Goodman, S. M. and Benstead, J. P. (eds.), *The Natural History of Madagascar.* University of Chicago Press, Chicago, pp. 1235–1242.
Porter, C. A., Sampaio, I., Schneider, M. P. C., Czelusniak, C., and Goodman, M. (1997). Evidence on primate phylogeny from ?-globin gene sequences and flanking regions. *J. Mol. Evol.* 40: 30–55.
Poux, C., Madsen, O., Marquard, E., Vieites, D. R., deJong, W. W., and Vences, M. (2005). Asynchronous colonization of Madagascar by four endemic clades of primates, insectivores, carnivores and rodents as inferred from nuclear genes. *Syst. Biol.* 54: 719–730.
Rabinowitz, P. D., Coffin, M. C., and Falvey, D. (1983). The separation of Madagascar and Africa. *Science* 220: 67–69.
Raxworthy, C. J., Forstner, M. J. R., and Nussbaum, R. A. (2002). Chameleon radiation by oceanic dispersal. *Nature* 415: 784–787.
Sanmartín, I. and Ronquist, F. (2004). Southern hemisphere biogeography inferred by event-based models: Plant versus animal patterns. *Syst. Biol.* 53: 216–243.
Schwartz, J. H. and Tattersall, I. (1985). Evolutionary relationships of living lemurs and lorises (Mammalia, Primates) and their potential affinity with European Eocene Adapidae. *Anthropol. Pap. Am. Mus.* 60: 1–100.
Seiffert, E. R. (2007). Early Evolution and biogeography of Lorisiform Strepsirrhines. *Folia Primatol.* 69: 27–35.
Seiffert, E. and Simons, E. L. (2000). *Widanelfarasia*, a diminutive placental from the Late Eocene of Egypt. *P. Natl. Acad. Sci. USA* 97: 2646–2651.
Seiffert, E., Simons, E. L., and Attiya, Y. (2003). Fossil evidence for an ancient divergence of lorises and galagos. *Nature* 422: 421–424.

Seiffert, E., Simons, E. L., Ryan, T. M., and Attiya, Y. (2005). Additional remains of *Wadilemur elegans*, a primitive stem galagid from the late Eocene of Egypt. *P. Natl. Acad. Sci. USA* 102: 11396–11401.

Seiffert, E. R., Simons, E. L., Ryan, T. M., Bown, T. M., and Attia, Y. (in press). New remains of Eocene and Oligocene Afrosoricida (Afrotheria) and the origin(s) of afrosoricid zalambdodonty. *J. Vert. Paleontol.*

Simpson, G. G. (1967). The Tertiary lorisiform primates of Africa. *B. Mus. Comp. Zool.* 136: 39–62.

Storey, B. C. (1995). The role of mantle plumes in continental breakup: Case histories from Gondwanaland. *Nature* 377: 301–308.

Storey, B. C., Mahoney, J. J., and Saunders, A. D. (1997). Cretaceous basalts in Madagascar and the transition between plume and continental lithosphere mantle sources. *Geophysical Monographs, American Geophysical Union* 100: 95–122.

Tattersall, I. (2005). Mechanisms of faunal origin and diversity in island environments: The case of Madagascar's mammals. *Annales Géologiques des Pays Helléniques* vol TK.

Van Couvering, J. A. and Harris, J. A. (1991). Late Eocene age of Fayum mammal fauna. *J. Hum. Evol.*, 21: 241–260.

Wells, N. A. (2003). Some hypotheses on the Mesozoic and Cenozoic paleoenvironmental history of Madagascar. In: Goodman, S. M. and Benstead, J. P. (eds.), *The Natural History of Madagascar*. University of Chicago Press, Chicago, pp. 16–34.

Yoder, A. and Flynn, J. J. (2003). Origin of Malagasy carnivora. In Goodman, S. M. and Benstead, J. P. (eds.), *The Natural History of Madagascar*. University of Chicago Press, Chicago, pp. 1253–1256.

Yoder, A. and Yang, Z. (2004). Divergence dates for Malagasy lemurs estimated from multiple gene loci: Geological and evolutionary context. *Mol. Ecol.* 13: 757–773.

Yoder, A., Cartmill, M., Ruvolo, M. K., Smith K., and Vilgalys, R. (1996). Ancient single origin for Malagasy primates. *P. Natl Acad. Sci. USA* 93:122–5126.

Yoder, A., Burns, M. M., Zehr, S., Delefosse, T., Véron, G., Goodman, S. M., and Flynn, J. J. (2003). Single origin of Malagasy carnivora from an African ancestor. *Nature* 421:734–737.

On the Brink of Extinction

Research for the Conservation of the Tonkin Snub-nosed Monkey (*Rhinopithecus avunculus*)

Herbert H. Covert, Le Khac Quyet and Barth W. Wright

Introduction

The Tonkin snub-nosed monkey (*Rhinopithecus avunculus*) is endemic to Vietnam and restricted to a small area in the northern portion of this country. This species was named by Dollman (1912), but very few sightings occurred in the following decades, leading Mittermeier and Cheney (1986:488) to state "the Vietnamese snub-nosed monkey (*Rhinopithecus avunculus*) from Tonkin may already be extinct. It is known from only a handful of museum specimens collected earlier in this century, and there are no recent reports of it from the wild." While not extinct, it is critically endangered. Nadler et al. (2003) and Mittermeier et al. (2005) report that only three small populations of Tonkin snub-nosed (TSN) monkeys are known, and they consist of less than 300 individuals. Its critical conservation status is reflected by the fact that Cowlishaw and Dunbar (2000), in their influential *Primate Conservation Biology* text, use the snub-nosed monkeys as the first example of primates in peril of extinction, stating that of the four *Rhinopithecus* species, the TSN monkey is the most endangered. Populations in Na Hang and Cham Chu were first identified in the early 1990s and as outlined below, are severely threatened by human activities. The third population occurs in Khau Ca adjacent to the Du Gia Nature Reserve, Ha Giang Province and was initially confirmed to exist in this region by one of us in 2002 (Le Khac Quyet, 2002). Here we provide the most detailed review of the research and conservation history to date, and the efforts that we are making to conserve and study the Tonkin snub-nosed monkey in Khau Ca.

This is an excellent topic to include in a volume dedicated to the work of Dr. Elwyn Simons. While Dr. Simons is best known for his work on the evolutionary history of primates, he has also been influential in primate conservation; particularly in Madagascar as noted by Oakley (2008) and Wright (2008). His work in Madagascar helped to "rediscover" the geographic

Herbert H. Covert
Department of Anthropology UCB 233, University of Colorado at Boulder, Boulder, CO 80309-0233, 303 492-1167, 303 492-1871
Covert@colorado.edu

J. G. Fleagle, C. C. Gilbert (eds.), *Elwyn Simons: A Search for Origins.*

distribution of the aye–aye (*Daubentonia madagascariensis*), an unusual primate that was thought to have a population less than a hundred individuals by many in the 1970s (Simons, 1993, 1995). Today the aye–aye is recognized as endangered but is known to occur throughout a wide geographic range, including the entire east coast of Madagascar (Mittermeier et al., 2006). As with the aye–aye, we hope that the "rediscovery" of the TSN monkey will lead to an increase in their numbers and geographic range.

Discovery of the TSN Monkey

The snub-nosed monkeys were first recognized in western science by Milne-Edwards in 1870 and these animals usually have been placed in the genus *Rhinopithecus* since 1872 (Milne-Edwards, 1872). Dollman named a new species of snub-nosed monkey in 1912, *Rhinopithecus avunculus,* based on two specimens: an adult female and an immature individual from Yen Bai Province. While this initial description focused primarily on the pelage of the adult specimen, it was also noted that the Vietnamese snub-nosed monkey was a bit smaller in body size and had a much longer tail than *R. bieti*, and in some features of the skull it was more similar to *R. roxellana* than to *R. bieti*. Pocock (1924) added to Dollman's discussion stating that it was similar to other snub-nosed monkeys "in cranial and facial characters, having a small upturned nose and a long upper lip, thus differing from *Nasalis*" (p. 330). He noted, however, that it differed from the other snub-nosed monkeys "in the structure of the hands and feet, which have digits remarkably long, as in *Nasalis*, the tip of the hallux, when turned forwards, reaching to the distal end of the first phalange of the 2nd digit" (p. 330–31). He concluded that these distinct aspects of its extremities were significant enough to place the Tonkin monkey into its own genus *Presbytiscus*. Thomas (1928) reviewed 12 additional specimens that had been collected in the winter of 1926 and 1927 by Delacour and Willoughby Lowe in Bac Kan, and supported Pocock's generic allocation noting that the new specimens of the TSN monkey were characterized by longer and more slender hands and digits than the Chinese snub-nosed monkeys. In addition, he noted that a number of the specimens had a brighter pelage on the crown and back of the neck. He concluded that the type specimen (an adult female) had an "unusually dull coloured, dark napped, and whitish throat" (p. 141). Additional surveys in this area during the early part of the last century did not observe individuals nor collect specimens of this species; for example in Osgood's (1932) report on the Kelley-Roosevelt and Delacour Asiatic expeditions it is noted that (p. 208) "although Delacour and Lowe obtained a fine series of twelve specimens of this monkey [*R. avunculus*] ... it is not represented in subsequent collections, and is otherwise known only from the type and one immature paratype." Following these initial descriptions, little additional research focused on the TSN monkey. For example, Zuckerman (1932) in his comprehensive discussion of primate behavior did not make any mention of snub-nosed monkeys, and a decade later, Hooton (1942) included a single paragraph discussion on the Chinese snub-nosed taxa. Napier and Napier (1967) provided a brief review of the genus *Rhinopithecus*, and repeated that *R. avunculus* has a long tail and long and slender hands and feet, citing Dolman and Pocock. Finally, Fleagle (1988) in his review of *Rhinopithecus* briefly noted that "one species is from the monsoon forests of Vietnam" (p. 190).

As mentioned above, at about the same time that Fleagle provided this brief statement, Mittermeier and Cheney (1986) were lamenting that the TSN monkey may be extinct and that little information had been collected on this species since its initial description. One of the primary reasons for the lack of attention to this rare creature was the fact that Vietnam was at war for much of the 20th century, and this sad truth precluded normal science, such as faunal surveys and ecological studies, from taking place. For example, Ratajszczak (1988:134) stated that "nearly nothing has been added to our knowledge of this species since its discovery. It was briefly observed in the wild by a team of Vietnamese zoologists in 1969, at which time it was still numerous in Ha Tuyen Province" (Fig. 1, Table 1). He added that no surveys had occurred between 1969 and 1988.

Ratajszczak et al. (1990) conducted a survey for WWF in the vicinity of Ba Be National Park where one TSN monkey was observed at Lung Luong Mountain in Cho Don District, Bac Kan Province, documenting their existence after Mittermeier's and Cheney's (1986) postulated extinction. They also

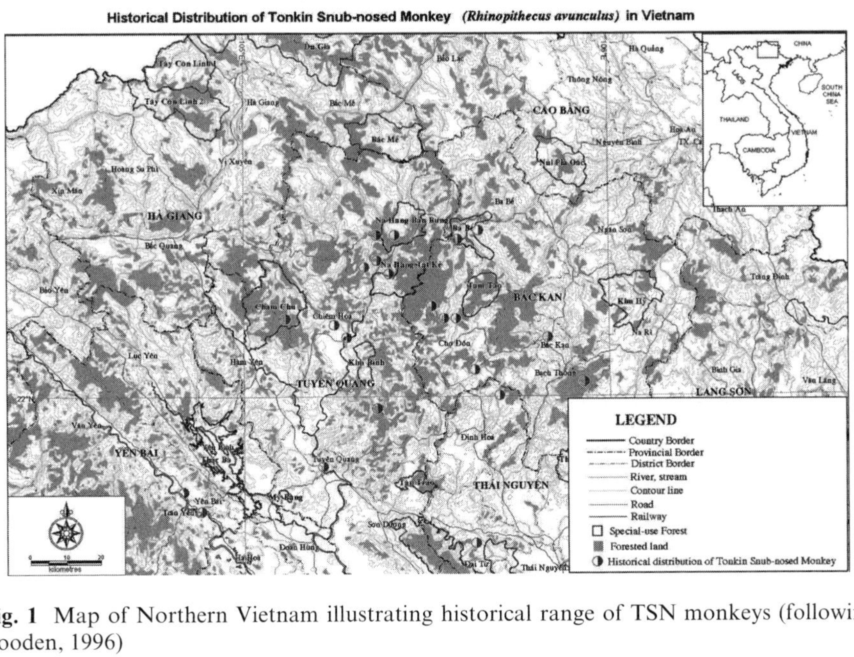

Fig. 1 Map of Northern Vietnam illustrating historical range of TSN monkeys (following Fooden, 1996)

Table 1 Present and past names of provinces where Tonkin snub-nosed monkeys have occurred in historical times

Present Names	Past Names
Yen Bai	Yen Bai, Yen Bay
Tuyen Quang	Ha Tuyen
Ha Giang	Ha Tuyen
Bac Kan	Bac Thai
Thai Nguyen	Bac Thai
Cao Bang	Cao Bang
Quang Ninh	Quang Ning

collected reports on the presence of TSN monkeys in three other locations in this district (Ban Thi, Ra Ban, and Pho Huan). This was followed by a project focusing on the TSN monkey (Ratajszczak et al., 1992). Surveys were conducted in Kien Thiet (Yen Son District), Cham Chu Mountain (Ham Yen and Chiem Hoa Districts), Tat Ke and Ban Bung (Na Hang District) of Ha Tuyen (now belonging to Tuyen Quang) Province, and Tam Dao National Park. They reported the presence of TSN monkeys in Tat Ke (estimated 40–50 individuals based on interviews and collected one skull), in Ban Bung (observed one group of 12 individuals, estimated 90–110 individuals based on interviews, and collected two adult female and one infant specimens), in Cham Chu (estimated a population of approximately15–30 based on interviews and collected two skulls), and in Ken Thiet (estimated 40 based on interviews). They concluded that a total of 190–250 individuals were present in Ha Tuyen Province (now a part of Tuyen Quang) and about 100 individuals in Bac Thai Province (comprised of Bac Kan and Thai Nguyen Provinces). Finally, they predicted that this species might also occur in Bac Quang District, now a portion of Ha Giang Province. Coincidentally, Nisbett and Ciochon (1993) provided a review of the status and ecology of primates in Northern Vietnam in the following year where they stated (p. 780): "although no data are available concerning behavioral ecology, *P. avunculus* is said to live in tropical rain forest." They also acknowledged that L. Lippold had communicated to them in 1988 that "remnant" populations of the TSN monkey existed in the area of Na Hang (presumably the populations at Tat Ke and Ban Bung).

Research in the Na Hang Area

Ratajszczak et al. (1992) and Wirths' (1992) "rediscovery" of TSN monkeys in the Na Hang area proved Lippold's information to be correct. Shortly after this Boonratana and Le Xuan Canh (1994) provided the initial details of the ecology and behavior of these primates based on their observations at two areas in Na Hang District. At about the same time, Pham Nhat (1994) provided information on the diet of this species based on stomach contents of animals that had been poached.

Boonratana and Le Xuan Canh (1998a, b) reviewed results from their six month study in Na Hang District (September 1993–February 1994). They split their time between the Ban Bung and Tat Ke areas. At this time, Ban Bung and Tat Ke were a part of the 20,000 ha Chiem Hoa – Na Hang Nature Reserve. These authors described this area as consisting of steep and mountainous terrain, a difficult landscape in which to observe arboreal animals. During this study they were in contact with the animals a total of 122 hours and they were able to observe them for a total of 47 hours. They characterized the social organization of this population to include single-male/multi-female groups (with an average of 15.2 individuals) and all male groups. They reported there were at least three single-male/multi-female groups in Ban Bung and two in Tat Ke. Further, there were two all male groups at Tat Ke. Finally, regarding social organization, Boonratana and Le Xuan Canh noted that a number of groups may join together at sleeping sites and feeding trees, and that occasionally, they may travel together. In contrast to the Chinese snub-nosed monkeys, these authors described the TSN monkey to be totally arboreal and their locomotion to include "quadrupedal walking, climbing and leaping" (p. 212). These animals were also noted to occasionally brachiate. Boonratana and Le Xuan Canh observed 34 feeding bouts, and the diet consisted of young leaves (38%), unripe fruit (47%), and ripe fruit and seeds (15%). It was estimated that one group ranged over an area of at least 10 km^2, but because this is based on observations of less than three months, Boonratana and Le Xuan Canh (1998a) were unable to determine if this area included the entire home range.

Boonratana and Le Xuan Canh (1998b) focused on the challenges of conserving the TSN monkey at Na Hang. First, they estimated that Tat Ke and Ban Bung were home to approximately 130 monkeys. The two most serious threats to these creatures in this area during the late 1990s were habitat disturbance and hunting. They noted that significant illegal harvesting of timber, bamboo, and fruit were taking place in the Na Hang Reserve. In addition, people had attempted to mine gold in this area, an activity that often includes significant deforestation. This mining appeared to be a non-commercial endeavor and one that had ceased by 1998 (Boonratana, pers. com.). Boonratana and Le Xuan Canh reported that TSN monkey meat is considered to be "bad-tasting," yet these creatures are hunted throughout their range. In addition to being hunted for the dinner table they are also "made into a medicinal stock as a cure for fatigue" (p. 319). These authors were able to convince local commune leaders to ban hunting of all wild animals in this area, and they also had helped to initiate a patrol system by rangers from the Forest Protection Unit. In essence, oonratana and Le Xuan Canh were able to mobilize the local community to become involved in the conservation of TSN monkeys. They concluded in recommending that the Vietnamese government establish a national park in Na Hang.

This pioneering research by Boonratana and Le Xuan Canh led to significant actions in the Na Hang Region. In 1994 Na Hang Nature Reserve was established and in 1998 the reserve was given Special Use Forest status (Martin, 2004). In December 1997 an internationally funded project "the Tonkin

Snub-nosed monkey Conservation Project in Vietnam" (TCP) was implemented by Bettina Martin, with funding from "Allwetter Zoo Münster, the Zoological Society for the Conservation of Species and Populations (ZSCSP), and Primate Conservation Inc. (PCI)" (Martin, 2004:103). Martin also partnered with the Forestry Protection Department (FPD), the People's Committee of Tuyen Quang Province, and the Institute of Ecology and Biological Resources (IEBR). It appeared up until 2002 that the efforts of Martin and coworkers would be sufficient to protect the populations of TSN monkeys in the Na Hang Nature Reserve. Accomplishments at that time included the training of a patrol group of 50 individuals; establishing a wildlife and forest monitoring system; building an office at Na Hang and seven guard posts; developing a nature awareness program for the communes and schools in the area; conducting scientific surveys with IEBR staff; establishing a range of cooperative projects with UNICEF, Protected Areas and the Resource Conservation Project (PARC), and others; and the development of alternative income programs for local villages (Martin, 2004). Nadler et al. (2003), citing communication with Boonrantana, noted that it is not possible to estimate the TSN monkey population size in Na Hang but suggest that it is less than 100 animals. Martin (2004) estimated the population to be around 110 based on TCP ranger reports.

The Damn Dam at Na Hang

Unfortunately, it appears that what was recently a significant conservation success at Na Hang is now a terrible tragedy, or in the words of Martin (2004:104) "even though the situation [conservation of TSN monkeys] improved during the last six years, a huge development activity surpassed conservation efforts: the hydropower dam at Na Hang." According to Martin, the dam construction that began in 2003 has been devastating. For example, during the first year of construction 7,000 workers moved to the town of Na Hang, which previously had a population of about 5,000 human inhabitants. Three-hundred and sixty trucks are passing daily through the town of Na Hang, causing significant noise and air pollution, not to mention creating serious traffic problems. And while the lake created by this dam will flood only a small portion of the Tat Ke section of the protected area, a significant number of people are being resettled away from the new lake and it is feared that they will damage the surrounding forests. Wolters (2004) added that the blasting at the dam site is yielding significant noise pollution, construction of housing for the dam builders has increased the illegal harvesting of timber in the protected areas, and the increase in human population has "accelerated demand for wildlife products and firewood as well as non timber forest products" (p. 9). In addition, the improvement of roads for dam construction has dramatically increased access to the Ban Bung section of the protected area – making it easier to enter for illegal extraction of resources. Finally, it is also quite evident from photos of the construction site that a tremendous amount of sediment has been dislodged and spread onto foliage by three primary sources; blasting, a dramatic increase in traffic, and clearing of the area for dam construction. This has certainly impacted the local ecology by covering all plants in the general vicinity with dirt, and thus affected the diet of the local fauna. The silica in silt and dust has been shown to increase dental wear. In primates such as leaf monkeys that are dependent on well developed dental shear this silt may significantly impact longevity and mortality because it increases the rate of wear (Ungar et al., 1995).

Wolters (2004) provided a slightly more recent review of the TCP and an assessment of the effects of the dam construction on the conservation of TSN monkeys at Na Hang. First, due to recent funding cuts the TCP reduced staff from 50 to 30 rangers. The outcome is that rangers are only patrolling areas that have been utilized by the TSM monkeys during the past few years. In addition, about 1/3rd of the rangers are being dedicated to patrolling and protecting the TSN monkeys in the Cham Chu area (see below). Second, it is becoming more and more difficult to locate the monkeys in the Na Hang area. Wolters reported that the last sighting of TSN monkeys in Na Hang were on June 26, 2004 in the Tat Ke section–after a six month period of no observations. She also noted that little science had been conducted in the Na Hang region since the original work of Boonratana and Le Xuan Canh in 1994 but beginning in 2005 a new study focusing on the ecology of the TSN monkey was to be initiated. Towards the end of her report Wolters (2004:23) stated "neither details about general behavior, breeding ecology, inter- and intra-group relationships and home ranges, nor other conservation related subjects are known but are urgently needed to evaluate current population status and to plan necessary steps and approaches for conserving the species." This highlights the importance of the recently initiated research of Dong Thanh Hai, of the Forestry University of Vietnam, at Na Hang (and our work at Khau Ca). However, he has had less than three hours of observations on the monkeys during the first six months of work in Na Hang (Dong Thanh Hai, personal communication). It appears that the population at Nha Hang has decreased significantly in the past few years.

Nguyen Nga (2000), in a brief report, described two months of survey (August 5 to October 4, 1998) in Na Hang Nature Reserve. While Nguyen Nga does not add any new information to that of Boonratana and Le Xuan Canh, it is worth noting that she found these animals to be rare and difficult to observe, stating that the TSN monkeys were not seen in Ban Bung and were encountered four times for a total of 1.5 hours in Tat Ke. These observations are certainly consistent with those of other researchers in this area during the past few years.

TSN monkeys have been known to be present in Cham Chu of Tuyen Quang Province since 1992 (Ratajszczak et al., 1992) and have been observed there as recently as 2001 (Long and Le Khac Quyet, 2001). Nadler et al. (2003), based on interviews of locals by Long and Le Khac Quyet, estimated the population here to include approximately 70 animals. While this area has been identified as a critical location for conserving the TSN monkey, it has never been approved formally as a protected area by the Vietnamese government. As noted by Wolters (2004), this should be done immediately to guarantee that Cham Chu is sufficiently protected.

The Khau Ca TSN Monkeys

As mentioned above, Ratajszczak et al. (1992) suggested that TSN monkeys might also exist in Ha Giang, Vietnam's most northern province. Yet, Kirkpatrick's (1998a) exhaustive review of all reported observations on snub-nosed monkeys does not include any observations of TSN monkeys in Ha Giang. In late 2001, one of us (Le Khac Quyet), working for Fauna and Flora International (FFI), conducted surveys in Ha Giang Province with the goal of confirming the existence of TSN monkey in this region. In December, a recently poached adult male specimen was given to Le Khac Quyet by a local hunter. This led to surveys of the Khau Ca forest just outside of Du Gia Nature Reserve where a group of TSN monkeys were sighted in January 2002 (Fig. 2). This discovery is exciting, not only does it extend the known present day distribution of this species to the north (and in fact, beyond the known historical range provided by Boonratana and Le Xuan Canh, 1998b), this area appears to be home to at least 50 TSN monkeys, and thus of critical importance for conserving this species.

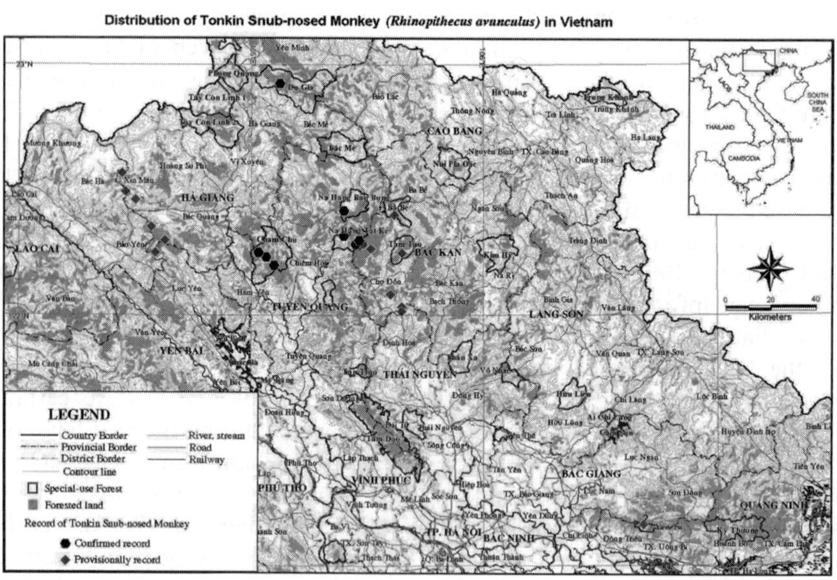

Fig. 2 Map of Northern Vietnam illustrating modern range of TSN monkeys

A Brief Review of *Rhinopithecus*

Today four species of snub-nosed monkeys are recognized with three living in mountainous regions of southern China (*Rhinopithecus roxellana, R. bieti,* and *R. brelichi*), and the TSN monkey restricted to a small area in northern Vietnam. A significant amount of attention has been focused on the Chinese snub-nosed monkeys during the past 15 years (Jablonski, 1998), and these animals are noteworthy for a number of reasons. First, along with the proboscis monkey, they are the largest extant colobines with *R. roxellana* males having a body mass of about 18 kg and females about 12 kg. The males in the other three species have body masses of about 15 kg and females about 9 kg with the TSN monkeys possibly having the smallest body mass (Fleagle, 1999) . Second, *R. roxellana* inhabits the coldest environments of any nonhuman primate, withstanding temperatures as cold as −30°C (Rowe, 1996). Third, their diets are quite unusual compared to other colobines as both *R. roxellana* and *R. bieti* consume a substantial amount of lichen. In addition, the diet of *R. roxellana* is better known than that of the other snub-nosed monkeys and varies significantly through out the year. Kirkpatrick (1998b:171) reported that "the diet is dominated by young leaves in the Spring, and by a mix of mature leaves, fruits and seeds, beard lichens and bryophytes in Summer and Autumn." Their Winter diet varies from site to site but often includes lichens, bark, and pine nuts. Fourth, their social organization is quite unusual for a colobine – some species are characterized by fission-fusion patterns, and all species have occasionally been observed in large bands. *R. roxellana* has been seen in bands as large as 300 in the Qinling Mountains and Shennongjia and 600 in Choushuigou, whereas *R. brelichi* has been seen in a band as large of 400 in Fanjingshan (Kirkpatrick, 1998b). These are the largest groupings ever reported for colobine primates.

Given the paucity of data on the TSN monkey we presently do not know how similar they are in aspects of their dietary and behavioral ecology to the Chinese snub-nosed monkeys. However, ecological study has begun at Khau Ca where an important comparative data set is being compiled. In the following sections we detail the ongoing research at this site.

Research at Khau Ca

Since 2002, FFI Indochina Programme has been actively involved in developing a conservation strategy for the TSN monkeys of Khau Ca, and one of us has been the leader of this effort (Le Khac Quyet, 2002). After initial discussions with provincial officers in Ha Giang City, contacts were made with the leaders of Tung Ba Commune. This commune is adjacent to Khau Ca and residents have historically extracted timber, non-timber products, and hunted in this area. During the past three years a professional working relationship has been developed with commune residents and provincial officers, and the local communities are active partners in developing a sustainable conservation plan for the TSN monkey at Khau Ca. For example, since April 2002, FFI has implemented an emergency conservation action plan in the form of an intensive awareness campaign. FFI has also assisted in the development of a core protection zone with participation from local communities, development of community patrol groups under leadership of Ha Giang Provincial Forestry Protection Department, and development of regulations for protection of the TSN monkey and its habitat by the local community with endorsement from the provincial level. These activities will continue for the foreseeable future, along with additional capacity building for the patrol group and patrol group leaders.

Fig. 3 Photograph of a pair of adult TSN monkeys. Note the sexual dimorphism (female on the left) and tail length

Study of the ecology of the TSN monkey in Khau Ca has been initiated (by LKQ) and more observations of these animals have been made to date than in any previous study, with more than 75 hours of visual contact. For example, initial field surveys were conducted in March–April of 2002, November 2003, and December 2004 through February of 2005. During these surveys data have been collected on feeding behavior, habitat preferences, and social behavior. In addition, a series of trails have been developed and a base camp for research and protection established. A number of photographs have been taken as well as some professional quality video of the monkeys (Fig. 3). During these surveys, all observations on TSN monkeys occurred in Khau Ca, a small patch of limestone forest located in three communes: Tung Ba (Vi Xuyen District), Yen Dinh and Minh Son (Bac Me District) of Ha Giang Province. Its total area is about 1,500 ha and it is mainly comprised of primary evergreen forest. According to the Investment Plan of Du Gia Nature Reserve documents, the Khau Ca area is located in the nature reserve's buffer zone; and thus, receives less conservation attention than the core zone of the reserve.

With the feasibility of conservation and study of the TSN monkeys at Khau Ca now documented we conducted an initial ecological viability study here in June and July of 2005. This study period provided further census data, as well as general dietary information and measurement of the toughness of foods previously observed to have been eaten by the TSN monkeys at this site. Toughness data were collected using a portable mechanical tester designed by Lucas et al. (2001) at Hong Kong University. General survey continued in August and September and in October we started to establish phenological transects and plots, and detailed behavioral data collection was initiated.

Initial Results

Community Based Conservation

Conservation efforts often depend on the cooperation and support of local communities. The lead community assistants at Khau Ca come from the nearby Tung Ba commune. Prior hunting experience has provided these field assistances with knowledge of the Khau Ca forest and the TSN monkeys. A lifetime of work in this area has also provided the local team with the physical skills necessary to navigate the rough rocky terrain of the karst formations where the TSN monkeys are found. The ability of these assistants to work in the forest and maintain an effective camp in close proximity is, indeed, the key to success for this project as revealed to both H. Covert and B. Wright during our initial visit to the area. Training in census techniques and the collection of plant phenology data is ongoing with these participants.

Trails

A trail system of 10 km has been established in the Khau Ca area since the initiation of this project in 2002. This trail system is thought to encapsulate the TSN monkeys' day and home ranges; and it travels through three forest types (lowland evergreen, montane evergreen, and limestone forests) covering a large altitudinal gradient from approximately 640 m to 1,300 m (Fig. 4).

Phenology Transects and Plots

We established four one kilometer long phenology transects, each of which is four meters wide and was selected to cover both altitudinal and soil gradients These transects follow established trails and did not require much foliage cutting, thus limiting forest disturbance. As evidence of this, no trees over 5 cm DBH have been damaged in anyway by trail cutting. This method of phenology data collection was employed to permit local field assistants to simultaneously

Fig. 4 Aerial photograph of illustrating protected areas in Ha Giang. Khau Ca is in the southern boundary of Du Gia Protected Area

collect phenological data and census information bimonthly by walking the established trails. It also reduces the amount of additional brush cutting needed to establish plots, which is of concern to the local populous (pers. com. Tung Ba head ranger) and could influence phenological patterns over the long run. Though these are not true transects (i.e. running in straight lines), they are treated as random sections of straight line transects, and with randomization statistics analyze these data for differences in species composition, tree density and diversity, forest structure, crop yield, and seasonal shifts in phenology.

Trees of 10 cm DBH or larger were sampled within two meters of either side of the one kilometer transects. The size of the trees in this forest and the large size of the study area suggest the use of a 10 cm DBH cutoff and relatively large plots (see Forest Mensuration, 3rd Edition, B. Husch et al., Krieger Publishing Company, Malabar, 1993). Standard phonological data will be collected including taxon, DBH, bole height, tree height, canopy shape, canopy diameters (both long and short), % leaf flush, % flower/inflorescence, % fruit, % horizontal substrates, % up 45% substrates, % down 45% substrates, and associated climbers and epiphytes.

All four transects were completed during October, 2006, at which time the first set of phenological data were collected. The DBH and canopy dimensions of sampled trees will be remeasured annually. Aluminum tags were used to identify the trees and to mark the height at which DBH was measured. Two botanists from Vietnam National University (VNU), Hanoi, Mr. Vu Anh Tai and Mr. Nguyen Anh Duc, provided tree family and species names and were included on the tree tags. GPS waypoints were taken every 50 m the altitude provided at the 50 m waypoints can be used to correlate altitudinal gradient changes with changes in tree density and species composition.

Eight 20 × 50 m plots have also been established in different eco-regions throughout the site. The data from these plots will be compared to those from the transects. The sampling methods used in these plots have been long established by botanists at VNU and will provide a data set that can be directly compared to other studies conducted by VNU researchers. The only addition to the established plot methods was the grade and direction of the plot slope. These data, in combination with information on feeding behavior that is being collected during the next 18 months will generate a much better understanding of the feeding ecology of this species including the identification of keystone and fall back foods, critical information for planning long-term conservation strategies.

Census

Bimonthly census walks of the entire trail system were initiated in December 2005. Although the rugged limestone terrain does not permit the establishment of straight-line transects, temporally systematic walks of these trails will accumulate a reasonable census data set. The most efficient use of the field assistants' time and available resources is to combine census and phenology walks. Distance and angle to sighted animals and animal counts will be included. Group sightings may possibly occur more than once during a given walk. However, during other walks the animals will not be seen. Over time this will provide a more reasonable estimate than intermittent opportunistic group counts. Two teams consisting of at least two field assistants will work together, permitting completion of the census and phenology walks in a reasonable amount of time.

Population Size

The TSN monkeys have been observed numerous times in Khau Ca during the past few years by members of our research team and an estimated of about 50 animals has been made but this is simply based on opportunistic counts rather than traditional census methods. In June, 2005 we had the opportunity to count a single group of TSN monkeys that were approximately 200–300 m across a ravine from the base camp. We had an unobstructed view of a number of individuals for approximately 2 hours of continuous observation. We counted 26 individuals as they crossed a single forest gap. Three of these individuals were females with dependent young, bringing the total count to 29 individuals. Smaller groups ranging from two to seven individuals were seen on seven other occasions in June. It is likely that these smaller groups include one adult male and a number of adult females and their young, while the larger observed group probably included at least two of these one male groups. Thus, we can say with confidence that there is a group of at least 29 individuals at Khau Ca. The relative ease with which individuals were encountered throughout this first period of ecological study strongly suggests that a larger population exists at this site. It is also of note that arm swinging was used by all observed individuals to cross a tree gap that was encountered by the group (see "Locomotion and Posture" below).

Diet

Our initial dietary findings reveal that the TSN monkeys at Khau Ca exploit a broad range of plants and plant parts. Since 2004, 18 food species have been identified (Table 2). Leaf stems are the most common part of the TSN monkeys' diet (31%), followed by ripe fruit (23%), unripe fruit (15%), inflorescences (15%) and buds (8%) (Table 3). Leaf lamina only contributes 8% to the total dietary profile. Among items tested for toughness, cambium (1577.5 Jm^{-2}), leaf lamina (1370.6 Jm^{-2}) and leaf stems (1230.9 Jm^{-2}) are the toughest tissues ingested by the TSN monkeys, followed by whole fruit (1147.53 Jm^{-2}), seed coats (1129.6 Jm^{-2}), bark (981.6 Jm^{-2}), mesocarp (382.9 Jm^{-2}), endosperms (351.9 Jm^{-2}), and exocarps (231.4 Jm^{-2}). B. Wright, L. Ulibarri and J. O'Brien have found that the toughness of the cambium, stems, and leaf lamina ingested by the TSN monkeys is greater than that of tissues selected by two *Pygathrix*

Table 2 List of food plants of TSNM at Khau Ca as of July, 2005

Latin name	Consumed items
Aceraceae	
Acer tonkinensis	Stems
Annonaceae	
Polyalthia suberosa	Flower
Unknown species 1	Ripe fruit
Araliaceae	
Brassaiopsis stellata	Ripe fruit
Schefflera palmiformis	Stems
Asclepiadaceae	
Goniostema punctatum	Stems
Clusiaceae	
Garcinia tinctoria	Stems
Euphorbiaceae	
Bridelia retusa	Ripe fruit
Sapium rotundifolium	Seeds
Icacinaceae	
Iodes seguin	Stems, ripe fruit
Rubiaceae	
Unknown species 2	Young leaves (buds)
Tiliaceae	
Excentrodendron tonkinensis	Young leaves (buds)

and two *Trachypithecus* colobine species studied at the Cuc Phuong National Park Endangered Primate Rescue Center.

Locomotion and Posture

TSN monkeys appear to be habitually arboreal in that they only rarely come to the ground in Khau Ca. The following is a list of locomotor modes that we observed these monkeys to use. The most frequently used locomotion is a symmetrical gait

Table 3 Counts and percentages for food types consumed by the TSN monkeys at Khau Ca

Items	Numbers	% of diet
Flowers	2	10
Fruit	8	40
Stems	6	30
Leaves	3	15
Seeds	1	5
Total	20	100

walk and other common modes include quadrupedal running, vertical climbing, rump-first descent, brachiation, brachiating leap, arrested drops, pronograde leaping, pumping leaping, and quadrupedal drop. Common postures include: sit out, sit/forelimb suspend, chair sit, bimanual cling, cling/forelimb suspend, stand/forelimb-suspend, forelimb-suspend/stand, and lie. In the near future a study of their positional behavior that will yield quantitative data will be started. It is worth noting here that forelimb suspension is commonly used in both locomotion and postures. Covert et al. (2004) have argued that suspensory behaviors may be an important aspect of the positional repertoire of all odd-nosed monkeys.

Perspective

We are confident that Khau Ca is an ideal location to protect the critically endangered TSN monkey and that we are well on our way to implementing a research and conservation program that will be successful in the short and long-term. Habitat loss and hunting continue to seriously threaten the survival of the TSN monkey throughout their range, and as noted by Le Khac Quyet (2004), the Khau Ca area until recently suffered habitat loss or destruction due to exploitation of timber and of non-timber forest products for commercial purposes, shifting cultivation, livestock feeding, fuel wood collecting and hunting. During the past three years, there appears to have been significant reductions in all of these activities, with only the exploitation of non-timber products remaining a common practice in this area. The most important task at hand to continue this trajectory of reduced habitat disturbance and hunting is the immediate establishment of a Species/Habitat Conservation Area for the TSN monkey at Khau Ca so that this area will be fully protected under Vietnamese law. Ultimately, this area, along with the Du Gia Nature Reserve, should be placed into a national park. Ongoing scientific research at this site is essential. It will provide quantitative data on those aspects of the habitat that are most important for the TSN monkeys' survival. These data will enable us to identify the best areas for development of expansion corridors. Presently it appears that areas to the north of Khau Ca are the most likely candidates for such activities, but forest surveys in the near future will aid in identifying the viability of these tracts.

Tonkin Snub-nosed Monkeys and Elwyn Simons

We are pleased to be able to provide this information on the ecology and conservation of the TSN monkey, and the work by the senior author on this project is a direct outcome of studying with Elwyn Simons between 1978 and 1985 at Duke University. Elwyn shared (and continues to share) with his students the importance of conducting field research in primate evolutionary biology. Our project at Khau Ca fits comfortably into Elwyn's vision that the most critical advances in our discipline occur when scientists invest the necessary time in the field to make observations that significantly add to our understanding of primate natural history.

Acknowledgment We thank the leaders of Ha Giang People's Committee and the Ha Giang Forestry Protection Department for allowing us to conduct this research. We also thank the citizens of Tung Ba Commune for providing accommodations before and after fieldwork at Khau Ca. Special thanks go to Mr. Khoan, Mr. Chuyen, and Mr. Khoán for their assistance in the field. Conversations with Dr. Le Xuan Canh, Mr. Vu Ngoc Thanh, Dr. Ramesh Boonratana, Ms. Sonya Wolters, Ms. Bettina Martin, and Mr. Tilo Nadler helped us to better understand the challenges for successfully conserving the Tonkin snub-nosed monkey. We thank Mr. Jonathan O'Brien and Mr. Larry Ulibarri for their assistance with data collection and for commenting on an earlier version of this manuscript. This research has been funded by Fauna and Flora International, Vodafone Group Plc., FFI/Flagship Species Fund, the Singapore Zoo, Margot Marsh Biodiversity Foundation, Primate Conservation Inc, the Zoological Society of San Diego, and National Geographic.

References

Boonratana, R., and Le Xuan Canh (1994). A report on the ecology, status and conservation of the Tonkin snub-nosed monkeys (*Rhinopithecus avunculus*) in Northern Vietnam. WCS, New York and IEBR, Hanoi.

Boonratana, R., and Le Xuan Canh (1998a). Preliminary observations of the ecology and behavior of the Tonkin snub-nosed monkey (*Rhinopithecus [Presbytiscus] avunculus*) in Northern Vietnam. In: Jablonski, N.G. (ed.), *The Natural History of the Doucs and Snub-nosed Monkeys*. World Scientific Publishing, Singapore, pp. 207–215.

Boonratana, R., and Le Xuan Canh (1998b). Conservation of Tonkin Snub-nosed Monkeys (*Rhinopithecus [Presbytiscus] avunculus*) in Vietnam. In: Jablonski, N.G. (ed.), *The Natural History of the Doucs and Snub-nosed Monkeys*. World Scientific Publishing, Singapore, pp. 315–322.

Cowlishaw, G., and Dunbar, R. (2000). *Primate Conservation Biology*. University of Chicago Press, Chicago.

Covert, H.H., Workman, C., and Byron, C. (2004). The EPRC as an important research center: Ontogeny of locomotor differences among Vietnamese colobines. In: Nadler, T., Streicher, U., and Ha Thang Long (eds.), *Conservation of Primates in Vietnam*. Haki Publishing, Hanoi, pp. 121–129.

Dollman, G. (1912). A new snub-nosed monkey. *Proc. Zool. Soc. London*. 106:503–504.

Fleagle, J.G. (1988). *Primate Adaptation and Evolution*, 1st ed. Academic Press, San Diego.

Fleagle, J.G. (1999). *Primate Adaptation and Evolution*, 2nd ed. Academic Press, New York.

Fooden, J. (1996). Zoogeography of Vietnamese primates. *Int. J. Primatol*. 17:845–899.

Groves, C. P. (2001). *Primate Taxonomy*. Smithsonian Institution Press, Washington.

Hooton, E. (1942). *Man's Poor Relations*. Doubleday, Doran and Company, Garden City, NY.

Jablonski, N.G. (ed.). (1998). *The Natural History of the Doucs and Snub-nosed Monkeys*. World Scientific Publishing, Singapore.

Kirkpatrick, P.C. (1998a). Toward a gazetteer of the snub-nosed and douc langurs. In: Jablonski, N.G. (ed.), *The Natural History of the Doucs and Snub-nosed Monkeys*. World Scientific Publishing, Singapore, pp. 337–372.

Kirkpatrick, P.C. (1998b). Ecology and behavior in snub-nosed and douc langurs. In: Jablonski, N.G. (ed.), *The Natural History of the Doucs and Snub-nosed Monkeys*. World Scientific Publishing, Singapore, pp. 155–190.

Le Khac Quyet (2002). Distribution and conservation of Tonkin snub-nosed monkey (*Rhinopithecus avunculus*) in Du Gia Nature Reserve, Ha Giang Province, Northeast Vietnam. Fauna and Flora International – Vietnam Programmes, Hanoi.

Le Khac Quyet (2004). Distribution and conservation of Tonkin snub-nosed monkey (*Rhinopithecus avunculus*) in Du Gia Nature Reserve, Ha Giang Province, Northeast Vietnam. In Nadler, T. Streicher, U., and Ha Thang Long (eds.), *Conservation of Primates in Vietnam*. Haki Publishing, Hanoi, pp 58–62.

Long, B., and Le Khac Quyet (2001). An initial assessment of conservation requirements for Cham Chu, Tuyen Quang Province including mammal and bird diversity surveys. Unpubl. Report Fauna amp; Flora International, Indochina-Programme, Hanoi.

Lucas, P.W., Beta, T., Darvell, B.W., Dominy, N.J., Essackjee, H.C., Lee, P.K.D., Osorio, D., Ramsden, L., Yamashita, N., and Yuen, T.D.B. (2001). Field kit to characterize physical, chemical and spatial aspects of potential primate foods. *Folia Primatol.* 72:11–25.

Martin, B. (2004). Conservation of the Tonkin snub-nosed monkey (*Rhinopithecus avunculus*) and its habitat at Ha Hang Nature Reserve during dam construction. In: Nadler, T., Streicher, U., and Ha Thang Long (eds.), *Conservation of Primates in Vietnam*. Haki Publishing, Hanoi, pp. 103–105.

Milne-Edwards, A. (1872). Memoire sur la Faune mammalogique du Tibet Oriental et Principalement de la principauté de Moupin. In: Milne-Edwards, E. and Milne-Edwards, A., (eds.), *Recherches pour Servir à l'Histoire Naturelle des Mammifères*. Masson, Paris, pp. 231–379.

Mittermeier, R.A., and Cheney, D.L. (1986). Conservation of primates and their habitats. In: Smuts, B.B., Cheney, D.L., Seyfarth, R.M., Wrangham, R.W., and Strusaker, T.T. (eds.), *Primate Societies*. University of Chicago Press: Chicago, pp. 477–490.

Mittermeier, R.A., Konstant, W.R., Hawkins, F., Louis, E.E., Langrand, O., Ratsimbazafy, J.,Rasoloarison, R. Ganzhorn, J.U., Rajaobelina, S., Tattersall, I., and Meyers, D.M. (2006). *Lemurs of Madagascar*. Conservation International, Washington, D.C.

Mittermeier, R. A., Valladares-Pádua, C., Rylands, A.B., Eudey, A.A., Butynski, T.M., Ganzhorn, J.U., Kormos, R. Aguiar, J.M., and Walker, S. (2005). *Primates in Peril. The World's 25 most endangered primates 2004–2006*. Conservation International, Washington D.C.

Nadler, T., Momberg, F., Nguyen Xuan Dang, and Lormée (2003). *Vietnam Primate Conservation Status Review 2002. Part 2: Leaf Monkeys*. FFI Vietnam Program and Frankfurt Zoological Society, Hanoi, Vietnam.

Napier, J.R., and Napier, P.H. (1967). *A Handbook of the Living Primates*. Academic Press, London.

Nguyen Nga (2000). A survey of Tonkin snub-nosed monkeys (*Rhinopithecus avunculus*) in northern Vietnam. *Folia Primatol.* 71:157–160.

Nisbett, R.A., and Ciochon, R.L. (1993). Primates in Northern Vietnam: A review of the ecology and conservation status of extant species, with notes on Pleistocene localities. *Int. J. Primatol.* 14:765–795.

Oakley, F. (2008). Introduction to the Festschrift. In: Fleagle, J. G. and Gilbert, C. C. (eds.), Elwyn Simons: A Search for Origins. Springer, New York, pp. 3–5.

Osgood, W. H. (1932). Mammals of the Kelley-Roosevelts and Delacour Asiatic expeditions. *Field Museum of Natural History, Zoological Series* 18(10):193–339.

Pham Nhat (1994). Some data on the food of the Tonkin snub-nosed monkey (*Rhinopithecus avunculus*). *Asian Primates* 4:9–11.

Pocock, R.I. (1924). A new genus of monkeys. *Proc. Zool. Soc. London*. 1924:330–331.

Ratajszczak, R. (1988). Notes on the current status and conservation of primates in North Vietnam. *Primate Conservation* 9:134–136.

Ratajszczak, R., Cox, R, and Ha Hinh Duc (1990). A preliminary survey of primates in North Vietnam. WWF Report, Project 3869.

Ratajszczak, R. Ngoc Can, and Pham Nhat (1992). A survey for Tonkin snub-nosed monkey (*Rhinopithecus avunculus*) in North Vietnam. FFI Preservation Society, London and WWF.

Rowe, N. (1996). *The Pictorial Guide to the Living Primates*. Pogonias Press, NY.

Simons, E.L. (1993). Discovery of the western aye–aye. *Lemur News* 1:6.

Simons, E.L. (1995). History, anatomy, subfossil record and management of *Daubentonia madagascariensis*. In: Alterman, L., Doyle, G.A., and Izard, M.K. (eds.) *Creatures of the Dark: The Nocturnal Prosimians*. Plenum Press, New York pp. 133–140.

Thomas, O. (1928). The Delacour exploration of French Indo-China. Mammals II: Mammals collected during the winter of 1926–27. *Proc. Zool. Soc. London* 1928(1):139–150.

Ungar, P.S., Teaford, M.F., Glander K.E., and Pastor R.F. (1995) Dust accumulation in the canopy: A potential cause of dental microwear in primates. *Am. J. Phys. Anthropol.* 97:93–99.

Wirth, R. (1992). Tonkin snub-nosed (*Rhinopithecus avunculus*) rediscovered. *Asian Primates* 2:1–2.

Wolters, S. (2004). Technical report (1) – Reassignment and restructuring Tonkin Snub-nosed Monkey Conservation Project, Na Hang, Vietnam.

Wright, P.C. (2008). Decades of lemur research and conservation: the Elwyn L. Simons influence. In: Fleagle, J. G. and Gilbert, C. C. (eds.), Elwyn Simons: A Search for Origins. Springer, New York, pp. 283–310.

Zuckerman, S. (1932). *The Social Life of Monkeys and Apes*. Harcourt, Brace and Company, New York.

Appendix

Publications by Elwyn Simons Through 2006

1 Patterson B, and Simons EL (1958) A new barylambdid pantodont from the late Paleocene. Breviora *93:*1–8.
2 Simons EL (1959) An anthropoid frontal bone from the Fayum Oligocene of Egypt: the oldest skull fragment of a higher primate. Am. Mus. Novitates *1976:*1–16.
3 Simons EL (1960) *Apidium* and *Oreopithecus*. Nature *186:*824–826.
4 Simons EL (1960) New fossil primates: a review of the past decade. American Scientist *48:*179–192.
5 Simons EL (1960) The Paleocene Pantodonta. Trans. Am. Philos. Soc. (N.S.) *50:*1–99.
6 Simons EL (1959) Book Review: Mankind in the Making (The Story of Human Evolution) by William Howells. American Scientist *48:*405A.
7 Simons EL, and Russell DE (1960) Notes on the cranial anatomy of *Necrolemur*. Breviora *127:*1–14.
8 Simons EL (1961) Notes on Eocene tarsioids and a revision of some Necrolemurinae. Bull. Brit. Mus. (Nat. Hist.), Geol. Ser *5:*45–69.
9 Simons EL (1961) An anthropoid mandible from the Oligocene Fayum beds of Egypt. Am. Mus. Novitates *2051:*1–5.
10 Simons EL (1961) The dentition of *Ourayia*. Its bearing on relationships of omomyid prosimians. Postilla *54:*1–20.
11 Simons EL (1961) The phyletic position of *Ramapithecus*. Postilla *57:*1–9.
12 Simons EL (1962) A new Eocene primate, *Cantius*, and a revision of some allied European lemuroids. Bull. Brit. Mus. (Nat. Hist.) Geol. Ser *7:*1–36.
13 Simons EL (1962) An expedition to the Egyptian desert: Yale Scientific Magazine, pp. 1–4.
14 Simons EL (1962) Two new primate species from the African Oligocene. Postilla *64:*1–12.
15 Simons EL (1962) Fossil evidence relating to the early evolution of primate behavior. Ann. New York Acad. Sci *102:*282–294.
16 Simons EL (1963) A critical reappraisal of Tertiary primates. In J Buettner-Janusch (ed.): Genetic and Evolutionary Biology of the Primates. New York: Academic Press, pp. 65–129.
17 Simons EL (1962) Primates (Comparative Anatomy and Taxonomy), Vol. V: Cebidae, part B by W.C. Osman Hill. American Scientist *51:*207A–207B.
18 Simons EL (1963) David Baldwin, O.C. Marsh, and the discovery of the first continental Paleocene faunas of the New World. Postilla *75:*1–11.
19 Simons EL (1963) Some Fallacies in the Study of Hominid Phylogeny. Science *141:*879–889.
20 Simons EL (1963) The Yale Collection of Fossil Primates. A brief survey of an extensive and world famous collection at Peabody Museum: Yale Scientific Magazine, pp. 22–23.
21 Simons EL, and Alexander HL (1964) Age of the Shasta ground sloth from Aden Crater, New Mexico. American Antiquity *29:*390–391.
22 Simons EL (1964) On the mandible of *Ramapithecus*. Proc. Nat. Acad. Sci *51:*528–535.
23 Simons EL (1964) Book Review: Old World Higher Primates: Classification and Taxonomy. Science *144:*709–710.
24 Simons EL (1964) The early relatives of man. Scientific American *211:*50–62.
25 Simons EL (1964) Book Review: The Geology of Egypt. American Journal of Science *262:*1237–1238.
26 Simons EL (1965) New fossil apes from Egypt and the initial differentiation of Hominoidea. Nature *205:*135–139.
27 Simons EL, and Pilbeam DR (1965) Preliminary revision of Dryopithecinae (Pongidae, Anthropoidea). Folia Primat *3:*81–152.

28 Simons EL (1965) Symposium Remarks on Pongid and Hominid Evolution. In PLD Vore (ed.): The Origin of Man, A Symposium. New York: The Wenner-Gren Foundation for Anthropological Research, Inc, pp. 43–45, 65, 68.
29 Simons EL, and Pilbeam DR (1965) Some problems of hominid classification. Am. Sci *53:*237–259.
30 Simons EL (1965) The hunt for Darwin's third ape: Medical Opinion and Review, pp. 74–81.
31 Simons EL (1966) In search of the missing link. Discovery, Yale Peabody Museum *1:*24–30.
32 Simons EL (1967) Unraveling the age of earth and man: Natural History, pp. 53–59.
33 Simons EL (1967) Fossil primates and the evolution of some primate locomotor systems. Am. J. Phys. Anth *26:*241–253.
34 Simons EL (1967) A fossil *Colobus* skull from the Sudan (Primates, Cercopithecidae). Postilla *111:*1–12.
35 Simons EL (1967) Order Primates, Order Pantodonta. The Fossil Record*:*763–787.
36 Simons EL (1967) The significance of primate paleontology for anthropological studies. Am. J. Phys. Anth *27:*307–332.
37 Simons EL (1967) The earliest apes. Scientific American *217:*28–35.
38 Simons EL (1967) Review of the phyletic interrelationships of Oligocene and Miocene Old World Anthropoidea. Problèmes actuels de Paléontologie (Evolution des vertébrés). Colloques Internationaux du Centre National de la Recherche Scientifique, Paris *163:*597–602.
39 Simons EL (1967) New evidence on the anatomy of earliest Catarrhine primates. In RS D. Starck, H.J. Kuhn (ed.): Neue Ergebnisse der Primatologie: Progress in Primatology. Stuttgart: Fischer, pp. 15–18.
40 Simons EL (1968) A source for dental comparison of *Ramapithecus* with *Australopithecus* and *Homo*. S. Af. Joun. Sci *64:*92–112.
41 Simons EL (1968) Primate evolution: International Encyclopedia of the Social Sciences. New York: MacMillan Co, pp. 210–215.
42 Simons EL (1968) Assessment of a Fossil Hominid (review). Science *160:*672–675.
43 Simons EL (1968) New Fossil Primates: A review.
44 Simons EL (1968) On the mandible of *Ramapithecus*. In G Kurth (ed.): Evolution and Hominization, Second Edition. Stuttgart: Fischer, pp. 139–149.
45 Simons EL (1968) Early Cenozoic mammalian faunas, Fayum Province, Egypt; Part I. African Oligocene Mammals: Introduction, history of study and faunal succession, pp. 1–21. Bulletin, Peabody Museum *28:*1–105.
46 Simons EL (1968) Hunting the "Dawn Apes" of Africa. Discovery *4:*19–32.
47 Simons EL (1968) *Gigantopithecus* in India. Newsletter, Yerkes Reg. Primate Res. Cent., Emory University*:*14–18.
48 Tattersall IM, and Simons EL (1969) Notes on some little-known primates from India. Folia Primat *10:*146–153.
49 Simons EL (1969) Late Miocene hominid from Fort Ternan, Kenya. Nature *211:*448–451.
50 Simons EL (1969) Doctor Leakey and the Dawn of Man. American Anthropologist *71:*577–578.
51 Simons EL (1969) In pursuit of man's pedigree: Yale Alumni Mag, pp. 24–27.
52 Simons EL (1969) Recent Advances in Paleoanthroplogy. Yearbook of Physical Anthropology.
53 Simons EL (1969) Miocene monkey (*Prohylobates*) from northern Egypt. Nature *223:*687–689.
54 Simons EL, and Chopra SRK (1969) A preliminary announcement of a new *Gigantopithecus* species from India. Proc. 2nd Int. Cong. Primat *2:*135–43.
55 Simons EL, and Chopra SRK (1969) *Gigantopithecus* (Pongidae, Hominoidea) A new species from North India. Postilla *138:*18 pages.

56 Simons EL, Pilbeam DR, and Ettel PC (1969) Controversial taxonomy of fossil hominids. Science *166:*258–259.
57 Simons EL (1969) The origin and radiation of the Primates. Ann. New York Acad. Sci *167:*319–331.
58 Simons EL, and Ettel PC (1970) *Gigantopithecus*. Scientific American *222:*76–85.
59 Simons EL (1970) The deployment and history of Old World monkeys (Cercopithecidae, Primates). In JF Napier and RP Napier (eds.): Old World Monkeys: Evolution, Systematics, and Behavior. New York: Academic Press, pp. 99–138.
60 Simons EL (1971) A current review of the interrelationships of Oligocene and Miocene Catarrhini. In AA Dahlberg (ed.): Dental Morphology and Evolution: University of Chicago Press, pp. 193–208.
61 Simons EL (1971) Prehuman primates: Anthropology Today. Del Mar, California: CRM Books, pp. 148–161.
62 Pilbeam DR, and Simons EL (1971) Humerus of *Dryopithecus* from Saint Gaudens, France. Nature *229:*406–407.
63 Simons EL, Pilbeam DR, and Boyer SJ (1971) Appearance of Hipparion in the Tertiary of the Siwalik Hills of north India, Kashmir and west Pakistan. Nature *229:*408–409.
64 Simons EL (1971) Tertiary Period: Encyclopedia Americana, pp. 524–576.
65 Simons EL (1971) Primates, fossil: *Gigantopithecus*. In DN Lapedes (ed.): McGraw-Hill Yearbook of Science and Technology. New York: McGraw-Hill, pp. 347–348.
66 Simons EL (1971) Relationships of *Amphipithecus* and *Oligopithecus*. Nature *232:*489–491.
67 Simons EL, and Pilbeam DR (1971) A gorilla-sized ape from the Miocene of India. Science *173:*23–37.
68 Simons EL, and Tattersall IM (1971) Origin of the family of Man: Ventures, pp. 47–55.
69 Simons EL, and D. R. Pilbeam (1972) Hominoid paleoprimatology. In R Tuttle (ed.): The Functional and Evolutionary Biology of Primates. Chicago: Aldine Press, pp. 24–27.
70 Simons EL (1972) Primate Evolution: An Introduction to Man's Place in Nature. NY: MacMillan Company.
71 Simons EL, and Fleagle J (1973) The history of extinct gibbon-like primates. In DM Rumbaugh (ed.): Gibbon and Siamang, Volume 2. Basel and New York: S. Karger, pp. 121–148.
72 Simons EL (1974) The relationships of *Aegyptopithecus* to other primates. Ann. Geol. Surv. Egypt *4:*149–156.
73 Simons EL, and Gingerich PD (1974) New carnivorous mammals from the Oligocene of Egypt. Ann. Geol. Surv. Egypt *4:*157–166.
74 Simons EL (1974) Notes on early Tertiary prosimians. In RD Martin, GA Doyle and A Walker (eds.): Prosimian Biology. London: Duckworth, pp. 415–433.
75 Simons EL (1974) *Parapithecus grangeri* (Parapithecidae, Old World Higher Primates): New species from the Oligocene of Egypt and the initial differentiation of Cercopithecoidea. Postilla *166:*1–12.
76 Fleagle JG, Simons EL, and Conroy GC (1975) Ape limb bone from the Oligocene of Egypt. Science *189:*135–137.
77 Conroy GC, Schwartz JH, and Simons EL (1975) Dental eruption patterns in Parapithecidae (Primates, Anthropoidea). Folia Primat *24:*275–281.
78 Simons EL, and Gingerich PD (1976) A new species of *Apterodon* (Mammalia,Creodonta) from the upper Eocene Qasr el-Sagha formation of Egypt. Postilla *168:*9.
79 Simons EL (1976) Primate radiations and the origins of hominoids. In RB Masterson (ed.): Evolution of the Nervous System and Behavior. Washington, D.C: V. H. Winston and Sons, Inc, pp. 383–391.
80 Simons EL (1976) The nature of the transition in the dental mechanism from pongids to hominids. Journ. Human Evol *5:*511–528.

81 Rose KD, and Simons EL (1977) Dental function in the Plagiomenidae: Origin and relationship of the mammalian order Dermoptera. Contrib. Mus. Paleont., Univ. of Michigan *24:*221–236.
82 Simons EL (1977) The fossil record of primate phylogeny. In M Goodman and R Tashian (eds.): Molecular Anthropology. New York: Plenum Press, pp. 35–62.
83 Simons EL (1977) *Ramapithecus*. Scientific American *236:*28–35.
84 Rose KD, Bown TM, and Simons EL (1977) An unusual new mammal from the early Eocene of Wyoming. Postilla *172:*1–12.
85 Gingerich PD, and Simons EL (1977) Systematics, phylogeny, and evolution of early Eocene Adapidae (Mammalia, Primates) in North America. Contrib. Mus. Paleont., Univ. Michigan *24:*245–279.
86 Simons EL (1978) Relationships between *Dryopithecus, Sivapithecus and Ramapithecus* and their bearing on hominid origins. Les plus anciens hominidés, Proceedings of the IX Congrès International des Sciences Préhistoriques et Protohistoriques, Nice, September 1976.
87 Simons EL, and Delson E (1978) Chapter 7. Cercopithecidae and Parapithecidae. In VJ Maglio and HBS Cooke (eds.): Evolution of African Mammals. Cambridge: Harvard University Press, pp. 100–119.
88 Simons EL, Andrews P, and Pilbeam DR (1978) Chapter 8. Cenozoic Apes. In VJ Maglio and HBS Cooke (eds.). Cambridge: Harvard University Press, pp. 120–146.
89 Simons EL, and Pilbeam DR (1978) Chapter 9. *Ramapithecus* (Hominidae, Hominoidea). In VJ Maglio and HBS Cooke (eds.). Cambridge: Harvard University Press, pp. 147–153.
90 Rose KD, Bown TM, and Simons EL (1978) *Alocodontulum*, a New Name for *Alocodon* Rose, Bown and Simons 1977, Non Thulborn, 1973. Journal of Paleontology *52:*1162–1162.
91 Simons EL (1978) Diversity among the early hominids: A vertebrate paleontologist's viewpoint. In CJ Jolly (ed.): Early Hominids of Africa. New York: St. Martin's Press, pp. 543–566.
92 Fleagle JG, and Simons EL (1978) *Micropithecus clarki*, a Small Ape from Miocene of Uganda. American Journal of Physical Anthropology *49:*427–440.
93 Fleagle JG, and Simons EL (1978) Humeral Morphology of Earliest Apes. Nature *276:*705–707.
94 Simons EL (1979) L'origine des hominidés. La Recherche, Paris *10:*260–267.
95 Fleagle JG, and Simons EL (1979) Anatomy of the Bony Pelvis in Parapithecid Primates. Folia Primatologica *31:*176–186.
96 Romero-Herrera AE, Lieska N, Goodman M, and Simons EL (1979) Use of Amino-Acid Sequence Analysis in Assessing Evolution. Biochimie *61:*767–779.
97 Simons EL, and Kay RF (1979) Comments on the adaptive strategy of the first African anthropoids. Proc. VII'th Cong. Int. Primat. Soc., Bangalore, India, Zeitschrift f. Morph. u. Anthropol *71:*143–148.
98 Kay RF, and Simons EL (1979) Ecology of Oligocene African Anthropoidea. American Journal of Physical Anthropology *50:*453–453 (abstract).
99 Kay RF, and Simons EL (1979) Ecology of Oligocene African Anthropoidea. International Journal of Primatology *1:21–37.*
100 Simons EL, and Kay RF (1980) Dawn Ape Provides Clue to Social-Life. Geotimes *25:*18–18.
101 Eaglen RH, and Simons EL (1980) Notes on the Breeding Biology of Thick-Tailed and Silvery Galagos in Captivity. Journal of Mammalogy *61:*534–537.
102 Fleagle JG, Kay RF, and Simons EL (1980) Sexual Dimorphism in Early Anthropoids. Nature *287:*328–330.

103 Simons EL (1980) Origine du genre Homo à partir d'Australopithecus, de Ramapithecus, ou d'une forme encore inconnue. Les processus de l'hominisation, Colloque international, CNRS, France *599:*1–310.
104 Simons EL, and Kay RF (1981) *Aegyptopithecus* and *Propliopithecus*. In S Parker (ed.): McGraw-Hill Yearbk. Sci. and Tech, pp. 77–80.
105 Simons EL, and Kay RF (1981) *Apidium* and *Parapithecus*. In S Parker (ed.): McGraw Hill Yearbk. Sci. and Tech, pp. 103–105.
106 Simons EL (1981) Mans Immediate Forerunners. Philosophical Transactions of the Royal Society of London Series B-Biological Sciences *292:*21–41.
107 Kay RF, Fleagle JG, and Simons EL (1981) A Revision of the Oligocene Apes of the Fayum Province, Egypt. American Journal of Physical Anthropology *55:*293–322.
108 Cartmill M, Macphee RDE, and Simons EL (1981) Anatomy of the Temporal Bone in Early Anthropoids, with Remarks on the Problem of Anthropoid Origins. American Journal of Physical Anthropology *56:*3–21.
109 Simons EL, and Covert HH (1981) Paleoprimatological research of the last fifty years–foci and trends. Am. J. Phys. Anthrop *56:*373–382.
110 Fleagle JG, Kay RF, and Simons EL (1981) Sexual Dimorphism in Early Anthropoids - Reply. Nature *290:*609.
111 Simons EL (1982) History of the primates: half-lemurs, sub-monkeys, monkeys and the archaic and progressive apes. In PV Dooren (ed.): De Evolutie van de mens. Maastricht, Holland: De Evolutie van de Primaten, Chapter 1, pp. 12–39.
112 Fleagle JG, and Simons EL (1982) The Humerus of *Aegyptopithecus zeuxis* - a Primitive Anthropoid. American Journal of Physical Anthropology *59:*175–193.
113 Bown TM, Kraus MJ, Wing SL, Fleagle JG, Tiffney BH, Simons EL, and Vondra CF (1982) The Fayum Primate Forest Revisited. Journal of Human Evolution *11:*603–632.
114 Fleagle JG, and Simons EL (1982) Skeletal Remains of *Propliopithecus chirobates* from the Egyptian Oligocene. Folia Primatologica *39:*161–177.
115 Kay RF, and Simons EL (1983) A re-assessment of the relationships between later Miocene and subsequent Hominoidea. In RL Ciochon and RS Corruchini (eds.): New Interpretations of Ape and Human Ancestry. New York and London: Plenum Press, pp. 577–624.
116 Simons EL (1983) Recent advances in knowledge of the earliest catarrhines of the Egyptian Oligocene (including the most ancient known presumed ancestors of man). Pontifical Academy of Sciences, Vatican City, Rome, S. V *50:*11–27.
117 Fleagle JG, and Simons EL (1983) The Tibio-Fibular Articulation in *Apidium phiomense*, an Oligocene Anthropoid. Nature *301:*238–239.
118 Simons EL, and Kay RF (1983) *Qatrania*, New Basal Anthropoid Primate from the Fayum, Oligocene of Egypt. Nature *304:*624–626.
119 Simons EL, and Meinel W (1983) Mandibular Ontogeny in the Miocene Great Ape *Dryopithecus*. International Journal of Primatology *4:*331–337.
120 Khashab B, Simons EL, and Fleagle JG (1983) Annotated bibliography of Egyptian vertebrate fossils up to the end of 1980. Min. of Industry and Mineral Resources, Geol. Surv. Egypt *Paper 65:*1–111.
121 Kay RF, and Simons EL (1983) Dental Formulas and Dental Eruption Patterns in Parapithecidae (Primates, Anthropoidea). American Journal of Physical Anthropology *62:*363–375.
122 Izard MK, and Simons EL (1983) A breeding colony of bushbabies. Lab. Animal *12:*12.
123 Macphee RDE, and Simons EL (1983) Team Finds Subfossil Lemur. Earth Science *36:*18–19.
124 Bown TM, and Simons EL (1984) First Record of Marsupials (Metatheria, Polyprotodonta) from the Oligocene in Africa. Nature *308:*447–449.

125 Macphee RDE, Simons EL, Wells NA, and Vuillaumerandriamanantena M (1984) Team Finds Giant Lemur Skeleton. Geotimes *29:*10–11.
126 Simons EL (1984) Dawn Ape of the Fayum. Natural History *93:*18–20.
127 Simons EL, and Bown TM (1984) A New Species of *Peratherium* (Didelphidae, Polyprotodonta) - the 1st African Marsupial. Journal of Mammalogy *65:*539–548.
128 Bown TM, and Simons EL (1984) African Marsupials - Vicariance or Dispersion - Reply. Nature *312:*379–380.
129 Gebo DL, and Simons EL (1984) Puncture Marks on Early African Anthropoids. American Journal of Physical Anthropology *65:*31–35.
130 Yoder AD, Simons EL, and Pollock J (1984) Conservation at Duke University Primate Center. IUCN/SSC Primate Specialist Group Newsletter *4:*48–50.
131 Izard MK, and Simons EL (1984) Management of Reproduction in a Breeding Colony of Bushbabies. International Journal of Primatology *5:*350.
132 Wright PC, and Simons EL (1984) Calls of the Mindanao Tarsier (*Tarsius syrichta*). American Journal of Physical Anthropology *63:*236.
133 Simons EL, and Bown TM (1985) *Afrotarsius chatrathi*, 1st Tarsiiform Primate Questionable Tarsiidae) from Africa. Nature *313:*475–477.
134 Simons EL, Wells NA, MacPhee RDE, Burney D, Chatrath PS, Dewar R, and Villaume-Randriamanantena M (1985) Sedimentology of *Ampasambazimba marsh*, one of Madagascar's best Holocene fossil deposits. Geol. Soc. Am. Annual Meeting (abstract).
135 Simons EL, Wells NA, MacPhee RDE, Burney D, Chatrath PS, Dewar R, and Villaume-Randriamanantena M (1985) Geology of several Holocene fossil sites in Madagascar. Geol. Soc. Am. Annual Meeting (abstract).
136 Simons EL (1985) Duke Primate Center: A home for endangered *Sifaka*. On the Edge *26:*8–11.
137 Simons EL (1985) Did Tarsiers Arise in Africa. American Journal of Physical Anthropology *66:*231.
138 Simons EL (1985) African origin, characteristics, and context of earliest higher primates. In PV Tobias (ed.): Hominid Evolution: Past, Present and Future. New York: Alan P. Liss, pp. 101–106.
139 Haring DM, Wright PC, and Simons EL (1985) Social Behaviors of *Tarsius syrichta* and *Tarsius bancanus*. American Journal of Physical Anthropology *66:*179.
140 Simons EL (1985) Origins and characteristics of the first hominoids. In E Delson (ed.): Ancestors: The Hard Evidence. New York: Alan R. Liss, pp. 37–41.
141 Jenkins FA, Simons EL, McKenna MC, and Gingerich PD (1985) Princeton Intellectual Trust. Science *229:*330.
142 Freed BZ, Wright PC, and Simons EL (1985) Infant Development and Parental Care in *Lemur mongoz* and *Lemur coronatus*. American Journal of Primatology *8:*338.
143 Izard MK, Wright PC, and Simons EL (1985) Gestation Length in *Tarsius bancanus*. American Journal of Primatology *9:*327–331.
144 Fleagle JG, Bown TM, Obradovich JD, and Simons EL (1986) How old are the Fayum primates? In Else and Lee (eds.): Primate Evolution: Cambridge Univ. Press, pp. 3–17.
145 Izard MK, and Simons EL (1986) Infant Survival and Litter Size in Primigravid and Multigravid *Galagos*. Journal of Medical Primatology *15:*27–35.
146 Simons EL, and Izard MK (1986) Management of reproduction in a breeding colony of bushbabies. In Else and Lee (eds.): Primate Ecology and Conservation: Cambridge Univ. Press, pp. 315–323.
147 Izard MK, and Simons EL (1986) Isolation of Females Prior to Parturition Reduces Neonatal Mortality in *Galago*. American Journal of Primatology *10:*249–255.
148 Simons EL (1986) Tarsiers Found in Fayum Quarries. Geotimes *31:*13.

149 Simons EL, Scheffrahn W, Welker C, Izard MK, Stanyon R, and Njuguna J (1986) Morphological, cytogenetic and blood group genetic findings in different species of the genus *Galago*: XIth Congr. Int. Primat. Soc. Göttingen, Germany.
150 Wright PC, Toyama LM, and Simons EL (1986) Courtship and Copulation in *Tarsius bancanus*. Folia Primatologica *46:*142–148.
151 Wright PC, Izard MK, and Simons EL (1986) Reproductive cycles in *Tarsius bancanus*. American Journal of Primatology *11:*207–215.
152 Simons EL (1986) Lemurs in a natural habitat in North Carolina: A crucial step in reintroduction to the wild: Primate Conservation: The Newsletter and Journal of the IUCN/SSC Primate Specialist Group, pp. 60–62.
153 Simons EL (1986) *Parapithecus grangeri* of the African Oligocene - an Archaic Catarrhine without Lower Incisors. Journal of Human Evolution *15:*205–213.
154 Fleagle JG, Bown TM, Obradovich JD, and Simons EL (1986) Age of the Earliest African Anthropoids. Science *234:*1247–1249.
155 Pereira ME, and Simons EL (1986) Sexually-differentiated responses to potential immigrant males in semi-free-ranging ringtailed lemurs. Primate Report *14:*113 (abstract).
156 Simons EL, and Bown TM (1987) New Oligocene Ptolemaiidae (Mammalia: ?Pantolesta) from the Jebel Qatrani Formation, Fayum Depression, Egypt. J. Vert. Paleo *7:*311–324.
157 Pereira ME, Klepper A, and Simons EL (1987) Tactics of Care for Young Infants by Forest-Living Ruffed Lemurs (*Varecia variegata variegata*) Ground Nests, Parking, and Biparental Guarding. American Journal of Primatology *13:*129–144.
158 Izard MK, and Simons EL (1987) Lactation and Interbirth Interval in the Senegal Galago (*Galago senegalensis moholi*). Journal of Medical Primatology *16:*323–332.
159 Simons EL, Rasmussen DT, and Gebo DL (1987) A New Species of *Propliopithecus* from the Fayum, Egypt. American Journal of Physical Anthropology *73:*139–147.
160 Rasmussen DT, Olson SL, and Simons EL (1987) Fossil birds from the Oligocene Jebel Qatrani Formation, Fayum Province, Egypt. Smithsonian Contributions to Paleobiology *62:*1–20.
161 Simons EL (1987) Protection des lémuriens Malgaches par la captivité. In RA Mittermeier, LH Rakotovao, V Randrianasolo, EJ Sterling and D Divetre (eds.): Priorités en matière de Conservation des espèces au Madagascar. Occasional papers on the IUCN Species Survival Commission No. 2.
162 Simons EL, Bown TM, and Rasmussen DT (1986) Discovery of Two Additional Prosimian Primate Families (Omomyidae, Lorisidae) in the African Oligocene. Journal of Human Evolution *15:*431–437.
163 Simons EL, Bown TM, and Rasmusssen DT (1987) Discovery of two additional prosimian primate families (Omomyidae, Lorisidae) in the African Oligocene. J. Hum. Evol *15:*431–437.
164 Cherry JA, Izard MK, and Simons EL (1987) Description of Ultrasonic Vocalizations of the Mouse Lemur (*Microcebus murinus*) and the Fat-Tailed Dwarf Lemur (*Cheirogaleus medius*). American Journal of Primatology *13:*181–185.
165 Gebo DL, and Simons EL (1987) Morphology and Locomotor Adaptations of the Foot in Early Oligocene Anthropoids. American Journal of Physical Anthropology *74:*83–101.
166 Simons EL (1987) Update on the primate colony at the Duke University Primate Center. Primate Conservation: The Newsletter and Journ. of the IUCN/SSC Primate Specialist Group *8:*51.
167 Wright PC, Haring DM, Simons EL, and Andau P (1987) Tarsiers: A conservation perspective. Primate Conservation: The Newsletter and Journal of the IUCN/SSC Primate Specialist Group *8:*51–54.
168 Gagnon M, Simons EL, Godfrey L, and Vuillaumerandriamanantana M (1988) Preliminary Report of Giant Fossil Lemur Findings in the Ankarana Region of Madagascar. American Journal of Physical Anthropology *75:*211.

169 Simons EL (1988) New faces of *Aegyptopithecus* from the Oligocene of Egypt. J. Human Evol *16:*273–289.

170 Rasmussen DT, and Simons EL (1988) New Oligocene hyracoids from Egypt. J. Vert. Paleo *8:*67–83.

171 Simons EL, and Rumpler Y (1988) *Eulemur* - New Generic Name for Species of Lemur Other Than *Lemur catta.* Comptes Rendus De L'Academie Des Sciences Serie III-Sciences De La Vie-Life Sciences *307:*547–551.

172 Simons EL, and Kay RF (1988) New Material of *Qatrania* from Egypt with Comments on the Phylogenetic Position of the Parapithecidae (Primates, Anthropoidea). American Journal of Primatology *15:*337–347.

173 Simons EL (1988) A New Species of *Propithecus* (Primates) from Northeast Madagascar. Folia Primatologica *50:*143–151.

174 Koop BF, Siemieniak D, Slightom JL, Goodman M, Dunbar J, Wright PC, and Simons EL (1989) *Tarsius* Delta-Globin and Beta-Globin Genes - Conversions, Evolution, and Systematic Implications. Journal of Biological Chemistry *264:*68–79.

175 Rasmussen DT, Tilden CD, and Simons EL (1989) New Specimens of the Giant Creodont *Megistotherium* (Hyaenodontidae) from Moghara, Egypt. Journal of Mammalogy *70:*442–447.

176 Simons EL, and Rasmussen DT (1989) Cranial Morphology of *Aegyptopithecus* and *Tarsius* and the Question of the Tarsier-Anthropoidean Clade. American Journal of Physical Anthropology *79:*1–23.

177 Pereira ME, Macedonia J, Haring DM, and Simons EL (1989) Maintenance of Primates in Captivity for Research: bancanusrd MK, and Simons EL (1989) Psychological Well-Being of Nocturnal Primates in Captivity. In E Segal (ed.): Psychological Well-Being of Captive Primates. Park Ridge, New Jersey: Noyes Publ, pp. 61–74.

178 Rasmussen DT, and Simons EL (1988) New Specimens of *Oligopithecus savagei*, Early Oligocene Primate from the Fayum, Egypt. Folia Primatologica *51:*182–208.

179 Simons EL (1989) Human Origins. Science *245:*1343–1350.

180 Simons EL (1989) Description of Two Genera and Species of Late Eocene Anthropoidea from Egypt. Proceedings of the National Academy of Sciences of the United States of America *86:*9956–9960.

181 Godfrey LR, Simons EL, Chatrath PJ, and Rakotosamimanana B (1990) A New Fossil Lemur (*Babakotia*, Primates) from Northern Madagascar. Comptes Rendus De L'Academie Des Sciences Serie II *310:*81–87.

182 Simons EL, and Rasmussen DT (1990) Vertebrate paleontology of Fayum: History of research, faunal review and future prospects. In R Said and CH Squires (eds.): Treatise on the Geology of Egypt. Rotterdam, Holland: Balkema, pp. 627–638.

183 Simons EL, Godfrey LR, Vuillaumerandriamanantena M, Chatrath PS, and Gagnon M (1990) Discovery of New Giant Subfossil Lemurs in the Ankarana Mountains of Northern Madagascar. Journal of Human Evolution *19:*311–319.

184 Simons EL (1990) Discovery of the Oldest Known Anthropoidean Skull from the Paleogene of Egypt. Science *247:*1567–1569.

185 Rasmussen DT, Gagnon M, and Simons EL (1990) Taxeopody in the Carpus and Tarsus of Oligocene *Pliohyracidae* (Mammalia, Hyracoidea) and the Phyletic Position of Hyraxes. Proceedings of the National Academy of Sciences of the United States of America *87:*4688–4691.

186 Gingerich PD, Smith BH, and Simons EL (1990) Hind Limbs of Eocene *Basilosaurus* - Evidence of Feet in Whales. Science *249:*154–157.

187 Bown TM, Holroyd PA, and Simons EL (1990) A new elephant-shrew from the Fayum Depression, Egypt, and the origin of Macroscelidea. J. Vert. Paleo *10:*15A (abstract).

188 Tilden CD, Holroyd PA, and Simons EL (1990) Phyletic affinities of *Apterodon* (Hyaenodontidae, Creodonta). J. Vert. Paleo *10:*46A (abstract).

189 Simons EL (1990) Discovery and Conservation of the Golden Crowned Sifaka, *Propithecus tattersali*, from Northeastern Madagascar. AAZPA Regional Conference Proceedings.
190 Leakey MG, Leakey RE, Richtsmeier JT, Simons EL, and Walker AC (1991) Similarities in *Aegyptopithecus* and *Afropithecus* Facial Morphology. Folia Primatologica *56:*65–85.
191 Simons EL, and Rasmussen DT (1991) The Generic Classification of Fayum Anthropoidea. International Journal of Primatology *12:*163–178.
192 Rasmussen DT, and Simons EL (1991) The Oldest Egyptian Hyracoids (Mammalia: Pliohyracidae): New Species of *Saghatherium* and *Thyrohyrax* from the Fayum. Neues Jahrbuch fur Geologie und Palaontologie Abhandlungen *182:*187–209.
193 Holroyd PA, and Simons EL (1991) The phyletic relationships of *Azibius*. Am. J. Phys. Anth *Suppl. 12:*94 (abstract).
194 Simons EL (1991) Endocranial molds of the brain of *Aegyptopithecus*. Am. J. Phys. Anth *Suppl. 12:*102 (abstract).
195 Simons EL, Holroyd PA, and Bown TM (1991) Early Tertiary Elephant-Shrews from Egypt and the Origin of the Macroscelidea. Proceedings of the National Academy of Sciences of the United States of America *88:*9734–9737.
196 Jungers WL, Godfrey LR, Simons EL, Chatrath PS, and Rakotosamimanana B (1991) Phylogenetic and Functional Affinities of *Babakotia* (Primates), a Fossil Lemur from Northern Madagascar. Proceedings of the National Academy of Sciences of the United States of America *88:*9082–9086.
197 Izard MK, Heath SJ, Hayes Y, and Simons EL (1991) Hematology, Serum Chemistry Values, and Rectal Temperatures of Adult Greater Galagos (*Galago garnettii* and *G. crassicaudatus*). Journal of Medical Primatology *20:*117–121.
198 Rasmussen DT, Bown TM, and Simons EL (1992) The Eocene-Oligocene Transition in Continental Africa. In DR Prothero and WA Berggren (eds.): Eocene-Oligocene Climatic and Biotic Evolution: Princeton Univ. Press, pp. 548–566.
199 Simons EL (1992) Fossil history of primates: Cambridge Encyclopedia of Human Evolution. England: Cambridge University Press, pp. 199–208.
200 Rasmussen DT, and Simons EL (1992) Paleobiology of the Oligopithecines, the Earliest Known Anthropoid Primates. International Journal of Primatology *13:*477–508.
201 Simons EL, Godfrey LR, Jungers WL, Chatrath PS, and Rakotosamimanana B (1992) A New Giant Subfossil Lemur, *Babakotia*, and the Evolution of the Sloth Lemurs. Folia Primatologica *58:*197–203.
202 Kappelman J, Simons EL, and Swisher CC (1992) New Age-Determinations for the Eocene-Oligocene Boundary Sediments in the Fayum Depression, Northern Egypt. Journal of Geology *100:*647–667.
203 Simons EL (1992) Diversity in the Early Tertiary Anthropoidean Radiation in Africa. Proceedings of the National Academy of Sciences of the United States of America *89:*10743–10747.
204 Simons EL (1993) Diversity in the Early Tertiary Anthropoidean Radiation in Africa. Proceedings of the National Academy of Sciences of the United States of America *90:*1634–1634. (correction)
205 Vuillaumerandriamanantena M, Godfrey LR, Jungers WL, and Simons EL (1992) Morphology, Taxonomy and Distribution of Megaladapis - Giant Subfossil Lemur from Madagascar. Comptes Rendus De L'Academie Des Sciences Serie II *315:*1835–1842.
206 Tattersall I, Simons EL, and Vuillaume-Randriamanantena M (1992) *Paleopropithecus ingens* G. Grandidier 1899 (Mammalia, Primates): Proposed conservation of both generic and specific names. Bull. Zool. Nomenclature *49:*55–57.
207 Rasmussen DT, and Simons. EL (1993) An antelope-like hyrax from the African Eocene. Journal of Vertebrate Paelontology (abstract).
208 Simons EL (1993) Egypt's Simian Spring. Natural History *102:*58–59, 104.

209 Stanyon R, Weinberg J, Simons EL, and Izard MK (1992) A Third Karyotype for *Galago demidovii* suggests the Existence of Multiple Species. Folia Primat. *59:*33–38.

210 B.S. Coffman WR, Hess KE, Glander PE, Feeser, and Simons EL (1993) Management of a Breeding Colony of Aye-Ayes (*Daubentonia madagascariensis*) at the Duke University Primate Center. AAZPA *1993 Regional Proceedings:*161–167.

211 Simons EL (1993) New Endocasts of *Aegyptopithecus* - Oldest Well-Preserved Record of the Brain in Anthropoidea. American Journal of Science *293A:*383–390.

212 Izard MK, Coffman B, Katz A, and Simons EL (1993) Reproduction in the collared lemur (*Eulemur fulvus collaris*). Am. J. Phys. Anth. *30:*320 (abstract).

213 Wunderlich RE, Jungers WL, and Simons EL (1993) New pedal remains of *Megaladapis* and their functional significance. Am. J. Phys. Anth. *Suppl. 16:*213 (abstract).

214 Simons EL (1993) Lost Lemurs of the Crocodile Caves: The Sciences, pp. 6–8.

215 Simons EL (1993) Discovery of the Western Aye-Aye. Lemur News *1:*6.

216 Bown TM, Rose KD, Simons EL, and Wing SL (1994) Distribution and Stratigraphic Correlation of Upper Paleocene and Lower Eocene Fossil Mammal and Plant Localities of the Fort Union, Willwood, and Tatman Formations, Southern Bighorn Basin, Wyoming: U. S. Geol. Surv. Prof. Paper 1540, pp. 103 pages.

217 Simons EL (1994) The Giant Aye-Aye *Daubentonia robusta.* Folia Primatologica *62:*14–21.

218 Shapiro L, Jungers WL, Godfrey LR, and Simons. EL (1994) Vertebral morphology of extinct lemurs. Am. Journ. Phys. Anth. *Suppl. 18:*179–180 (abstract).

219 Jungers WL, Simons EL, and Godfrey LR (1994) Phalangeal curvature and locomotor adaptations in subfossil lemurs. Am. Journ. Phys. Anth. *Suppl. 18:*117–118 (abstract).

220 Wunderlich RE, Jungers WL, Godfrey LR, and Simons EL (1994) Pedal form and function in subfossil indroids. Am Journ.Phys. Anth. *Suppl. 18:*211–212 (abstract).

221 Ravosa MJ, and Simons EL (1994) Mandibular Growth and Function in *Archaeolemur.* American Journal of Physical Anthropology *95:*63–76.

222 Wunderlich RE, Jungers WL, Godfrey LR, Simons EL, and Chatrath PS (1994) Functional morphology of subfossil Malagasy primate feet. Am. Soc. Phys. Anth. Annual Meeting (abstract).

223 Domning DP, Gingerich PD, Simons EL, and Ankel-Simons FA (1994) A New Early Oligocene Dugongid (Mammalia, Sirenia) from Fayum Province, Egypt. Contrib. Mus. Paleont. Univ. Michigan *29:*89–108.

224 Simons EL (1994) New monkeys (*Prohylobates*) and an ape humerus from the Miocene Moghara Formation of Northern Egypt. Proc. XIV Int. Primatol. Conf. Strassbourg, France, 1993*:*247–253.

225 Simons EL, Rasmussen DT, Brown TM, and Chatrath PS (1994) The Eocene Origin of Anthropoid primates: Adaptation, Evolution, and Diversity. In JG Fleagle and RF Kay (eds.): Anthropoid Origins. N.Y.: Plenum Press, pp. 179–202.

226 Gebo DL, Simons EL, Rasmussen DT, and Dagosto M (1994) Eocene anthropoid postcrania from the Fayum, Egypt. In JG Fleagle and RF Kay (eds.): Anthropoid Origins. N.Y.: Plenum Press, pp. 203–234.

227 Simons EL, and Rasmussen DT (1994) A Remarkable Cranium of *Plesiopithecus teras* (Primates, Prosimii) from the Eocene of Egypt. Proceedings of the National Academy of Sciences of the United States of America *91:*9946–9950.

228 Simons EL (1994) New evidence concerning the earliest African higher primates. Annals Egyptian Geol. Surv. and Mining Auth. Proc. Inter. Conf. 30 Years Cooper *20:*101–113.

229 Haring DM, Hess WR, Coffman BS, Simons EL, and Owens TM (1994) Natural history and captive management of the Aye-Aye at the Duke University Primate Center, Durham. International Zoo Yearbook*:*201–219.

230 Simons EL, and Rasmussen DT (1995) A Whole New World of Ancestors: Eocene Anthropoideans from Africa. Evolutionary Anthropology *3:*128–139.
231 Simons EL, and Bown TM (1995) Ptolemaiida, a New Order of Mammalia - with Description of the Cranium of *Ptolemaia grangeri*. Proceedings of the National Academy of Sciences of the United States of America *92:*3269–3273.
232 Izard MK, Epps B, and Simons EL (1995) Reproduction in the brown lemur (*Eulemur fulvus fulvus*). American Journal of Primatology *36:*129 (abstract).
233 Fleagle JG, and Simons EL (1995) Limb Skeleton and Locomotor Adaptations of *Apidium phiomense*, an Oligocene Anthropoid from Egypt. American Journal of Physical Anthropology *97:*235–289.
234 Simons EL (1995) Skulls and Anterior Teeth of *Catopithecus* (Primates, Anthropoidea) from the Eocene and Anthropoid Origins. Science *268:*1885–1888.
235 Hamrick MW, Meldrum DJ, and Simons EL (1995) Anthropoid Phalanges from the Oligocene of Egypt. Journal of Human Evolution *28:*121–145.
236 Simons EL (1995) Crania of *Apidium*: Primitive Anthropoidean (Primates, Parapithecidae) from the Egyptian Oligocene. American Museum Novitates *3124:*1–10.
237 Simons EL, Godfrey LR, Jungers WL, Chatrath PS, and Ravaoarisoa J (1995) A New Species of *Mesopropithecus* (Primates, Palaeopropithecidae) from Northern Madagascar. International Journal of Primatology *16:*653–682.
238 Jungers WL, Godfrey LR, Simons EL, and Chatrath PS (1995) Subfossil *Indri indri* from the Ankarana Massif of Northern Madagascar. American Journal of Physical Anthropology *97:*357–366.
239 Simons EL (1995) History, Anatomy, Subfossil Record and Management of *Daubentonia madagascariensis*. In L Alterman, GA Doyle and MK Izard (eds.): Creatures of the Dark: The Nocturnal Prosimians. New York: Plenum Press, pp. 133–140.
240 Simons EL, Rasmussen DT, and Gingerich PD (1995) New cercamoniine adapid from Fayum, Egypt. Journal of Human Evolution *29:*577–589.
241 Simons EL (1995) Egyptian Oligocene Primates: A review. Yearbook of Physical Anthropology *38:*199–238.
242 Simons EL, Burney DA, Chatrath PS, Godfrey LR, Jungers WL, and Rakotosamimanana B (1995) Ams C-14 Dates for Extinct Lemurs from Caves in the Ankarana Massif, Northern Madagascar. Quaternary Research *43:*249–254.
243 Shoshani J, Groves CP, Simons EL, and Gunnell GF (1996) Primate phylogeny: Morphological vs molecular results. Molecular Phylogenetics and Evolution *5:*102–154.
244 Godfrey LR, Wilson JM, Simons EL, Stewart PD, and Vuillaume-Randriamanantena M (1996) Ankarana: Window to Madagascar's Past. Lemur News *1:*16–17.
245 Wilson J, Godfrey LR, Simons EL, Stewart P, and Vuillaume-Randriamanantena M (1996) Past and present lemur fauna at Ankarana, Northern Madagascar. Primate Conservation *16:*47–52.
246 Wunderlich RE, Simons EL, and Jungers WL (1996) New pedal remains of *Megaladapis* and their functional significance. American Journal of Physical Anthropology *100:*115–138.
247 Simons EL, and Rasmussen DT (1996) Skull of *Catopithecus browni*, an early tertiary catarrhine. American Journal of Physical Anthropology *100:*261–292.
248 Teaford MF, Maas MC, and Simons EL (1996) Dental microwear and microstructure in early Oligocene primates from the Fayum, Egypt: Implications for diet. American Journal of Physical Anthropology *101:*527–543.
249 Holroyd PA, Simons EL, Bown TM, Polly PD, and Kraus MJ (1996) New records of terrestrial mammals from the upper Eocene Qasr El Sagha Formation, Fayum Depression, Egypt. Paleovertebrata *25:*175–192.
250 Simons EL (1997) Discovery of the smallest Fayum Egyptian primates (Anchomomyini, Adapidae). Proceedings of the National Academy of Sciences of the United States of America *94:*180–184.

251 Simons EL (1997) Lemurs: Old and New. In SM Goodman and BD Patterson (eds.): Natural Change and Human Impact in Madagascar. Washington, D.C.: Smithsonian Institution Press, pp. 142–156.
252 Godfrey LR, Jungers WL, Reed KE, Simons EL, and Chatrath PS (1997) Subfossil Lemurs: Inferences about the Past and Present Primate Communities in Madagascar. In SM Goodman and BD Patterson (eds.): Natural Change and Human Impact in Madagascar. Washington, D.C.: Smithsonian Institution Press, pp. 218–256.
253 Simons EL, and Miller ER (1997) An upper dentition of *Aframonius dieides* (Primates) from the Fayum, Egyptian Eocene. Proceedings of the National Academy of Sciences of the United States of America *94:*7993–7996.
254 Miller ER, and Simons EL (1997) Dentition of *Proteopithecus sylviae*, an archaic anthropoid from the Fayum, Egypt. Proceedings of the National Academy of Sciences of the United States of America *94:*13760–13764.
255 Jungers WL, Godfrey LR, Simons EL, and Chatrath PS (1997) Phalangeal curvature and positional behavior in extinct sloth lemurs (Primates, Palaeopropithecidae). Proceedings of the National Academy of Sciences of the United States of America *94:*11998–12001.
256 Bloch JI, Fisher DC, Gingerich PD, Gunnell GF, Simons EL, and Uhen MD (1997) Cladistic analysis and anthropoid origins. Science *278:*2134–2135.
257 Simons EL (1997) Preliminary description of the cranium of *Proteopithecus sylviae*, an Egyptian late Eocene anthropoidean primate. Proceedings of the National Academy of Sciences of the United States of America *94:*14970–14975.
258 Miller ER, Rasmussen DT, and Simons EL (1997) Fossil storks (Ciconiidae) from the Late Eocene and Early Miocene of Egypt. Ostrich *68:*23–26.
259 Simons EL (1998) The prosimian fauna of the fayum Eocene/Oligocene deposits of Egypt. Folia Primatologica *69:*286–294.
260 Fleagle JG, Richmond BF, Ankel-Simons F, Chatrath PS, and Simons EL (1998) *Aegyptopithecus zeuxis* and the Evolution of Old World Higher Primates. Proceedings of the Geological Survey of Egypt Centennial*:*277–287.
261 Miller ER, and Simons EL (1998) Relationships between the Mammalian Fauna from Wadi Moghara, Quattara Depression, Egypt, and other early Miocene Faunas. Proceedings of the Geological Survey of Egypt Centennial*:*547–580.
262 Simons EL, and Chatrath PS (1998) Eocene Mammalian Faunas of Africa with Particular Reference to the Age Correlation of Primates at Locality 41. Proceedings of the Geological Survey of Egypt Centennial*:*775–783.
263 Simons EL, Cornero S, and Bown TM (1998) The Taphonomy of Fossil Vertebrate Quarry L-41, Upper Eocene, Fayum Province, Egypt. Proceedings of the Geological Survey of Egypt Centennial*:*785–791.
264 Schwartz JH, Shoshani J, Tattersall I, Simons EL, and Gunnell G (1998) *Lorisidae* Grey, 1821 and *Galagidae* Grey, 1825 (Mammalia, Primates): proposed conservation as the correct original spellings. Bulletin of Zoological Nomenclature *55:*165–168.
265 Simons EL (1998) Chapter 5, Prosimians: The Psychological Well-being of Nonhuman Primates: a Report of the Committee on Well-being of Nonhuman Primates, Institute for Laboratory Animal Research, National Research Council. Washington, D.C.: National Academy Press, pp. 55–67.
266 Rasmussen DT, Conroy GC, and Simons EL (1998) Tarsier-like locomotor specializations in the Oligocene primate *Afrotarsius*. Proceedings of the National Academy of Sciences of the United States of America *95:*14848–14850.
267 Simons EL, Plavcan JM, and Fleagle JG (1999) Canine sexual dimorphism in Egyptian Eocene anthropoid primates: *Catopithecus* and *Proteopithecus*. Proceeding of the National Academy of Sciences of the United States of America *96:*2559–2562.

268 DeBlieux DD, and Simons EL (1999) Cranial and dental anatomy of the late Eocene hyracoid *Antilohyrax pectidens* (Mammalia) from the Fayum, Egypt. Journal of Vertebrate Paleontology 19(3) (abstract).
269 Rasmussen DT, and Simons EL (2000) Ecomorphological diversity among Paleogene hyracoids (Mammalia): A new cursorial browser from the Fayum, Egypt. Journal of Vertebrate Paleontology *20:*167–176.
270 Godfrey LR, Jungers WL, Simons EL, Chatrath PS, and Rakotosamimanana B (1999) Past and Present Distributions of Lemurs in Madagascar. In H Rasamiminana, B Rakotosamimanana, J Ganzhorn and S Goodman (eds.): New Directions in Lemur Studies. New York: Kluwer Academic/Plenum Press, pp. 19–53.
271 Simons EL, and Seiffert ER (1999) A partial skeleton of *Proteopithecus sylviae* (Primates, Anthropoidea): first associated dental and postcranial remains of an Eocene anthropoidean. Comptes Rendus De L'Academie Des Sciences Serie II Fascicule a- Sciences De La Terre Et Des Planetes *329:*921–927.
272 Seiffert ER, and Simons EL (2000) *Widanelfarasia*, a diminutive placental from the late Eocene of Egypt. Proceedings of the National Academy of Sciences of the United States of America *97:*2646–2651.
273 Hamrick MW, Simons EL, and Jungers WL (2000) New wrist bones of the Malagasy giant subfossil lemurs. Journal of Human Evolution *38:*635–650.
274 Jungers WL, Godfrey LR, Simons EL, Wunderlich RE, Richmond BG, Chatrath PS and Rakotosamimanana B. (2002) Ecomorphology and behavior of giant extinct lemurs from Madagascar. In JM Plavcan, R Kay, C Van Schaik and WL Jungers. (ed.): Reconstructing Behavior in the Primate Fossil Record. New York: Kluwer Academic/ Plenum Press pp. 371–411.
275 Simons EL (2000) A view on the science: Physical anthropology at the millennium. American Journal of Physical Anthropology *112:*441–446.
276 Seiffert ER, Simons EL, and Fleagle JG (2000) Anthropoid humeri from the late Eocene of Egypt. Proceedings of the National Academy of Sciences of the United States of America *97:*10062–10067.
277 Thewissen JGM, and Simons EL (2001) Skull of *Megalohyrax eocaenus* (Hyracoidea, Mammalia) from the Oligocene of Egypt. Journal of Vertebrate Paleontology *21:*98–106.
278 Rasmussen DT, Simons EL, Hertel F, and Judd A (2001) Hindlimb of a giant terrestrial bird from the upper Eocene, Fayum, Egypt. Palaeontology *44:*325–337.
279 Seiffert ER, and Simons EL (2001) Astragalar morphology of Late Eocene anthropoids from the Fayum Depression (Egypt) and the origin of catarrhine primates Journal of Human Evolution *41:* 577–606.
280 Kirk EC, and Simons EL (2001) Diets of fossil primates from the Fayum Depression of Egypt: a quantitative analysis of molar shearing. Journal of Human Evolution *40:*203–229.
281 Simons EL (2001) The cranium of *Parapithecus grangeri*, an Egyptian Oligocene anthropoidean primate. Proceedings of the National Academy of Sciences of the United States of America *98:*7892–7897.
282 Ciochon RL, Gingerich PD, Gunnell GF, and Simons EL (2001) Primate postcrania from the late middle Eocene of Myanmar. Proceedings of the National Academy of Sciences of the United States of America *98:*7672–7677.
283 Simons EL, and Meyers DM (2001) Folklore and Beliefs about the Aye aye (*Daubentonia madagascariensis*). Lemur News *6:*11–16.
284 King SJ, Godfrey LR, and Simons EL (2001) Adaptive and phylogenetic significance of ontogenetic sequences in *Archaeolemur*, subfossil lemur from Madagascar. Journal of Human Evolution *41:*545–576.
285 DeBlieux DD, and Simons EL (2002) Cranial anatomy of *Antilohyrax pectidens* a late Eocene hyracoid (Mammalia) from the Fayum, Egypt. Journal of Vertebrate Paleontology *22:*121–135.

286 Simons EL, Seiffert ER, Chatrath PS, and Attia Y. (2002) Earliest record of a parapithecid anthropoid from the Jebel Qatrani Formation, northern Egypt. Folia Primatologica *72:*316–331.
287 Schwartz GT, Samonds KE, Godfrey LR, Jungers WL, Simons EL. (2002) Dental microstructure and life history in subfossil Malagasy lemurs. Proceedings of the National Academy of Sciences *99:*6124–6129.
288 Gaffney ES, Deblieux DD, Simons EL, Sanchez-Villagra MR, Meylan PA (2002) Redescription of the skull of Dacquemys Williams, 1954, a podocnemidid side-necked turtle from the late Eocene of Egypt. American Museum Novitates *3372:* 1–16.
289 Rossie JB, Simons EL, Gauld SC, Rasmussen DT. (2002) Paranasal sinus anatomy of Aegyptopithecus: Implications for hominoid origins. Proceedings of the National Academy of Sciences *99(12):* 8454–8456
290 Simons EL (2002) The Fossil Record of Human Origins among the Anthropoidea. In B Sherwood (ed.): New Perspectives in Primate Evolution and Behavior. West Yorkshire, UK: Westbury Academic and Scientific Publishing, 13–28.
291 Seiffert ER, Simons EL, Attia Y (2003) Fossil evidence for an ancient divergence of lorises and galagos. Nature *422:* 421–424.
292 Simons EL (2003) The Fossil Record of Tarsier Evolution. In PC Wright, EL Simons, and S Gursky (eds.): Tarsiers: Past, Present, and Future. New Brunswick: Rutgers University Press, 9–34.
293 Wright PC, Pochron ST, Herring EH, and Simons EL (2003) Can We Predict Seasonal Behavior and Social Organization from Sexual Dimorphism and Testes Measurements? In PC Wright, EL Simons, and S Gursky (eds.): Tarsiers: Past, Present, and Future. New Brunswick: Rutgers University Press, 260–273.
294 Jernvall J, Wright PC, Ravoavy, FL, Simons EL (2003). Report on Findings of Subfossils at Ampoza and Ampanihy in Southwestern Madagascar. Lemur News 8, 21–23.
295 Godfrey L, Jungers W, Simons EL (2003). "Box 1: Early Descriptions, Early Discoveries," in Godfrey L, Jungers W, The Extinct Sloth Lemurs of Madagascar. Evolutionary Anthropology 12, 255.
296 Seiffert ER, Simons EL and CVM Simons (2004) Phylogenetic, biogeographic, and adaptive implications of new fossil evidence bearing on early stem catarrhine evolution. In CF Ross and RF Kay (eds.) Anthropoid Origins: New Visions. New York: Kluwer Academic/Plenum Publishers, Chapter 7,157–181.
297 Bush EC, Simons EL, Dubowitz D, Allman, JM (2004) Endocranial volume and optic foramen size in *Parapithecus grangeri.* In CF Ross and RF Kay (eds.) Anthropoid Origins: New Visions. New York: Kluwer Academic/Plenum Publishers, Chapter 21, 603–614.
298 Simons EL (2004) The cranium and the adaptations of *Parapithecus grangeri*, a stem catarrhine. In CF Ross and RF Kay (eds.) Anthropoid Origins: New Visions. New York: Kluwer Academic/Plenum Publishers, Chapter 8, 183–204.
299 Simons EL, Simons VFH, Chatrath PS, Muldoon KM, Oliphant M, Pistole N, and C. Savvas. (2004) Research on subfossils in Southwestern Madagascar and Ankilitelo Cave. Lemur News 9, 12–16.
300 Godfrey LR, Simons EL, Jungers WL, DeBlieux D, Chatrath PS. (2004) New discovery of subfossil *Hapalemur simus*, the greater bamboo lemur, in western Madagascar. Lemur News 9, 9–11.
301 Godfrey LR, Semprebon G, Jungers WL, Sutherland M, Simons EL, Solounias N. (2004) Dental use wear in extinct lemurs: evidence of diet and niche differentiation. Journal of Human Evolution *47:*145–169.
302 Bush EC, Simons EL, Allman JM. (2004) High-Resolution Computed Tomography Study of the Cranium of a Fossil Anthropoid Primate, *Parapithecus grangeri*: New Insights Into the Evolutionary History of Primate Sensory Systems. The Anatomical Record Part A *281A:*1083–1087.

303 Seiffert ER, Simons EL, Ryan TM, and Attia Y. (2005) Additional remains of *Wadilemur elegans*, a primitive stem galagid from the late Eocene of Egypt. Proceedings of the National Academy of Sciences *102(32)*:11396–11401.

304 Jungers WL, Lemelin P, Godfrey LR, Wunderlich RE, Burney DA, Simons EL, Chatrath, PS, James, HF. (2005) The hands and feet of *Archaeolemur*: metrical affinities and their functional significance. Journal of Human Evolution *49(1)*:36–55.

305 Seiffert ER, Simons EL, Clyde WC, Rossie JB, Attia Y, Bown TM, Chatrath PS, Mathison ME. (2005) Basal anthropoids from Egypt and the antiquity of Africa's higher primate radiation. Science *310*:300–304.

306 Murray AM, Simons EL, and Attia,YS. (2005) A new clupeid fish (Clupeomorpha) from the Oligocene of the Fayum, Egypt, with notes on some other fossil Clupeomorphs. Journal of Vertebrate Paleontology *32(2)*:300–308.

307 Shapiro LJ, Seiffert CVM, Godfrey LR, Jungers WL, Simons EL, Randria GFN. (2005) Morphometric Analysis of Lumbar Vertebrae in Extinct Malagasy Strepsirrhines. American Journal of Physical Anthropology *1285(4)*:823–839.

308 Muldoon KM, and Simons EL (2005) New late Holocene micromammals from Ankilitelo Cave, SW Madagascar. Journal of Vertebrate Paleontology 25:94A–95A (abstract).

309 Simons, EL. (2005) Eocene and Oligocene mammals from the Fayum, Egypt. First International Conference on the Geology of the Tethys, Cairo University, Egypt, 439–450.

310 DeBlieux DD, Baumrind MR, Simons EL, Chatrath PS, Meyer GE, and Attia Y. (2006) Sexual dimorphism of the internal mandibular chamber in Fayum Pliohyracidae (mammalia). Journal of Vertebrate Paleontology *26(1)*:160–169.

311 Muldoon KM and Simons EL. (2006) Ecogeographic size variation in small-bodied subfossil primates from Ankilitelo, SW Madagascar. American Journal of Physical Anthropology *129(S42)*:135–135 (abstract).

312 Muldoon KM, and Simons EL. (2006) Environmental change and human impact: evidence from Ankilitelo Cave, SW Madagascar. Journal of Vertebrate Paleontology *26(S3)*:104A (abstract).

313 Simons EL, Seiffert ER, Ryan TM, and Attia Y. (2007) A remarkable female cranium of Aegyptopithecus zeuxis (Catarrhini, Propliopithecidae). Proceedings of the National Academy of Sciences *104*:8731–8736.

314 Muldoon KM, and Simons EL. (2007) What can small mammals tell us about giant lemurs?: Paleoenvironment of the Late Holocene fauna at Ankilitelo Cave, SW Madagascar. American Journal of Physical Anthropology *315(S)*:175 (abstract).

315 Scott JR, Ungar PS, Jungers WL, Godfrey LR, Scott RS, Simons EL, Teaford MF, and Walker A. (2007) Dental microwear texture analysis of megaladapids and archaeolemurids. American Journal of Physical Anthropology *315(S)*:212 (abstract).

Subject Index

Printed in the United States of America